C. O. Kappe, A. Stadler

Microwaves in Organic and Medicinal Chemistry

Methods and Principles in Medicinal Chemistry

Edited by R. Mannhold, H. Kubinyi, G. Folkers

Editorial Board
H.-D. Höltje, H. Timmerman, J. Vacca, H. van de Waterbeemd, T. Wieland

Previous Volumes of this Series:

C. Oliver Kappe, Alexander Stadler

Microwaves in Organic and Medicinal Chemistry

WILEY-VCH

WILEY-VCH Verlag GmbH & Co. KGaA

Series Editors

Prof. Dr. Raimund Mannhold
Biomedical Research Center
Molecular Drug Research Group
Heinrich-Heine-Universität
Universtätsstrasse 1
40225 Düsseldorf
Germany
Raimund.mannhold@uni-duesseldorf.de

Prof. Dr. Hugo Kubinyi
Donnersbergstrasse 9
67256 Weisenheim am Sand
Germany
kubinyi@t-online.de

Prof. Dr. Gerd Folkers
Collegium Helveticum
STW/ETH Zentrum
8092 Zürich
Switzerland
folkers@collegium.ethz.ch

Authors

Prof. Dr. C. Oliver Kappe
Institute of Chemistry
Karl-Franzens-University Graz
Heinrichstrasse 28
8010 Graz
Österreich
Oliver.Kappe@uni-graz.at

Dr. Alexander Stadler
Institute of Chemistry
Karl-Franzens-University Graz
Heinrichstrasse 28
8010 Graz
Österreich
alexander.stadler@anton-paar.com

■ All books published by Wiley-VCH are
carefully produced. Nevertheless, authors,
editors, and publisher do not warrant the
information contained in these books,
including this book, to be free of errors.
Readers are advised to keep in mind that
statements, data, illustrations, procedural
details or other items may inadvertently
be inaccurate.

Library of Congress Card No.: applied for
British Library Cataloguing-in-Publication Data
A catalogue record for this book is available
from the British Library.

**Bibliographic information published by
Die Deutsche Bibliothek**
Die Deutsche Bibliothek lists this publication
in the Deutsche Nationalbibliografie; detailed
bibliographic data is available in the Internet at
<http://dnb.ddb.de>.

© 2005 WILEY-VCH Verlag GmbH & Co. KGaA,
Weinheim

Printed in the Federal Republic of Germany.
Printed on acid-free paper.

Typesetting   Kühn & Weyh, Satz und Medien,
Freiburg
Printing   betz-druck GmbH, Darmstadt
Bookbinding   Buchbinderei J. Schäffer GmbH,
Grünstadt

ISBN-13: 978-3-527-31210-8
ISBN-10: 3-527-31210-2

Contents

Microwaves in Organic and Medicinal Chemistry. C. Oliver Kappe, Alexander Stadler
Copyright © 2005 WILEY-VCH Verlag GmbH & Co. KGaA, Weinheim
ISBN: 3-527-31210-2

Preface

Until recently, the application of microwaves in organic synthesis was a curiosity. Starting in 1986, but increasing in use only from the mid-nineties onwards, microwave heating has now become a wide-spread technique in organic chemistry. On the one hand, this development was enabled and assisted by the availability of commercial microwave equipment, which produces much more homogeneous heating conditions than the formerly used domestic microwave ovens. On the other hand, organic chemists realized that microwave heating not only significantly speeds up chemical reactions from hours or days to minutes but also enables "new" chemistry.

Medicinal chemistry has the need for large sets of new compounds, with druglike character and structural diversity. Originally, combinatorial chemistry was expected to generate a vast amount of new drug candidates, due to the sheer numbers of analogs that can be produced in parallel. Unfortunately, these early expectations were not met. Large series of chemically related analogs were produced, in libraries of up to millions of compounds, but most of these compounds were completely inactive. Design was dominated by synthetic accessibility, not by medicinal chemistry know-how. Thus, in recent years combinatorial chemistry developed from mixtures of impure compounds to libraries of purified, single compounds, from mere chemicals to druglike structures, and from large libraries to much smaller libraries with different scaffolds. Microwave applications aid to produce such libraries within extremely short time – this application possibly being the most important use of this fascinating new approach.

Since long C. Oliver Kappe, Karl Franzens University, Graz, Austria, is an internationally leading expert in microwave chemistry. In 2000, he created the "microwaves in organic chemistry" (MAOS) Website as information source for organic and medicinal chemists (www.maos.net), informing there about recent literature and providing links to the Websites of instrument vendors and other MAOS-related organizations. Correspondingly, the Editors of the series "Methods and Principles of Medicinal Chemistry" asked Oliver Kappe and his former coworker Alexander Stadler, now at Anton Paar GmbH, Graz, to contribute to our series "Methods and Principles in Medicinal Chemistry" a book on practical applications of microwave chemistry.

"Microwaves in Organic and Medicinal Chemistry" is the very first monograph on this topic, which deals not only with mere synthetic applications of microwave

Microwaves in Organic and Medicinal Chemistry. C. Oliver Kappe, Alexander Stadler
Copyright © 2005 WILEY-VCH Verlag GmbH & Co. KGaA, Weinheim
ISBN: 3-527-31210-2

chemistry but primarily with typical applications in medicinal chemistry. After a brief introduction into microwave synthesis and its history, theory and differences to classical heating are discussed, followed by a chapter on different commercial equipment for microwave heating. Microwave processing techniques, the use of microwave reactors and general comments on reaction optimization are discussed in the next two chapters. Two literature surveys, Part A, General organic synthesis, and Part B, Combinatorial and high-throughput synthesis methods, constitute the major part of the book. Organic reactions, from Heck to Pauson-Khand reactions and from Diels-Alder reactions to Michael additions, are discussed in a systematic manner. Afterwards syntheses of N-, O-, and S-containing five- and six-membered heterocyclic ring systems are presented, followed by sections on, e.g., peptide synthesis, multicomponent reactions, and the use of polymer-supported reagents, catalysts, and scavengers. The final chapter presents an outlook and conclusions.

This book is a treasure trove for every organic and medicinal chemist. From personal experience we confirm: once being familiar with microwave heating one would not like to miss it any longer! Oliver Kappe and Alexander Stadler aid with their monograph to the further broad distribution of microwave-supported organic synthesis. We are very grateful for their excellent contribution, as well as we thank the publisher Wiley-VCH, especially Dr. Frank Weinreich, for ongoing support of the series "Methods and Principles in Medicinal Chemistry".

May 2005

Raimund Mannhold, Düsseldorf
Hugo Kubinyi, Weisenheim am Sand
Gerd Folkers, Zurich

Personal Foreword

We are currently witnessing an explosive growth in the general field of "microwave chemistry". The increase of interest in this technology stems from the realization that microwave-assisted synthesis, apart from many other enabling technologies, actually provides significant practical and economic advantages. Although microwave chemistry is currently used in both academic and industrial contexts, the impact on the pharmaceutical industry especially, has led to the development of microwave-assisted organic synthesis (MAOS) from a laboratory curiosity in the 1980s and 1990s to a fully accepted technology today. The field has grown such that nearly every pharmaceutical company and more and more academic laboratories now actively utilize this technology for their research.

One of the main barriers facing a synthetic chemist contemplating the use of microwave synthesis today is – apart from access to suitable equipment – obtaining education and information on the fundamental principles and possible applications of this new technology. Thus, the aim of this book is to give the reader a well-structured, up-to-date, and exhaustive overview of known synthetic procedures involving the use of microwave technology, and to illuminate the "black box" stigma that microwave chemistry still has.

Our main motivation for writing *"Microwaves in Organic and Medicinal Chemistry"* derived from our experience in teaching microwave chemistry in the form of short courses and workshops to researchers from the pharmaceutical industry. In fact, the structure of this book closely follows a course developed for the American Chemical Society and can be seen as a compendium for this course. It is hoped that some of the chapters of this book are sufficiently convincing as to encourage scientists not only to use microwave synthesis in their research, but also to offer training for their students or co-workers.

We would like to thank Hugo Kubinyi for his encouragement and motivation to write this book. Thanks are also due to Mats Larhed, Nicholas E. Leadbeater, Erik Van der Eycken and scientists from Anton Paar GmbH, Biotage AB, CEM Corp., and Milestone s.r.l., who have been kind enough to read various sections of the manuscript and to provide valuable suggestions. Foremost we would like to thank Doris Dallinger, Bimbisar Desai, Toma Glasnov, Jenny Kremsner and the other members of the Kappe research group for spending their time searching the "microwave literature", and for tolerating this distraction. We are particularly indebted to

Microwaves in Organic and Medicinal Chemistry. C. Oliver Kappe, Alexander Stadler
Copyright © 2005 WILEY-VCH Verlag GmbH & Co. KGaA, Weinheim
ISBN: 3-527-31210-2

Doris Dallinger for carefully proofreading the complete manuscript, and to Jenny Whedbee for providing the cover art. We are very grateful to Frank Weinreich and his colleagues at Wiley-VCH for their assistance during the preparation of the manuscript and for the preparation of the finished book.

This book is dedicated to Rajender S. Varma, a pioneer in the field of microwave synthesis, who inspired us to enter this exciting research area in the 1990s.

Graz, Austria, April 2005

C. Oliver Kappe
Alexander Stadler

1
Introduction: Microwave Synthesis in Perspective

1.1
Microwave Synthesis and Medicinal Chemistry

Improving research and development (R&D) productivity is one of the biggest tasks facing the pharmaceutical industry. In the next 10 years, the pharmaceutical industry will see many patents of drugs expire. In order to remain competitive, pharma companies need to pursue strategies that will offset the sales decline and see robust growth and shareholder value. The impact of genomics and proteomics is creating an explosion in the number of drug targets. Today's drug therapies are based solely on approximately 500 biological targets, while in 10 years from now the number of targets could well reach 10000. In order to identify more potential drug candidates for all of these targets, pharmaceutical companies have made major investments in high-throughput technologies for genomic and proteomic research, combinatorial chemistry, and biological screening. However, lead compound optimization and medicinal chemistry remain the bottlenecks in the drug discovery process. Developing chemical compounds with the desired biological properties is time-consuming and expensive. Consequently, increasing interest is being directed toward technologies that allow more rapid synthesis and screening of chemical substances to identify compounds with functional qualities.

Medicinal chemistry has benefited tremendously from the technological advances in the field of combinatorial chemistry and high-throughput synthesis. This discipline has been the innovative machine for the development of methods and technologies which accelerate the design, synthesis, purification, and analysis of compound libraries. These new tools have had a significant impact on both lead identification and lead optimization in the pharmaceutical industry. Large compound libraries can now be designed and synthesized to provide valuable leads for new therapeutic targets. Once a chemist has developed a suitable high-speed synthesis of a lead, it is now possible to synthesize and purify hundreds of molecules in parallel to discover new leads and/or to derive structure–activity relationships (SAR) in unprecedented timeframes.

The bottleneck of conventional parallel/combinatorial synthesis is typically optimization of reaction conditions to afford the desired products in suitable yields and purities. Since many reaction sequences require at least one or more heating steps for extended time periods, these optimizations are often difficult and time-consum-

Microwaves in Organic and Medicinal Chemistry. C. Oliver Kappe, Alexander Stadler
Copyright © 2005 WILEY-VCH Verlag GmbH & Co. KGaA, Weinheim
ISBN: 3-527-31210-2

ing. Microwave-assisted heating under controlled conditions has been shown to be an invaluable technology for medicinal chemistry and drug discovery applications since it often dramatically reduces reaction times, typically from days or hours to minutes or even seconds. Many reaction parameters can be evaluated in a few hours to optimize the desired chemistry. Compound libraries can then be rapidly synthesized in either a parallel or (automated) sequential format using this new, enabling technology. In addition, microwave synthesis allows for the discovery of novel reaction pathways, which serve to expand "chemical space" in general, and "biologically relevant, medicinal chemistry space" in particular.

Specifically, microwave synthesis has the potential to impact upon medicinal chemistry efforts in at least three major phases of the drug discovery process: lead generation, hit-to-lead efforts, and lead optimization. Medicinal chemistry addresses what are fundamentally biological and clinical problems. Focusing first on the preparation of suitable molecular tools for mechanistic validation, efforts ultimately turn to the optimization of biochemical, pharmacokinetic, pharmacological, clinical, and competitive properties of drug candidates. A common theme throughout this drug discovery and development process is speed. Speed equals competitive advantage, more efficient use of expensive and limited resources, faster exploration of structure–activity relationships (SAR), enhanced delineation of intellectual property, more timely delivery of critically needed medicines, and can ultimately determine positioning in the marketplace. To the pharmaceutical industry and the medicinal chemist, time truly does equal money, and microwave chemistry has become a central tool in this fast-paced, time-sensitive field.

Chemistry, like all sciences, consists of never-ending iterations of hypotheses and experiments, with results guiding the progress and development of projects. The short reaction times provided by microwave synthesis make it ideal for rapid reaction scouting and optimization, allowing very rapid progress through the "hypotheses-experiment-results" iterations, resulting in more decision points per unit time. In order to fully benefit from microwave synthesis, one has to "be prepared to fail in order to succeed". While failure could cost a few minutes, success would gain many hours or even days. The speed at which multiple variations of reaction conditions can be performed allows a morning discussion of "What should we try?" to become an after lunch discussion of "What were the results?" (the "let's talk after lunch" mantra) [1]. Not surprisingly, therefore, most pharmaceutical, agrochemical, and biotechnology companies are already heavily using microwave synthesis as frontline methodology in their chemistry programs, both for library synthesis and for lead optimization, as they realize the ability of this enabling technology to speed chemical reactions and therefore the drug discovery process.

1.2
Microwave-Assisted Organic Synthesis (MAOS) – A Brief History

While fire is now rarely used in synthetic chemistry, it was not until Robert Bunsen invented the burner in 1855 that the energy from this heat source could be applied to a reaction vessel in a focused manner. The Bunsen burner was later superseded by the isomantle, the oil bath or the hot plate as a means of applying heat to a chemical reaction. In the past few years, heating and driving chemical reactions by microwave energy has been an increasingly popular theme in the scientific community [1, 2].

Microwave energy, originally applied for heating foodstuffs by Percy Spencer in the 1940s, has found a variety of technical applications in the chemical and related industries since the 1950s, in particular in the food-processing, drying, and polymer industries. Other applications range from analytical chemistry (microwave digestion, ashing, extraction) [3] to biochemistry (protein hydrolysis, sterilization) [3], pathology (histoprocessing, tissue fixation) [4], and medical treatments (diathermy) [5]. Somewhat surprisingly, microwave heating has only been implemented in organic synthesis since the mid-1980s. The first reports on the use of microwave heating to accelerate organic chemical transformations (MAOS) were published by the groups of Richard Gedye (Scheme 1.1) [6] and Raymond J. Giguere/George Majetich [7] in 1986. In those early days, experiments were typically carried out in sealed Teflon or glass vessels in a domestic household microwave oven without any temperature or pressure measurements. The results were often violent explosions due to the rapid uncontrolled heating of organic solvents under closed-vessel conditions. In the 1990s, several groups started to experiment with solvent-free microwave chemistry (so-called dry-media reactions), which eliminated the danger of explosions [8]. Here, the reagents were pre-adsorbed onto either an essentially microwave-transparent (i.e., silica, alumina or clay) or strongly absorbing (i.e., graphite) inorganic support, that additionally may have been doped with a catalyst or reagent. Particularly in the early days of MAOS, the solvent-free approach was very popular since it allowed the safe use of domestic microwave ovens and standard open-vessel technology. While a large number of interesting transformations using "dry-media" reactions have been published in the literature [8], technical difficulties relating to non-uniform heating, mixing, and the precise determination of the reaction temperature remained unresolved, in particular when scale-up issues needed to be addressed.

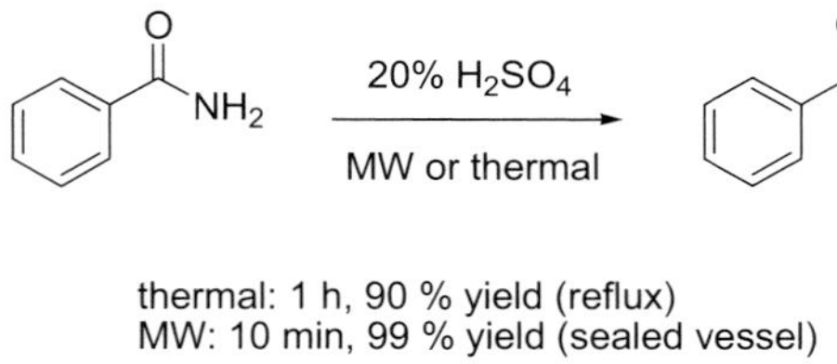

thermal: 1 h, 90 % yield (reflux)
MW: 10 min, 99 % yield (sealed vessel)

Scheme 1.1 Hydrolysis of benzamide. The first published example (1986) of microwave-assisted organic synthesis.

Alternatively, microwave-assisted synthesis has been carried out using standard organic solvents under open-vessel conditions. If solvents are heated by microwave irradiation at atmospheric pressure in an open vessel, the boiling point of the solvent typically limits the reaction temperature that can be achieved. In order to nonetheless achieve high reaction rates, high-boiling microwave-absorbing solvents have been frequently used in open-vessel microwave synthesis [9]. However, the use of these solvents presented serious challenges in relation to product isolation and recycling of the solvent. Because of the recent availability of modern microwave reactors with on-line monitoring of both temperature and pressure, MAOS in dedicated sealed vessels using standard solvents – a technique pioneered by Christopher R. Strauss in the mid-1990s [10] – has been celebrating a comeback in recent years. This is clearly evident surveying the recently published (since 2001) literature in the area of controlled microwave-assisted organic synthesis (MAOS). It appears that the combination of rapid heating by microwaves with sealed-vessel (autoclave) technology will most likely be the method of choice for performing MAOS on a laboratory scale in the future. Importantly, recent innovations in microwave reactor technology now allow controlled parallel and automated sequential processing under sealed-vessel conditions, and the use of continuous- or stop-flow reactors for scale-up purposes.

Since the early days of microwave synthesis, the observed rate accelerations and sometimes altered product distributions compared to oil-bath experiments have led to speculation on the existence of so-called "specific" or "non-thermal" microwave effects [11]. Historically, such effects were claimed when the outcome of a synthesis performed under microwave conditions was different from that of the conventionally heated counterpart at the same apparent temperature. Reviewing the present literature [12], it appears that today most scientists agree that in the majority of cases the reason for the observed rate enhancements is a purely thermal/kinetic effect, i.e., a consequence of the high reaction temperatures that can rapidly be attained when irradiating polar materials in a microwave field, although effects that are caused by the unique nature of the microwave dielectric heating mechanism ("specific microwave effects") clearly also need to be considered. While for the medicinal chemist in industry this discussion may seem largely irrelevant, the debate on "microwave effects" is undoubtedly going to continue for many years in the academic world. Regardless of the nature of the observed rate enhancements (for further details on microwave effects, see Section 2.5), microwave synthesis has now truly matured and has moved from a laboratory curiosity in the late 1980s to an established technique in organic synthesis, heavily used in both academia and industry.

The initially slow uptake of the technology in the late 1980s and 1990s has been attributed to its lack of controllability and reproducibility, coupled with a general lack of understanding of the basics of microwave dielectric heating. The risks associated with the flammability of organic solvents in a microwave field and the lack of available dedicated microwave reactors allowing for adequate temperature and pressure control were major concerns. Important instrument innovations (see Chapter 3) now allow for careful control of time, temperature, and pressure profiles, paving the way for reproducible protocol development, scale-up, and transfer from laboratory to laboratory and from scientist to scientist. Today, microwave chemistry is as reliable as the vast arsenal of synthetic methods that preceded it. Since 2001, therefore, the number of publications related to MAOS has increased dramatically (Fig. 1.1), to such a level that it might be assumed that, in a few years, most chemists

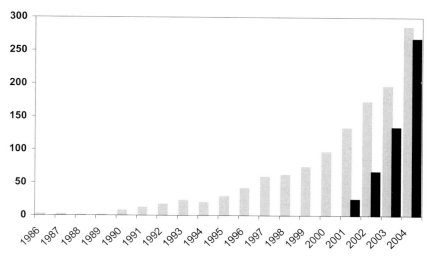

Fig. 1.1 Publications on microwave-assisted organic synthesis (1986–2004). Gray graphs: Number of articles involving MAOS for seven selected synthetic organic chemistry journals (*J. Org. Chem*, *Org. Lett.*, *Tetrahedron*, *Tetrahedron Lett.*, *Synth. Commun.*, *Synthesis*, *Synlett*; SciFinder scholar search, keyword: "microwave not spectroscopy"). The black graphs represent the number of publications (2001–2004) reporting MAOS experiments in dedicated reactors with adequate process control (ca. 50 journals, full text search: microwave).

will probably use microwave energy to heat chemical reactions on a laboratory scale [1, 2]. Not only is direct microwave heating able to reduce chemical reaction times significantly, but it is also known to reduce side reactions, increase yields, and improve reproducibility. Therefore, many academic and industrial research groups are already using MAOS as a technology for rapid reaction optimization, for the efficient synthesis of new chemical entities, or for discovering and probing new chemical reactivity.

1.3
Scope and Organization of the Book

Today, a large body of work on microwave-assisted synthesis exists in the published and patent literature. Many review articles [8–20], several books [21–23], and information on the world-wide-web [24] already provide extensive coverage of the subject. The goal of the present book is to present carefully scrutinized, useful, and practical information for both beginners and advanced practitioners of microwave-assisted organic synthesis. Special emphasis is placed on concepts and chemical transformations that are of importance to medicinal chemists, and that have been reported in the most recent literature (2002–2004). The extensive literature survey is limited to reactions that have been performed using controlled microwave heating conditions, i.e., where dedicated microwave reactors for synthetic applications with adequate temperature and pressure measurements have been employed. After a discussion of microwave dielectric heating theory and microwave effects (Chapter 2), a review of the existing equipment for performing MAOS is presented (Chapter 3). This is followed by a chapter outlining the different processing techniques in a microwave-heated experiment (Chapter 4) and a chapter on "how to get started" with microwave synthesis, including safety aspects (Chapter 5). Finally, a literature survey with more than 600 references is presented in Chapters 6, 7, and 8.

References

[1] N. Leadbeater, *Chemistry World* **2004**, *1*, 38–41.

[2] D. Adam, *Nature* **2003**, *421*, 571–572.

[3] H. M. Kingston, S. J. Haswell (Eds.), *Microwave-Enhanced Chemistry. Fundamentals, Sample Preparation and Applications*, American Chemical Society, Washington DC, **1997**.

[4] R. T. Giberson, R. S. Demaree (Eds.), *Microwave Techniques and Protocols*, Humana Press, Totowa, New Jersey, **2001**.

[5] W. E. Prentice, *Therapeutic Modalities for Physical Therapists*, McGraw-Hill, New York, **2002**.

[6] R. Gedye, F. Smith, K. Westaway, H. Ali, L. Baldisera, L. Laberge, J. Rousell, *Tetrahedron Lett.* **1986**, *27*, 279–282.

[7] R. J. Giguere, T. L. Bray, S. M. Duncan, G. Majetich, *Tetrahedron Lett.* **1986**, *27*, 4945–4958.

[8] A. Loupy, A. Petit, J. Hamelin, F. Texier-Boullet, P. Jacquault, D. Mathé, *Synthesis* **1998**, 1213–1234; R. S. Varma, *Green Chem.* **1999**, 43–55; M. Kidawi, *Pure Appl. Chem.* **2001**, *73*, 147–151; R. S. Varma, *Pure Appl. Chem.* **2001**, *73*, 193–198; R. S. Varma, *Tetrahedron* **2002**, *58*, 1235–1255; R. S. Varma, *Advances in Green Chemistry: Chemical Syntheses Using Microwave Irradiation*, Kavitha Printers, Bangalore, **2002**.

[9] A. K. Bose, B. K. Banik, N. Lavlinskaia, M. Jayaraman, M. S. Manhas, *Chemtech* **1997**, *27*, 18–24; A. K. Bose, M. S. Manhas, S. N. Ganguly, A. H. Sharma, B. K. Banik, *Synthesis* **2002**, 1578–1591.

[10] C. R. Strauss, R. W. Trainor, *Aust. J. Chem.* **1995**, *48*, 1665–1692; C. R. Strauss, *Aust. J. Chem.* **1999**, *52*, 83–96; C. R. Strauss, in *Microwaves in Organic Synthesis* (Ed.: A. Loupy), Wiley-VCH, Weinheim, **2002**, pp 35–60 (Chapter 2).

[11] L. Perreux, A. Loupy, *Tetrahedron* **2001**, *57*, 9199–9223; N. Kuhnert, *Angew. Chem. Int. Ed.* **2002**, *41*, 1863–1866; C. R. Strauss, *Angew. Chem. Int. Ed.* **2002**, *41*, 3589–3590.

[12] C. O. Kappe, *Angew. Chem. Int. Ed.* **2004**, *43*, 6250–6284.

[13] General organic synthesis: R. A. Abramovitch, *Org. Prep. Proced. Int.* **1991**, *23*, 685–711; S. Caddick, *Tetrahedron* **1995**, *51*, 10403–10432; P. Lidström, J. Tierney, B. Wathey, J. Westman, *Tetrahedron* **2001**, *57*, 9225–9283; B. L. Hayes, *Aldrichim. Acta* **2004**, *37*, 66–77.

[14] For more technical reviews, see: M. Nüchter, B. Ondruschka, W. Bonrath, A. Gum, *Green Chem.* **2004**, *6*, 128–141; M. Nüchter, U. Müller, B. Ondruschka, A. Tied, W. Lautenschläger, *Chem. Eng. Technol.* **2003**, *26*, 1207–1216.

[15] Cycloaddition reactions: A. de la Hoz, A. Díaz-Ortis, A. Moreno, F. Langa, *Eur. J. Org. Chem.* **2000**, 3659–3673.

[16] Heterocycle synthesis: J. Hamelin, J.-P. Bazureau, F. Texier-Boullet, in *Microwaves in Organic Synthesis* (Ed.: A. Loupy), Wiley-VCH, Weinheim, **2002**, pp 253–294 (Chapter 8); T. Besson, C. T. Brain, in *Microwave-Assisted Organic Synthesis* (Eds.: P. Lidström, J. P. Tierney), Blackwell Publishing, Oxford, **2005** (Chapter 3); Y. Xu, Q.-X. Guo, *Heterocycles* **2004**, *63*, 903–974.

[17] Radiochemistry: N. Elander, J. R. Jones, S.-Y. Lu, S. Stone-Elander, *Chem. Soc. Rev.* **2000**, 239–250; S. Stone-Elander, N. Elander, *J. Label. Compd. Radiopharm.* **2002**, *45*, 715–746.

[18] Homogeneous transition metal catalysis: M. Larhed, C. Moberg, A. Hallberg, *Acc. Chem. Res.* **2002**, *35*, 717–727; K. Olofsson, M. Larhed, in *Microwave-Assisted Organic Synthesis* (Eds.: P. Lidström, J. P. Tierney), Blackwell Publishing, Oxford, **2005** (Chapter 2).

[19] Medicinal chemistry: J. L. Krstenansky, I. Cotterill, *Curr. Opin. Drug Discovery Dev.* **2000**, *4*, 454–461; M. Larhed, A. Hallberg, *Drug Discovery Today* **2001**, *6*, 406–416; B. Wathey, J. Tierney, P. Lidström, J. Westman, *Drug Discovery Today* **2002**, *7*, 373–380; N. S. Wilson, G. P. Roth, *Curr. Opin. Drug Discovery Dev.* **2002**, *5*, 620–629; C. D. Dzierba, A. P. Combs, in *Ann. Rep. Med. Chem.* (Ed.: A. M. Doherty), Academic Press, **2002**, vol. 37, pp 247–256.

[20] Combinatorial chemistry: A. Lew, P. O. Krutznik, M. E. Hart, A. R. Chamberlin, *J. Comb. Chem.* **2002**, *4*, 95–105; C. O. Kappe, *Curr. Opin. Chem. Biol.* **2002**, *6*, 314–320; P. Lidström, J. Westman, A. Lewis, *Comb. Chem. High Throughput Screen.* **2002**, *5*, 441–458; H. E. Blackwell, *Org. Biomol. Chem.* **2003**, *1*, 1251–1255; F. Al-Obeidi, R. E. Austin, J. F. Okonya, D. R. S. Bond, *Mini-Rev. Med. Chem.* **2003**, *3*, 449–460; K. M. K. Swamy, W.-B. Yeh, M.-J. Lin, C.-M. Sun, *Curr. Med. Chem.* **2003**, *10*, 2403–2423; *Microwaves in Combinatorial and High-Throughput Synthesis* (Ed.: C. O. Kappe), Kluwer, Dordrecht, **2003** (a special issue of *Mol. Diversity* **2003**, *7*, pp 95–307); A. Stadler, C. O. Kappe, in *Microwave-Assisted Organic Synthesis* (Eds.: P. Lidström, J. P. Tierney), Blackwell Publishing, Oxford, **2005** (Chapter 7).

[21] A. Loupy (Ed.), *Microwaves in Organic Synthesis*, Wiley-VCH, Weinheim, **2002**.

[22] B. L. Hayes, *Microwave Synthesis: Chemistry at the Speed of Light*, CEM Publishing, Matthews, NC, **2002**.

[23] P. Lidström, J. P. Tierney (Eds.), *Microwave-Assisted Organic Synthesis*, Blackwell Publishing, Oxford, **2005**.

[24] For online resources on microwave-assisted organic synthesis (MAOS), see: *www.maos.net*.

2
Microwave Theory

The physical principles behind and the factors determining the successful application of microwaves in organic synthesis are not widely familiar to chemists, possibly
because electric field theory is generally taught in engineering or physics rather
than in chemistry. Nevertheless, it is essential for the synthetic chemist involved in
microwave-assisted organic synthesis to have at least a basic knowledge of the
underlying principles of microwave–matter interactions and on microwave effects.
The basic understanding of macroscopic microwave interactions with matter was
formulated by von Hippel in the mid-1950s [1]. In this chapter, a brief summary on
the current understanding of microwaves and their interactions with matter is
given. For more in depth discussion of this quite complex field, the reader is
referred to recent review articles [2–5].

2.1
Microwave Radiation

Microwave irradiation is electromagnetic irradiation in the frequency range of 0.3 to
300 GHz, corresponding to wavelengths of 1 cm to 1 m. The microwave region of
the electromagnetic spectrum (Fig. 2.1) therefore lies between infrared and radio frequencies. Wavelengths between 1 cm and 25 cm are extensively used for RADAR
transmissions and the remaining wavelength range is used for telecommunications.
All domestic "kitchen" microwave ovens and all dedicated microwave reactors for
chemical synthesis that are commercially available today operate at a frequency of
2.45 GHz (corresponding to a wavelength of 12.25 cm) in order to avoid interference
with telecommunication and cellular phone frequencies. There are other frequency
allocations for microwave heating applications (ISM frequencies, see Table 2.1) [6],
but these are not generally employed in dedicated reactors for synthetic chemistry.
Indeed, published examples of organic syntheses carried out with microwave heating at frequencies other than 2.45 GHz are extremely rare [7].

Microwaves in Organic and Medicinal Chemistry. C. Oliver Kappe, Alexander Stadler
Copyright © 2005 WILEY-VCH Verlag GmbH & Co. KGaA, Weinheim
ISBN: 3-527-31210-2

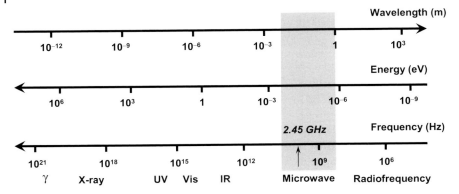

Fig. 2.1 The electromagnetic spectrum.

Table 2.1 ISM microwave frequencies (for industrial, scientific, and medical use) (data from [6]).

Frequency (MHz)	Wavelength (cm)
433.92 ± 0.2	69.14
915 ± 13	32.75
2450 ± 50	12.24
5800 ± 75	5.17
24125 ± 125	1.36

From comparison of the data presented in Table 2.2 [8], it is obvious that the energy of the microwave photon at a frequency of 2.45 GHz (0.0016 eV) is too low to cleave molecular bonds and is also lower than Brownian motion. It is therefore clear that microwaves cannot "induce" chemical reactions by direct absorption of electromagnetic energy, as opposed to ultraviolet and visible radiation (photochemistry).

Table 2.2 Comparison of radiation types and bond energies (data from [6, 8]).

Radiation type	Frequency (MHz)	Quantum energy (eV)	Bond type	Bond energy (eV)
Gamma rays	3.0×10^{14}	1.24×10^6	CC single bond	3.61
X-rays	3.0×10^{13}	1.24×10^5	CC double bond	6.35
Ultraviolet	1.0×10^9	4.1	CO single bond	3.74
Visible light	6.0×10^8	2.5	CO double bond	7.71
Infrared light	3.0×10^6	0.012	CH bond	4.28
Microwaves	2450	0.0016	OH bond	4.80
Radiofrequencies	1	4.0×10^{-9}	Hydrogen bond	0.04–0.44

2.2
Microwave Dielectric Heating

Microwave-enhanced chemistry is based on the efficient heating of materials by "microwave dielectric heating" effects [4, 5]. Microwave dielectric heating is dependent on the ability of a specific material (for example, a solvent or reagent) to absorb microwave energy and convert it into heat. Microwaves are electromagnetic waves which consist of an electric and a magnetic field component (Fig. 2.2). For most practical purposes related to microwave synthesis it is the electric component of the electromagnetic field that is of importance for wave–material interactions, although in some instances magnetic field interactions (for example with transition metal oxides) can also be of relevance [9].

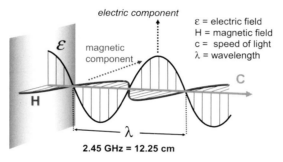

Fig. 2.2   Electric and magnetic field components in microwaves.

The electric component of an electromagnetic field causes heating by two main mechanisms: dipolar polarization and ionic conduction. The interaction of the electric field component with the matrix is called the dipolar polarization mechanism (Fig. 2.3.a) [4, 5]. For a substance to be able to generate heat when irradiated with microwaves it must possess a dipole moment. When exposed to microwave frequencies, the dipoles of the sample align in the applied electric field. As the applied field oscillates, the dipole field attempts to realign itself with the alternating electric field and, in the process, energy is lost in the form of heat through molecular friction and dielectric loss. The amount of heat generated by this process is directly related to the ability of the matrix to align itself with the frequency of the applied field. If the dipole does not have enough time to realign (high-frequency irradiation) or reorients too quickly (low-frequency irradiation) with the applied field, no heating occurs. The allocated frequency of 2.45 GHz used in all commercial systems lies between these two extremes and gives the molecular dipole time to align in the field, but not to follow the alternating field precisely. Therefore, as the dipole reorientates to align itself with the electric field, the field is already changing and generates a phase difference between the orientation of the field and that of the dipole. This phase difference causes energy to be lost from the dipole by molecular friction and collisions, giving rise to dielectric heating. In summary, field energy is transferred to the medium and electrical energy is converted into kinetic or thermal energy, and ultimately into heat. It should be emphasized that the interaction between microwave radiation and the polar solvent which occurs when the frequency of the radiation approximately matches the frequency of the rotational relaxation process is not a quantum mechanical resonance phenomenon. Transitions between quantized rotational bands are not involved and the energy transfer is not a property of a specific molecule, but the result of a collective phenomenon involving the bulk [4, 5]. The heat is generated by frictional forces occurring between the polar molecules, the rotational velocity of which has been increased by the coupling with the microwave irradiation. It should also be noted that gases cannot be heated under microwave irradiation, since the distance between the rotating molecules is too great. Similarly, ice is also (nearly) microwave transparent, since the water dipoles are constrained in a crystal lattice and cannot move as freely as in the liquid state.

The second major heating mechanism is the ionic conduction mechanism (Fig. 2.3.b) [4, 5]. During ionic conduction, as the dissolved charged particles in a sample (usually ions) oscillate back and forth under the influence of the microwave field, they collide with their neighboring molecules or atoms. These collisions cause agitation or motion, creating heat. Thus, if two samples containing equal amounts of distilled water and tap water, respectively, are heated by microwave irradiation at a fixed radiation power, more rapid heating will occur for the tap water sample due to its ionic content. Such ionic conduction effects are particularly important when considering the heating behavior of ionic liquids in a microwave field (see Section 4.3.3.2). The conductivity principle is a much stronger effect than the dipolar rotation mechanism with regard to the heat-generating capacity.

Fig. 2.3 (a) Dipolar polarization mechanism. (b) Dipolar molecules try to align with an oscillating electric field. Ionic conduction mechanism. Ions in solution will move in the electric field.

2.3
Dielectric Properties

The heating characteristics of a particular material (for example, a solvent) under microwave irradiation conditions are dependent on the dielectric properties of the material. The ability of a specific substance to convert electromagnetic energy into heat at a given frequency and temperature is determined by the so-called loss tangent, $\tan \delta$. The loss factor is expressed as the quotient $\tan \delta = \varepsilon''/\varepsilon'$, where ε'' is the dielectric loss, indicative of the efficiency with which electromagnetic radiation is

Table 2.3 Loss tangents (tan δ) of various solvents (2.45 GHz, 20 °C) (data from [10]).

Solvent	tan δ	Solvent	tan δ
Ethylene glycol	1.350	N,N-dimethylformamide	0.161
Ethanol	0.941	1,2-dichloroethane	0.127
Dimethyl sulfoxide	0.825	Water	0.123
2-propanol	0.799	Chlorobenzene	0.101
Formic acid	0.722	Chloroform	0.091
Methanol	0.659	Acetonitrile	0.062
Nitrobenzene	0.589	Ethyl acetate	0.059
1-butanol	0.571	Acetone	0.054
2-butanol	0.447	Tetrahydrofuran	0.047
1,2-dichlorobenzene	0.280	Dichloromethane	0.042
1-methyl-2-pyrrolidone	0.275	Toluene	0.040
Acetic acid	0.174	Hexane	0.020

converted into heat, and ε' is the dielectric constant describing the polarizability of the molecules in the electric field. A reaction medium with a high tan δ is required for efficient absorption and, consequently, for rapid heating. Materials with a high dielectric constant such as water (ε' at 25 °C = 80.4) may not necessarily also have a high tan δ value. In fact, ethanol has a significantly lower dielectric constant (ε' at 25 °C = 24.3), but heats much more rapidly than water in a microwave field due to its higher loss tangent (tan δ: ethanol = 0.941, water = 0.123). The loss tangents for some common organic solvents are summarized in Table 2.3 [10]. In general, solvents can be classified as high (tan δ > 0.5), medium (tan δ 0.1–0.5), or low microwave-absorbing (tan δ < 0.1). Other common solvents without a permanent dipole moment, such as carbon tetrachloride, benzene, and dioxane, are more or less microwave-transparent. It has to be emphasized that a low tan δ value does not preclude a particular solvent from being used in a microwave-heated reaction. Since either the substrates or some of the reagents/catalysts are likely to be polar, the overall dielectric properties of the reaction medium will in most cases allow sufficient heating by microwaves. Furthermore, polar additives such as alcohols or ionic liquids can be added to otherwise low-absorbing reaction mixtures in order to increase the absorbance level of the medium (see Section 4.3.3.2).

The loss tangent values are both frequency- and temperature-dependent. Fig. 2.4 shows the dielectric properties of distilled water as a function of frequency at 25 °C [1, 4, 5]. It is apparent that the dielectric loss ε'' has an appreciable value over a wide frequency range. The dielectric loss ε'' goes through a maximum as the dielectric constant ε' falls. The heating, as measured by ε'', reaches its maximum at around 18 GHz, while all domestic microwave ovens and dedicated reactors for chemical synthesis operate at a much lower frequency, 2.45 GHz. The practical reason for the lower frequency is the necessity to heat food efficiently throughout its interior. If the frequency is optimal for a maximum heating rate, the microwaves are absorbed in the outer regions of the food, and penetrate only a short distance [4].

According to definition, the penetration depth is the point where 37% (1/e) of the initially irradiated microwave power is still present [6]. The penetration depth is in-

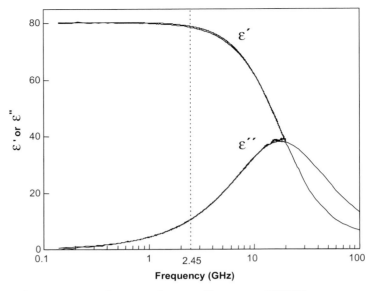

Fig. 2.4 Dielectric properties of water as a function of frequency at 20 °C [11].

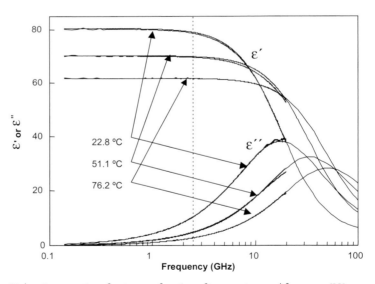

Fig. 2.5 Dielectric properties of water as a function of temperature and frequency [11].

versely proportional to tan δ and therefore critically depends on factors such as temperature and irradiation frequency. For a solvent such as water, the penetration depth at room temperature is only of the order of a few centimeters. The dielectric loss and loss tangent of water and most other organic solvents decrease with increasing temperature (Fig. 2.5), hence the absorption of microwave radiation in water decreases at higher temperatures. In turn, the penetration depth of microwaves increases. Issues relating to the penetration depth are critically important when considering the scale-up of MAOS (see Section 4.5).

The interaction of microwave irradiation with matter is characterized by three different processes: absorption, transmission and reflection. Highly dielectric materials such as polar organic solvents strongly absorb microwaves and consequently rapid heating of the medium ensues. Non-polar materials exhibit only small interactions with penetrating microwaves and can thus be used as construction materials for reactors. If microwave radiation is reflected by the material surface, there is little or no coupling of energy in the system. The temperature of the material increases only marginally. This holds true in particular for metals with high conductivity.

2.4
Microwave Versus Conventional Thermal Heating

Traditionally, organic synthesis is carried out by conductive heating with an external heat source (for example, an oil bath or heating mantle). This is a comparatively slow and inefficient method for transferring energy into the system since it depends on convection currents and on the thermal conductivity of the various materials that must be penetrated, and results in the temperature of the reaction vessel being higher than that of the reaction mixture. In addition, a temperature gradient can develop within the sample and local overheating can lead to product, substrate or reagent decomposition.

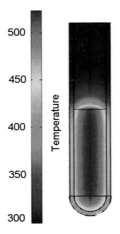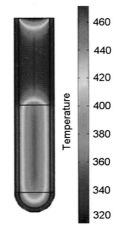

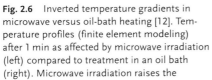

Fig. 2.6 Inverted temperature gradients in microwave versus oil-bath heating [12]. Temperature profiles (finite element modeling) after 1 min as affected by microwave irradiation (left) compared to treatment in an oil bath (right). Microwave irradiation raises the temperature of the whole volume simultaneously (bulk heating), whereas in the oil-heated tube the reaction mixture in contact with the vessel wall is heated first. Temperature scales in Kelvin. Reproduced with permission from [12].

In contrast, microwave irradiation produces efficient internal heating (in core volumetric heating) by direct coupling of microwave energy with the molecules (solvents, reagents, catalysts) that are present in the reaction mixture. Since the reaction vessels employed are typically made out of (nearly) microwave-transparent materials such as borosilicate glass, quartz or Teflon, the radiation passes through the walls of the vessel and an inverted temperature gradient as compared to conventional thermal heating results (Fig. 2.6). If the microwave cavity is well designed, the temperature increase will be uniform throughout the sample. The very efficient internal heat transfer results in minimized wall effects (no hot vessel surface), which may lead to the observation of so-called specific microwave effects (see Section 2.5.2), for example in the context of diminished catalyst deactivation. It should be emphasized that microwave dielectric heating and thermal heating by convection are totally different processes, and that any comparison between the two is inherently difficult.

2.5
Microwave Effects

Despite the relatively large body of published work on microwave-assisted chemistry (see Chapters 6 and 7), and the basic understanding of high-frequency electromagnetic irradiation and microwave–matter interactions, the exact reasons why and how microwaves enhance chemical processes are still not fully understood. Several groups have speculated on the existence of so-called "microwave effects" [13]. Such "microwave effects" could be the consequence of specific wave–material interactions, leading to a decrease in activation energy or an increase in the pre-exponential factor in the Arrhenius law due to orientation effects of polar species in an electromagnetic field [13]. Other researchers strictly denounce the existence of non-thermal effects and rationalize all rate enhancements in terms of the rapid heating and the high temperatures that are attained in a microwave-heated chemical reaction [14], the formation of microscopic or macroscopic hotspots, or the selective heating of a specific component in the reaction mixture [15]. The controversy about microwave effects has led to heated debates at microwave chemistry conferences and in the chemical literature [13–17]. Although the present chapter cannot provide a definitive answer on the issue of microwave effects, the basic concepts will be illustrated in the following sections.

It should be obvious from a scientific standpoint that the question of microwave effects needs to be addressed in a serious manner, given the rapid increase in the use of microwave technology in chemical sciences, in particular organic synthesis. There is an urgent need to remove the "black box" stigma of microwave chemistry and to provide a scientific rationalization for the observed effects. This is even more important if one considers safety aspects once this technology moves from small-scale laboratory work to pilot- or production-scale instrumentation.

Since the early days of microwave synthesis, the observed rate accelerations and sometimes altered product distributions compared to oil-bath experiments have led to speculation on the existence of so-called "specific" or "non-thermal" microwave effects. Historically, such effects were claimed when the outcome of a synthesis performed under microwave conditions was different from that of the conventionally heated counterpart at the same apparent temperature. An extreme example is highlighted in Scheme 2.1. Here, Soufiaoui and coworkers [18] synthesized a series of 1,5-diazepin-2-ones in high yield in only 10 min by the condensation of *ortho*-aryldiamines with β-ketoesters in xylene under microwave irradiation in an open vessel at reflux temperature, utilizing a conventional domestic microwave oven. Surprisingly, they observed that no reaction occurred when the same reaction mixtures were heated conventionally for 10 min at the same temperature. In their publication, the authors specifically point to the involvement of "specific effects (which are not necessarily thermal)" in rationalizing the observed product yields. These results could be taken as clear evidence for a specific microwave effect. Interestingly, however, Gedye and Wei later reinvestigated the exact same reaction under thermal and microwave conditions and found there to be virtually no difference in the rates of the microwave and the conventionally heated reactions, with both leading to similar product yields [7]. The literature is full of examples like the one highlighted above, with conflicting reports on the involvement or non-involvement of "specific" or "non-thermal" microwave effects for a wide variety of different types of chemical reactions [13–17]. Microwave effects are the subject of considerable current debate and controversy and it is evident that extensive research efforts are necessary in order to truly understand these and related phenomena.

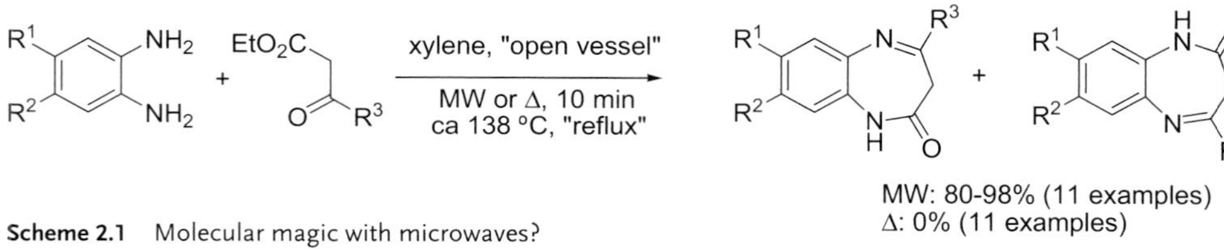

MW: 80-98% (11 examples)
Δ: 0% (11 examples)

Scheme 2.1 Molecular magic with microwaves?

Essentially, one can envisage three different possibilities for rationalizing rate enhancements observed in a microwave-assisted chemical reaction [12]:

- thermal effects (kinetics),
- specific microwave effects,
- non-thermal (athermal) microwave effects.

Clearly, a combination of two or all three contributions may be responsible for the observed phenomena, which makes the investigation of microwave effects an extremely complex subject.

2.5.1
Thermal Effects (Kinetics)

Reviewing the present literature, it appears that today many scientists suggest that in the majority of cases the reason for the observed rate enhancements is a purely thermal/kinetic effect, that is, a consequence of the high reaction temperatures that

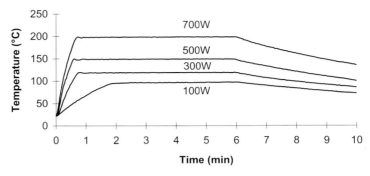

Fig. 2.7 Temperature profiles for a 30 mL sample of 1-methyl-2-pyrrolidone heated under open-vessel microwave irradiation conditions [19]. Multimode microwave heating at different maximum power levels for 6 min with temperature control using the feedback from a fiber-optic probe. After the set temperatures of 200 °C (700 W), 150 °C (500 W), 120 °C (300 W), and 100 °C (100 W) are reached, the power regulates itself down to an appropriate level (not shown). Reproduced with permission from [19].

can rapidly be attained when irradiating polar materials in a microwave field [14]. As shown in Fig. 2.7, even a moderately strong microwave-absorbing solvent such as 1-methyl-2-pyrrolidone (NMP, bp 202–204 °C, tan δ = 0.275, see Table 2.3) can be heated very rapidly ("microwave flash heating") in a microwave cavity. As indicated in Fig. 2.7, a sample of NMP can be heated to 200 °C within ca. 40 s, depending on the maximum output power of the magnetron [19].

A recent trend in microwave-assisted organic synthesis is to perform reactions under sealed-vessel conditions in relatively small, so-called single-mode microwave reactors with high power density (see Chapter 3) [12]. Under these autoclave-type conditions, microwave-absorbing solvents with a comparatively low boiling point such as methanol (tan δ = 0.659) can be rapidly superheated to temperatures of more than 100 °C in excess of their boiling points when irradiated under microwave conditions (Fig. 2.8). The rapid increase in temperature can be even more pronounced for media with extreme loss tangents, such as ionic liquids (see Section 4.3.3.2), for which temperature jumps of 200 °C within a few seconds are not uncommon. Naturally, such temperature profiles are very difficult if not impossible to reproduce by standard thermal heating. Therefore, comparisons with conventionally heated processes are inherently troublesome.

Dramatic rate enhancements when comparing reactions that are performed at room temperature or under standard oil-bath conditions (heating under reflux) with high-temperature microwave-heated processes have frequently been observed. As Mingos and Baghurst [4] have pointed out based on simply applying the Arrhenius law [$k = A \exp(-E_a/RT)$], a transformation that requires 68 days to reach 90% conversion at 27 °C will show the same degree of conversion within 1.61 seconds (!) when performed at 227 °C (Table 2.4). Due to the very rapid heating and extreme temperatures observable in microwave chemistry, it appears obvious that many of the reported rate enhancements can be rationalized in terms of purely thermal/kinetic effects. In the absence of any "specific" or "non-thermal" effects, however, one would expect reactions carried out under open-vessel, reflux conditions to proceed at the same reaction rate, regardless of whether heated by microwaves or by a thermal process [18]. It should be emphasized that for these strictly thermal effects, the pre-exponential factor A and the energy term (activation energy E_a) in the Arrhenius equation are not affected, and only the temperature term changes (see Section 2.5.3).

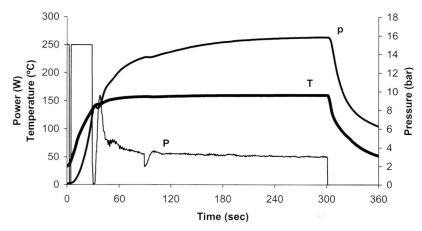

Fig. 2.8 Temperature (T), pressure (p), and power (P) profiles for a 3 mL sample of methanol heated under sealed-vessel microwave irradiation conditions [12]. Single-mode microwave heating (250 W, 0–30 s), temperature control using the feedback from IR thermography (40–300 s), and active gas-jet cooling (300–360 s). The maximum pressure in the reaction vessel was ca. 16 bar. After the set temperature of 160 °C is reached, the power regulates itself down to ca. 50 W. Reproduced with permission from [12].

Table 2.4 Relationship between temperature and time for a typical first-order reaction ($A = 4 \times 10^{10}$ mol^{-1}s^{-1}, $E_a = 100$ kJ mol^{-1}) (data from [4]).

Temperature (°C)	Rate constant, k (s^{-1})	Time (90% conversion)
27	1.55×10^{-7}	68 days
77	4.76×10^{-5}	13.4 h
127	3.49×10^{-3}	11.4 min
177	9.86×10^{-2}	23.4 s
227	1.43	1.61 s

2.5.2
Specific Microwave Effects

In addition to the above mentioned thermal/kinetic effects, microwave effects that are caused by the unique nature of the microwave dielectric heating mechanisms (see Section 2.2) must also be considered. These effects should be termed "specific microwave effects" and shall be defined as *accelerations of chemical transformations in a microwave field that cannot be achieved or duplicated by conventional heating, but essentially are still thermal effects*. In this category falls, for example, the superheating effect of solvents at atmospheric pressure [21–23]. The question of the boiling point of a liquid undergoing microwave irradiation is one of the basic problems facing microwave heating. Several groups have established that the enthalpy of vaporization is the same under both microwave and conventional heating conditions [24]. These studies have also shown that the rate of evaporation, as well as the temperature of both vapor and liquid at the interface, strongly depends on the experimental conditions. Initial studies of boiling phenomena related to microwave chemistry were reported in 1992 by Baghurst and Mingos [21], and later by Saillard and co-workers [22]. It was established that microwave-heated liquids boil at temperatures above the equilibrium boiling point at atmospheric pressure. For many solvents, the superheating temperature can be up to 40 °C above the classical boiling point [23]. Therefore, in a microwave-heated reactor, the average temperature of the solvent can be significantly higher than the atmospheric boiling point. This is because the microwave power is dissipated over the whole volume of the solvent. The most significant way to lose excess thermal energy is by boiling. However, this will only occur at the existing liquid–gas interfaces, in contrast to a thermally heated solvent where boiling typically occurs at nucleation points (cavities, pits, and scratches) on the glass reactor surface [21]. The bulk temperature of a microwave-irradiated solvent under boiling depends on many factors, such as the physical properties of the solvent, reactor geometry, mass flow, heat flow, and electric field distribution. It

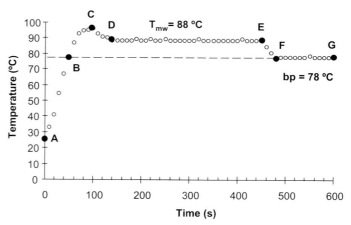

Fig. 2.9 Microwave heating (single-mode reactor) of ethanol under open-vessel conditions. Initially, the temperature rises during the heating phase (AB), above the normal boiling point of ethanol (78 °C), to a point C at which the solvent bumps and starts to boil at the vapor/liquid interface. At this point, the temperature drops to a plateau region (D) and it can be maintained at this temperature for many hours to within ± 1 °C. Addition of a boiling stone (E) brings the temperature to the normal boiling point of ethanol (FG). Reproduced with permission from [23].

should be emphasized that practically all superheating can be removed by adding boiling stones or stirring (Fig. 2.9) [23]. Importantly, however, the kinetics of homogeneous organic reactions shows an extension of Arrhenius behavior into the superheated temperature region [23]. Therefore, reaction rate enhancements of the order of 10–100 can be achieved, which is normally only possible under pressure.

Closely related to the superheating effect under atmospheric pressure are wall effects, more specifically the elimination of wall effects caused by inverted temperature gradients (Fig. 2.6). With microwave heating, the surface of the wall is generally not heated since the energy is dissipated inside the bulk liquid. Therefore, the temperature at the inner surface of the reactor wall is lower than that of the bulk liquid. It can be assumed that while in a conventional oil-bath experiment (hot vessel surface, Fig. 2.6) temperature-sensitive species, for example catalysts, may decompose at the hot reactor surface (wall effects), the elimination of such a hot surface will increase the lifetime of the catalyst and therefore will lead to better conversions in a microwave-heated as compared to a conventionally heated process.

Another phenomenon characteristic of microwave dielectric heating is mass heating, that is, the rapid and even heating of the whole reaction mixture by microwaves (volumetric heating). An example illustrating this effect, involving the decomposition of urea to cyanuric acid (Scheme 2.2), was studied by Berlan [16]. Cyanuric acid is obtained by heating urea to temperatures of around 250 °C. Under conventional heating the reaction is sluggish, and chemical yields are low due to the formation of various side-products. The reason for this is that cyanuric acid, which is first formed as a solid at the walls of the reactor, is a poorly heat-conductive material (it decomposes without melting at 300 °C), and it forms an insulating crust which prevents heat transfer to the rest of the reaction mixture. Increasing the temperature of the wall (that is, the oil-bath temperature) results in partial decomposition and does not improve the chemical yield of cyanuric acid significantly. In contrast, very good yields of cyanuric acid (83%) can be obtained by volumetric microwave heating on a 2 gram scale for 2 min without any urea or biuret side-product being detected at the end of the reaction [14].

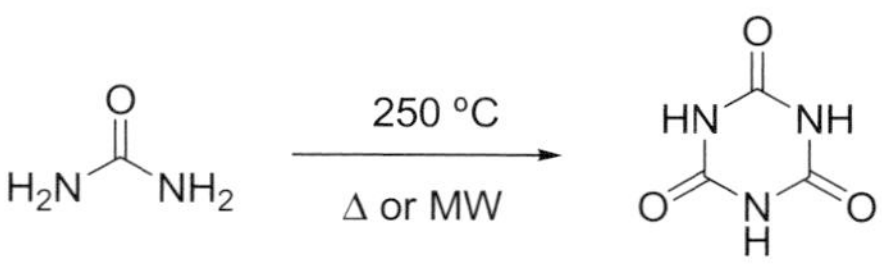

Scheme 2.2   Decomposition of urea to cyanuric acid.

The same concept of volumetric *in situ* heating by microwaves was also exploited by Larhed and coworkers in the context of scaling-up a biochemical process such as the polymerase chain reaction (PCR) [25]. In PCR technology, strict control of temperature in the heating cycles is essential in order not to deactivate the enzymes involved. With classical heating of a milliliter-scale sample, the time required for heat transfer through the wall of the reaction tube and to obtain an even temperature in the whole sample is still substantial. In practice, the slow distribution of heat (temperature gradients), together with the importance of short processing times and reproducibility, limits the volume for most PCR transformations in conventional thermocyclers to 0.2 mL. With microwave heating, the thermal gradients are eliminated since the full volume is heated simultaneously. Therefore, microwave heating under strict temperature control has been shown to be an extremely valuable tool for carrying out large-scale PCR processing at a volume of up to 15 mL [25].

Probably one of the most important "specific microwave effects" results from the selective heating of strongly microwave-absorbing heterogeneous catalysts or reagents in a less polar reaction medium. Selective heating generally means that in a sample containing more than one component, only that component which couples with microwaves is selectively heated. The non-absorbing components are thus not heated directly, but only by heat transfer from the heated component. For homogeneous mixtures, for example polar reagents in a microwave-transparent solvent (liquid/liquid systems), some authors have speculated about the formation of "molecular radiators" (microscopic hot spots) by direct coupling of microwave energy to specific microwave-absorbing reagents or substrates in solution [26]. The existence of such "molecular radiators" is difficult to prove experimentally. It should also be noted that it is not possible to selectively "activate" polar functional groups (so-called antenna groups [27]) within a larger molecule by microwave irradiation. It is tempting for a chemist to give a chemical significance to the fact that localized rotations of such antenna groups are indeed possible [5], and to speculate that microwave dielectric heating of molecules containing these groups may result in an enhancement of reaction rates specifically at these groups. However, the dielectric heating process involves rapid energy transfer from these groups to neighboring molecules and it is not possible to store the energy in a specific part of the molecule [5].

For heterogeneous mixtures, in particular for gas/solid systems involved in heterogeneous gas-phase catalysis [15, 28], selective heating of the catalyst bed is of importance and here the sometimes observed rate enhancements and changes in selectivities have been attributed to the formation of localized (macroscopic) hot spots having temperatures 100–150 °C above the measured bulk temperature [29]. The measurement or estimation of temperature distributions induced by microwave heating in solid materials is, however, very difficult. Consequently, most local temperature fluctuations are greater than those measured. Under stronger microwave irradiation it is, therefore, very easy to obtain local temperature gradients. Temperature measurements usually yield an average temperature, because temperature gradients induce convective motions. Despite these difficulties, some methods, for example IR thermography, can reveal surface temperature distributions without any contact with the sample under study.

Of greater importance for the organic and medicinal chemist are microwave-assisted transformations in organic solvents catalyzed by a heterogeneous catalyst (liquid/solid systems) such as palladium-on-charcoal (Pd/C). Since here the catalyst is a very strong absorber of microwave energy (see Section 4.1), it can be assumed that the reaction temperature on the catalyst surface is significantly higher than the bulk temperature of the solvent, in particular when a solvent with a low tan δ value is chosen (Table 2.3). This can be proven indirectly by irradiating samples of Pd/C in carbon tetrachloride using microwave-transparent quartz vessels and measuring the temperature profiles of the bulk solvent [30]. More direct proof of selective heating effects involving solid/liquid systems was recently obtained by Bogdal and coworkers [31]. In the example shown in Scheme 2.3, primary and secondary alcohols were oxidized with a chromium dioxide reagent under microwave conditions. Irradiation of a neat sample of chromium dioxide for 2 min led to surface temperatures of up to 360 °C, as measured with an IR thermovision camera. When the oxidant was suspended in the weakly microwave-absorbing toluene, the temperature of the chromium dioxide reached ca. 140 °C. Importantly, even though the temperature of the solid material was higher than the boiling point of toluene, no boiling was observed in the reaction vessel.

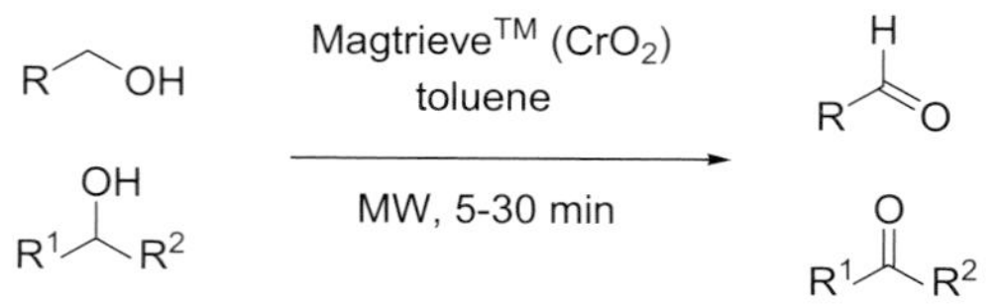

R $\diagup$ OH

OH
R^1 $\diagup$ R^2

Magtrieve^{TM} (CrO$_2$)
toluene
————————————→
MW, 5–30 min

H
R $\diagdown$ O

O
R^1 $\diagup$ R^2

Scheme 2.3 Oxidation of primary and secondary alcohols with chromium dioxide.

These results provide clear evidence for the existence of selective heating effects in MAOS involving heterogeneous mixtures. It should be stressed that the standard methods for determining the temperature in microwave-heated reactions, namely with an IR pyrometer from the outside of the reaction vessel, or with a fiber-optic probe on the inside, would only allow measurement of the average bulk temperature of the solvent, not the "true" reaction temperature on the surface of the solid reagent.

A selective heating in liquid/liquid systems was exploited by Strauss and coworkers in a Hofmann elimination reaction using a two-phase water/chloroform system (Fig. 2.10) [32]. The temperatures of the aqueous and organic phases under microwave irradiation were 110 and 55 °C, respectively, due to the different dielectric properties of the solvents (Table 2.3). This temperature differential prevented decomposition of the final product. Comparable conditions would be difficult to obtain using traditional heating methods. A similar effect has been observed by Hallberg and coworkers in the preparation of β,β-diarylated aldehydes by hydrolysis of enol ethers in a two-phase toluene/aqueous hydrochloric acid system [33].

In summary, all potential rate enhancements discussed above falling under the category of "specific microwave effects", such as the superheating effect of solvents at atmospheric pressure, the selective heating of strongly microwave-absorbing heterogeneous catalysts or reagents in a less polar reaction medium, the formation of "molecular radiators" by direct coupling of microwave energy to specific reagents in homogeneous solution (microscopic hotspots), and the elimination of wall effects caused by inverted temperature gradients, are essentially still a result of a *thermal* effect (that is, a change in temperature compared to heating by standard convection methods), although it may be difficult to determine the exact reaction temperature experimentally.

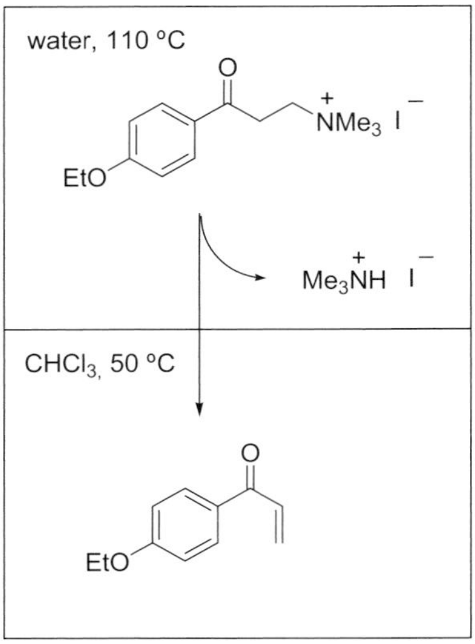

Fig. 2.10 Selective dielectric heating of water/chloroform mixtures.

2.5.3
Non-Thermal (Athermal) Microwave Effects

In contrast to the so-called "specific microwave effects" described in Section 2.5.2, some authors have suggested the possibility of "non-thermal microwave effects" (also referred to as athermal effects). These should be classified as *accelerations of chemical transformations in a microwave field that cannot be rationalized in terms of either purely thermal/kinetic or specific microwave effects*. Essentially, most non-thermal effects result from a proposed direct interaction of the electric field with specific molecules in the reaction medium. It has been argued, for example, that the presence of an electric field leads to orientation effects of dipolar molecules and hence changes the pre-exponential factor A [34] or the activation energy (entropy term) [35] in the Arrhenius equation. Furthermore, a similar effect should be observed for polar reaction mechanisms, where the polarity is increased on going from the ground state to the transition state, resulting in an enhancement of reactivity through a lowering of the activation energy. Many publications in the literature use arguments like this to explain the outcome of a chemical reaction carried out under microwave irradiation [13]. A recent study by Loupy and coworkers illustrates this point [36]. In the example shown in Scheme 2.4, two irreversible Diels–Alder cycloaddition processes were compared. In the first example (Scheme 2.4a), no difference in either yield or selectivity was observed between the conventionally and the microwave-heated reactions. Detailed ab initio calculations on the cycloaddition process revealed that here a synchronous, isopolar (concerted) mechanism is operational, in which no charges are developed on going from the ground state to the transition state. On the contrary, in the second example (Scheme 2.4b), a significant difference in product yield was observed on comparing the thermally- and micro-wave-heated runs. Here, ab initio calculations on transition-state geometries and dipole moments revealed a significant degree of charge development on going from the ground state to the transition state. The authors take these experimental results as clear evidence for the involvement of non-thermal microwave effects arising from electrostatic interactions of polar molecules with the electric field, that is, for the stabilization of the transition state and thereby a decrease of the activation energy [36]. However, other scientists [2, 10, 14] denounce the existence of such dipolar ori-entation effects in electric fields on the grounds of overriding disorientation phe-nomena (thermal agitation) that should prevent any statistically significant orienta-tion (alignment) of dipoles. Several authors have used quantum mechanical meth-ods to study the interaction between reactant molecules and the electromagnetic field [37, 38]. Regardless of the rationalization, in many instances microwave irradia-tion leads to a change in selectivity (chemo-, regio-, and stereoselectivity) when com-pared to conventional heating [39].

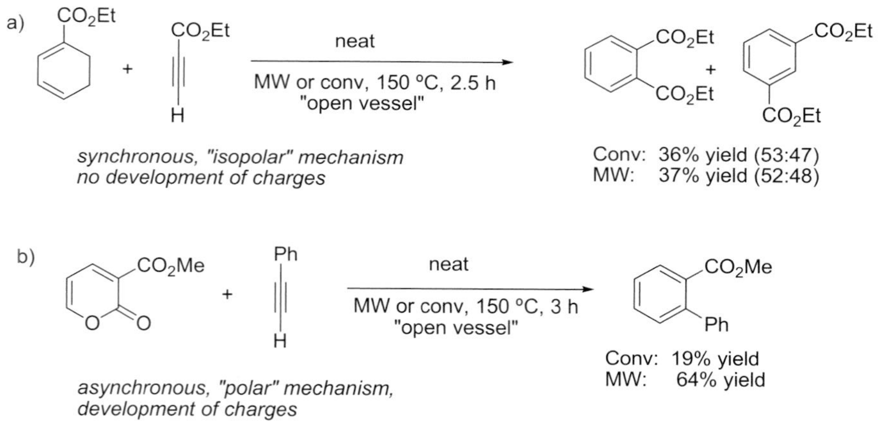

Scheme 2.4   Involvement of non-thermal microwave effects in Diels–Alder cycloaddition reactions.

Related to the issue of non-thermal or specific microwave effects is the recent con-cept that simultaneous external cooling of the reaction mixture (or maintaining sub-ambient reaction temperatures) while heating with microwaves leads to an enhance-ment of the overall process (PowerMax, "Enhanced Microwave Synthesis" [10, 40, 41]). Here, the reaction vessel is cooled from the outside with compressed air while being irradiated by microwaves. This allows a higher level of microwave power to be directly administered to the reaction mixture, but will prevent overheating by con-tinuously removing latent heat. Some authors [10, 41] have made the argument that since microwave energy is transferred into the sample at a much faster rate than that at which molecular kinetic relaxation can occur, non-equilibrium conditions will result in a microwave-heated reaction system. Therefore, the more power that is applied, the higher the "instantaneous" temperature will be relative to the measured bulk temperature [10, 41]. Published applications of the simultaneous cooling technology are rare [41–43, 47]. In the example shown in Scheme 2.5, the authors compared the outcome of the condensation of an acid chloride with an isonitrile followed by hydrolysis of the intermediate to an α-ketoamide under microwave heating with and without external cooling [42]. While irradiation at a constant 100 W for 1 min (step 1) without cooling (measured temperature 150 °C) led only to a black tar-like product, the same reaction with the cooling feature turned on (100 °C) provided a 69% isolated yield of the target compound [42]. While it is tempting to speculate about the apparent benefits of this approach, one must be aware that the actual reaction temperature for the chemistry described in Scheme 2.5 has not been determined. In the set-up utilized by the investigators [42], a standard external IR pyrometer (see Section 3.5.2) was used to determine the reaction temperature. Since the IR pyrometer will only provide the surface temperature of the reaction vessel (and not the "true" reaction temperature inside the reactor [6, 47]), cooling of the reaction vessel from the outside with compressed air will not afford a reliable temperature measurement. Therefore, without knowing the actual reaction temperature, care must be taken not to misinterpret the results obtained with the simultaneous cooling approach [41–43]. For this type of experimental set-up, the use of a fiber-optic probe measuring the internal temperature is advised [47].

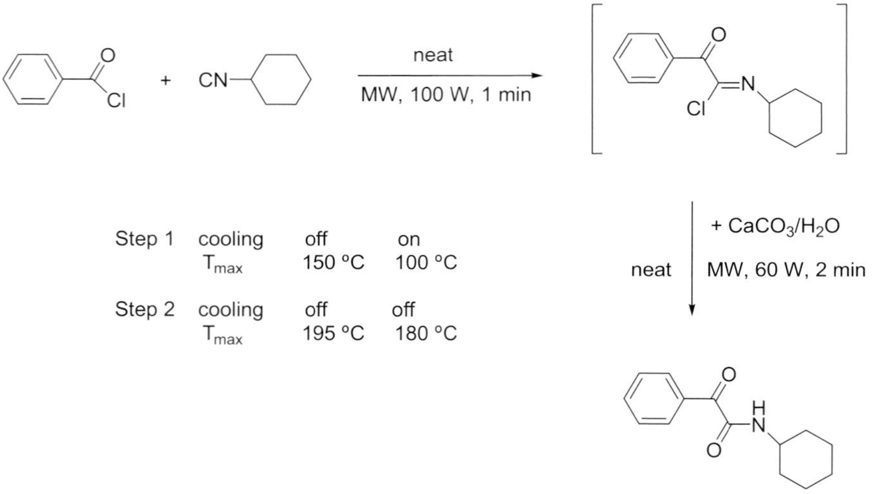

Step 1	cooling	off	on
	T_{max}	150 °C	100 °C
Step 2	cooling	off	off
	T_{max}	195 °C	180 °C

Scheme 2.5 Application of the microwave heating–simultaneous cooling approach.

Recently, Hajek and coworkers have reported results on microwave-assisted chemistry performed by cooling of a reaction mixture to as low as –176 °C. Reaction rates were recorded under microwave and conventional conditions. The higher reaction rates under microwave heating at sub-ambient temperatures were attributed to a superheating of the heterogeneous K10 catalyst [44].

Another unusual phenomenon in microwave synthesis was described by Ley and coworkers. In several examples [45, 46], the authors found that pulsed microwave irradiation gave higher conversions than irradiating the reaction mixture continuously for the same amount of time. In the example shown in Scheme 2.6, the highest conversion in the Claisen rearrangement was obtained by exposing the reaction to three 15 min spells of microwave heating at 220 °C (with intermittent cooling). Continuous microwave irradiation for 45 min at the same temperature provided a somewhat lower yield [45].

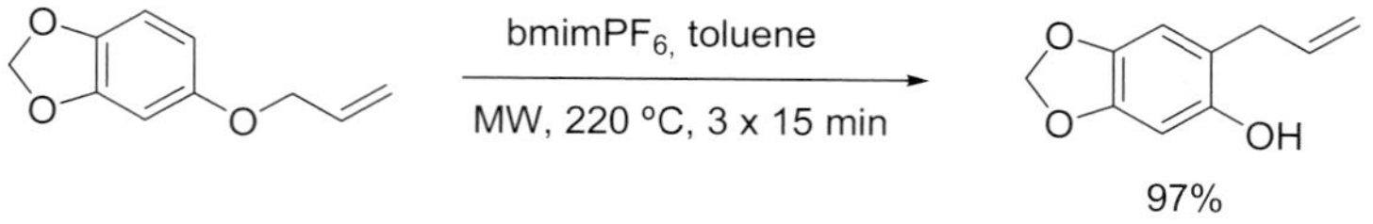

MW (220 °C)	Conversion (%)
3 x 15 min	97
30 min continuous	78
45 min continuous	86
2+2+2+1+1+15+15+15	85

Scheme 2.6 Pulsed versus continuous microwave irradiation ($bmimPF_6$ = 1-butyl-3-methylimidazolium hexafluorophosphate).

Microwave effects are still the subject of considerable current debate and controversy, and the reader should be aware that there is no agreement in the scientific community on the role that "microwave effects" play, not even on a definition of terms.

References

[1] A. R. von Hippel, *Dielectric Materials and Applications*, MIT Press, Cambridge, MA, USA, **1954**.

[2] D. Stuerga, M. Delmotte, in *Microwaves in Organic Synthesis* (Ed.: A. Loupy), Wiley-VCH, Weinheim, **2002**, pp 1–34 (Chapter 1).

[3] D. M. P. Mingos, in *Microwave-Assisted Organic Synthesis* (Eds.: P. Lidström, J. P. Tierney), Blackwell Publishing, Oxford, **2004** (Chapter 1).

[4] D. R. Baghurst, D. M. P. Mingos, *Chem. Soc. Rev.* **1991**, *20*, 1–47.

[5] C. Gabriel, S. Gabriel, E. H. Grant, B. S. Halstead, D. M. P. Mingos, *Chem. Soc. Rev.* **1998**, *27*, 213–223.

[6] M. Nüchter, B. Ondruschka, W. Bonrath, A. Gum, *Green Chem.* **2004**, *6*, 128–141.

[7] R. N. Gedye, J. B. Wei, *Can. J. Chem.* **1998**, *76*, 525–532.

[8] E. Neas, M. Collins, in *Introduction to Microwave Sample Preparation: Theory and Practice* (Eds.: H. M. Kingston, L. B. Jassie), Ameri-

can Chemical Society, Washington, DC, **1988**.

[9] D. V. Stass, J. R. Woodward, C. R. Timmel, P. J. Hore, K. A. McLauchlan, *Chem. Phys. Lett.* **2000**, *329*, 15–22; C. R. Timmel, P. J. Hore, *Chem. Phys. Lett.* **1996**, *257*, 401–408; J. R. Woodward, R. J. Jackson, C. R. Timmel, P. J. Hore, K. A. McLauchlan, *Chem. Phys. Lett.* **1997**, *272*, 376–382.

[10] B. L. Hayes, *Microwave Synthesis: Chemistry at the Speed of Light*, CEM Publishing, Matthews, NC, **2002**.

[11] D. D. Grice, P. D. I. Fletcher, S. J. Haswell, Department of Chemistry, University of Hull, UK, unpublished data.

[12] C. O. Kappe, *Angew. Chem. Int. Ed.* **2004**, *43*, 6250–6284.

[13] L. Perreux, A. Loupy, *Tetrahedron* **2001**, *57*, 9199–9223; A. de la Hoz, A. Díaz-Ortiz, A. Moreno, *Chem. Soc. Rev.* **2005**, *34*, 164–178.

[14] N. Kuhnert, *Angew. Chem. Int. Ed.* **2002**, *41*, 1863–1866; C. R. Strauss, *Angew. Chem. Int.*

Ed. **2002**, *41*, 3589–3590; M. Panunzio,
E. Campana, G. Martelli, P. Vicennati,
E. Tamanini, *Mat. Res. Innovat.* **2004**, *8*, 27–31.

[15] M. Hajek, in *Microwaves in Organic Synthesis*
(Ed.: A. Loupy), Wiley-VCH, Weinheim,
2002, pp 345–378 (Chapter 10).

[16] J. Berlan, *Radiat. Phys. Chem.* **1995**, *45*,
581–589.

[17] F. Langa, P. de la Cruz, A. de la Hoz,
A. Díaz-Ortiz, E. Díez-Barra, *Contemp. Org.
Synth.* **1997**, *4*, 373–386.

[18] K. Bougrin, A. K. Bannani, S. F. Tetouani,
M. Soufiaoui, *Tetrahedron Lett.* **1994**, *35*,
8373–8376.

[19] A. Stadler, C. O. Kappe, *Eur. J. Org. Chem.*
2001, 919–925.

[20] R. N. Gedye, in *Microwaves in Organic Synthesis* (Ed.: A. Loupy), Wiley-VCH, Weinheim, **2002**, pp 115–146 (Chapter 10).

[21] D. R. Baghurst, D. M. P. Mingos, *J. Chem.
Soc., Chem. Commun.* **1992**, 674–677.

[22] R. Saillard, M. Poux, J. Berlan, M. Audhuy-
Peaudecerf, *Tetrahedron* **1995**, *51*,
4033–4042.

[23] F. Chemat, E. Esveld, *Chem. Eng. Technol.*
2001, *24*, 735–744.

[24] E. Abtal, M. Lallemant, G. Bertrand,
G. Watelle, *J. Chim. Phys. Phys.-Chim. Biol.*
1985, *82*, 381–389; G. Roussy, J. M. Thibaut,
P. Collin, *Thermochim. Acta* **1986**, *98*, 57–62.

[25] K. Orrling, P. Nilsson, M. Gullberg,
M. Larhed, *Chem. Commun.* **2004**, 790–791.

[26] N.-F. K. Kaiser, U. Bremberg, M. Larhed,
C. Moberg, A. Hallberg, *Angew. Chem. Int.
Ed.* **2000**, *39*, 3596–3598; A. Steinreiber,
A. Stadler, S. F. Mayer, K. Faber, C. O. Kappe,
Tetrahedron Lett. **2001**, *42*, 6283–6286.

[27] R. Laurent, A. Laporterie, J. Dubac, J. Berlan,
S. Lefeuvre, M. Audhuy, *J. Org. Chem.* **1992**,
57, 7099–7102.

[28] H. Will, P. Scholz, B. Ondruschka, *Chem.
Ing. Tech.* **2002**, *74*, 1057–1067.

[29] X. Zhang, D. O. Hayward, D. M. P. Mingos,
Catal. Lett. **2003**, *88*, 33–38.

[30] S. Garbacia, B. Desai, O. Lavastre, C. O.
Kappe, *J. Org. Chem.* **2003**, *68*, 9136–9139.

[31] D. Bogdal, M. Lukasiewicz, J. Pielichowski,
A. Miciak, Sz. Bednarz, *Tetrahedron* **2003**, *59*,
649–653; M. Lukasiewicz, D. Bogdal,
J. Pielichowski, *Adv. Synth. Catal.* **2003**, *345*,
1269–1272.

[32] K. D. Raner, C. R. Strauss, R. W. Trainor,
J. S. Thorn, *J. Org. Chem.* **1995**, *60*,
2456–2460.

[33] P. Nilsson, M. Larhed, A. Hallberg, *J. Am.
Chem. Soc.* **2001**, *123*, 8217–8225.

[34] J. Jacob, L. H. L. Chia, F. Y. C. Boey, *J. Mater.
Sci.* **1995**, *30*, 5322–5327; J. G. P. Binner,
N. A. Hassine, T. E. Cross, *J. Mater. Sci.*
1995, *30*, 5389–5322; C. Shibata, T. Kashima,
K. Ohuchi, *Jpn. J. Appl. Phys.* **1996**, *35*,
316–319.

[35] J. Berlan, P. Giboreau, S. Lefeuvre,
C. Marchand, *Tetrahedron Lett.* **1991**, *32*,
2363–2366; D. A. Lewis, J. D. Summers,
T. C. Ward, J. E. McGrath, *J. Polym. Sci. Part
A* **1992**, *30*, 1647–1653.

[36] A. Loupy, F. Maurel, A. Sabatié-Gogová,
Tetrahedron **2004**, *60*, 1683–1691.

[37] S. Kalhori, B. Minaev, S. Stone-Elander,
N. Elander, *J. Phys. Chem. A.* **2002**, *106*,
8516–8524.

[38] A. Miklavc, *Chem. Phys. Chem.* **2001**,
553–555.

[39] A. de la Hoz, A. Díaz-Ortiz, A. Moreno, *Curr.
Org. Chem.* **2004**, *8*, 903–918.

[40] B. L. Hayes, M. J. Collins, Jr., World Patent
2004, WO 04002617.

[41] B. L. Hayes, *Aldrichim. Acta* **2004**, *37*, 66–77.

[42] J. J. Chen, S. V. Deshpande, *Tetrahedron Lett.*
2003, *44*, 8873–8876.

[43] F. Mathew, K. N. Jayaprakash, B. Fraser-
Reid, J. Mathew, J. Scicinski, *Tetrahedron
Lett.* **2003**, *44*, 9051–9054; K. Crawford,
S. K. Bur, C. S. Straub, A. Padwa, *Org. Lett.*
2003, *5*, 3337–3340; C. E. Humphrey,
M. A. M. Easson, J. P. Tierney, N. J. Turner,
Org. Lett. **2003**, *5*, 849–852; A. R. Katritzky,
Y. Zhang, S. K. Singh, P. J. Steel, *ARKIVOC*
2003 (xv), 47–65; C. J. Bennett, S. T. Caldwell,
D. B. McPhail, P. C. Morrice, G. G. Duthie,
R. C. Hartley, *Bioorg. Med. Chem.* **2004**, *12*,
2079–2098; A.-L. Villard, B. Warrington,
M. Ladlow, *J. Comb. Chem.* **2004**, *6*, 611–622;
M. Bejugam, S. L. Flitsch, *Org. Lett.* **2004**, *6*,
4001–4004.

[44] J. Kurfürstová, M. Hajek, *Res. Chem.
Intermed.* **2004**, *30*, 673–681.

[45] I. R. Baxendale, A.-I. Lee, S. V. Ley, *J. Chem.
Soc., Perkin Trans.* **2002**, 1850–1857.

[46] T. Durand-Reville, L. B. Gobbi, B. L. Gray,
S. V. Ley, J. S. Scott, *Org. Lett.* **2002**, *4*, 3847–
3850; I. R. Baxendale, S. V. Ley, M. Nessi,
C. Piutti, *Tetrahedron* **2002**, *58*, 6285–6304.

[47] N. E. Leadbeater, S. J. Pillsbury, E. Shana-
han, V. A. Williams, *Tetrahedron* **2005**, *61*,
3565–3585.

3
Equipment Review

3.1
Introduction

Although many of the early pioneering experiments in microwave-assisted organic synthesis were carried out with domestic microwave ovens, the current trend is undoubtedly to use dedicated instruments for chemical synthesis (see Fig. 1.1). In a domestic microwave oven, the irradiation power is generally controlled by on-off cycles of the magnetron (pulsed irradiation), and it is typically not possible to monitor the reaction temperature in a reliable way. Combined with the inhomogeneous field produced by the low-cost multimode designs and the lack of safety controls, the use of such equipment cannot be recommended. In contrast, all of today's commercially available dedicated microwave reactors for synthesis feature built-in magnetic stirrers, direct temperature control of the reaction mixture with the aid of fiber-optic probes or IR sensors, and software that enables on-line temperature/pressure control by regulation of the microwave power output. Currently, two different philosophies with respect to microwave reactor design are emerging: multimode and monomode (also referred to as single-mode) reactors. In the so-called multimode instruments (conceptually similar to a domestic oven), the microwaves that enter the cavity are reflected by the walls and the load throughout the typically large volume. In most instruments, a mode stirrer ensures that the field distribution is as homogeneous as possible. In the much smaller monomode cavities, only one mode is present and the electromagnetic irradiation is directed through a precision-designed and built rectangular or circular wave guide to the reaction vessel mounted at a fixed distance from the radiation source, creating a standing wave. The key difference between the two types of reactor systems is that whereas in multimode cavities several reaction vessels can be irradiated simultaneously in multi-vessel rotors (parallel synthesis), in monomode systems typically only one vessel at a time can be irradiated. In the latter case, however, high throughput can be achieved by means of integrated robotics that move individual reaction vessels in and out of the microwave cavity. Most instrument companies offer a variety of diverse reactor platforms with different degrees of sophistication with respect to automation, database capabilities, safety features, temperature and pressure monitoring, and vessel design. Importantly, single-mode reactors processing comparatively small volumes also have a

Microwaves in Organic and Medicinal Chemistry. C. Oliver Kappe, Alexander Stadler
Copyright © 2005 WILEY-VCH Verlag GmbH & Co. KGaA, Weinheim
ISBN: 3-527-31210-2

built-in cooling feature that allows for rapid cooling of the reaction mixture by compressed air after completion of the irradiation period (see Fig. 2.8). The dedicated single-mode instruments available today can process volumes ranging from 0.2 to ca. 50 mL under sealed-vessel conditions (250 °C, ca. 20 bar), and somewhat higher volumes (ca. 150 mL) under open-vessel reflux conditions. In the much larger multi-mode instruments, several liters can be processed under both open- and closed-vessel conditions. For both single- and multimode cavities, continuous-flow reactors are nowadays available that already allow the preparation of kilograms of materials using microwave technology.

This chapter provides a detailed description of the various commercially available microwave reactors that are dedicated for microwave-assisted organic synthesis. A comprehensive coverage of microwave oven design, applicator theory, and a description of waveguides, magnetrons, and microwave cavities lies beyond the scope of this book. Excellent coverage of these topics can be found elsewhere [1–4]. An overview of experimental, non-commercial microwave reactors has recently been presented by Stuerga and Delmotte [4].

3.2
Domestic Microwave Ovens

At its beginning in the mid-1980s, microwave-assisted organic synthesis was carried out exclusively in conventional multimode domestic microwave ovens [5, 6]. Gedye and his group used sealed Teflon vessels for their first synthesis under microwave irradiation (see Scheme 1.1) [5]. The main drawback of these household appliances is the lack of control systems. In general, it is not possible to determine the reaction temperature in an accurate way using domestic microwave ovens. The lack of pressure control and of a means of stirring the reaction mixture additionally makes performing chemical syntheses in domestic microwave ovens troublesome. Furthermore, the pulsed irradiation (on-off duty cycles) and the resulting inhomogeneity of the microwave field may lead to problems of reproducibility. Another concern is safety. Heating organic solvents in open vessels in a microwave oven can lead to violent explosions induced by electric arcs inside the cavity or sparking resulting from the switching of the magnetrons. On the other hand, working with sealed vessels under pressure without real-time monitoring of the pressure can lead to unexpected vessel failures in the case of a thermal runaway and to serious accidents. Early on, simple modifications of the available ovens with self-made accessories such as mechanical stirrers or reflux condensers mounted through holes in the cavity were attempted in order to generate instrumentation useful for chemical synthesis (Fig. 3.1). However, some safety risks remained as these instruments were not explosion-proof and leakage of microwaves harmful to the operator could occur. Clearly, the use of any microwave equipment not specifically designed for organic synthesis cannot be recommended and is generally banned from most industrial and academic laboratories today.

Fig. 3.1 Modified domestic household microwave oven. Inlets for temperature measurement by IR pyrometer (left side) and for attaching reflux condensers (top) are visible. A magnetic stirrer is situated below the instrument.

3.3
Dedicated Microwave Reactors for Organic Synthesis

The growing interest in MAOS during the mid-1990s led to increased demand for more sophisticated microwave instrumentation, offering, for example, stirring of the reaction mixture, temperature measurement, and power control features. For scientifically valuable, safe, and reproducible work, the microwave instruments utilized should offer the following features:

- built-in magnetic or mechanical stirring,
- accurate temperature measurement,
- pressure control,
- continuous power regulation,
- efficient post-reaction cooling,
- computer-aided method programming,
- explosion-proof cavities.

A particularly difficult problem in microwave processing is the correct measurement of the reaction temperature during the irradiation phase. Classical temperature sensors (thermometers, thermocouples) will fail since they will couple with the electromagnetic field. Temperature measurement can be achieved either by means of an immersed temperature probe (fiber-optic or gas-balloon thermometer) or on the outer surface of the reaction vessels by means of a remote IR sensor. Due to the volumetric character of microwave heating, the surface temperature of the reaction vessel will not always reflect the actual temperature inside the vessel [7].

Since the early applications of microwave-assisted synthesis were based on the use of domestic multimode microwave ovens, the primary focus in the development of dedicated microwave instruments was inevitably the improvement of multimode

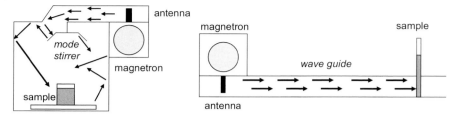

Fig. 3.2 Multimode (left) versus single-mode cavities (right).

reactors. In general, one or two magnetrons create the microwave irradiation, which is typically directed into the cavity through a waveguide and distributed by a mode stirrer (Fig. 3.2). The microwaves are reflected from the walls of the cavity, thus interacting with the sample in a chaotic manner. Multimode cavities may therefore show multiple energy pockets with different levels of energy intensity, thus resulting in hot and cold spots. To provide a more even energy distribution, the samples are continuously rotated within the cavity. Consequently, multimode instruments conveniently allow for an increase in reaction throughput by the use of multi-vessel rotors for parallel synthesis or scale-up. A general problem associated with multimode instruments is a poor level of performance in small-scale experiments (< 3 mL). While the generated microwave power is high (1000–1400 W), the power density of the field is generally rather low. This makes the heating of small individual samples rather difficult, a major drawback especially for research and development purposes. Therefore, the use of multimode instruments for small-scale synthetic organic research applications has become much less extensive as compared to the use of the much more popular single-mode cavities (see Chapters 6 and 7).

Monomode (also known as single-mode) instruments generate a single, highly homogeneous energy field of high power intensity. Thus, these systems couple efficiently with small samples and the maximum output power is typically limited to 300 W. The microwave energy is created by a single magnetron and directed through a rectangular waveguide to the sample, which is positioned at a maximized energy point (Fig. 3.2). To create optimum conditions for different samples, a tuning device allows adjustment of the microwave field for variations in the samples (0.2–20 mL). Thus, the highly homogeneous field generally enables excellent reproducibility.

In addition to the rectangular waveguide applicator, instruments with a self-tuning circular waveguide are also available (Fig. 3.3). This cavity features multiple entry points for introducing the microwave energy into the cavity (slots), compensating for variations in the coupling characteristics of the sample.

Due to the cavity design, it is suitable for vessels of different types and sizes, such as sealed 10–80 mL vials or 125 mL round-bottomed flasks. With this system, pressurized reactions in sealed vessels as well as traditional refluxes at atmospheric pressure can be performed.

Recent advances and further improvements have led to a broad variety of applications for single-mode microwave instruments, offering flow-through systems as well as special features such as solid-phase peptide synthesis or *in situ* on-line analy-

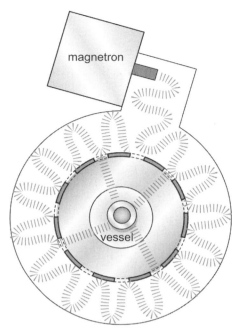

Fig. 3.3   Circular single-mode cavity.

tics. Thus, the use of single-mode reactors has tremendously increased since the year 2000 and these types of reactors have become very popular in many synthetic laboratories, both in industry and academia. However, it should be pointed out that which type of instrumentation is to be used is more a matter of the desired application and scale than of the kind of chemistry to be performed. Both multi-mode and single-mode reactors can be used to carry out chemical reactions efficiently and to improve classical heating protocols.

The following sections give a comprehensive description of all commercially available multimode and single-mode microwave reactors as of December 2004, including various accessories and special applications. Essentially, there are currently four instrument manufacturers that produce microwave reactors for laboratory-scale organic synthesis, namely Anton Paar GmbH (Graz, Austria) [8], Biotage AB (Uppsala, Sweden) [9], CEM Corporation (Matthews, NC, USA) [10], and Milestone s.r.l. (Sorisole, Italy) [11].

3.4
Multimode Instruments

The development of multimode reactors for organic synthesis occurred mainly from already available microwave acid-digestion/solvent-extraction systems. Instruments for this purpose were first designed in the 1980s and with the growing demand for synthesis systems, these reactors were subsequently adapted for organic synthesis applications. Therefore, there is still a close relationship between multimode microwave digestion systems and synthesis reactors.

3.4.1
Milestone s.r.l.

Based on their microwave digestion system, Milestone offers the MicroSYNTH labstation (also known as ETHOS series) multimode instrument (Fig. 3.4 and Table 3.1), which is available with various accessories. Two magnetrons deliver 1000 W microwave output power and a patented pyramid-shaped microwave diffuser ensures homogeneous microwave distribution within the cavity [12].

The modular MicroSYNTH platform offers the diversity of different rotor and vessel systems, enabling reactions in volumes ranging from 3 to 500 mL under open- and sealed-vessel conditions in batch and parallel manners at up to 50 bar. The START package offers simple laboratory glassware for reactions at atmospheric pressure under reflux conditions (Fig. 3.4). This system can be upgraded (see Section 3.4.1.1) by a teaching kit (16-vessel rotor for reactions at < 1.5 bar, Fig. 3.5) or a research lab kit, equipped with the so-called MonoPREP module (Fig. 3.6) for single small-scale experiments (3–30 mL) in the multimode cavity. In addition, equipment for combinatorial chemistry approaches (CombiCHEM kit, microtiter well-plates up to 96 × 1 mL), extraction, UV photoexcitation, and flow-through techniques is available (see Section 3.4.1.2).

Post-reaction cooling of the reaction mixture is achieved by means of a constant air-flow through the cavity (Table 3.1). Due to the design of the thick plastic segments, the cooling of the high-pressure rotors is not very efficient, and external cooling by immersing the rotor in a water bath is recommended. Thus, vessels under

Fig. 3.4   Milestone MicroSYNTH/START labstation.

Table 3.1 MicroSYNTH Platform – General Features

Feature	Description
Cavity size (volume)	$35 \times 35 \times 35$ cm (43 L)
Delivered power	1600 W (2 magnetrons)
Max. output power	1000 W
Temperature control	Immersed fiber-optic probe (max. 300 °C);
	Outside IR remote sensor (optional)
Pressure measurement	Pneumatic pressure sensor (optional)
Cooling system	Air flow 1.8 m^3 min^{-1}, external water-cooling recommended
Magnetic stirring	ASM-400 (infinitely manual variable)
External PC	Standard equipment, touch screen optional

pressure have to be handled outside the cavity, but a special cooling rack is available for this purpose.

Temperature measurement is achieved by means of a fiber-optic probe immersed in a single reference vessel. An available option is an IR sensor for monitoring the outside surface temperature of each vessel, mounted in the sidewall of the cavity about 5 cm above the bottom. The reaction pressure is measured by a pneumatic sensor connected to one reference vessel. Therefore, the parallel rotors should be filled with identical reaction mixtures to ensure homogeneity.

For all MicroSYNTH systems, reactions are monitored through an external control terminal utilizing the EasyWAVE software packages. The runs can be controlled by adjusting either the temperature, the pressure, or the microwave power output in a defined program of up to ten steps. The software enables on-line modification of any method parameter and the reaction process is monitored through an appropriate graphical interface. An included solvent library and electronic lab journal feature simplifies the experimental documentation.

3.4.1.1 Accessories

As mentioned above, numerous rotors and vessel types for individual applications are available for the MicroSYNTH/ETHOS platform. Most of them are derived from the original digestion system and have been adapted for synthesis purposes.

For reactions at atmospheric pressure, standard laboratory glassware such as round-bottomed flasks or simple beakers from 0.25 to 2 L can be used. A protective mount in the ceiling of the cavity enables the connection of reflux condensers or distillation equipment. An additional mount in the sidewall allows for sample withdrawal, flushing with gas to create inert atmospheres, or live monitoring of the reaction with a video camera. Most of the published results in controlled MAOS have been obtained from reactions in sealed vessels, and thus in the following mostly accessories for sealed-vessel reaction conditions are described.

- *Teaching Lab Kit* (Fig. 3.5): This is a basic rotor for standard organic reactions allowing an introduction to microwave-mediated chemistry in teaching laboratories. It is designed for 16 × 20 mL glass vessels with operation limits of 1.5 bar and ca. 150 °C.

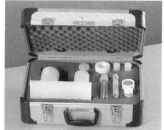
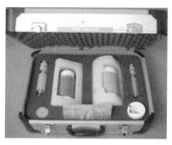

Fig. 3.5 Milestone Teaching Kit, Research Kit, and MedChem Kit (left to right).

Fig. 3.6 Milestone MonoPREP module (left) and its parallel set-up rotor PRO-6 (right).

- *Research Lab Kit* (Fig. 3.5): Package containing a MonoPREP module with a 12 mL and a 50 mL glass vessel.
- *MedChem Kit* (Fig. 3.5): An accessory especially designed for medicinal chemistry laboratories to cover working volumes of 2–140 mL. The package contains a 12 mL glass vial, a 50 mL quartz vessel, and single-pressure reactor segments (100 mL and 270 mL, respectively).
- *MonoPREP module* (Fig. 3.6): This tool is designed for single optimization runs at elevated pressure on a comparatively small scale in multimode ovens.

Fig. 3.7 Milestone MultiPREP-50/80 with PFA vessels (left) and PRO 16/24 with PTFE-TFM pressure vessels (right).

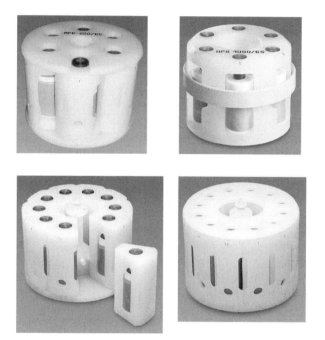

Fig. 3.8 Milestone high-pressure rotors: MPR-6, HPR monobloc, HPR-10 and MPR-12 (left to right).

It comes with two glass vessels suitable for volumes of 3–30 mL at operation limits of 200 °C and 15 bar. A cooling mechanism returns the reaction mixture to 30 °C after the irradiation. For enhanced optimization, the PRO-6 rotor, a parallel version of the MonoPREP module with six identical vessels, can be applied.

- *MultiPREP rotors* (Fig. 3.7): For high-throughput synthesis, Milestone offers a series of different parallel rotors for 36, 50 or 80 glass or Teflon-PFA vessels (35/50 mL) to perform reactions at atmospheric pressure up to 230 °C.

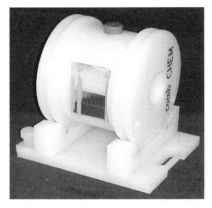
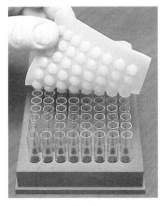

Fig. 3.9   Milestone CombiCHEM system: Rotor with wellplates made of Weflon™.

Fig. 3.10   Milestone special accessory tools for parallel reflux/extraction, protein hydrolysis, and UV-induced reactions (left to right).

- *PRO 16/24 rotor* (Fig. 3.7): Advanced rotor for high-throughput purposes under elevated temperature and pressure conditions utilizing 16 or 24 reaction containers. The 70 mL PTFE-TFM vessels offer 35 mL working volume at 200 °C up to 20 bar.
- *MPR-6 large-volume rotor* (Fig. 3.8): Dedicated for parallel scale-up, this rotor comes with six segments for 270 mL PTFE-TFM vessels. Reactions can be performed on a 15–140 mL scale at operation limits of 200 °C up to 10 bar.
- *HPR monobloc* (Fig. 3.8): For applications at high pressure, the HPR monobloc is available. This tool features six 100 mL PTFE-TFM vessels fixed in a strong casing, serving operation limits of 300 °C and 100 bar.
- *High-pressure rotor* (Fig. 3.8): Two variations of parallel high-pressure rotors are available. The MPR-12 comes with 12 segmented 100 mL PTFE-TFM vessels for reactions at up to 260 °C and 35 bar. For more forceful conditions in volumes of 8–50 mL, the HPR-10 provides for 10 segmented 100 mL PTFE-TFM vessels for reactions at up to 250 °C at 55 bar.

- *CombiCHEM System* (Fig. 3.9): For small-scale combinatorial chemistry applications, this barrel-type rotor is available. It can hold two 24- to 96-well microtiter plates utilizing glass vials (0.5–4 mL) at up to 4 bar at 150 °C. The plates are made of Weflon™ (graphite-doped Teflon) to ensure uniform heating and are sealed by an inert membrane sheet. Axial rotation of the rotor tumbles the microwell plates to admix the individual samples. Temperature measurement is achieved by means of a fiber-optic probe immersed in the center of the rotor.

Additionally, advanced tools for special applications are offered, including provisions for parallel reflux, solvent extraction, and hydrolysis, as well as electrodeless discharge lamps for photochemistry (Fig. 3.10). A detailed description of these accessories can be found on the Milestone website [11].

3.4.1.2 Scale-Up Systems

As already mentioned, the scale-up of microwave-assisted reactions is of specific interest in many industrial laboratories. For this purpose, Milestone offers two different continuous-flow systems (Fig. 3.11).

The FlowSYNTH equipment employs quartz or ceramic flow-through cells of various sizes (10–60 mm diameter) inside the regular ETHOS cavity. The reagents are pumped through the microwave field from the bottom to the top of the cavity under maximum operating conditions of 250 °C or 40 bar. The flow rate is dependent on the cell used. The cell design is such that suspensions and inhomogeneous mixtures can also be used. Temperature and pressure control throughout the entire course of the process is achieved by an in-line thermal sensor and an in-line pressure-control valve, respectively. Additionally available 380 mL MRS-batch tubes made of fused silica or ceramic are also applicable for conventional batch synthesis without flow-through of the sample (stop-flow technique) at operation limits of 200 °C or 14 bar.

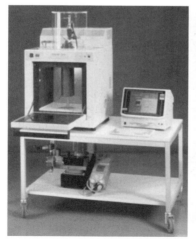

Fig. 3.11 Milestone scale-up reactors: FlowSYNTH (left) and ETHOSpilot 4000 (right).

Stirring bars or Weflon™ heating chips can be inserted to enable sufficient admixing and heating of weakly absorbing samples.

The ETHOS pilot 4000 labstation is a prototype reactor designed for scale-up to the kilogram scale, built from two regular Milestone cavities. The reaction mixtures can be heated in either a continuous-flow or a batch-type manner. The delivered microwave output power is 2500 W, which can be boosted to 5000 W if required. The reaction tubes (quartz or ceramic) are custom built, with several diameters and lengths being available, covering a broad range of flow rates and pressure conditions. For reaction control, the temperature is monitored over the whole length of the reaction cell, as in the FlowSYNTH system.

A recent addition to the spectrum of available reactors for scale-up is the Ultra-CLAVE batch reactor. Microwave power settings of 0–1000 W may be selected. Continuous unpulsed microwave energy from the system's magnetron is introduced into the 3.5 L high-pressure vessel through a special microwave-transparent port. The internal geometry of the vessel is optimized for direct microwave coupling with zero reflectance, ensuring maximum sample heating efficiency. Reaction conditions of up to 100 bar and 250 °C are possible.

3.4.2
CEM Corporation

The MARS™ Microwave Synthesis System (Fig. 3.12 and Table 3.2) is based on the related MARS 5™ digestion instrument and offers different sets of rotor systems with several vessel designs and sizes for various synthesis applications.

For reactions at atmospheric pressure, standard laboratory glassware such as round-bottomed flasks from 0.5 to 3 L can be used. A protective mount in the ceil-

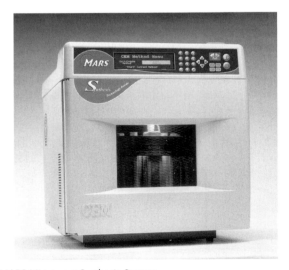

Fig. 3.12 CEM MARS Microwave Synthesis System.

ing of the cavity allows the connection of a reflux condenser or of distillation equipment, as well as providing access for the addition of reagents and sample withdrawal. For parallel reactions at atmospheric pressure, turntables for up to 120 vials are available. Furthermore, a turntable for conventional 96-well titer plates is offered, allowing reactions to be performed in a minimum volume of 0.1 mL per vessel (Table 3.3 and Fig. 3.13). Reactions under pressure can be carried out in the HP-500 Plus rotor (14 × 100-mL vessels), in the XP-1500 Plus rotor (12 × 100-mL vessels), or in the Xpress rotor (40 × 55- or 75-mL vessels) for high-throughput synthesis at elevated pressure (Table 3.4 and Fig. 3.14).

Temperature measurement in the rotor systems is accomplished by means of an immersed fiber-optic probe in one reference vessel or by an IR sensor on the surface of the vessels positioned at the bottom of the cavity. Pressure measurement in HP-

Table 3.2 MARS Microwave Synthesis System – General Features

Feature	Description
Cavity size (volume)	Approx. 50 L
Delivered power	1500 W
Max. output power	1200 W
Temperature control	Outside IR remote sensor; Immersed fiber-optic probe (optional)
Pressure measurement	Pneumatic pressure sensor (optional)
Cooling system	Air flow through cavity 100 m^3 h^{-1}
External PC	Optional; not required as integrated key panel is standard equipment

Table 3.3 MARS-S – Features of Atmospheric Pressure Parallel Rotors.

	Combichem	Parallel synthesis	High throughput
Turntable	3 × 96-well plates	52 ×	120 ×
Vessel volume	1 mL	50 mL	15 mL
Operation volume	min. 0.1 mL	max. 30 mL	max. 10 mL
Vessel material	glass	glass or PFA	glass or PFA
Max. temperature	150 °C	150 °C	150 °C
Temp. measurement	fiber optic	fiber optic	fiber optic

Table 3.4 MARS Microwave Synthesis System – Features of Parallel Pressure Rotors.

	HP-500+	XP-1500+	MARSXpress
Turntable	14 ×	12 ×	40 ×
Vessel volume	100 mL	100 mL	55 or 75 mL
Operation volume	max. 70 mL	max. 70 mL	max. 30 or 50 mL
Vessel material	PTFE-PFA	PTFE-TFM	PFA
Operation pressure	30 bar	100 bar	30 bar
Max. temp.	260 °C	300 °C	260 °C
Temp. measurement	fiber optic	fiber optic	IR

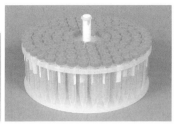

Fig. 3.13 CEM MARS parallel rotors: CombiChem, parallel synthesis, high throughput (left to right).

Fig. 3.14 CEM MARS pressure rotors: XP1500, HP500, Xpress (clockwise).

and XP-rotors is achieved by means of an electronic sensor in one reference vessel. Therefore, parallel reactions should only be carried out with identical reaction mixtures in the individual vessels. For reactions at high pressures, the HP- and XP-rotor vessels offer a choice of covers: fully sealed or self-venting. The temperature and pressure feedback control of the MARS System monitors and carefully regulates the amount of power being supplied to the reactions to provide optimum reaction control. The system will automatically shut down the microwave power if the temperature in the control vessel rises too high or if the vessel starts to over-pressurize. For the MARSXpress rotor, no internal pressure measurement is available as the vessels are "self-regulating" to prevent over-pressure. The MARSXpress maintains reaction control by means of the self-regulating pressure vessels and the temperature feed-back control. The temperature control sensor monitors the temperature in each ves-sel and adjusts the power output accordingly to maintain the user-defined tempera-ture set point. The MARSXpress offers real-time display of the temperature in each vessel. All of CEM's high-pressure vessels have a unique open-architecture design that allows air-flow within the cavity to cool the vessels quickly.

The general maximum output power of the instrument is 1200 W, but the MARS' control panel offers two low-energy levels with unpulsed microwave output powers of 300 and 600 W, respectively. This feature avoids overheating of the reaction mix-ture and unit when just small amounts of reagents are used.

The MARS comes with a software package, operated via the integrated spill-proof key-pad. The instrument can be connected to an external PC, but this is not required for most common operations. Methods and reaction protocols can be designed as tem-perature/time profiles or with precise control of constant power during the reaction.

3.4.3
Biotage AB

For scale-up applications, Biotage offers the Emrys Advancer batch reactor (Fig. 3.15), providing a multimode cavity for operations with one 350–850 mL Teflon reaction vessel under high-pressure conditions. An operating volume of ca. 50–500 mL at a maximum of 20 bar enables the production of 10–100 g of product in a single run. Homogeneous heating is ensured by a precise field-tuning mechanism and vigorous magnetic and/or overhead stirring of the reaction mixture. Direct scal-ability allows translation of the optimized reaction conditions from the Emrys/Initiator system (see Section 3.5.1) to a larger scale.

Fig. 3.15 Biotage Emrys Advancer scale-up instrument with multifunctional chamber head (right).

The maximum output power of the Emrys Advancer is 1100 W, generating a heating rate of 0.5–4 °C s^{-1}, which allows the maximum temperature of 250 °C for a reaction volume of 300 mL to be reached in comparable times as in the monomode experiments. Several connection ports in the chamber head (Fig. 3.15) enable the addition of reagents during the irradiation, sample removal for analysis, *in situ* monitoring by real-time spectroscopy, or the creation of inert/reactant gas atmospheres. Cooling is achieved by means of an effective definite gas-expansion mechanism, which ensures drastically shortened cooling periods (200 mL of ethanol is cooled from 180 to 40 °C within 1 min). Due to the dimensions of the instrument (160 × 85 × 182 cm), extra laboratory space is required to make operations comfortable. This instrument is a custom-built, user-specified product, manufactured on request.

3.4.4
Anton Paar GmbH

The Anton Paar Synthos 3000 (Fig. 3.16 and Table 3.5) is the most recent multi-mode instrument to come onto the market. It is a microwave reactor dedicated for scaled-up synthesis in quantities of up to approximately 250 g per run and designed for chemistry under high-pressure and high-temperature conditions. The instrument enables direct scaling-up of already elaborated and optimized reaction protocols from single-mode cavities without changing the reaction parameters.

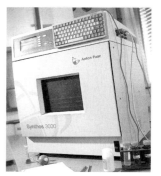

Fig. 3.16 Anton Paar Synthos 3000; rotors and vessel types.

Table 3.5 Anton Paar Synthos 3000 Features.

Feature	Description
Cavity size (volume)	45 × 42 × 35 cm (66 L)
Installed power	1700 W (2 magnetrons)
Max. output power	1400 W
Temperature control	Immersed gas balloon thermometer (max. 300 °C); outside IR remote sensor (max. 400 °C)
Pressure measurement	Hydraulic system (max. 86 bar)
Cooling system	190 m^3 h^{-1} (forced air flow through rotor, 4 steps adjustable)
Magnetic stirring	600 rpm (4 steps software adjustable)
Rotor speed	3 rpm
External PC	Optional, not required as key panel + keyboard is standard equipment

The use of two magnetrons (1400 W continuously delivered output power) allows mimicking of small-scale runs to produce large amounts of the desired compounds within a similar time frame. The homogeneous microwave field guarantees identical conditions at every position of the rotors (8 or 16 vessels), resulting in good reproducibility of experiments. Since it offers high operation limits (80 bar at 300 °C), this instrument facilitates the investigation of new reaction methods. The instrument can be operated with either an 8- or a 16-position rotor, equipped with various vessel types (Fig. 3.16) for different pressure and temperature conditions. Various

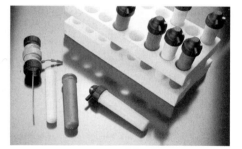

Fig. 3.17 Anton Paar accessories for special applications: gas loading system, filtration unit, UV lamp (left to right).

accessories (Fig. 3.17) allow for special applications, such as the creation of inert/reactive gas atmospheres, reactions in pre-pressurized vessels, and chemistry in near-critical water, as well as solid-phase synthesis or photochemistry.

3.4.4.1 Parameter Control

Temperature measurement is achieved by means of a remote IR sensor beneath the lower outer surface of the vessels. The operation limit of the IR sensor is 400 °C, but it is regulated by the software safety features to 280 °C as the operation limits of the materials used are around 300 °C. For additional control, temperature measurement in a reference vessel by means of an immersed gas-balloon thermometer is available. The operational limit of this temperature probe is 310 °C, making it suitable for reactions under extreme temperature and pressure conditions.

The pressure is measured by means of a hydraulic system, either in one reference vessel of the 16-vessel rotor or simultaneously for all vessels of the 8-vessel rotor. The operational limit is 86 bar, sufficient for synthetic applications. In addition, a pressure rate limit is set to 3.0 bar s^{-1} by the control software provided. Protection against sudden pressure peaks is provided by metal safety disks incorporated into the vessel caps (safety limits of 70 bar or 120 bar, respectively) and by software regulations, depending on the rotor used and the vessel type.

All parameters are transmitted in a wireless fashion by IR data transfer from the rotors to the system control computer of the instrument to eliminate disturbing cables and hoses from within the cavity.

3.4.4.2 Rotors and Vessels

For the individual rotor types, different vessels are available. The 16-vessel rotor, dedicated for the performance of standard reactions at up to 240 °C, offers 100 mL PTFE-TFM liners. Applying different pressure jackets allows continuous operation of these vessels to a maximum of 20 or 40 bar, respectively. The 8-vessel rotor is designed for high-pressure reactions, offering PTFE-TFM liners or quartz vessels, which enable reactions to be performed at 300 °C and 80 bar for several hours.

Table 3.6 Synthos 3000 Vessels and Operation Limits.

	MF100	HF100	XF100	XQ80
Volume	100 mL	100 mL	100 mL	80 mL
Operating volume	6–60 mL	6–60 mL	6–60 mL	6–60 mL
Max. temp.	200 °C	240 °C	260 °C	300 °C
Operation pressure	20 bar	40 bar	60 bar	80 bar
Safety limit	70 bar	70 bar	120 bar	120 bar
Liner material	PTFE-TFM	PTFE-TFM	PTFE-TFM	quartz
Pressure jacket	PEEK	ceramics	ceramics	none
Pre-pressurizing	no	no	20 bar	20 bar

3.5
Single-Mode Instruments

The first microwave instrument company offering single-mode cavities ("focused microwaves" [7]) was the French company Prolabo (Fig. 3.18) [13]. In the early 1990s, the Synthewave 402 was released, followed by the Synthewave 1000. The instruments were designed with a rectangular waveguide, providing "focused micro-waves" with a maximum output power of 300 W. The cavity was designed for the use of cylindrical glass or quartz tubes of several diameters for reactions at atmo-spheric pressure only. Temperature measurement was accomplished by means of an IR sensor at the bottom of the vessels, which required calibration by a fiber-optic probe. In 1999, all patents and microwave-based product lines were acquired by CEM Corporation. However, several instruments are still in use, mainly within the French scientific community, and several publications per year describing the use of this equipment appear in the literature.

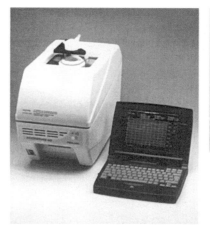

Fig. 3.18 Prolabo single-mode microwave instruments Synthewave 402 (left) and Synthewave 1000 (right). Reproduced with permission from [14].

3.5.1
Biotage AB

3.5.1.1 Emrys™ Platform (2000–2004)
Until 2004, Biotage (formerly Personal Chemistry) offered the Emrys™ monomode reactor series of instruments (Fig. 3.19). Although no longer commercially available, many instruments are currently still in use. Therefore, this line of products is dis-cussed in detail in this chapter [15].

The Emrys Creator is the basic application tool for reactions on a 0.5–5.0 mL scale. The Emrys Optimizer allows for automation with integrated robotics so that

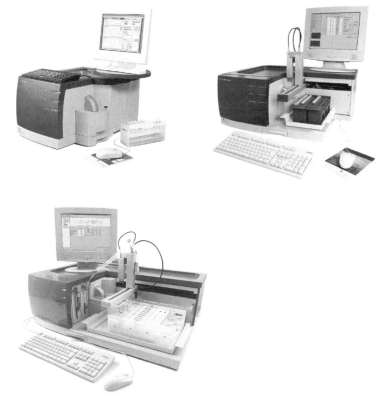

Fig. 3.19 Personal Chemistry (Biotage) Emrys product series: Creator, Optimizer, and Liberator (clockwise).

up to 60 reactions can be run in series. The EXP upgrade offers a larger microwave cavity and additional vessel sizes from very small scale (0.2–0.5 mL) to a low-level scale-up vessel (20 mL). The Emrys Liberator is a fully automated instrument with robotic sample handling and a liquid dispenser for high-throughput library synthesis. Up to 120 reactions can be operated sequentially with this equipment. Additionally available reaction kits enable standard reactions to be carried out under different conditions at one time. In contrast to other single-mode reactors, all Emrys instruments only operate with an external computer control.

The Emrys microwave unit consists of a closed rectangular waveguide tube combined with a deflector device that physically maximizes the energy absorption by the reaction mixture by means of a power sensor (Dynamic Field Tuning™). For this reason, the operation volume is limited (0.2–20 mL), but the uniform and high density heating process results in fast and highly reproducible synthesis results. The output power is maximized to continuously deliver 300 W, which is sufficient for rapid heating of most reaction mixtures. The Emrys series is equipped with built-in magnetic stirring at a fixed level. Temperature measurement is achieved by means of an IR sensor perpendicular to the position of the vial in the waveguide, working in a measuring range of 60–250 °C. This arrangement requires a minimum filling height in each vessel type in order to obtain accurate temperature values. Thus, the temperature is measured on the outer surface of the reaction vessels, and no internal temperature measurement is available. The pressure limit for the Emrys™ instruments is 20 bar, imposed by the sealing mechanism which utilizes Teflon-coated silicon seals in aluminum crimp tops. Pressure control is achieved by means of a sensor integrated into the closing lid of the cavity, which senses the deformation of the seal due to the pressure that is built-up. At 20 bar, the instrument switches off and the cooling mechanism is activated as the reaction is aborted. Efficient cooling is accomplished by means of a pressurized air supply at a rate of approximately 60 L min^{-1}, enabling cooling from 250 °C to 40 °C within approximately one minute, depending on the heat capacity of the solvent used.

Reaction control is temperature-based, with the system trying to attain the adjusted maximum temperature as rapidly as possible. When employing polar reaction mixtures with a high level of absorption, the output power can be limited to a maximum of 150 W to avoid overheating of the sample. The instruments of the Emrys series come with a comprehensive software package for the creation of protocols, including an ISIS draw surface for graphical description of the reactions.

A very useful tool is the Emrys Knowledge Database, representing a collection of detailed protocols for reactions performed with Biotage (Personal Chemistry) instruments. To date, this web-based tool contains around 4000 entries.

3.5.1.2 Initiator™ and Initiator Sixty

The currently available instrumentation from Biotage is the Initiator™ reactor for small-scale reactions in a single-mode cavity (Fig. 3.20). This instrument is closely related to the former Creator, but is now equipped with a touch-screen for "on-the-fly" control or changes of parameters, and no external PC is needed. The enhanced version, the Initiator™ Sixty, in succession to the Optimizer EXP, is equipped with a

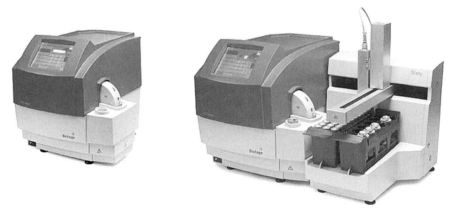

Fig. 3.20 Biotage Initiator and Initiator Sixty (right).

Fig. 3.21 Biotage Initiator vessel types (0.2–0.5 mL, 0.5–2.5 mL, 2.0–5.0 mL, 5.0–20.0 mL (left to right)) and schematic depiction of maximum volume.

robotic gripper and allows small range scale-up using different vessels with operating volumes of 0.2–20 mL (Fig. 3.21). With this automated upgrade, 24 to 60 reactions can be carried out sequentially and unattended. In March 2005 the Initiator Eight was released.

Similar to its predecessors of the Emrys series, the operation limits for the Initiator system are 60–250 °C at a maximum pressure of 20 bar. Temperature control is achieved in the same way by means of an IR sensor perpendicular to the sample position. Thus, the temperature is measured on the outer surface of the reaction vessels, and no internal temperature measurement is available. Pressure measurement is accomplished by a non-invasive sensor integrated into the cavity lid, which measures the deformation of the Teflon seal of the vessels. Efficient cooling is accomplished by means of a pressurized air supply at a rate of approximately 60 L min^{-1}, which enables cooling from 250 °C to 40 °C within one minute.

As no external PC is now needed, the characteristics of the software package have changed compared to the Emrys series. Via the touch-screen, the user is now able to change the reaction protocols "on-the-fly" without aborting the experiment. Temperature and time changes can be made immediately and furthermore the power can be adjusted to a defined value over the process time. In addition, so-called "cooling-while-heating" can be applied, whereby more microwave energy is introduced into the reaction system as the cooling is activated during the irradiation process (see Section 2.5.3 for details).

The above mentioned Knowledge Database has been adapted for use with the Initiator System as the technical features of the new instruments are very similar to those of the Emrys series.

3.5.2
CEM Corporation

The CEM Discover® platform, introduced in 2001, offers a single-mode instrument based on the self-tuning circular waveguide technique (see Fig. 3.3). This concept allows for derivatization of the reactor to accommodate additional application-specific modules to address other common laboratory objectives. While the Discover®

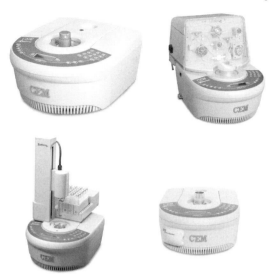

Fig. 3.22 Available variations of the CEM Discover platform: BenchMate, Voyager, Investigator, and Explorer$_{PLS}$ (clockwise).

reactor covers a variety of reaction conditions in open- (up to 125 mL) and closed-vessel systems (max. 7 or 50-mL working volumes) (Fig. 3.22) [16], it can be easily upgraded for automation (Explorer$_{PLS}$®, with robotic gripper and up to 24-position autosampler racks) or for scale-up by a flow-through approach (Voyager™). Offering a very economical choice in terms of cost of ownership and in footprint, the Discover BenchMate provides a fully featured entry-level system with reaction temperature and pressure management. The chemist has the ability to instantaneously change any reaction parameter during the reaction process, leading to improved instrument control and better reaction optimization, another valuable feature introduced by this line of reactors. The Investigator module™ brings to the platform real-time, "*in situ*" analysis of the reaction using an integrated Raman spectroscopy system. As with the automation and scale-up modules, the reactor core includes all of the necessary capabilities to accommodate the Raman module with minimal modification of the system as a whole. The analytical technique includes a proprietary fluorescence correction that overcomes the typical limitations that accompany Raman spectroscopy. The cavity design allows for the easy insertion of a quartz light pipe with focusing optics to enhance the Raman signal. The system software interrogates the Raman spectra and uses this information as a feedback mechanism to assist in the optimization of the reaction conditions.

The Voyager System converts the standard reactor into a flow-through system (see Section 4.5), designed to allow the scale-up of reactions while still maintaining the advantages of single-mode energy transfer. While the technology accommodates both continuous- and stop-flow formats, the stop-flow technique better accommodates the majority of scale-up applications encountered in today's synthesis laboratory. The Voyager system in stop-flow mode is operated with a special 80-mL vessel (see Fig. 3.23), imposing reaction limits of 225 °C or 15 bar, and is applicable even for heterogeneous mixtures, slurries, and solid-phase reactions.

Reaction scale-up using the Voyager system in genuine continuous-flow format is achieved by the use of special coiled flow-through cells. The reaction coils are made of glass or Teflon (Fig. 3.24) with a maximum flow rate of 20 mL min^{-1} and operational limits of 250 °C or 17 bar. The continuous-flow format should only be used for homogeneous solution-phase chemistry, as slurried mixtures may cause prob-

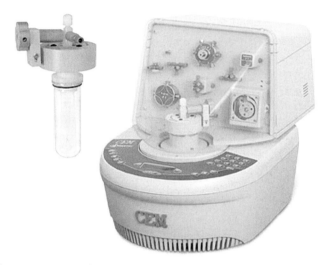

Fig. 3.23   The CEM Voyager and its 80-mL reaction vessel.

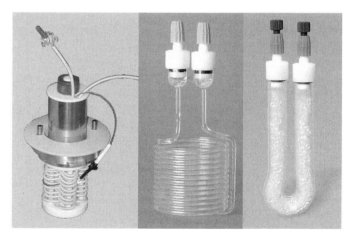

Fig. 3.24   Applicable flow-through cells for the CEM Voyager: 5 mL Kevlar-reinforced Teflon coil (left), 10 mL glass coil (center), active flow cell (right).

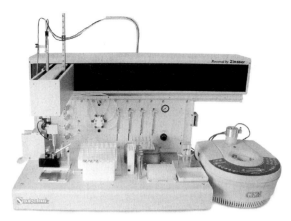

Fig. 3.25 Advanced applications: CEM Navigator (left) and Liberty (right).

lems with the pumping system. In addition, active flow cells (Fig. 3.24), which can be equipped with catalysts or scavengers on solid supports, are also available.

For advanced high-throughput synthesis, the fully automated Navigator™ Microwave Compound Factory is available (Fig. 3.25). This equipment offers dedicated XYZ robotics, including but not limited to the following options: capping/decapping station, automated weighing, liquid handling (up to 8 substrates) and solid reagent addition, as well as solid/liquid sample withdrawal. A number of integrated analytics (SPE, HPLC, flash, Raman) can be delivered; the system is built according to customer requirements.

In addition, for solid-phase peptide synthesis, the Liberty™ reaction system is available (Fig. 3.25). This instrument, based on the same Discover® reactor core, enables the synthesis of up to 12 peptides in an unattended manner using a fluidics module to enable the controlled addition of resins, amino acids, coupling, deprotection, and washing reagents, as well as cleavage cocktails. Vessels with volumes of 10–100 mL allow the synthesis of peptides on a 0.1 to 5-mmol scale, with up to 25 amino acids, accommodating both naturally occurring and non-natural amino acids. Typical cycle times with this system are around 15 min, including all washings, for each residue addition. The Liberty also allows programmable cleavage from the resin, either immediately after the synthesis or at a later programmed time. Potential issues related to racemization have been addressed through the use of precise temperature control.

Routine temperature measurement within the Discover series is achieved by means of an IR sensor positioned beneath the cavity below the vessel. This allows accurate temperature control of the reaction even when using minimal volumes of materials (0.2 mL). The platform also accepts an optional fiber-optic temperature sensor system that addresses the need for temperature measurement where IR technology is not suitable, such as with sub-zero temperature reactions or with specialized reaction vessels. Pressure regulation is achieved by means of the IntelliVent™ pressure management technology. If the pressure in the vial exceeds 20 bar, the

IntelliVent sensor allows for a controlled venting of the pressure and then reseals to maintain optimum safety and extend application scope. All Discover instruments are equipped with a built-in keypad for programming the reaction procedures and facilitating "on-the-fly" changes. All vessel types are equipped with corresponding stirring bars, ensuring optimum admixing of the reagents. Flow cells facilitate mixing as the reagents travel through a circuitous path. The patented enhanced cooling system (PowerMAX™) can be operated during irradiation, thereby achieving simultaneous cooling for "enhanced microwave synthesis" (EMS), which results in a more efficient energy transfer into the sample and leads to improved results (see Section 2.5.3 for details).

The latest extension in this context is the Discover® CoolMate™ (Fig. 3.26), a microwave system for performing sub-ambient temperature chemistry. The reactor is equipped with a jacketed low-temperature vessel, and the system's microwave-transparent cooling medium and chilling technology keep the bulk temperature low (–80 to +35 °C). Thus, thermal degradation of compounds is prevented while microwave energy is introduced to the reaction mixture.

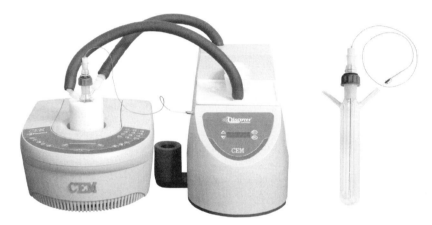

Fig. 3.26 CEM Discover CoolMate: low-temperature microwave system.

3.6
Discussion

Since the first appearance of industrially designed microwave instruments for organic synthesis in the early 1990s, the interest in microwave-assisted synthesis has grown tremendously.

Whereas in the last decade microwave irradiation was mainly applied to accelerate and optimize well-known and established reactions, current trends are indicative of the future use of microwave technology for the development of completely new reaction pathways in organic synthesis. Limited by vessel and cavity size, microwave-assisted synthesis has hitherto been focused predominantly on reaction optimization and method development on a small scale (<10 mmol). Today, there is increasing demand for the large-scale microwave production (>100 g per run) of valuable intermediates. To this end, the development of large flow-through systems is currently under investigation.

A limiting factor in attempting to introduce microwave instrumentation to a large number of academic and industrial laboratories is certainly the price of the equipment. Although prices have come down dramatically since the first introduction of single-mode microwave instruments in 2000, the average microwave reactor remains a rather expensive piece of equipment [17]. Academic laboratories in particular often cannot afford large budgets for new instrumentation, and thus market forces are constantly pushing instrument manufacturers to produce and offer cheaper reactors. This will certainly help to realize the intent over time to replace classical oil baths by microwave reactors as conventional heating sources in chemical laboratories. Up-to-date information about available instruments for microwave-assisted organic synthesis can be found on the websites of the instrument manufacturers [8–11].

References

[1] R. V. Decareau, R. A. Peterson, *Microwave Processing and Engineering*, VCH, Weinheim, **1986**.

[2] A. C. Metaxas, R. J. Meredith, *Industrial Microwave Heating*, P. Peregrinus, London, UK, **1983**.

[3] C. A. Balanis, *Advanced Engineering Electromagnetics*, Wiley, New York, USA, **1989**.

[4] D. Stuerga, M. Delmotte, in *Microwaves in Organic Synthesis* (Ed.: A. Loupy), Wiley-VCH, Weinheim, **2002**, pp 1–34 (Chapter 1).

[5] R. Gedye, F. Smith, K. Westaway, H. Ali, L. Baldisera, L. Laberge, J. Rousell, *Tetrahedron Lett.* **1986**, *27*, 279–282.

[6] R. J. Giguere, T. L. Bray, S. M. Duncan, G. Majetich, *Tetrahedron Lett.* **1986**, *27*, 4945–4958.

[7] M. Nüchter, B. Ondruschka, W. Bonrath, A. Gum, *Green Chem.* **2004**, *6*, 128–141.

[8] Anton Paar GmbH, Anton-Paar-Str. 20, A-8054 Graz, Austria; phone: (internat.) +43–316–257180; fax: (internat.) +43–316–257918; http://www.anton-paar.com.

[9] Biotage AB, Kungsgatan 76, SE-753 18 Uppsala, Sweden; phone: (internat.) +46–18565900; fax: (internat.) +46–18591922; http://www.biotage.com.

[10] CEM Corporation, P.O. Box 200, Matthews, NC 28106, USA; Toll-free: (inside US) +1–800–7263331; phone: (internat.) +1–704–8217015; fax: (internat.) +1–704–8217894. http://www.cemsynthesis.com (US website for CEM synthesis equipment); www.cem.com (main website), www.mikrowellen-synthese.de (German website).

[11] Milestone s.r.l., Via Fatebenefratelli, 1/5, 24010 Sorisole, BG, Italy; phone: (internat.) +39–035–573857; fax: (internat.) +39–035–575498; http://www.milestonesci.com (US website), www.milestonesrl.com (Italian website), www.mls-mikrowellen.de (German website).

[12] L. Favretto, *Mol. Div.* **2003**, *6*, 287–291.

[13] R. Commarmont, R. Didenot, J. F. Gardais (Prolabo), French Patent 84/03496, 1986.

[14] J. Cleophax, M. Liagre, A. Loupy, A. Petit, *Org. Proc. Res. Dev.* **2000**, *4*, 498–504.

[15] J.-S. Schanche, *Mol. Div.* **2003**, *6*, 293–300.

[16] J. D. Ferguson, *Mol. Div.* **2003**, *6*, 281–286.

[17] J. R. Minkel, *Drug Discov. Dev.* **2004**, March issue (special report).

4
Microwave Processing Techniques

In modern microwave synthesis, a variety of different processing techniques can be utilized, aided by the availability of diverse types of dedicated microwave reactors. While in the past much interest was focused on, for example, solvent-free reactions under open-vessel conditions [1], it appears that nowadays most of the published examples in the area of controlled microwave-assisted organic synthesis (MAOS) involve the use of organic solvents under sealed-vessel conditions [2] (see Chapters 6 and 7). Despite this fact, a brief summary of alternative processing techniques is presented in the following sections.

4.1
Solvent-Free Reactions

A frequently used processing technique employed in microwave-assisted organic synthesis since the early 1990s involves solvent-less ("dry media") procedures [1], where the reagents are pre-adsorbed onto either an essentially microwave-transparent (silica, alumina or clay) or a strongly absorbing (graphite) inorganic support, that additionally can be doped with a catalyst or reagent. Particularly in the early days of MAOS the solvent-free approach was very popular since it allowed the safe use of domestic household microwave ovens and standard open-vessel technology. While a large number of interesting transformations using "dry media" reactions have been published in the literature [1], technical difficulties relating to non-uniform heating, mixing, and the precise determination of the reaction temperature remain unresolved [3], in particular when scale-up issues need to be addressed.

One of the simplest methods involves the mixing of the neat reagents and subsequent irradiation with microwaves. In general, pure, dry solid organic substances do not absorb microwave energy, and therefore almost no heating will occur. Thus, small amounts of a polar solvent (e.g., N,N-dimethylformamide, water) often need to be added to the reaction mixture in order to allow for dielectric heating by microwave irradiation [3]. A case study demonstrates this point: according to a literature procedure, the reaction of benzoin with urea yields 4,5-diphenyl-4-imidazolin-2-one within 4 min of microwave irradiation in a domestic oven (Scheme 4.1) [4]. The reaction mixture is prepared by thoroughly mixing the two solid reactants. Interest-

Microwaves in Organic and Medicinal Chemistry. C. Oliver Kappe, Alexander Stadler
Copyright © 2005 WILEY-VCH Verlag GmbH & Co. KGaA, Weinheim
ISBN: 3-527-31210-2

ingly, this reaction could not be reproduced in a dedicated reactor [3]. As it turned out, the glass rotary plate utilized in a domestic oven is distinctly microwave-absorbing and warms to temperatures of 120–140 °C upon microwave irradiation, at which point benzoin melts through convective heating and the "microwave reaction" displayed in Scheme 4.1 is initiated. Only with the addition of small amounts of water could this transformation eventually be performed in a Teflon reactor inside a dedicated microwave instrument [3].

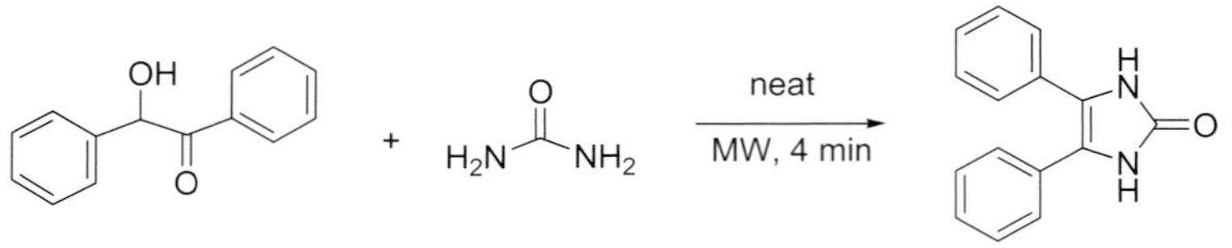

Scheme 4.1   Synthesis of imidazolin-2-one.

An alternative technique utilizes microwave-transparent or only weakly absorbing inorganic supports such as silica, alumina, or clay materials [1]. These reactions are effected by the reagents/substrates immobilized on the porous solid supports and have advantages over the conventional solution-phase reactions because of their good dispersion of active reagent sites, associated selectivity, and easier work-up. The recyclability of some of these solid supports and the avoidance of the waste disposal problems associated with the use of solvents renders these processes eco-friendly "green" protocols. In general, the substrates are pre-adsorbed onto the surface of the solid support and then exposed to microwave irradiation. In the example shown in Scheme 4.2, the solid support (in this case, neutral alumina) serves to catalyze the deprotection of aldehyde diacetates. Brief exposure of diacetate derivatives of aromatic aldehydes to microwave irradiation (domestic microwave oven) on a neutral alumina surface rapidly regenerated the corresponding aldehydes [5]. Selectivity in these deprotection reactions could be achieved by merely adjusting the irradiation time. For molecules bearing an acetoxy functionality (R = OAc), the aldehyde diacetate was selectively removed in 30 s, whereas an extended period of 2 min was required to cleave both the diacetate and ester groups. Unfortunately, the reaction temperature during the irradiation process could not be recorded, hence re-optimization of the conditions using temperature-controlled dedicated microwave reactors will be necessary.

Apart from examples where the inorganic support merely acts as a catalyst, there are many instances where a solid-supported reagent can be used very effectively in

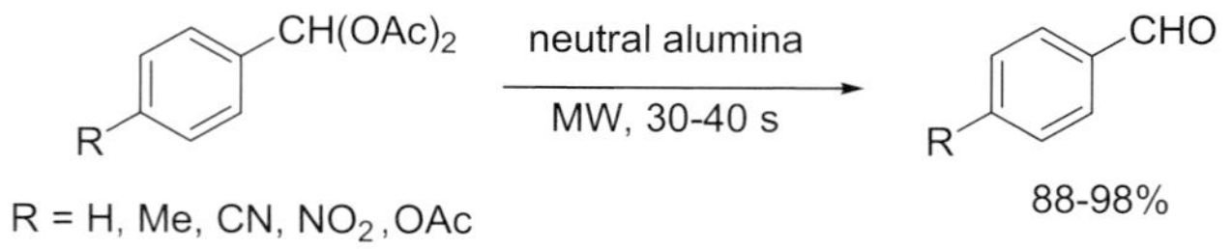

Scheme 4.2   Deprotection of aldehyde diacetates on neutral alumina.

the process. This is particularly true for oxidation reactions with metal-based reagents. For example, Varma and Dahiya have developed a method in which montmorillonite K 10 clay-supported iron(III) nitrate (so-called clayfen) is used under solvent-free conditions for the oxidation of alcohols to carbonyl compounds [6]. This process (Scheme 4.3) is significantly accelerated by exposure to microwave irradiation and simply involves mixing of the neat alcohols with clayfen and brief irradiation of the reaction mixtures with microwaves for 15–60 s. Remarkably, no carboxylic acids were formed in the oxidation of primary alcohols (R^1 = H). Similar microwave-assisted solvent-free oxidations have also been carried out with manganese- [7], copper- [8], and chromium-based [9] oxidation reagents adsorbed on suitable inorganic supports.

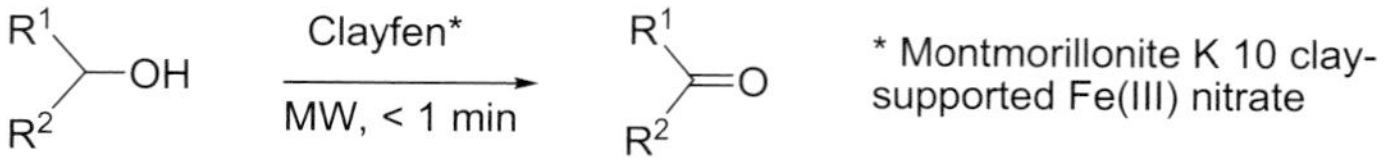

Scheme 4.3   Oxidation of alcohols by iron(III) nitrate supported on clay (clayfen).

In addition to cases where the inorganic support itself acts as a catalyst, or where a reagent has been impregnated on the solid support, it is also possible to additionally dope the support material with, for example, metal catalysts. Scheme 4.4 illustrates a Sonogashira coupling that was performed on a strongly basic potassium fluoride/alumina support, doped with a palladium/copper(I) iodide/triphenylphosphine mixture. The resulting aryl alkynes were synthesized in very high yields (82–97%) [10]. Many more examples of solvent-free microwave-assisted transformations can be found in review articles [1] and in recent books on microwave synthesis [11, 12].

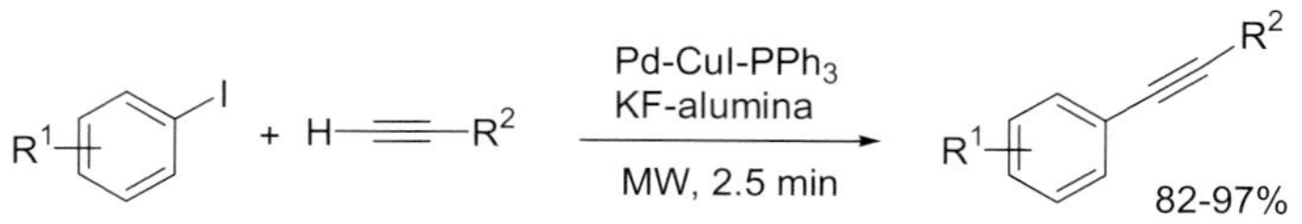

Scheme 4.4   Sonogashira coupling on palladium-doped alumina.

In contrast to solvent-free ("dry media") microwave processing involving (when properly dried) weakly microwave-absorbing supports such as silica, alumina, clays, and zeolites, an alternative is to use strongly microwave-absorbing supports such as graphite. For reactions which require high temperatures, the idea of using a reaction support which takes advantage of both strong microwave coupling and strong adsorption of organic molecules has received increasing attention in recent years [13]. Since many organic compounds do not interact appreciably with microwave irradiation, such a support could be an ideal "sensitizer", able to absorb and convert the energy provided by a microwave source and to transfer it to the chemical reagents.

Most forms of carbon interact strongly with microwaves. When irradiated at 2.45 GHz, amorphous carbon and graphite in powdered form rapidly reach ca. 1000 °C within 1 min of irradiation. An example of a solvent-free Diels–Alder reaction performed on a graphite support is shown in Scheme 4.5. Here, diethyl fumarate and anthracene adsorbed on graphite reacted within 1 min of microwave irradiation under open-vessel conditions to provide the corresponding cycloadduct in 92% yield [14]. The maximum temperature recorded by an IR-pyrometer was 370 °C. In other cases, it was necessary to reduce the microwave power and therefore the reaction temperature in order to avoid retro-Diels–Alder reactions [13].

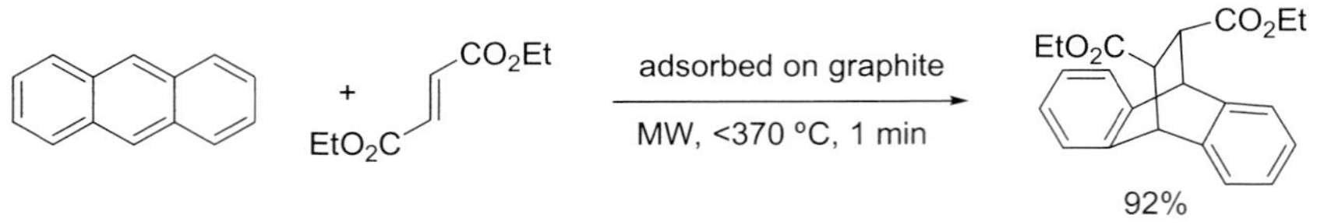

Scheme 4.5 Diels–Alder cycloaddition using graphite as sensitizer.

In addition to graphite being used as a "sensitizer" (energy converter), there are several examples in the literature where the catalytic activity of metal inclusions in graphite has been exploited ("graphimets"). The Friedel–Crafts acylation of anisole is a case in point, where the presence of catalytic amounts of iron oxide magnetite crystallites (Fe_3O_4) allowed efficient acylation with benzoyl chloride within 5 min of irradiation (Scheme 4.6) [13]. In contrast, the same reaction did not proceed when attempted in the presence of high-purity graphite without appreciable Fe_3O_4 content.

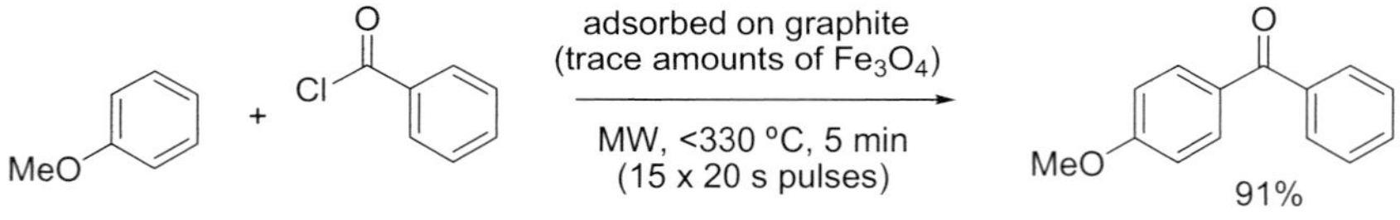

Scheme 4.6 Friedel–Crafts acylation using a "graphimet" as sensitizer and catalyst.

Since graphite is a very strong absorber of microwave heating, the temperature must be carefully controlled to avoid melting of the reactor. The use of a quartz reactor is highly preferable.

4.2
Phase-Transfer Catalysis

In addition to solvent-free processing, phase-transfer catalytic conditions (PTC) have also been widely employed as a processing technique in MAOS [15]. In phase-transfer catalysis, the reactants are situated in two separate phases, for example liquid-

liquid or solid-liquid. In liquid-liquid PTC, because the phases are mutually insoluble, ionic reagents are typically dissolved in the aqueous phase, while the substrate
remains in the organic phase. In solid-liquid PTC, on the other hand, ionic reagents
may be used in their solid state as a suspension in the organic medium. Transport
of the anions from the aqueous or solid phase to the organic phase is facilitated by
phase-transfer catalysts, typically quaternary onium salts or cation-complexing
agents. Phase-transfer catalytic reactions are perfectly tailored for microwave activation, and the combination of solid-liquid PTC and microwave irradiation typically
gives the best results in this area [15]. Numerous transformations in organic synthesis can be achieved under solid-liquid PTC and microwave irradiation in the absence
of solvent, generally at atmospheric pressure in open vessels.

Potassium acetate, for example, can be readily alkylated by the use of an equivalent amount of an alkylating reagent (for example, an alkyl halide) in the presence
of the phase-transfer catalyst Aliquat 336 (10 mol%) (Scheme 4.7) [16]. Yields are
always near quantitative within a few minutes of microwave irradiation, irrespective
of the chain length and the nature of the leaving group. This procedure has been
scaled-up from 50 mmol to 2 mol scale in a large batch reactor [17].

$$CH_3COOK + nC_8H_{17}Br \xrightarrow[\text{MW, 160 °C, 5 min}]{\text{neat, Aliquat 336}} CH_3COOnC_8H_{17}$$
98%

Scheme 4.7 *O*-Alkylation of potassium acetate under phase-transfer-catalysis conditions.

A related example involves the *N*-alkylation of azaheterocycles such as imidazole
(Scheme 4.8). These reactions are typically performed using 1.5 equivalents of the
alkyl halide and a catalytic amount of tetrabutylammonium bromide (TBAB). The
reactants are adsorbed either on a mixture of potassium carbonate and potassium
hydroxide or on potassium carbonate alone and then irradiated with microwaves
(domestic oven) for 0.5 to 1 min. With benzyl chloride, a yield of 89% was achieved
after 40 s [18].

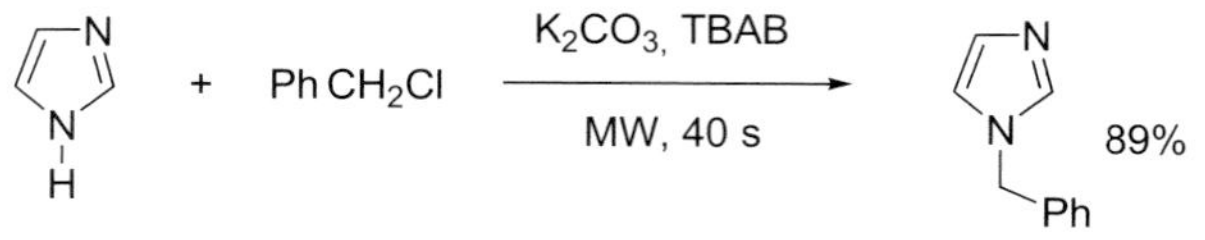

Scheme 4.8 *N*-Alkylation of imidazole with alkyl halides under phase-transfer-catalysis conditions.

4.3
Reactions Using Solvents

4.3.1
Open- versus Closed-Vessel Conditions

Microwave-assisted syntheses can be carried out using standard organic solvents under either open- or sealed-vessel conditions. If solvents are heated by microwave irradiation at atmospheric pressure in an open vessel, the boiling point of the solvent (as in an oil-bath experiment) typically limits the reaction temperature that can be achieved. In the absence of any specific or non-thermal microwave effects (such as the superheating effect at atmospheric pressure, see Section 2.5.2), the expected rate enhancements would be comparatively small. In order to nonetheless achieve high reaction rates, high-boiling microwave-absorbing solvents such as dimethyl sulfoxide, 1-methyl-2-pyrrolidone, 1,2-dichlorobenzene, or ethylene glycol (see Table 2.3) have frequently been used in open-vessel microwave synthesis [19]. However, the use of these solvents presents serious challenges during product isolation. The approach has been adapted for lower-boiling solvents, such as toluene, by periodic interruption of the heating. However, this modification not only precludes attainment of the high temperatures that are advantageous in microwave synthesis, but also generates a potentially serious fire hazard when unmodified domestic microwave instruments are used [19]. Because of the recent availability of modern microwave reactors with on-line monitoring of both temperature and pressure (see Chapter 3), MAOS in sealed vessels has been celebrating a comeback in recent years. This is clearly evident surveying the recently published literature in the area of MAOS (see Chapters 6 and 7), and it appears that the combination of rapid dielectric heating by microwaves with sealed-vessel technology (autoclaves) will most likely be the method of choice for performing MAOS in the future.

A typical example of open-vessel microwave technology is illustrated in Scheme 4.9. Here, poly(ethylene glycol)-200 (PEG-200) was used as a high-boiling solvent (boiling point > 250 °C) for the synthesis of furans from 2-butene-1,4-diones and 2-butyne-1,4-diones [20]. The high dielectric constant, high boiling point, and high water miscibility of PEG-200 make it an ideal protic solvent for microwave-assisted organic reactions at atmospheric pressure. Microwave irradiation in a domestic oven for 1–5 min produced excellent yields of the desired furans in the one-pot palladium-mediated reduction–dehydrative cyclization sequence of ene-1,4-diones or 2-yne-1,4-diones (Scheme 4.9) [20]. In a closely related approach, polyaryl-pyrroles were obtained from the same starting materials and ammonium formate [21]. Another area where high-boiling solvents are frequently used is microwave-assisted solid-phase organic synthesis (see Section 7.1). In solid-phase synthesis, the synthesized compounds are attached to an insoluble polymer support, which can easily be removed by filtration. Therefore, high-boiling solvents do not represent a problem during work-up. More examples of microwave synthesis under atmospheric pressure conditions, both in solution and on solid phases, are described in Chapters 6 and 7.

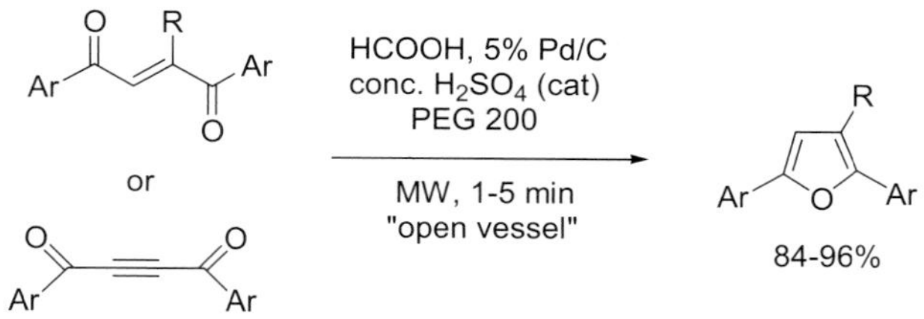

Scheme 4.9 Catalytic transfer hydrogenations in PEG 200 under open-vessel conditions.

An example of solid-phase microwave synthesis where the use of open-vessel technology is essential is shown in Scheme 4.10. The transesterification of β-keto esters with a supported alcohol (Wang resin) is carried out in 1,2-dichlorobenzene (DCB) as a solvent under controlled microwave heating conditions [22]. The temperature is kept constant at 170 °C, ca. 10 degrees below the boiling point of the solvent, thereby allowing safe processing in the microwave cavity. In order to achieve full conversion to the desired resin-bound β-keto ester, it is essential that the methanol formed can be removed from the equilibrium [22].

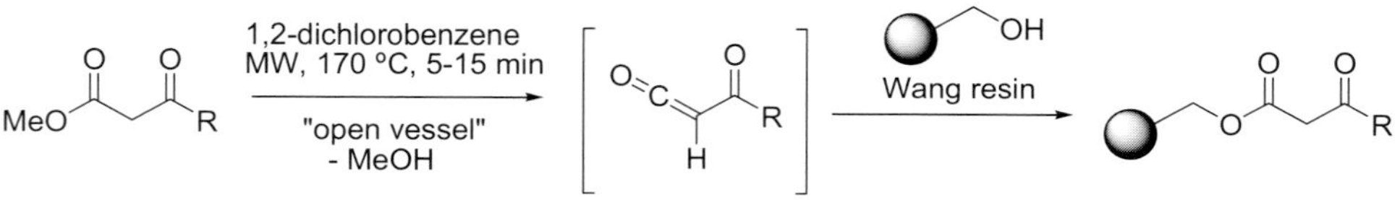

Scheme 4.10 Microwave-assisted solid-phase transesterifications.

Another example of the necessity to perform open-vessel microwave synthesis in specific cases is outlined in Scheme 4.11, namely the formation of 4-hydroxyquinolin-2-(1H)-ones from anilines and malonic esters. The corresponding conventional,

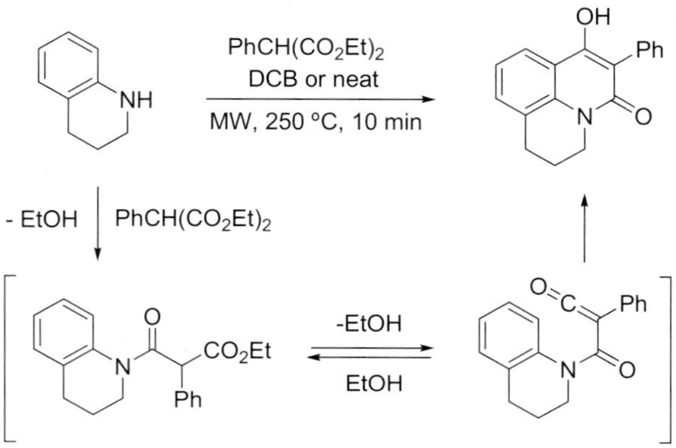

Scheme 4.11 Formation of 4-hydroxyquinolin-2-(1H)-ones from anilines and malonic esters.

thermal protocol involves heating of the two components in equimolar amounts in an oil bath at 220–300 °C for several hours (without solvent) [23], whereas similar high yields can be obtained by microwave heating at 250 °C for 10 min using 1,2-dichlorobenzene (DCB) as solvent [24]. Again, it was essential to use open-vessel technology here, since two equivalents of a volatile by-product (ethanol) are formed that under normal (atmospheric pressure) conditions are simply distilled off and therefore removed from the equilibrium. Preventing the removal of this ethanol from the reaction mixture, for example by using a standard closed-vessel microwave system that would result in a pressure build-up in the reaction vial, leads to significantly lower yields (Table 4.1). These experiments highlight the importance of choosing appropriate experimental conditions when using microwave heating technology. In the present example, scale-up of the synthesis shown in Scheme 4.11 would clearly only be feasible using open-vessel technology [25]. Loupy and co-workers have described a set-up for open-vessel microwave reactors whereby water formed during the reaction can be removed from the reaction medium [26].

Table 4.1 Dependence of the yield of quinolone product (Scheme 4.11) on the use of closed- or open-vessel microwave heating[a] (data from [24]).

Reagents (mmol)	Solvent (mL)	Yield (%)	Pressure (bar)
1	2	76	3.6
2	2	67	5.3
4	2	60	7.4
1	0.5	91	2.0
2[b]	neat	92	
4[b]	neat	90	

a) Microwave heating, 250 °C, 10 min, 1,2-dichlorobenzene (DCB) or neat.
b) Open vessel.

4.3.2
Pre-Pressurized Reaction Vessels

Relatively little work has been performed with gaseous reagents in sealed-vessel microwave experiments. Although several publications describe this technique in the context of heterogeneous gas-phase catalytic reactions important for industrial processes [27], the use of pre-pressurized reaction vessels in conventional microwave-assisted organic synthesis (MAOS) involving solvents is rare. Due to the design of modern single-mode microwave reactors and their reaction vessels, pre-pressurization is not possible. Several authors have, however, described the use of reactive gases in such experiments, along with experimental techniques for applying a slight overpressure (2–3 bar). Examples are summarized in Scheme 4.12. The Diels–Alder cycloaddition reaction of the heterodiene pyrazinone with ethene (Scheme 4.12a) furnished the corresponding bicyclic cycloadduct [28]. Under conventional conditions, these cycloaddition reactions have to be carried out in an autoclave applying an ethene pressure of 25 bar before the set-up is heated at 110 °C for 12 h. In the current example, Diels–Alder addition of ethene in a sealed vessel that had been flushed with the gas prior to sealing was completed after irradiation for 140 min at 190 °C. It was, however, not possible to further increase the reaction rate by raising the temperature. At temperatures above 200 °C, an equilibrium between the cycloaddition and the competing retro-Diels–Alder fragmentation process was observed [28]. Only by using a multimode microwave reactor that allowed pre-pressurization of the reaction vessel with 10 bar of ethene (see Fig. 3.18) could the Diels–Alder addition be carried out much more efficiently at 220 °C within 10 min [29].

The second example involves the synthesis of *ortho*-dipropynylated arenes (Scheme 4.12b), which serve as precursors to tribenzocyclyne by way of an alkyne metathesis reaction (see also Scheme 6.31). Here, a Sonogashira reaction was carried out in a pre-pressurized (propyne at ca. 2.5 bar) sealed microwave vessel in a standard single-mode microwave reactor. Double-Sonogashira coupling of the dibromodiiodobenzene was completed within 20 min at 110 °C [30].

Another example involves dioxygen-promoted regioselective oxidative Heck arylations of electron-rich alkenes with arylboronic acids (Scheme 4.12c). For this, two types of microwave reactors have been used. In a single-mode instrument (1 mmol run; 25 mL vessel), the Heck arylation was performed by first pre-pressurizing the

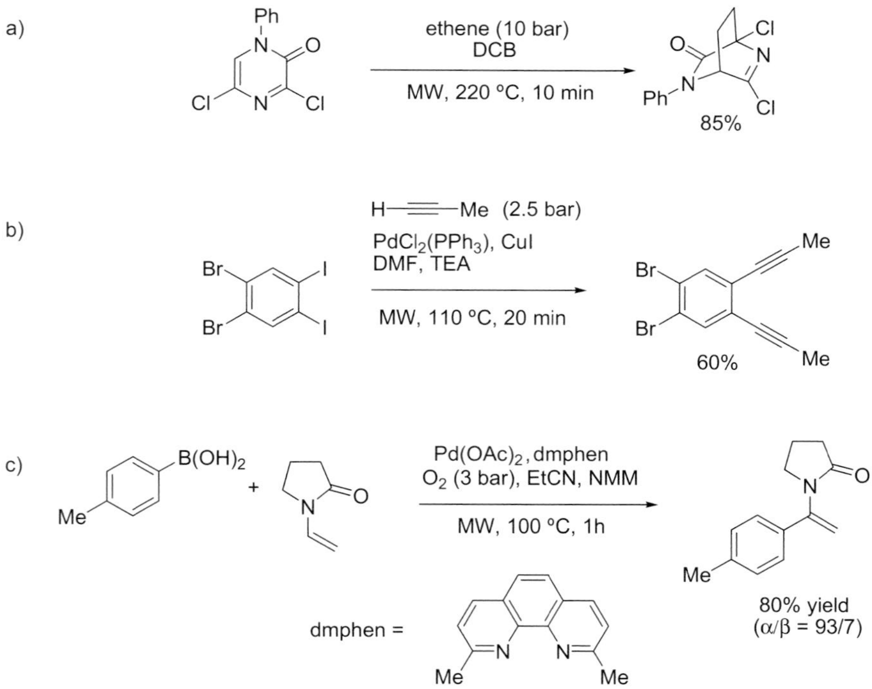

Scheme 4.12 Examples of microwave-assisted syntheses using pre-pressurized reactors.

vessel with oxygen to ca. 3 bar and subsequently heating under microwave conditions. The same reaction was also performed on a multigram scale (10 mmol) in a multimode autoclave reactor (see Fig. 3.16) using a continuous flow of oxygen (3 bar), providing nearly identical results [31].

4.3.3
Non-Classical Solvents

Besides using standard organic solvents in conjunction with microwave synthesis, the use of either water or so-called ionic liquids as "alternative reaction media" [32] has become increasingly popular in recent years.

4.3.3.1 Water as Solvent

Over the past decades, there has been increasing interest in the use of aqueous reaction media for performing organic reactions [33]. Temperatures of 100 °C and below have been employed for synthesis [33], often to exploit the so-called hydrophobic effect [34]. Conversely, at temperatures approaching the supercritical region (T_c = 374 °C), water has found degradative applications in addition to still being a solvent useful for synthetic purposes [35]. For microwave synthesis, the subcritical region (also termed near-critical) at temperatures in the range 150–300 °C is particularly attractive [35–38]. The dielectric constant, ε', of water decreases from 78 at 25 °C to 20 at 300 °C, this latter value being comparable with that of typical organic solvents such as acetone at ambient temperature (Fig. 4.1) [36]. Therefore, water behaves as a pseudo-organic solvent at elevated temperatures, allowing the dissolution of many organic substrates. In addition to the environmental advantages of using water in place of organic solvents, isolation of products is normally facilitated by the decrease in solubility of the organic reaction products upon post-reaction cooling. Furthermore, the ionic product (dissociation constant, pK_W) of water is greatly influenced by temperature, increasing by three orders of magnitude between 25 °C and 250 °C (Table 4.2) [39]. Water is therefore a significantly stronger acid and base in the subcritical region than under ambient conditions, which can be exploited for organic synthesis.

Numerous organic transformations have been carried out in subcritical water under conventional thermal conditions [35]. The application of microwave heating

Table 4.2 Properties of water under different conditions (data from [39]).

Fluid	Ordinary water (T < 150 °C) (p < 0.4 MPa)	Subcritical water (T = 150–350 °C) (p = 0.4–20 MPa)	Supercritical water (T > 374 °C) (p > 22.1 MPa)
Temperature (°C)	25	250	400
Pressure (bar)	1	50	250
Density (g cm^{-3})	1	0.8	0.17
Dielectric constant ε'	78.5	27.1	5.9
pK_W	14	11.2	19.4

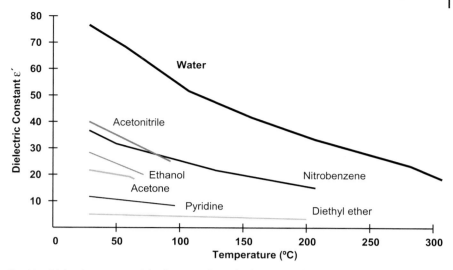

Fig. 4.1 Dielectric properties (ε') of water and standard organic solvents as a function of temperature (adapted from [36]).

in combination with the use of high-temperature water as a "green" solvent was pioneered by Strauss and co-workers in the 1990s [36, 37]. In an early example, the Fischer indole synthesis of 2,3-dimethylindole from 2-butanone and phenyl-hydrazine was accomplished in an aqueous medium at 222 °C within 30 min (Scheme 4.13) [36]. When 1 M sulfuric acid was used instead of water, the yield was comparable but the same transformation could be achieved within 1 min at the same temperature. In a related example, the decarboxylation of indole-2-carboxylic acid was investigated. At 255 °C, the decarboxylation of the acid was quantitative within 20 min (Scheme 4.13) [36].

More recently, Molteni and co-workers have described the three-component, one-pot synthesis of fused pyrazoles by reacting cyclic 1,3-diketones with N,N-dimethyl-

Scheme 4.13 Organic transformations utilizing near-critical water as solvent.

formamide dimethyl acetal (DMFDMA) and a suitable bidentate nucleophile such as a hydrazine derivative (Scheme 4.14) [40]. The reaction proceeds through initial formation of an enamino ketone as the key intermediate from the 1,3-diketone and DMFDMA precursors, followed by a tandem addition–elimination/cyclodehydration step. The authors were able to perform the multicomponent condensation by heating all three building blocks together with a small amount of acetic acid (2.6 equivalents) in water at 220 °C for 1 min. Upon cooling, the desired reaction products crystallized directly and were isolated in high purity by simple filtration. Although most of the starting materials are actually insoluble in water at room temperature, water at 220 °C behaves much like an organic solvent and is therefore able to dissolve the organic starting materials. The authors also successfully used other bidentate nucleophiles such as amidines and hydroxylamine for the synthesis of related heterocycles [40].

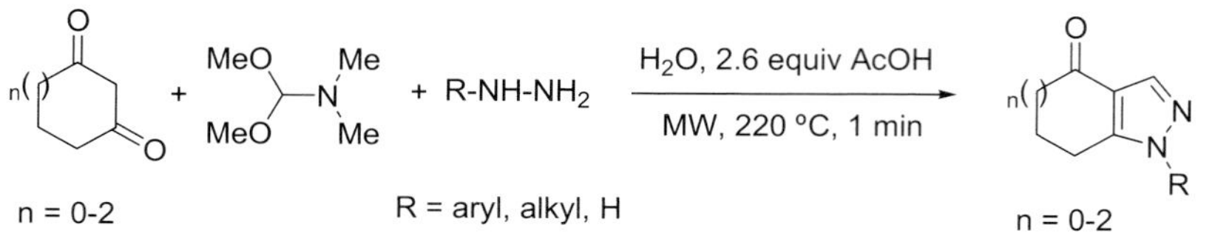

Scheme 4.14 Aqueous cyclodehydrations in the preparation of fused pyrazoles in superheated water.

Similarly, Vasudevan and Verzal have found that terminal alkynes can be hydrated under neutral, metal-free conditions using water as solvent (Scheme 4.15) [41]. While this reaction typically requires a catalyst such as gold(III) bromide, employing microwave-superheated distilled water allowed this chemistry to proceed without any catalyst. Extension of this methodology led to a one-pot conversion of alkynes to imines (hydroamination).

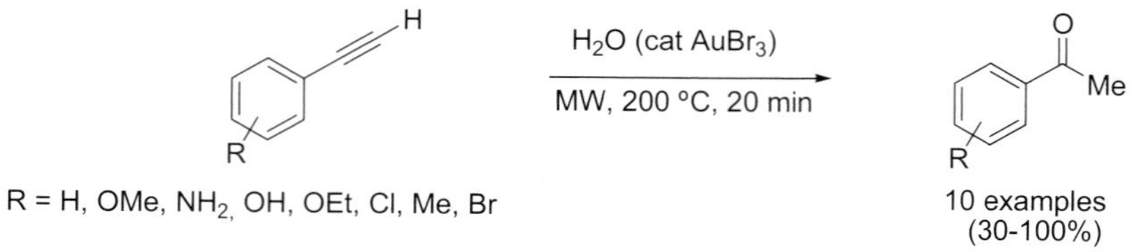

Scheme 4.15 Hydration of terminal alkynes in superheated water.

A recent publication by the group of Baran reports the total synthesis of ageliferin, an antiviral agent with interesting molecular architecture (Scheme 4.16) [42]. Just 1 min of microwave irradiation of sceptrin, another natural product, at 195 °C in water under sealed-vessel conditions provides ageliferin in 40% yield, along with 52% of recovered starting material. Remarkably, if the reaction is performed without

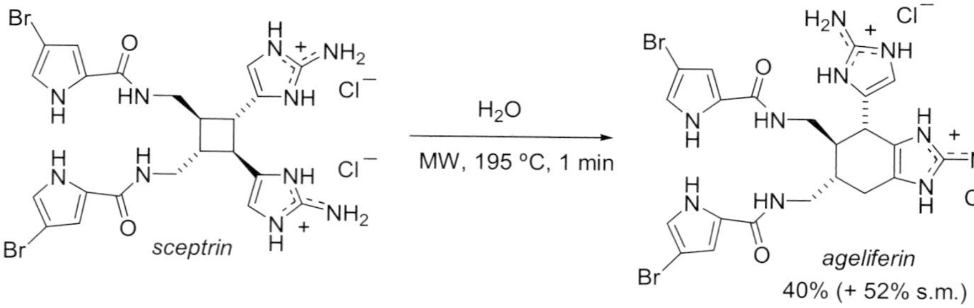

Scheme 4.16 Vinylcyclobutane–cyclohexene rearrangements in hot water.

microwaves at the same temperature, only starting material and decomposition products are observed.

There are many other examples in the literature where sealed-vessel microwave conditions have been employed to heat water as a reaction solvent well above its boiling point. Examples include transition metal catalyzed transformations such as Suzuki [43], Heck [44], Sonogashira [45], and Stille [46] cross-coupling reactions, in addition to cyanation reactions [47], phenylations [48], heterocycle formation [49], and even solid-phase organic syntheses [50] (see Chapters 6 and 7 for details). In many of these studies, reaction temperatures lower than those normally considered "near-critical" (Table 4.2) have been employed (100–150 °C). This is due in part to the fact that with single-mode microwave reactors (see Section 3.5) 200–220 °C is the current limit to which water can be safely heated under pressure since these instruments generally have a 20 bar pressure limit. For generating truly near-critical conditions around 280 °C, special microwave reactors able to withstand pressures of up to 80 bar have to be utilized (see Section 3.4.4).

4.3.3.2 Ionic Liquids

Room temperature ionic liquids (RTIL) are a new class of solvents that are entirely constituted of ions. Ionic liquids made up of organic cations and appropriate anions have attracted much recent attention as environmentally benign solvents for chemistry by virtue of the fact that they have melting points close to room temperature [51]. In some instances they have also been used as reagents. They have negligible vapor pressures and are immiscible with a range of organic solvents, meaning that organic products can be easily removed and the ionic liquid can be recycled. In addition, they have a wide accessible temperature range (typically > 300 °C), are nonflammable, and are relatively easy to use and to recycle. From the perspective of microwave chemistry one of the points of key importance is their high polarity and that this is variable depending on the cation and anion and hence can effectively be tuned to a particular application.

Ionic liquids interact very efficiently with microwaves through the ionic conduction mechanism (see Section 2.2) and are rapidly heated at rates easily exceeding $10\ °C\ s^{-1}$ without any significant pressure build-up [52]. Therefore, safety problems arising from over-pressurization of heated sealed reaction vessels are minimized.

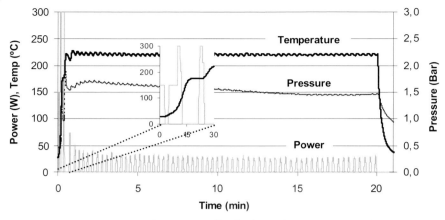

Fig. 4.2 Temperature, power, and pressure profiles for the Heck reaction presented in Scheme 4.17, performed in neat bmimPF$_6$ (set temperature 220 °C). The enlarged area shows the initial temperature response and the corresponding microwave power during the first 30 s of irradiation (reproduced with permission from [53]).

Figure 4.2 shows the heating profile for the Heck reaction shown in Scheme 4.17, in which 4 mol% of palladium(II) chloride/tri(*ortho*-tolyl)phosphine was used as the catalyst/ligand system [53]. Note the rapid increase of the measured reaction temperature during the first seconds of microwave irradiation. Full conversions were achieved within 5 min (X = I) and 20 min (X = Br). For transformations that were run without the phosphine ligand a reaction time of 45 min was required. A key feature of this catalytic/ionic liquid system is its recyclability: the phosphine-free ionic catalyst phase PdCl$_2$/bmimPF$_6$ (1-butyl-3-methylimidazolium hexafluorophosphate) was recyclable at least five times. After each cycle, the volatile product was directly isolated in high yield by rapid distillation under reduced pressure [53].

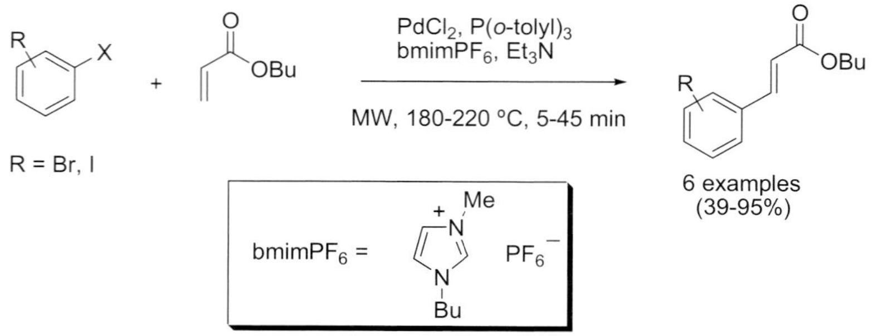

Scheme 4.17 Heck reactions in ionic liquids.

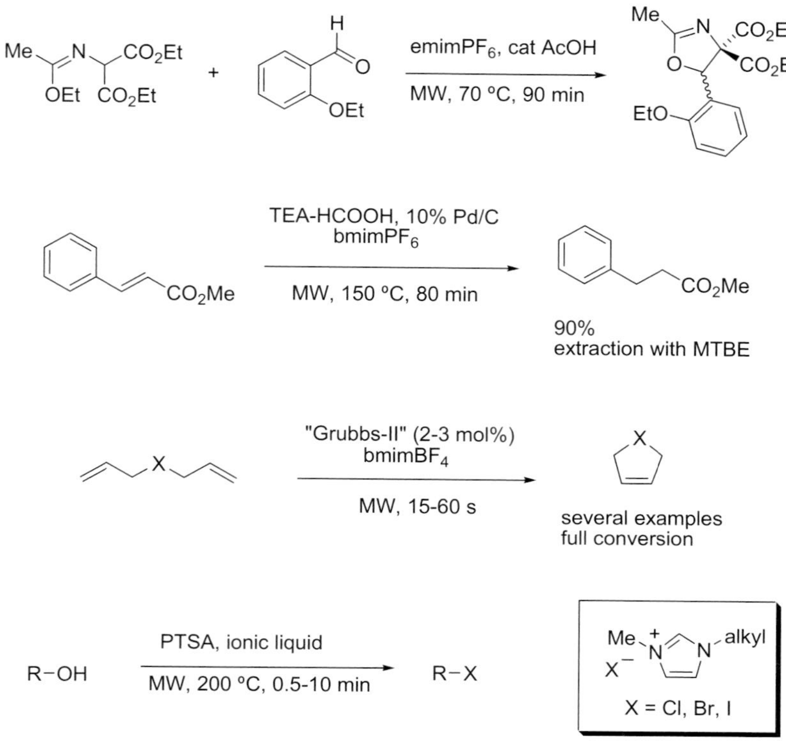

Scheme 4.18   Microwave syntheses in 1-alkyl-3-methylimidazolium
ionic liquids (see text for references).

   The concept of performing microwave synthesis in room temperature ionic liquids (RTIL) as reaction media has been applied to several different organic transformations (Scheme 4.18), such as 1,3-dipolar cycloaddition reactions [54], catalytic transfer hydrogenations [55], ring-closing metathesis [56], the conversion of alcohols to alkyl halides [57, 58], and several others [59–61].
   As an alternative to the use of the rather expensive ionic liquids as solvents, several research groups have used ionic liquids as "doping agents" for microwave heating of otherwise nonpolar solvents such as hexane, toluene, tetrahydrofuran, and dioxane. This technique, first introduced by Ley and co-workers in 2001 for the conversion of amides into thioamides using a polymer-supported reagent in a weakly microwave-absorbing solvent such as toluene (Scheme 4.19) [62], is becoming increasingly popular, as demonstrated by the many recently published examples (for details, see Chapter 6) [28, 63–71]. Systematic studies on temperature profiles and the thermal stability of ionic liquids under microwave irradiation conditions by the Leadbeater [63] and Ondruschka groups [52] have shown that the addition of a small amount of an ionic liquid (0.1 mmol/mL solvent) suffices to obtain dramatic changes in the heating profiles by changing the overall dielectric properties (tan δ

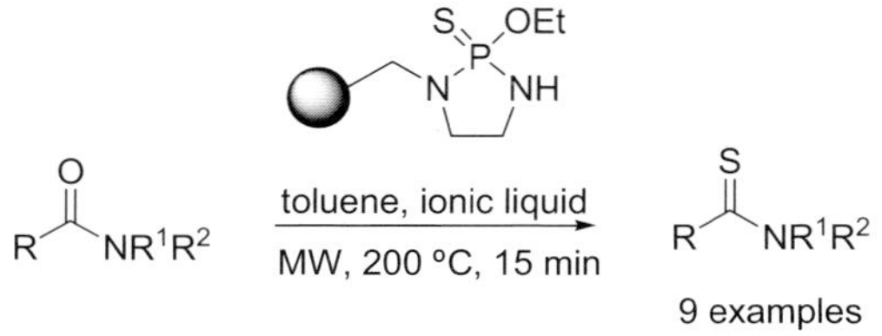

Scheme 4.19 Use of ionic liquid-doped toluene as a solvent for microwave synthesis.

value) of the reaction medium (Table 4.3). Other authors have shown that some substrates decompose in the presence of ionic liquids under microwave heating [69]. Reviews on microwave-promoted organic synthesis using ionic liquids were published in 2004 [70, 71].

It is worth noting that ionic liquids can also be very rapidly and efficiently prepared under microwave conditions. Much of the early work in this field was published by Varma and co-workers, applying open-vessel conditions [72]. Later, Khadlikar and Rebeiro demonstrated that the preparation of ionic liquids was also feasible by applying closed-vessel microwave conditions, eliminating the dangers of working with toxic and volatile alkyl halides in the open atmosphere [73]. A comprehensive study on the microwave-assisted preparation of ionic liquids was published by Deetlefs and Seddon in 2003. These authors synthesized a large number of ionic liquids based on the 1-alkyl-3-methylimidazolium, 1-alkyl-2-methylpyrazolium, 3-alkyl-4-methylthiazolium, and 1-alkylpyridinium cations by solvent-free alkylation of the

Table 4.3 Microwave heating effects of doping organic solvents with ionic liquids (IL) A and B (data from [63]).[a]

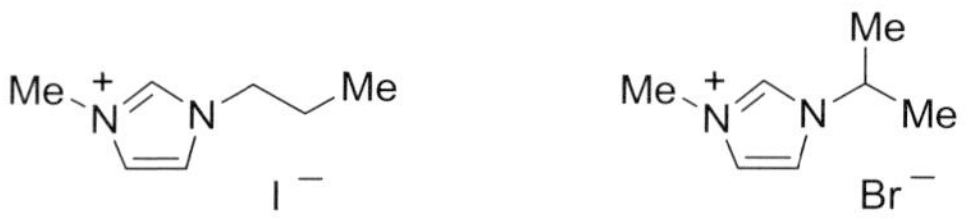

Solvent used	IL added	Temperature attained (°C)	Time taken (s)	Temperature attained without IL (°C) [b]
Hexane	A	217	10	46
	B	228	15	
Toluene	A	195	150	109
	B	234	130	
THF	A	268	70	112
	B	242	60	
Dioxane	A	264	90	76
	B	246	60	

a) Experiments run using a constant 200 W irradiation power (single-mode cavity) with 0.2 mmol IL/2 mL solvent under sealed-vessel conditions.
b) Temperature attained during the same irradiation time but without any IL added.

corresponding basic heterocyclic cores (Scheme 4.20) [74]. These methods offer dramatically reduced reaction times as compared to conventional methods, minimize the generation of organic waste, and also afford the ionic liquid products in excellent yields and with high purity. The syntheses were performed on flexible reaction scales ranging from 50 mmol to 2 mol in either sealed or open vessels, with a significant reduction of the large molar excess of haloalkane (often required in conventional ionic liquid syntheses).

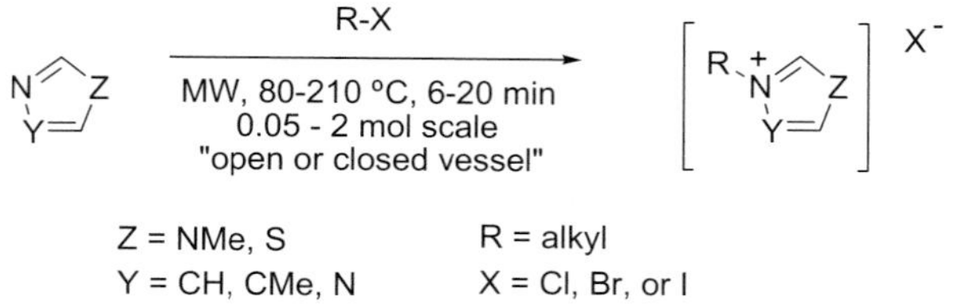

R-X

MW, 80-210 °C, 6-20 min
0.05 - 2 mol scale
"open or closed vessel"

Z = NMe, S   R = alkyl
Y = CH, CMe, N   X = Cl, Br, or I

Scheme 4.20 Preparation of ionic liquids under microwave conditions.

Using a similar protocol, Loupy and coworkers have reported the synthesis of chiral ionic liquids based on (1R,2S)-(–)-ephedrinium salts under microwave irradiation conditions (Scheme 4.21a) [75]. Importantly, the authors were also able to demonstrate that the desired hexafluorophosphate salts could be prepared in a one-pot protocol by *in situ* anion-exchange metathesis (Scheme 4.21b). The synthesis and transformation of so-called "task-specific ionic liquids" is discussed in more detail in Section 7.4.

a)

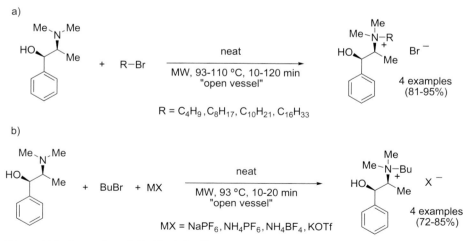

neat

MW, 93-110 °C, 10-120 min
"open vessel"

R = C$_4$H$_9$, C$_8$H$_{17}$, C$_{10}$H$_{21}$, C$_{16}$H$_{33}$

4 examples
(81-95%)

b)

neat

MW, 93 °C, 10-20 min
"open vessel"

MX = NaPF$_6$, NH$_4$PF$_6$, NH$_4$BF$_4$, KOTf

4 examples
(72-85%)

Scheme 4.21 Preparation of chiral ionic liquids.

4.4
Parallel Processing

Parallel processing of synthetic operations has been one of the cornerstones of medicinal and high-throughput synthesis for years. In the parallel synthesis of compound libraries, compounds are synthesized using ordered arrays of spatially separated reaction vessels adhering to the traditional "one vessel/one compound" philosophy. The defined location of the compound in the array provides the structure of the compound. A commonly used format for parallel synthesis is the 96-well microtiter plate, and today "combinatorial libraries" comprising hundreds to thousands of compounds can be synthesized by parallel synthesis, often in an automated fashion.

As pointed out in the previous chapters of this book, performing organic chemical transformations under microwave conditions often allows a reduction of reaction times from many hours or days to minutes or even seconds. For many organic or medicinal chemists in both academia and industry there is therefore little need to perform parallel synthesis any more, as carrying out rapid microwave chemistry in a sequential processing fashion becomes almost as efficient as performing traditional parallel synthesis. This is particularly true for the generation of small, focused libraries (50–100 compounds) utilizing fully automated sequential microwave synthesis platforms (see Chapter 3 for details). Despite the growing trend in the pharmaceutical industry to synthesize smaller, more focused libraries, medicinal chemists in a high-throughput synthesis environment often still need to generate large compound libraries using a parallel synthesis approach.

Microwave-assisted reactions allow rapid product generation in high yield under uniform conditions. Therefore, they should be ideally suited for parallel synthesis applications. The first example of parallel reactions carried out under microwave irradiation conditions involved the nucleophilic substitution of an alkyl iodide with 60 diverse piperidine or piperazine derivatives (Scheme 4.22) [76]. Reactions were carried out in a multimode microwave reactor in individual sealed polypropylene vials using acetonitrile as solvent. Screening of the resulting 2-aminothiazole library in a herpes simplex virus-1 (HSV-1) assay led to three confirmed hits, demonstrating the potential of this method for rapid lead optimization.

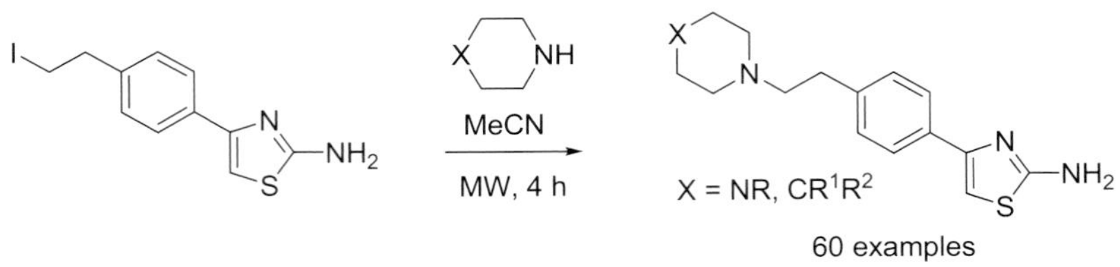

Scheme 4.22 Nucleophilic substitution reactions performed in parallel.

In a 1998 publication, the concept of microwave-assisted parallel synthesis in plate format was introduced for the first time. Using the three-component Hantzsch pyridine synthesis as a model reaction, libraries of substituted pyridines were prepared in a high-throughput parallel fashion. In this variation of the Hantzsch multi-component reaction, ammonium nitrate was used as the ammonium source as well as oxidizing agent, while bentonite clay served as an inorganic support (Scheme 4.23) [77]. Microwave irradiation was carried out in 96-well filter-bottom polypropylene plates, into which the corresponding eight 1,3-dicarbonyl compounds and twelve aldehyde building blocks had been dispensed using a robotic liquid handler. Microwave irradiation of the 96-well plate containing the aldehydes, the 1,3-dicarbonyl compounds, and ammonium nitrate/clay in a domestic microwave oven for 5 min produced the expected pyridine library directly, and the desired products were then extracted from the solid support using an organic solvent and collected in a receiving plate. HPLC/MS analysis showed that the reactions were uniformly successful across the 96-well reactor plate, without any residual starting material remaining. The diversity of this method was further extended when mixtures of two different 1,3-dicarbonyl compounds were used in the same Hantzsch synthesis, potentially leading to three distinct pyridine derivatives.

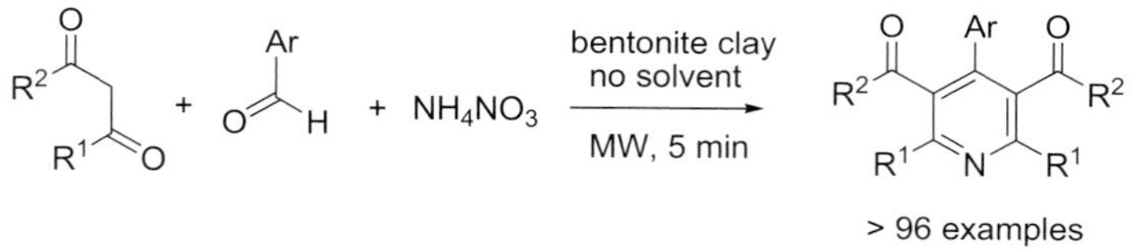

Scheme 4.23   Generation of Hantzsch-type pyridines in a 96-well plate.

Related applications of solvent-free microwave-enhanced parallel processes are summarized in Scheme 4.24. For example, dihydropyrimidines were obtained in a microwave-expedited version of the classical Biginelli three-component condensation (Scheme 4.24a) [78]. Neat mixtures of β-ketoesters, aryl aldehydes, and (thio)ureas with polyphosphate ester (PPE) as reaction mediator were irradiated in a domestic microwave oven for 1.5 min. The desired dihydropyrimidines were obtained in 61–95% yields after aqueous work-up. Similarly, imidazo-annulated pyridines, pyrazines, and pyrimidines were rapidly prepared in high yield by an Ugi-type multicomponent reaction using montmorillonite K-10 clay as an inorganic support (Scheme 4.24b) [79]. In the same way, a diverse set of 1,2,4,5-tetrasubstituted imidazoles was synthesized by condensation of 1,2-dicarbonyl compounds with aldehydes, amines, and ammonium acetate on acidic alumina as a support (Scheme 4.24c) [80]. A 25-member library of thioamides bearing additional basic amine functionalities (R^1R^2) was prepared by oxygen/sulfur exchange utilizing Lawesson's reagent. Microwave irradiation of the thoroughly mixed amide/thionation reagent mixture for 8 min produced the corresponding thioamides in good yields and high purities after solid-phase extraction (Scheme 4.24d) [81]. In another example, the regiospecific three-component cyclocondensation of equimolar amounts of aminopyrimidin-4-ones, benzoylacetonitrile, and arylaldehydes furnished the densely functionalized pyrido[2,3-d]pyrimidinones in 70–75% yield (Scheme 4.24e) [82]. The process was conducted in a domestic microwave oven using Pyrex glass vials in the absence of any solvent or support.

Most of the parallel reactions described in Schemes 4.23 and 4.24 were performed as dry-media reactions, in the absence of any solvent. In many cases, the starting materials and/or reagents were supported on an inorganic solid support, such as silica gel, alumina, or clay, that absorbs microwave energy or acts as a catalyst for the reaction (see also Section 4.1). In this context, an interesting method for the optimization of silica-supported reactions has been described [83]. The reagents were co-spotted neat or in solution onto a thin-layer chromatographic (TLC) plate.

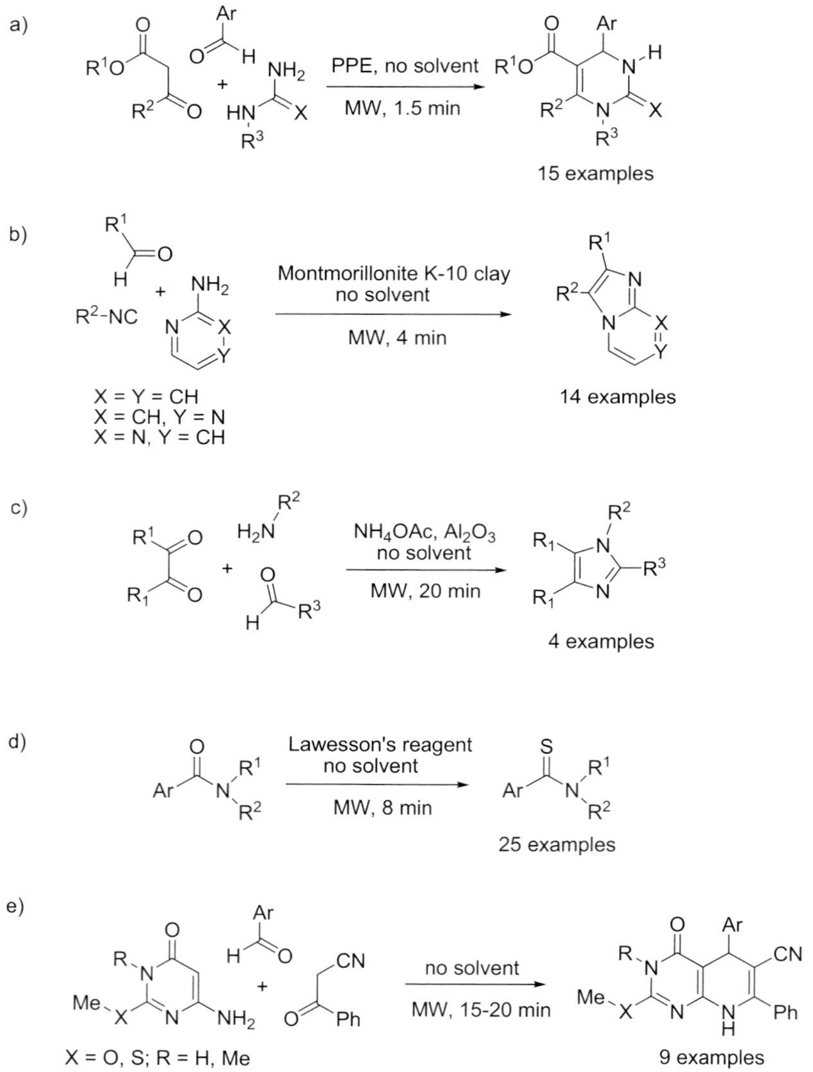

Scheme 4.24 Solvent-free parallel synthesis under microwave conditions (domestic oven).

The glass plate was exposed to microwave irradiation, eluted, and viewed by standard TLC visualization procedures to assess the results of the reaction. In this particular example, the synthesis of an arylpiperazine library (Scheme 4.25) was described, but the simplicity and general utility of the approach for the rapid screening of solvent-free microwave reactions may make this a powerful screening and reaction optimization tool. The synthesized compounds were later screened for their antimicrobial activity without their removal from the TLC plate utilizing bioautographical methods [84].

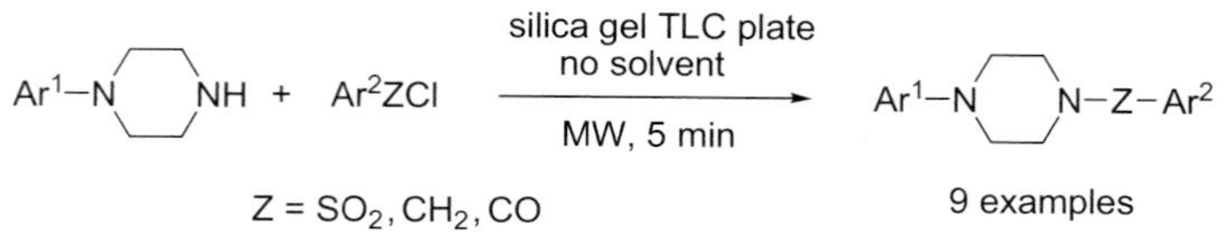

$$Ar^1-N \underset{\smile}{\overset{\frown}{N}}H \ + \ Ar^2ZCl \quad \xrightarrow[\text{MW, 5 min}]{\substack{\text{silica gel TLC plate} \\ \text{no solvent}}} \quad Ar^1-N \underset{\smile}{\overset{\frown}{N}}-Z-Ar^2$$

$$Z = SO_2, CH_2, CO$$

9 examples

Scheme 4.25 Microwave chemistry on a TLC plate.

Other microwave-assisted parallel processes, for example those involving solid-phase organic synthesis, are discussed in Section 7.1. In the majority of the cases described so far, domestic multimode microwave ovens were used as heating devices, without utilizing specialized reactor equipment. Since reactions in household multimode ovens are notoriously difficult to reproduce due to the lack of temperature and pressure control, pulsed irradiation, uneven electromagnetic field distribution, and the unpredictable formation of hotspots (Section 3.2), in most contemporary published methods dedicated commercially available multimode reactor systems for parallel processing are used. These multivessel rotor systems are described in detail in Section 3.4.

An important issue in parallel microwave processing is the homogeneity of the electromagnetic field in the microwave cavity. Inhomogeneities in the field distribution may lead to the formation of so-called hot and cold spots, resulting in different reaction temperatures in individual vessels or wells. Published examples of parallel microwave processing in dedicated multivessel rotor systems have included the generation of a 21-member library by parallel solid-phase Knoevenagel condensations under open-vessel conditions (see Scheme 7.39) [22]. Here, the temperature was monitored with the aid of a shielded thermocouple inserted into one of the reaction containers. It was confirmed by standard temperature measurements performed immediately after the irradiation period that the resulting end temperature in each vessel was the same to within ± 2 °C [22]. In another study involving a 36 sealed-vessel rotor system (see Fig. 3.7), the uniformity of the reaction conditions in such a parallel set-up was investigated. For this purpose, 36 identical Biginelli condensations using benzaldehyde, ethyl acetoacetate, and ureas as building blocks (see Scheme 4.24a) were run employing ethanol as solvent and hydrochloric acid as catalyst [85, 86]. All 36 individual vessels provided identical yields of dihydropyrimidine product (65–70%) within experimental error (no reaction details provided). In a subsequent experiment, six different aldehyde components (Fig. 4.3) were used to construct a small library of dihydropyrimidine analogues. Again, the yields of isolated products did not vary significantly depending on the position in the rotor, although slightly increased yields were obtained for mixtures that were placed in the inner circle, which would indicate a somewhat higher temperature in those reaction vessels (Fig. 4.3) [85, 86].

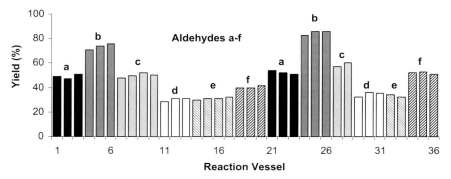

Fig. 4.3 Isolated yields of Biginelli dihydropyrimidine products (Scheme 4.24 a) in different reaction vessels of a 36 vessel rotor (Fig. 3.7). Outer ring, vessels 1–20; inner ring, vessels 21–36). Aldehydes: **a**, benzaldehyde; **b**, 2-hydroxybenzaldehyde; **c**, 3,4-dimethoxybenzaldehyde; **d**, 3-nitrobenzaldehyde; **e**, 2-chlorobenzaldehyde; **f**, 4-(*N,N*-dimethylamino)benzaldehyde. Adapted from [85].

Similar results were achieved when Biginelli reactions in acetic acid/ethanol (3:1) as solvent (120 °C, 20 min) were run in parallel in an eight-vessel rotor system (see Fig. 3.17) on an 8 × 80 mmol scale [87]. Here, the temperature in one reference vessel was monitored with the aid of a suitable probe, while the surface temperature of all eight quartz reaction vessels was also monitored (deviation less than 10 °C; Fig. 4.4). The yield in all eight vessels was nearly identical and the same set-up was also used to perform a variety of different chemistries in parallel mode [87]. Various other parallel multivessel systems are commercially available for use in different multimode microwave reactors. These are presented in detail in Chapter 3.

The construction of a custom-built parallel reactor with expandable reaction vessels that accommodate the pressure build-up during a microwave irradiation experiment has also been reported [88]. The system was used for the parallel synthesis of a 24-member library of substituted 4-sulfanyl-1*H*-imidazoles [88].

Several articles in the area of microwave-assisted parallel synthesis have described irradiation of 96-well filter-bottom polypropylene plates in conventional household microwave ovens for high-throughput synthesis. While some authors have not reported any difficulties in relation to the use of such equipment (see Scheme 4.24) [77], others have experienced problems in connection with the thermal instability of the polypropylene material itself [89], and with respect to the creation of temperature gradients between individual wells upon microwave heating [89, 90]. Figure 4.5 shows the temperature gradients after irradiation of a conventional 96-well plate for 1 min in a domestic microwave oven. For the particular chemistry involved (Scheme 7.45), the 20 °C difference between the inner and outer wells was, however, not critical.

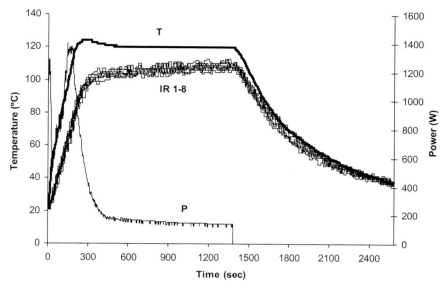

Fig. 4.4 Temperature and power profiles for a Biginelli condensation (Scheme 4.24.a) under sealed quartz vessel/microwave irradiation conditions (see Fig. 3.17). Linear heating ramp to 120 °C (3 min), temperature control using the feedback from the reference vessel temperature measurement (constant 120 °C, 20 min), and forced air cooling (20 min). The reaction was performed in eight quartz vessels each containing 40 mL of reaction mixture. The profiles show the temperature measured in one reference vessel by means of an internal gas balloon thermometer (T), the surface temperature monitoring of the eight individual vessels by IR thermography (IR 1–8), and the magnetron power (P, 0–1400 W). Reproduced with permission from [87].

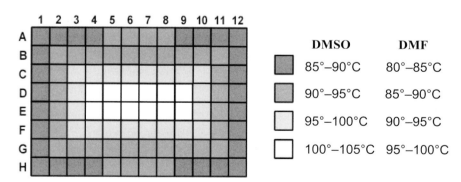

Fig. 4.5 Temperature gradients within a microwave-heated microtiter plate; 1 mL per well, heated continuously for 1 min at full power in a conventional microwave oven. Adapted from [90].

To overcome the problems associated with using conventional polypropylene deep-well plates in a microwave reactor, specifically designed well plates in 12, 24, 48, and 96 reaction well formats have been developed (see Section 3.4.1.1) [86]. These plates consist of a base of carbon-doped Teflon (Weflon®) for better heat distribution with glass inserts as reaction vessels. The vessels are sealed with Teflon-laminated silicon mats (Fig. 4.6). For such a large number of reaction vessels, mixing with magnetic stirrers is impracticable and therefore a new mixing method was developed using overhead shaking (Fig. 4.6). Here, temperature measurement is accomplished by means of a fiber-optic sensor inside the Weflon block, just underneath the reaction vessels [86]. Since several such devices can be mounted on top of one another, several hundred reactions may potentially be performed in one irradiation cycle. It is important to note that with this system the material used for the preparation of the plates (Weflon) absorbs microwave energy, which means that the sealed glass vials will be heated by microwave irradiation regardless of the dielectric properties of the reactants/solvents. The system is designed to interface with conventional liquid handler/dispensers to achieve a high degree of automation in the whole process.

Fig. 4.6 Ethos combCHEM system (Milestone, Inc.); left: 96 deep-well plates; right: overhead rotor with two plates. Operating limits: 200 °C, 5 bar.

The reaction homogeneity in a 24-well plate was investigated by monitoring the esterification of hexanoic acid with 1-hexanol at 120 °C for 30 min. The difference in conversion between the individual vessels was 4% with a standard deviation of 2.4% (Fig. 4.7). It thus appears that all individual reactions were irradiated homogeneously in the applied microwave field.

As already mentioned above, a different strategy to achieve high throughput in microwave-assisted reactions can be realized by performing automated sequential microwave synthesis in monomode microwave reactors. Since it is currently not feasible to have more than one reaction vessel in a monomode microwave cavity, a robotic system has been integrated into a platform that moves individual reaction

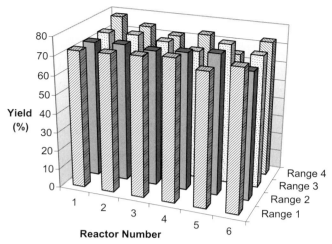

Fig. 4.7 Reaction homogeneity expressed in terms of conversion for the synthesis of hexyl hexanoate in a 24-well reaction plate. Reproduced with permission from [86].

vessels in and out of a specifically designed cavity. With some instruments, a liquid handler additionally allows the dispensing of reagents into sealed reaction vials, while a gripper moves each sealed vial in and then out of the microwave cavity after irradiation (see Chapter 3). Some instruments can process up to 120 reactions per run with a typical throughput of 12–15 reactions/hour in an unattended fashion. In contrast to the parallel synthesis application in multimode cavities, this approach allows the user to perform a series of optimization or library production reactions with each reaction separately programmed. A case study describing the preparation of Biginelli libraries by sequential processing is discussed in Section 5.3.

The issue of parallel versus sequential synthesis using multimode or monomode cavities, respectively, deserves special comment. While the parallel set-up allows for a considerably higher throughput achievable in the relatively short timeframe of a microwave-enhanced chemical reaction, the individual control over each reaction vessel in terms of reaction temperature/pressure is limited. In the parallel mode, all reaction vessels are exposed to the same irradiation conditions. In order to ensure similar temperatures in each vessel, the same volume of the identical solvent should be used in each reaction vessel because of the dielectric properties involved [86]. As an alternative to parallel processing, the automated sequential synthesis of libraries can be a viable strategy if small focused libraries (20–200 compounds) need to be prepared. Irradiating each individual reaction vessel separately gives better control over the reaction parameters and allows for the rapid optimization of reaction conditions. For the preparation of relatively small libraries, where delicate chemistries are to be performed, the sequential format may be preferable. This is discussed in more detail in Chapter 5.

4.5
Scale-Up in Batch and Continuous-Flow

Most examples of microwave-assisted chemistry published to date and presented in this book (see Chapters 6 and 7) were performed on a scale of less than 1 g (typically 1–5 mL reaction volume). This is in part a consequence of the recent availability of single-mode microwave reactors that allow the safe processing of small reaction volumes under sealed-vessel conditions by microwave irradiation (see Chapter 3). While these instruments have been very successful for small-scale organic synthesis, it is clear that for microwave-assisted synthesis to become a fully accepted technology in the future there is a need to develop larger scale MAOS techniques that can ultimately routinely provide products on a multi kg (or even higher) scale.

Bearing in mind some of the physical limitations of microwave heating technology, such as magnetron power or penetration depth (see Section 2.3), two different approaches for microwave synthesis on a larger scale (> 100 mL volume) have emerged. While some groups have employed larger batch-type multimode or monomode reactors (< 1000 mL processing volume), others have used continuous-flow techniques (multi- and monomode) to overcome the inherent problems associated with MAOS scale-up. An additional key point in processing comparatively large volumes under pressure in a microwave field is the safety aspect, as any malfunction or rupture of a large pressurized reaction vessel may have significant consequences. In general, one should note that published examples of MAOS scale-up experiments are rare, in particular those involving complex organic reactions.

Modern single-mode microwave technology allows the performance of MAOS in very small reaction volumes (< 0.2 mL). A detailed study by Takvorian and Combs highlights the advantage of performing microwave chemistry in very small reaction vessels [91]. The ultra-low volume vials utilized in this study (see Fig. 3.21) enabled the authors to run reactions in a very concentrated fashion, minimizing reaction times and utilizing only limited amounts of sometimes expensive scaffolds and reagents. With those single-mode reactors that do not require a minimum filling volume (since the temperature is measured from the bottom and not from the side by an external IR sensor), even volumes as low as 50 µL can be processed. With today's commercially available single-mode cavities, the largest volumes that can be processed under sealed-vessel conditions are ca. 50 mL, with different vessel types being available for scale-up in a linear fashion from 0.05 to 50 mL (see Chapter 3). Under open-vessel conditions, higher volumes (> 1000 mL) have been processed with microwave irradiation without presenting any technical difficulties [74].

In this chapter, microwave scale-up to volumes > 100 mL in sealed vessels is discussed. An important issue for the process chemist is the potential for direct scalability of microwave reactions, allowing rapid translation of previously optimized small-scale conditions to a larger scale. Several authors have reported independently on the feasibility of directly scaling reaction conditions from small-scale single-mode (typically 0.5–5 mL) to larger scale multimode batch microwave reactors (20–500 mL) without reoptimization of the reaction conditions [24, 87, 92–94].

The results of scale-up experiments carried out with a dedicated multimode scale-up reactor utilizing multivessel rotors (Synthos 3000) are summarized in Scheme 4.26 [87]. In all cases, the yields obtained in the small-scale single-mode experiments (1–4 mmol) could be reproduced on a larger scale (40–640 mmol) without the need to reoptimize the reaction conditions. It has to be noted, however, that in many cases the rapid heating and cooling profiles seen with a small-scale single-mode reactor with high power density ("microwave flash heating"; see Fig. 2.8) cannot be replicated on a larger scale. The heating profile for the transformation shown in Scheme 4.26a, for example, is reproduced in Fig. 4.4. Despite the somewhat longer heating and cooling periods, no appreciable difference was found in the outcome of the studied reactions.

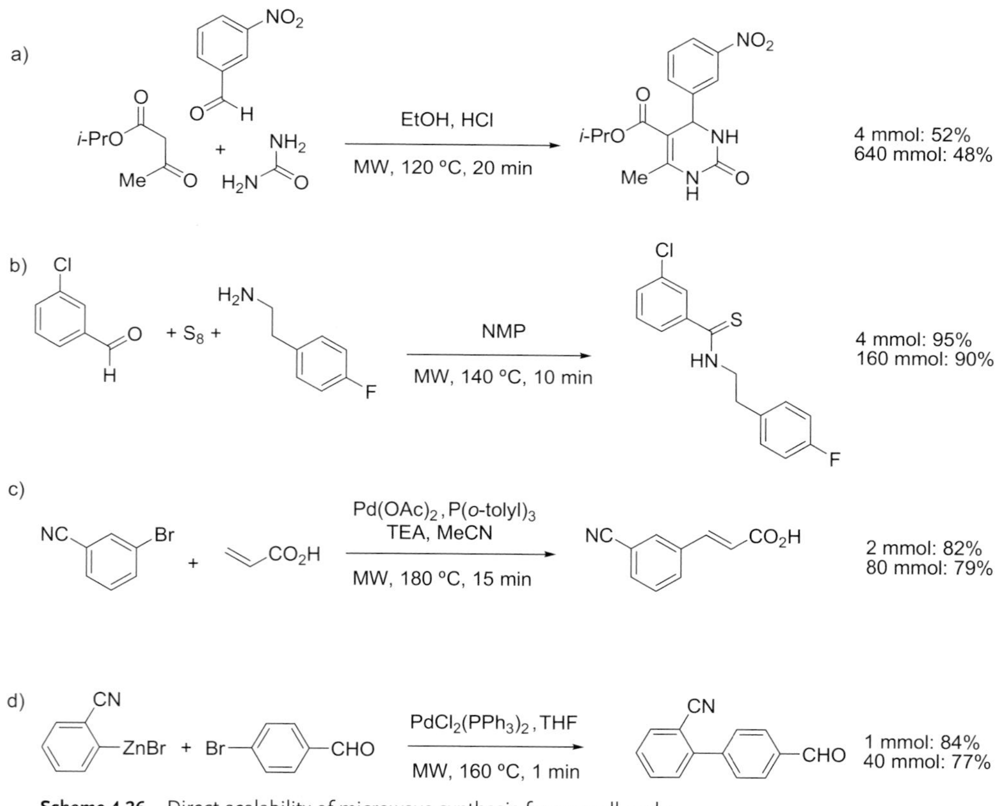

Scheme 4.26 Direct scalability of microwave synthesis from small-scale single-mode reactors (1–4 mmol) to large-scale multimode batch reactors (40–640 mmol).

Similar scale-up results were obtained for Mannich [93] and oxidative Heck processes [30] (Scheme 4.27) utilizing a single reaction vessel in a different multimode batch reactor (Emrys Advancer). Again, yields were comparable on going from a small-scale single-mode reactor to a larger multimode reactor. Here, rapid cooling after the microwave heating step is possible by means of a patented expansion cooling process. Earlier, comparatively large-scale microwave synthesis work was published by Strauss [95] and others [17, 96].

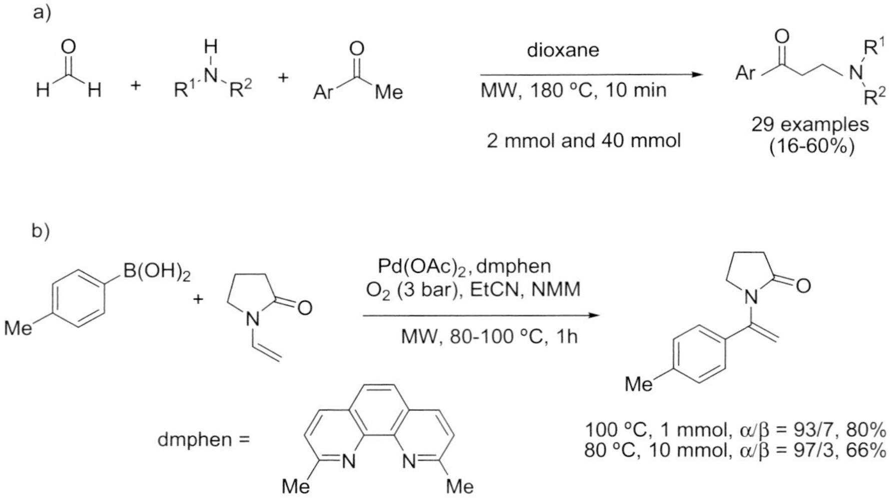

a)

b)

Scheme 4.27 Direct scalability of microwave synthesis from small-scale single-mode reactors (1–2 mmol) to large-scale multimode batch reactors (10–40 mmol).

Mainly because of safety concerns and issues related to the penetration depth of microwaves into absorbing materials such as organic solvents, the preferable option for processing volumes of > 1000 mL under sealed-vessel microwave conditions is a continuous-flow technique [68, 97–106], although the number of published examples using dedicated microwave reactors is limited [68, 102–106]. In such a system, the reaction mixture is passed through a microwave-transparent coil that is positioned in the cavity of a single- or multimode microwave reactor. The previously optimized reaction time under batch microwave conditions now needs to be related to a "residence time" (the time for which the sample stays in the microwave-heated coil) at a specific flow rate. While the early pioneering work in this area stems from the group of Strauss [97], others have since made notable contributions to this field, often utilizing custom-built microwave reactors or modified domestic microwave units [98–101].

Recently published examples of continuous-flow organic microwave synthesis include, for example, 1,3-dipolar cycloaddition chemistry in the CEM CF Voyager system (see Figs. 3.23 and 3.24). The cycloaddition of dimethyl acetylenedicarboxylate with benzyl azide in toluene was first carefully optimized with respect to solvent, temperature, and time under batch conditions. The best protocol was then translated to a continuous-flow procedure in which a solution 0.33 M in both building blocks was pumped through a Kevlar-reinforced Teflon coil (total capacity 10 mL) heated in the single-mode reactor at 110 °C (10 min residence time) (Scheme 4.28) [102]. This method provided a 91% conversion to the desired triazole product.

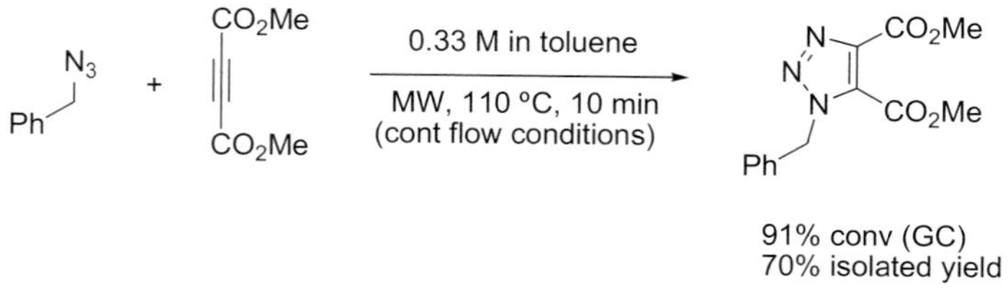

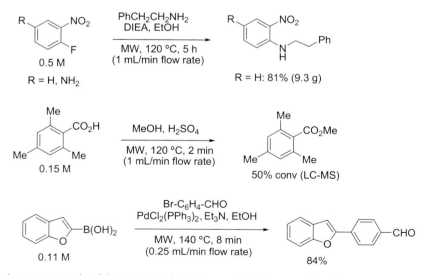

Scheme 4.28   1,3-Dipolar cycloaddition reactions under continuous-flow conditions.

In related work, Wilson and co-workers described a custom-made flow reactor based on the Biotage Emrys Synthesizer single-mode batch reactor (see Fig. 3.19) fitted with a coiled glass flow cell [103]. The flow cell was inserted into the cavity from the bottom of the instrument and the system was operated in either open- or closed-loop mode. The temperature was monitored and controlled by means of the instrument's internal infrared sensor (located in the cavity) and the instrument software. The different synthetic transformations investigated with this system included nucleophilic aromatic substitutions, esterifications, and Suzuki reactions, as summarized in Scheme 4.29 [103]. In all cases, product yields were equivalent to or exceeded those using conventional heating methods and were scalable to multigram

Scheme 4.29   Nucleophilic aromatic substitutions, esterifications, and Suzuki reactions under continuous-flow conditions.

quantities. Clogging of the lines and over-pressurization were some of the limitations observed.

In the context of elaborating a degradation method for hazardous and toxic halogenated aromatic hydrocarbons, Varma and co-workers reported the hydrodechlorination of chlorinated benzenes using a microwave-assisted continuous-flow process involving so-called "active flow cells" (Scheme 4.30) [106]. Here, a 15 mL quartz U-tube filled with spherical particles of 0.5% palladium-on-alumina (Pd/Al$_2$O$_3$) catalyst was fitted into the CEM Voyager CF reactor. Chlorobenzene was fed into the reactor at an adjustable flow rate along with a controlled flow of hydrogen from a cylinder. Comparison studies between the microwave and the thermally heated process indicated that the microwave method was both more efficient and more selective than the thermal process, in that the formation of cyclohexane was suppressed under microwave conditions.

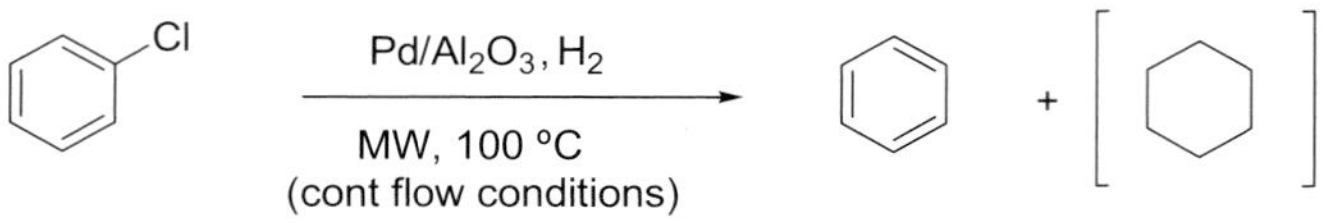

Scheme 4.30 Hydrodechlorination of chlorobenzene using active flow cells in a continuous-flow microwave process.

Current single-mode continuous-flow microwave reactors allow the processing of comparatively small volumes. Much larger volumes can be processed in continuous-flow reactors that are housed inside a multimode microwave system. In a 2001 publication, Shieh and coworkers described the methylation of phenols, indoles, and benzimidazoles with dimethyl carbonate under continuous-flow microwave conditions using a Milestone ETHOS-CFR reactor (see Fig. 3.11) [104]. In a typical procedure, a solution containing the substrate, dimethyl carbonate, 1,8-diazabicyclo[5.4.0]undec-7-ene (DBU) base, tetrabutylammonium iodide (TBAI), and a solvent was circulated by a pump through the microwave reactor, which was preheated to 160 °C and 20 bar by microwave irradiation (Scheme 4.31). Under these condi-

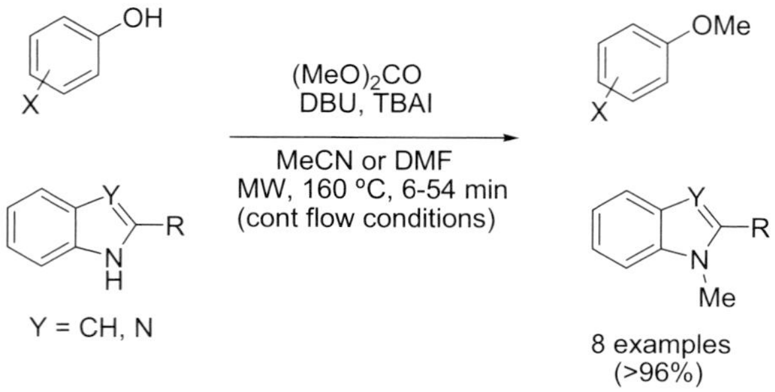

Scheme 4.31 Methylation of phenols, indoles, and benzimidazoles in a multimode continuous-flow microwave reactor.

tions, the time required for the methylation of phenols was reduced from hours to minutes, representing a nearly 2000-fold rate increase. Similar results were achieved for benzylations employing dibenzyl carbonate [68], and the same authors also reported the usefulness of this general method for the esterification of carboxylic acids [105]. Benzoic acid, for example, was converted to its methyl ester on a 100 g scale within 20 min (microwave residence time), utilizing the dimethyl carbonate/ DBU-mediated continuous-flow protocol [105].

As mentioned above, one remaining problem with continuous-flow reactors is clogging of the lines and the difficulties in processing heterogeneous mixtures. Since many organic transformations involve some form of insoluble reagent or catalyst, single-mode so-called stop-flow microwave reactors have been developed (Fig. 3.23), in which peristaltic pumps, capable of pumping slurries and even solid reagents, are used to fill a batch reaction vessel (80 mL) with the reaction mixture. After microwave processing in batch, the product mixture is pumped out of the system, which is then ready to receive the next batch of the reaction mixture. At the time of writing, no published applications of these microwave systems had yet been reported.

A rather unusual application of continuous-flow microwave processing has been described by the group of Haswell [107]. The authors used microwave energy to deliver heat locally to a heterogeneous palladium-supported catalyst (Pd/Al_2O_3, catalyst channel: $1.5 \times 0.08 \times 15$ mm) located within a micro reactor device. A 10–15 nm gold film patch, located on the outside surface of the base of a glass micro reactor, was found to efficiently assist in the heating of the catalyst, allowing Suzuki cross-coupling reactions to proceed very effectively (Scheme 4.32).

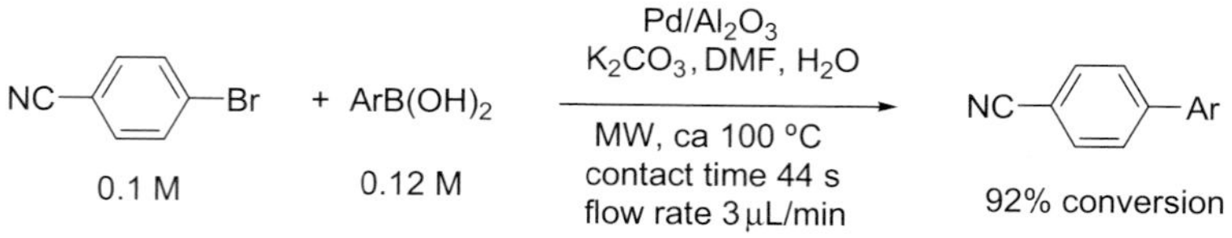

Scheme 4.32   Suzuki reactions in a micro reactor environment.

Critically evaluating the currently available instrumentation for microwave scale-up in batch and continuous-flow (Chapter 3), one may argue that for processing volumes of < 1000 mL a batch process may be preferable. By carrying out sequential runs in batch mode, kg quantities of product can easily be obtained. When larger quantities of a specific product need to be prepared on a regular basis, it may be worthwhile evaluating a continuous-flow protocol. Large-scale continuous-flow microwave reactors (flow rate 20 L h^{-1}) are currently under development [86, 108]. However, at the present time, there are no documented published examples of the use of microwave technology for organic synthesis on a production-scale level (> 1000 kg), which is a clear limitation of this otherwise so successful technology [109].

References

[1] A. Loupy, A. Petit, J. Hamelin, F. Texier-Boullet, P. Jacquault, D. Mathé, *Synthesis* **1998**, 1213–1234; R. S. Varma, *Green Chem.* **1999**, 43–55; M. Kidawi, *Pure Appl. Chem.* **2001**, *73*, 147–151; R. S. Varma, *Pure Appl. Chem.* **2001**, *73*, 193–198; R. S. Varma, *Tetrahedron* **2002**, *58*, 1235–1255; R. S. Varma, *Advances in Green Chemistry: Chemical Syntheses Using Microwave Irradiation*, Kavitha Printers, Bangalore, **2002**.

[2] C. O. Kappe, *Angew. Chem. Int. Ed.* **2004**, *44*, 6250–6284.

[3] M. Nüchter, U. Müller, B. Ondruschka, A. Tied, W. Lautenschläger, *Chem. Eng. Technol.* **2003**, *26*, 1207–1216.

[4] A. Shaabani, *J. Chem. Res. (S)* **1998**, 672–673.

[5] R. S. Varma, A. K. Chatterjee, M. Varma, *Tetrahedron Lett.* **1993**, *34*, 3207–3210.

[6] R. S. Varma, R. Dahiya, *Tetrahedron Lett.* **1997**, *38*, 2043–2044.

[7] R. S. Varma, R. K. Saini, R. Dahiya, *Tetrahedron Lett.* **1997**, *38*, 7823–7824.

[8] R. S. Varma, R. Dahiya, *Tetrahedron Lett.* **1998**, *39*, 1307–1308.

[9] R. S. Varma, R. K. Saini, *Tetrahedron Lett.* **1998**, *39*, 1481–1482.

[10] G. W. Kabalka, L. Wang, V. N. Namboodiri, R. M. Pagni, *Tetrahedron Lett.* **2000**, *41*, 5151–5154.

[11] A. Loupy (Ed.), *Microwaves in Organic Synthesis, Wiley-VCH, Weinheim*, **2002**.

[12] B. L. Hayes, *Microwave Synthesis: Chemistry at the Speed of Light*, CEM Publishing, Matthews, NC, **2002**.

[13] A. Laporterie, J. Marquié, J. Dubac, in *Microwaves in Organic Synthesis* (Ed.: A. Loupy), Wiley-VCH, Weinheim, **2002**, pp 219–252 (Chapter 7).

[14] B. Garrigues, C. Laporte, R. Laurent, A. Laporterie, J. Dubac, *Liebigs Ann.* **1996**, 739–741.

[15] S. Deshayes, M. Liagre, A. Loupy, J.-L. Luche, A. Petit, *Tetrahedron* **1999**, *55*, 10851–10870; A. Loupy, A. Petit, D. Bogdal, in *Microwaves in Organic Synthesis* (Ed.: A. Loupy), Wiley-VCH, Weinheim, **2002**, pp 147–180 (Chapter 5).

[16] A. Loupy, A. Petit, M. Ramdani, C. Yvanaef, M. Majdoub, B. Labiad, D. Villemin, *Can. J. Chem.* **1993**, *71*, 90–95.

[17] J. Cléophax, M. Liagre, A. Loupy, A. Petit, *Org. Proc. Res. Dev.* **2000**, *4*, 498–504.

[18] D. Bogdal, J. Pielichowski, K. Jaskot, *Heterocycles* **1997**, *45*, 715–722.

[19] A. K. Bose, B. K. Banik, N. Lavlinskaia, M. Jayaraman, M. S. Manhas, *Chemtech* **1997**, *27*, 18–24; A. K. Bose, M. S. Manhas, S. N. Ganguly, A. H. Sharma, B. K. Banik, *Synthesis* **2002**, 1578–1591.

[20] H. S. P. Rao, S. Jothilingam, *J. Org. Chem.* **2003**, *68*, 5392–5394.

[21] H. S. P. Rao, S. Jothilingam, H. W. Scheeren, *Tetrahedron* **2004**, *60*, 1625–1630.

[22] G. A. Strohmeier, C. O. Kappe, *J. Comb. Chem.* **2002**, *4*, 154–161.

[23] W. Stadlbauer, O. Schmut, T. Kappe, *Monatsh. Chem.* **1980**, *111*, 1005–1013.

[24] A. Stadler, S. Pichler, G. Horeis, C. O. Kappe, *Tetrahedron* **2002**, *58*, 3177–3183.

[25] J. H. M. Lange, P. C. Verveer, S. J. M. Osnabrug, G. M. Visser, *Tetrahedron Lett.* **2001**, *42*, 1367–1369.

[26] G. Vo-Tanh, H. Lahrache, A. Loupy, I.-J. Kim, D.-H. Chang, C.-H. Jun, *Tetrahedron* **2004**, *60*, 5539–5543.

[27] H. Will, P. Scholz, B. Ondruschka, *Chem. Ing. Tech.* **2002**, *74*, 1057–1067; D. D. Tanner, P. Kandanarachchi, Q. Ding, Q. H. Shao, D. Vizitiu, J. A. Franz, *Energy Fuels* **2001**, *15*, 197–204; X. Zhang, C. S.-M. Lee, D. M. P. Mingos, D. O. Hayward, *Catal. Lett.* **2003**, *88*, 129–139; X. Zhang, D. O. Hayward, D. M. P. Mingos, *Catal. Lett.* **2003**, *88*, 33–38.

[28] E. Van der Eycken, P. Appukkuttan, W. De Borggraeve, W. Dehaen, D. Dallinger, C. O. Kappe, *J. Org. Chem.* **2002**, *67*, 7904–7909.

[29] N. Kaval, W. Dehaen, C. O. Kappe, E. Van der Eycken, *Org. Biomol. Chem.* **2004**, *2*, 154–156.

[30] O. Š. Miljani$cacute$, K. P. C. Vollhardt, G. D. Whitener, *Synlett* **2003**, 29–34.

[31] M. M. S. Andappan, P. Nilsson, H. von Schenck, M. Larhed, *J. Org. Chem.* **2004**, *69*, 5212–5218.

[32] D. J. Adams, P. J. Dyson, S. J. Tavener, *Chemistry in Alternative Reaction Media*, Wiley, Chichester, UK, **2004**.

[33] R. Breslow, *Acc. Chem. Res.* **1991**, *24*, 159–164; P. A. Grieco, *Aldrichimica Acta* **1991**, *24*, 59–66; C.-J. Li, *Chem. Rev.* **1993**, *93*, 2023–

2035; C.-J. Li, T.-H. Chan, *Organic Reactions in Aqueous Media*, Wiley, New York, NY, **1997**; U. M. Lidström, *Chem. Rev.* **2002**, *102*, 2751–2772.

[34] W. Blokzijl, J. B. F. N. Engberts, *Angew. Chem. Int. Ed. Engl.* **1993**, *32*, 1545–1579; B. Widom, P. Bhimalapuram, K. Koga, *Phys. Chem. Chem. Phys.* **2003**, *5*, 3085–3093.

[35] D. Bröll, C. Kaul, A. Krämer, P. Krammer, T. Richter, M. Jung, H. Vogel, P. Zehner, *Angew. Chem. Int. Ed.* **1999**, *38*, 2998–3014; P. E. Savage, *Chem. Rev.* **1999**, *99*, 603–621; M. Siskin, A. R. Katritzky, **2001**, *101*, 825–836; A. R. Katritzky, D. A. Nichols, M. Siskin, R. Murugan, M. Balasubramanian, *Chem. Rev.* **2001**, *101*, 837–892; N. Akiya, P. E. Savage, *Chem. Rev.* **2002**, *102*, 2725–2750.

[36] C. R. Strauss, R. W. Trainor, *Aust. J. Chem.* **1995**, *48*, 1665–1692.

[37] C. R. Strauss, *Aust. J. Chem.* **1999**, *52*, 83–96.

[38] S. A. Nolen, C. L. Liotta, C. E. Eckert, R. Gläser, *Green Chem.* **2003**, *5*, 663–669.

[39] P. Krammer, H. Vogel, *J. Supercrit. Fluids* **2000**, *16*, 189–206.

[40] V. Molteni, M. M. Hamilton, L. Mao, C. M. Crane, A. P. Termin, D. M. Wilson, *Synthesis* **2002**, 1669–1674.

[41] A. Vasudevan, M. K. Verzal, *Synlett* **2004**, 631–634.

[42] P. S. Baran, D. P. O'Malley, A. L. Zografos, *Angew. Chem. Int. Ed.* **2004**, *43*, 2674–2677.

[43] N. E. Leadbeater, M. Marco, *J. Org. Chem.* **2003**, *68*, 888–892; L. Bai, J.-X. Wang, Y. Zhang, *Green Chem.* **2003**, *5*, 615–617.

[44] J.-X. Wang, Z. Liu, Y. Hu, B. Wei, L. Bin, *Synth. Commun.* **2002**, *32*, 1607–1614.

[45] N. E. Leadbeater, M. Marco, B. J. Tominack, *Org. Lett.* **2003**, *5*, 3919–3922; P. Appukkuttan, W. Dehaen, E. Van der Eycken, *Eur. J. Org. Chem.* **2003**, 4713–4716.

[46] N. Kaval, K. Bisztray, W. Dehaen, C. O. Kappe, E. Van der Eycken, *Mol. Diversity* **2003**, *7*, 125–133.

[47] R. K. Arvela, N. E. Leadbeater, H. M. Torenius, H. Tye, *Org. Biomol. Chem.* **2003**, *1*, 1119–1121.

[48] D. Villemin, M. J. Gomez-Escalonilla, J.-F. Saint-Clair, *Tetrahedron Lett.* **2001**, *42*, 635–637.

[49] T. A. Bryson, J. J. Stewart, J. M. Gibson, P. S. Thomas, J. K. Berch, *Green Chem.* **2003**, *5*, 174–176; T. A. Bryson, J. M. Gibson,

J. J. Stewart, H. Voegtle, A. Tiwari, J. H. Dawson, W. Marley, B. Harmon, *Green Chem.* **2003**, *5*, 177–180.

[50] J. Westman, R. Lundin, *Synthesis* **2003**, 1025–1030.

[51] P. Wasserscheid, T. Welton (Eds.), *Ionic Liquids in Synthesis*, Wiley-VCH, Weinheim, **2003**.

[52] J. Hoffmann, M. Nüchter, B. Ondruschka, P. Waserscheid, *Green Chem.* **2003**, *5*, 296–299.

[53] K. S. A. Vallin, P. Emilsson, M. Larhed, A. Hallberg, *J. Org. Chem.* **2002**, *67*, 6243–6246.

[54] J. F. Dubreuil, J. P. Bazureau, *Tetrahedron Lett.* **2000**, *41*, 7351–7355.

[55] H. Berthold, T. Schotten, H. Hönig, *Synthesis* **2002**, 1607–1610.

[56] K. G. Mayo, E. H. Nearhoof, J. J. Kiddle, *Org. Lett.* **2002**, *4*, 1567–1570.

[57] N. E. Leadbeater, H. M. Torenius, H. Tye, *Tetrahedron* **2003**, *59*, 2253–2258.

[58] H.-P. Nguyen, H. Matondo, M. Baboulene, *Green Chem.* **2003**, *5*, 303–305.

[59] X. Xie, J. Lu, B. Chen, J. Han, X. She, X. Pan, *Tetrahedron Lett.* **2004**, *45*, 809–812.

[60] R. Trotzki, M. Nüchter, B. Ondruschka, *Green Chem.* **2003**, *5*, 285–290.

[61] J. K. Lee, D.-C. Kim, C. E. Song, S.-G. Lee, *Synth. Commun.* **2003**, *33*, 2301–2307.

[62] S. V. Ley, A. G. Leach, R. I. Storer, *J. Chem. Soc. Perkin Trans. 1* **2001**, 358–361.

[63] N. E. Leadbeater, H. M. Torenius, *J. Org. Chem.* **2002**, *67*, 3145–3148.

[64] S. Garbacia, B. Desai, O. Lavastre, C. O. Kappe, *J. Org. Chem.* **2003**, *68*, 9136–9139.

[65] G. K. Datta, K. S. A. Vallin, M. Larhed, *Mol. Diversity* **2003**, *7*, 107–114.

[66] N. Srinivasan, A. Ganesan, *Chem. Commun.* **2003**, 916–917.

[67] N. E. Leadbeater, H. M. Torenius, H. Tye, *Mol. Diversity* **2003**, *7*, 135–144.

[68] W.-C. Shieh, S. Lozanov, O. Repič, *Tetrahedron Lett.* **2003**, *44*, 6943–6945.

[69] A. G. Glenn, P. B. Jones, *Tetrahedron Lett.* **2004**, *45*, 6967–6969.

[70] N. E. Leadbeater, H. M. Torenius, H. Tye, *Comb. Chem High Throughp. Screen.* **2004**, *7*, 511–528.

[71] J. Habermann, S. Ponzi, S. V. Ley, *Mini-Rev. Org. Chem.* **2005**, *2*, 125–137.

[72] R. S. Varma, V. V. Namboodiri, *Chem. Commun.* **2001**, 643–644; R. S. Varma, V. V. Namboodiri, *Tetrahedron Lett.* **2002**, *43*, 5381–5383; V. V. Namboodiri, R. S. Varma, *Chem. Commun.* **2002**, 342–343.

[73] B. M. Khadilkar, G. L. Rebeiro, *Org. Proc. Res. Dev.* **2002**, *6*, 826–828.

[74] M. Deetlefs, K. R. Seddon, *Green Chem.* **2003**, *5*, 181–186.

[75] G. Vo Thanh, B. Pegot, A. Loupy, *Eur. J. Org. Chem.* **2004**, 1112–1116.

[76] C. N. Selway, N. K. Terret, *Bioorg. Med. Chem.* **1996**, *4*, 645–654.

[77] I. C. Cotterill, A. Ya. Usyatinsky, J. M. Arnold, D. S. Clark, J. S. Dordick, P. C. Michels, Y. L. Khmelnitsky, *Tetrahedron Lett.* **1998**, *39*, 1117–1120.

[78] C. O. Kappe, D. Kumar, R. S. Varma, *Synthesis* **1999**, 1799–1803.

[79] R. S. Varma, D. Kumar, *Tetrahedron Lett.* **1999**, *40*, 7665–7669.

[80] A. Ya. Usyatinsky, Y. L. Khmelnitsky, *Tetrahedron Lett.* **2000**, *41*, 5031–5034.

[81] R. Olsson, H. C. Hansen, C.-M. Andersson, *Tetrahedron Lett.* **2000**, *41*, 7947–7950.

[82] J. Quiroga, C. Cisneros, B. Insuasty, R. Abonía, M. Nogueras, A. Sánchez, *Tetrahedron Lett.* **2001**, *42*, 5625–5627.

[83] L. Williams, *Chem. Commun.* **2000**, 435–436.

[84] L. Williams, O. Bergersen, *J. Planar Chromat.* **2001**, *14*, 318–321.

[85] M. Nüchter, W. Lautenschläger, B. Ondruschka, A. Tied, *LaborPraxis* **2001**, *25* (1), 28–31; M. Nüchter, B. Ondruschka, A. Tied, W. Lautenschläger, K. J. Borowski, *Am. Gen./Proteom. Techn.* **2001**, *1*, 34–39.

[86] M. Nüchter, B. Ondruschka, *Mol. Diversity* **2003**, *7*, 253–264.

[87] A. Stadler, B. H. Yousefi, D. Dallinger, P. Walla, E. Van der Eycken, N. Kaval, C. O. Kappe, *Org. Proc. Res. Dev.* **2003**, *7*, 707–716.

[88] C. M. Coleman, J. M. D. MacElroy, J. F. Gallagher, D. F. O'Shea, *J. Comb. Chem.* **2002**, *4*, 87–93.

[89] B. M. Glass, A. P. Combs, in *High-Throughput Synthesis. Principles and Practices* (Ed.: I. Sucholeiki), Marcel Dekker, Inc., New York, **2001**, Chapter 4.6, pp 123–128.

[90] B. M. Glass, A. P. Combs, Article E0027, «*Fifth International Electronic Conference on Synthetic Organic Chemistry*» (Eds.: C. O. Kappe, P. Merino, A. Marzinzik, H. Wennemers, T. Wirth, J. J. Vanden Eynde, S.-K. Lin), CD-ROM edition, ISBN 3–906980–06–5, MDPI, Basel, Switzerland, **2001**.

[91] A. G. Takvorian, A. P. Combs, *J. Comb. Chem.* **2004**, *6*, 171–174.

[92] M. Iqbal, N. Vyse, J. Dauvergne, P. Evans, *Tetrahedron Lett.* **2002**, *43*, 7859–7862.

[93] F. Lehmann, Å. Pilotti, K. Luthman, *Mol. Diversity* **2003**, *7*, 145–152.

[94] S. A. Shackelford, M. B. Anderson, L. C. Christie, T. Goetzen, M. C. Guzman, M. A. Hananel, W. D. Kornreich, H. Li, V. P. Pathak, A. K. Rabinovich, R. J. Rajapakse, L. K. Truesdale, S. M. Tsank, H. N. Vazir, *J. Org. Chem.* **2003**, *68*, 267–275.

[95] K. D. Raner, C. R. Strauss, R. W. Trainor, J. S. Thorn, *J. Org. Chem.* **1995**, *60*, 2456–2460.

[96] B. Perio, M.-J. Dozias, J. Hamelin, *Org. Proc. Res. Dev.* **1998**, *2*, 428–430.

[97] T. Cablewski, A. F. Faux, C. R. Strauss, *J. Org. Chem.* **1994**, *59*, 3408–3412.

[98] K. Kazba, B. R. Chapados, J. E. Gestwicki, J. L. McGrath, *J. Org. Chem.* **2000**, *65*, 1210–1214.

[99] B. M. Khadilkar, V. R. Madyar, *Org. Proc. Res. Dev.* **2001**, *5*, 452–455.

[100] E. Esveld, F. Chemat, J. van Haveren, *Chem. Eng. Technol.* **2000**, *23*, 279–283.

[101] G. Pipus, I. Plazl, T. Koloini, *Chem. Eng. J.* **2000**, *76*, 239–245.

[102] K. A. Savin, M. Robertson, D. Gernert, S. Green, E. J. Hembre, J. Bishop, *Mol. Diversity* **2003**, *7*, 171–174.

[103] N. S. Wilson, C. R. Sarko, G. Roth, *Org. Proc. Res. Dev.* **2004**, *8*, 535–538.

[104] W.-C. Shieh, S. Dell, O. Repič, *Org. Lett.* **2001**, *3*, 4279–4281.

[105] W.-C. Shieh, S. Dell, O. Repič, *Tetrahedron Lett.* **2002**, *43*, 5607–5609.

[106] U. R. Pillai, E. Sahle-Demessie, R. S. Varma, *Green Chem.* **2004**, *6*, 295–298.

[107] P. He, S. J. Haswell, P. D. Fletcher, *Lab Chip* **2004**, *4*, 38–41.

[108] R. Bierbaum, M. Nüchter, B. Ondruschka, *Chem. Ing. Techn.* **2004**, *76*, 961–965.

[109] M. Hajek, in *Microwaves in Organic Synthesis* (Ed.: A. Loupy), Wiley-VCH, Weinheim, **2002**, pp 345–378 (Chapter 10).

5
Starting with Microwave Chemistry

5.1
Why Use Microwave Reactors?

Since the initial experiments in the mid-1980s, the use of microwave energy for heating chemical reactions has shown tremendous benefits in organic synthesis. Significant rate enhancements, improved yields, and cleaner reaction profiles have been reported for many different reaction types over the past two decades. Thus, when planning a reaction protocol or designing novel synthetic pathways, the use of microwave irradiation as heating source should become a first choice and not the last resort! This chapter presents a general overview on the practical issues associated with performing organic synthesis under microwave irradiation.

Whilst in the early published reports in the area of microwave-assisted organic synthesis (MAOS) the utilization of simple, domestic microwave ovens without any parameter control was described, it has become evident that the use of these instruments was not only dangerous but also sometimes led to irreproducible results (see Chapter 2). To meet the increasing demand of innovative preparative chemists for precise reaction control, the development of dedicated microwave reactors for synthetic purposes was necessary in order to provide scientists with accurate measurements of the process-determining parameters. Modern microwave reactors enable the combination of rapid in-core heating with sealed-vessel (autoclave) conditions and an on-line monitoring of reaction parameters. Performing reactions at elevated pressure has therefore become vastly simplified as compared to conventional heating methods. Precise temperature and pressure sensors enable on-line reaction control at any stage of the process. Intelligent software packages allow stepwise method development and optimization. Additional tools for automation and special techniques simplify the procedures of new and improved compound syntheses (see Chapter 3).

The major advantages in using dedicated microwave reactors in organic synthesis can be summarized as follows:

- *Rate enhancement:* Due to the higher temperatures used, reaction times are often drastically reduced from hours to minutes or even seconds.

Microwaves in Organic and Medicinal Chemistry. C. Oliver Kappe, Alexander Stadler
Copyright © 2005 WILEY-VCH Verlag GmbH & Co. KGaA, Weinheim
ISBN: 3-527-31210-2

- *Increased yield:* In many cases the short reaction times minimize the occurrence of unwanted side reactions.
- *Improved purity:* Cleaner reactions due to less by-products lead to simplified purification steps.
- *Greater reproducibility:* The homogeneous microwave field present in dedicated single-mode reactors promises comparable results in every experimental run.
- *Expanded reaction conditions ("reaction space"):* Access to transformations and reaction conditions which cannot be easily achieved under conventional conditions.

5.2
Translating Conventionally Heated Methods

Microwave heating is often applied to already known conventional thermal reactions in order to accelerate the reaction and therefore to reduce the overall process time. When developing completely new reactions, the initial experiments should preferably be performed only on a small scale applying moderately enhanced temperatures to avoid exceeding the operational limits of the instrument (temperature, pressure). Thus, single-mode reactors are highly applicable for method development and reaction optimization.

5.2.1
Open or Closed Vessels?

Utilizing dedicated microwave reactors for organic synthesis allows a choice of open- or closed-vessel conditions. Microwave flash heating in open vessels enables reflux conditions to be reached in drastically reduced times as compared to an oil bath (see Fig. 2.7). In addition to the rapid dielectric heating using sealed vessels, reactions can be performed at temperatures far above the boiling point of the solvent used at elevated pressure (see Fig. 2.8). The combination of higher reaction temperatures and elevated pressure has led to impressive results with respect to reduced reaction times and improved yields. Most of the recently described microwave procedures (see Chapters 6 and 7) operate under sealed-vessel conditions and as a consequence all available microwave reactors are mainly designed for those conditions. Nevertheless, both multimode and monomode reactors can also be applied for reflux conditions at atmospheric pressure. For most transformations, sealed-vessel conditions provide satisfactory results, but sometimes it is necessary to utilize open-vessel conditions, for example in order to allow a volatile by-product to be removed from the reaction mixture (see Section 4.3). However, in order to accomplish significant rate enhancements (> 100 fold), the use of sealed vessels is preferred.

One major benefit of performing microwave-assisted reactions at atmospheric pressure is the possibility of using standard laboratory glassware (round-bottomed flasks, reflux condensers) in the microwave cavity to carry out syntheses on a larger scale. In contrast, pressurized reactions require special vessels and scale-up to more than one liter of processing volume can only be achieved using parallel rotors or flow-through systems (see Section 3.5). Another important aspect of open-vessel technology is safety, especially on a larger scale. It is inherently safer to perform reactions under atmospheric pressure conditions, rather than in a sealed vessel.

5.2.2
Choice of Solvent

The next step in any translation from an original thermal method is the choice of solvent. In general, microwave-mediated transformations can be performed using the same solvent as in a classical protocol. However, one needs to be aware that solvents interact differently with microwaves, according to their dielectric properties. The higher the loss tangent (tan δ), the better the conversion of microwave energy into heat and the more effective the microwave heating (see Section 2.2). Low absorbing solvents (Table 5.1) may also be treated with microwaves, but are poorly heated if they are used in pure form. However, as reaction mixtures are often composed of different (polar) reagents, there should always be enough potential for efficient microwave coupling. It has been argued that if only small amounts of polar reagents are used, the low absorbing solvent may act as a "heat sink", drawing away the thermal heat from the reagents, thereby protecting thermally labile compounds and keeping the reaction temperature low [1]. When employing nearly microwave-transparent reaction mixtures, the addition of strongly microwave-absorbing passive heating sources such as Weflon™-coated stir bars is recommended to achieve appropriate heat transfer within the mixture. These additives heat the reaction mixture by thermal convection and also help to protect parts of the microwave hardware equipment from absorbing too much microwave energy. Alternatively, low absorbing solvents may be doped with small amounts of strongly absorbing solvents such as ionic liquids (see Section 4.3.3.2).

Table 5.1 Physical Properties of Common Solvents (data from [1]).

Solvent	b.p. (°C)	ε'	ε''	tan δ	Microwave absorbance
Ethylene glycol	197	37.0	49.950	1.350	very good
Dimethyl sulfoxide	189	45.0	37.125	0.825	good
Ethanol	78	24.3	22.866	0.941	good
Methanol	63	32.6	21.483	0.659	good
Water	100	80.4	9.889	0.123	medium
1-methyl-2-pyrrolidone	204	32.2	8.855	0.275	medium
N,N-dimethylformamide	154	37.7	6.070	0.161	medium
1,2-dichlorobenzene	180	9.9	2.772	0.280	medium
Acetonitrile	81	37.5	2.325	0.062	medium
Dichloromethane	40	9.1	0.382	0.042	low
Tetrahydrofuran	66	7.4	0.348	0.047	low
Toluene	110	2.4	0.096	0.040	very low

When utilizing microwave heating in sealed vessels, it is no longer necessary to use high-boiling solvents, as in a conventional reflux set-up, to achieve a high reaction temperature. With modern instrumentation, it is easily possible to carry out a reaction in methanol at 160 °C (see Fig. 2.8), or a transformation in dichloromethane at 100 °C.

Utilizing low-boiling solvents often results in the development of a significant pressure at elevated temperatures. Therefore, it is crucial not to exceed the operational limits of the equipment used. With some instruments, a solvent library will help the user to estimate the expected pressure in the reaction vessel. The limiting factor is in general the pressure stability of the vessel used. Common pressure vessels for monomode reactors usually have a pressure limit of 20 bar, imposed by the sealing/capping technique used. For multimode instruments, different materials and vessel designs are available, enabling reactions to be performed at up to 100 bar. Thus, the physical properties of the reagents used may also influence the choice of the instrument to be utilized.

Microwave-mediated reactions can also be easily carried out without solvents (see Section 4.1). The requirements for these dry media reactions are different to those for reactions in solution. As no solvent is involved, the pressure built-up is rather low, and in most instances such reactions are performed under open-vessel conditions. On the other hand, these mixtures can easily be locally overheated, even though the overall bulk temperature may be comparatively low (macroscopic hot-spot formation). Stirring and accurate temperature measurement can prove rather difficult within such a matrix, impeding the investigation of certain reaction conditions. Thus, degradation or decomposition of reagents can be a severe problem for these kind of reactions.

5.2.3
Temperature and Time

The most crucial point for a successful microwave-mediated synthesis is the optimized combination of temperature and time. According to the Arrhenius equation, $k = A \exp(-E_a/RT)$, a halving of the reaction time with every temperature increase of 10 degrees can be expected. With this rule of thumb, many conventional protocols can be converted into an effective microwave-mediated process. As a simple example, the time for a reaction in refluxing ethanol can be reduced from 8 h to only 2 min by increasing the temperature from 80 °C to 160 °C (Fig. 5.1; see also Table 2.4).

For many transformations, the reaction times are in fact significantly shorter than the Arrhenius equation would predict, probably because of the additional pressure that is developed, or arguably due to the involvement of microwave effects (see Section 2.5).

When investigating completely new reactions for which no thermal protocol is available, a feasible starting point is approximately 30–40 °C above the boiling point of the solvent used [1]. Know-how on the way to proceed will improve with personal practical experience and varies with the type of chemistry to be accomplished. Per-

Increasing Temperature

80 °C	90 °C	100 °C	110 °C	120 °C	130 °C	140 °C	150 °C	160 °C
8 h	4 h	2 h	1 h	30 min	15 min	8 min	4 min	2 min

Decreasing reaction time

Fig. 5.1 Relationship between reaction time and temperature for a chemical reaction.

forming a reaction initially for 10 min generally gives a good overview of its progress. After monitoring the reaction mixture, further modifications to the procedure may be deemed necessary, according to the results obtained:

- *No reaction:* Temperature too low or non-suitable reaction system; increase temperature or change solvent.
- *Incomplete conversion:* Extend reaction time and/or increase temperature.
- *Decomposed substrates:* Temperature too high; decrease temperature.
- *Reaction complete:* Reduce time until conversion is still complete to maximize rate enhancement.

Usually, the instrument temperature limit of a monomode reactor is 250 °C, while multimode instruments can be subjected to up to 300 °C, at least for a limited time. This might allow the development of completely new reaction procedures at high temperatures and elevated pressures using sealed vessels, but one has to consider that the pressure limits of some vessel types decrease with increasing temperature. In particular, when employing PTFE or PFA vessels, these may be severely deformed under high-temperature/high-pressure conditions. For these high-temperature reactions, glass or quartz vessels should be employed.

No specific recommendations can be given about the optimum reaction time. As speeding up reactions is a key motive for employing microwave irradiation, the reaction should be expected to reach completion within a few minutes. On the other hand, a reaction should be run until full conversion of the substrates is achieved. In general, if a microwave reaction under sealed-vessel conditions is not completed within 60 min then it needs further reviewing of the reaction conditions (solvent, catalyst, molar ratios). The reported record for the longest microwave-mediated reaction is 22 h for a copper-catalyzed *N*-arylation (see Scheme 6.63). The shortest ever published microwave reaction requires a microwave pulse of 6 s to reach completion (ultra-fast carbonylation chemistry; see Scheme 6.49).

5.2.4
Microwave Instrument Software

All commercially available industrial microwave reactors have software packages which commonly allow for the creation of either a temperature-controlled or a power-controlled program. Using a temperature-controlled program, the system generally tries to reach the set temperature as fast as possible by applying the maximum microwave output power (see Figs. 2.7 and 2.8). To keep exothermic or highly reactive processes under control, dedicated heating ramps can be programmed, for example reaching 120 °C within 2 min, or a definite stepwise reaction progress may be designed (Fig. 5.2). With both variations, very good control over the reaction temperature is provided.

On the other hand, samples can be irradiated at constant microwave power over a certain fixed period, for example at 100 W for 10 min. As there is no control over the resulting temperature or pressure, care has to be taken not to exceed the operational limits of the system and this type of program should only be used for well-known reactions with non-critical limits, or under open-vessel (reflux) conditions. Since in this method only the applied energy and not the resulting temperature is controlled, the quality of reaction control is often superior employing a temperature-controlled program.

Some software packages additionally offer pressure-controlled method development, which relies on the resulting pressure as a limiting factor. The microwave power is regulated by the adjusted pressure limit, and thus there is no influence on the resulting temperature. Because the reaction temperature is the most crucial parameter for successful chemical synthesis, this program variation is used only rarely. For preliminary experiments, it is recommended that temperature programs

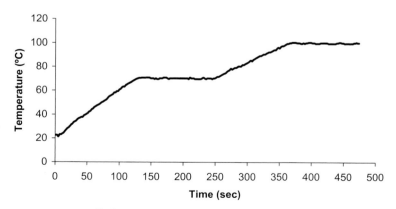

Fig. 5.2 Temperature profile for a 30 mL sample of water heated under sealed-vessel conditions. Multimode microwave heating with 100 W maximum power for 8 min with temperature control using the feedback from a fiber-optic probe: ramp within 120 s to 70 °C; hold for 120 s at 70 °C; ramp within 120 s to 100 °C; hold for 120 s at 100 °C.

are used in order to investigate the operational limits of the reaction. Once the maximum temperature and pressure are known, the variation of microwave output power can be a further means of optimizing the reaction outcome. The option of introducing more power into the reaction system by simultaneous external cooling of the reaction mixture ("heating-while-cooling") has been discussed in Section 2.5.3.

Consequently, which strategy is utilized in reaction optimization experiments is highly dependent on the type of instrument used. Whilst multimode reactors employ powerful magnetrons with up to 1500 W microwave output power, monomode reactors apply a maximum of only 300 W. This is due to the high density microwave field in a single-mode set-up and the smaller sample volumes that need to be heated. In principle, it is possible to translate optimized protocols from monomode to multimode instruments and to increase the scale by a factor of 100 without a loss of efficiency (see Section 4.5).

As a starting point for reaction optimization, Biotage offers a microwave reaction database (Emrys™ PathFinder) with ca. 4000 validated entries for microwave chemistry performed with Biotage single-mode instruments.

5.3
Reaction Optimization and Library Generation – A Case Study

As a suitable model reaction to highlight the steps necessary to successfully translate thermal conditions to microwave conditions, and to outline the general workflow associated with any microwave-assisted reaction sequence, in this section we describe the complete protocol from reaction optimization through to the production of an automated library by sequential microwave-assisted synthesis for the case of the Biginelli three-component dihydropyrimidine condensation (Scheme 5.1) [2, 3].

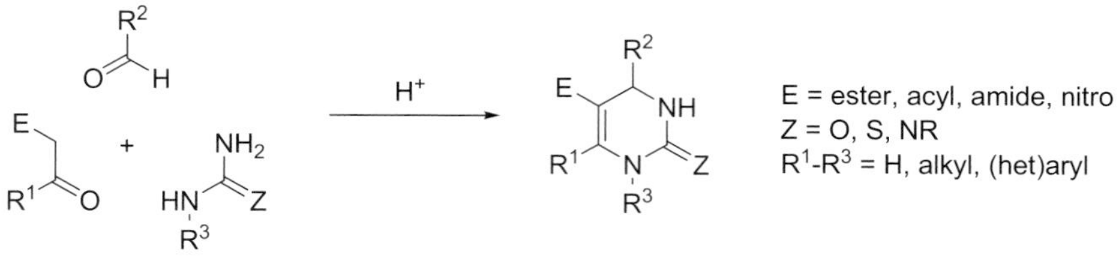

Scheme 5.1   The classic Biginelli dihydropyrimidine cyclocondensation.

The Biginelli multicomponent synthesis is particularly attractive, since the resulting dihydropyrimidine (DHPM) scaffold displays a wide range of biological effects, such as antiviral, antitumor, antibacterial, and anti-inflammatory activities, which has led to the development of a number of lead compounds based on this structural core [4]. Furthermore, functionalized DHPMs have emerged as orally active antihypertensive agents or α_{1a} adrenoceptor-selective antagonists [5]. Therefore, in the context of library generation and screening, a variety of different combinatorial protocols based on the classical Biginelli MCR have been advanced in recent years [6].

As the standard procedure for the Biginelli condensation is rather straightforward, involving one-pot condensation of the three building blocks in a solvent such as ethanol using a strongly acidic catalyst such as hydrochloric acid [7], small libraries of DHPMs are readily accessible by parallel synthesis [8]. One major drawback of the original method, besides the long reaction times involving reflux temperatures, are the moderate yields that are frequently obtained when using more complex building blocks. This has led to the development of a number of improved protocols, many of them involving Lewis acids instead of the traditional mineral acid catalysts [7].

Applying microwave irradiation towards the Biginelli protocols provides significant rate enhancements and frequently higher product yields. Most of these published microwave-assisted procedures involve solvent-free protocols using standard domestic microwave ovens, which do not allow the generation of high quality libraries in an automated fashion. In this case study, a dedicated single-mode microwave reactor (Emrys Synthesizer, see Section 3.5, Fig. 3.19), designed for performing automated chemical synthesis and optimization in a high-throughput/combinatorial chemistry environment, was utilized [2, 3].

5.3.1
Choice of Solvent

In order to make the Biginelli protocol amenable to an automated library generation format, utilizing the integrated robotic interface of the instrument, attempts were made to dissolve most of the building blocks used in solvents compatible with the reaction conditions. Since many of the published protocols employ either ethanol or acetic acid as solvents in Biginelli-type condensations, a 3:1 mixture of acetic acid

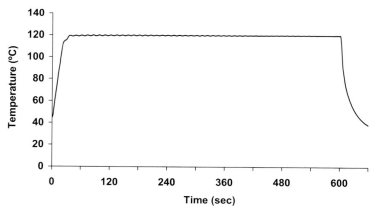

Fig. 5.3 Heating profile for a typical Biginelli condensation in AcOH/EtOH (3:1) under sealed-vessel microwave irradiation conditions: microwave flash heating (300 W, 0–40 s), temperature control using the feedback from IR thermography (constant 120 °C, 40–600 s), and active cooling (600–660 s).

(AcOH) and ethanol (EtOH) was used for the development of a microwave-assisted solution-phase protocol. Both solvents effectively couple with microwave radiation (see Table 2.3), and this allows the reaction mixture to heat up very rapidly under microwave irradiation conditions, leading to so-called microwave flash heating conditions (Fig. 5.3). Preliminary runs with other solvents, such as dioxane or THF, proved far less effective, both in terms of heating rates and product yields. Furthermore, the AcOH/EtOH solvent combination has the advantage that all building blocks are soluble under the reaction conditions at elevated temperatures, whereas the resulting DHPM are comparatively insoluble at room temperature facilitating product isolation.

5.3.2
Catalyst Selection

The next parameter to be considered was the selection of a suitable catalyst. The traditionally used hydrochloric acid proved not to be the most suitable reaction promoter due to inadvertent decomposition of the urea components to ammonia leading to unwanted by-products. Thus, the use of more tolerable Lewis acids such as Yb(OTf)$_3$, InCl$_3$, FeCl$_3$, and LaCl$_3$, which have all been reported to be very effective catalysts for Biginelli condensations [7], was considered. An initial screening of these catalysts for the model system ethyl acetoacetate (R^1 = Me, E = CO$_2$Et), benzaldehyde (R^2 = Ph), and urea (Z = O, R^3 = H) revealed that 10 mol% of Yb(OTf)$_3$ was the most effective catalyst with the AcOH/EtOH solvent system, providing the corresponding DHPM in 92% isolated yield (Fig. 5.4). The use of 5 mol% of the same catalyst furnished a significantly lower yield.

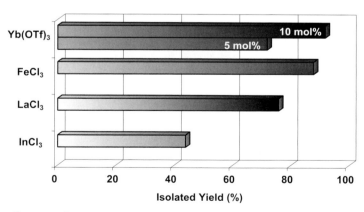

Fig. 5.4 Efficiencies of several Lewis acids in Biginelli cyclocondensation (AcOH/EtOH, 120 °C, 10 min).

5.3.3
Time and Temperature

Having identified an efficient solvent/catalyst combination (AcOH/EtOH, 3:1;
10 mol% Yb(OTf)$_3$), the next issue involved optimization of the reaction time and
temperature. Employing microwave flash heating in sealed vessels allows the application of temperatures far above the boiling points of the solvents as the process is
not limited by reflux. After a few optimization cycles, 120 °C proved to be a very
efficient reaction temperature (Fig. 5.5). While higher temperatures led to decreased
yields due to the formation of undesired by-products, lower reaction temperatures
necessitated longer reaction times for complete conversion. For the model system, a
total irradiation time of 10 min at 120 °C resulted in a 92% isolated yield of pure
product. The final DHPM product precipitated directly after the active cooling period (Fig. 5.3) and showed no traces of impurities when submitted to ^{1}H NMR and
HPLC analyses. Although an increase in reaction time to 15 min further increased
the yield in this example to 94%, longer reaction times were generally not used
unless warranted by the specific building block combinations (see below).

All reactions were run in 2–5 mL microwave process vials (Fig. 3.21), employing
equimolar 4.0 mmol amounts of the building blocks in a total reaction volume of
2.0 mL. At a temperature of 120 °C, this led to a pressure of ca. 3–4 bar in the vial,
well below the permissible limit of 20 bar. Although it has been shown that higher
yields of some DHPMs can be obtained by employing an excess of either the carbonyl or the urea building block, no attempt was made to optimize the molar ratio during this library generation.

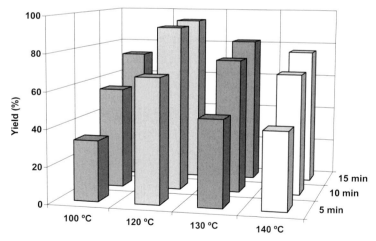

Fig. 5.5 Optimization of reaction temperature and time for the Biginelli
cyclocondensation in AcOH/EtOH (3:1) with 10 mol% of Yb(OTf)$_3$.

5.3.4
Reinvestigation by a "Design of Experiments" Approach

The above mentioned conditions can be further improved by applying a statistics-based "Design of Experiments" (DoE) approach [9]. DoE is a novel method for the optimization and screening of experimental parameters in organic synthesis. Experimental design and statistical tools for data analysis are used to provide key information in decision-making after only a few experiments. Tye recently reinvestigated the microwave-assisted Biginelli reaction conditions utilizing the DoE technique in order to find the optimal parameters [9]. Initially, a thermal protocol using copper(II) triflate as catalyst and acetonitrile as solvent was chosen. In a two-level fractional factorial screening design, concentration in mol% of the Lewis acid, temperature, time, equivalents of urea and concentration thereof were varied. After conducting 19 reactions and fitting the data using appropriate DoE software (MODDE), concentration, mol% of Lewis acid, and temperature were identified as the main factors [9]. The optimized protocol furnished 90% conversion employing 15 mol% of copper(II) triflate at 80 °C for 10 min with 1.0 equivalent of urea at a concentration of 0.5 M. In the next phase, a range of Lewis acids and solvents were screened under the optimized conditions. After a further 17 runs, copper(II) triflate was confirmed as the optimum catalyst when using acetonitrile as the solvent. However, when employing ethanol, two other catalysts (copper(II) bromide and vanadium(III) chloride) showed comparable results to copper(II) triflate. These results demonstrate the potential power of the DoE method, which can readily gauge interaction effects between the limiting factors of a reaction. Rapid microwave protocols are highly suitable for this statistical approach, since many more additional experiments can be carried out compared to standard thermal reactions [10].

5.3.5
Optimization for Troublesome Building Block Combinations

Having established an optimized set of reaction conditions for one model compound (conditions A, Fig. 5.6) of the planned DHPM library, potentially troublesome reagents and reagent combinations in the selection of building blocks can be identified. Thioureas, for example, are known to give significantly lower yields when employed in the Biginelli condensation. Thus, for the thioureas used, LaCl$_3$ turned out to be the catalyst of choice in this microwave-assisted protocol (conditions B, Fig. 5.6). Somewhat acid-sensitive substrates, such as furan-2-carbaldehyde, require a modified protocol using neat ethanol as a solvent at 100 °C (conditions C, Fig. 5.6). Some other examples of reaction conditions fine-tuned to specific building block combinations differing from the developed standard protocol are also presented Fig. 5.6.

The yields for the optimized microwave-assisted Biginelli condensations are in general comparable to or higher than those obtained using the standard reflux conditions. More importantly, however, reaction times have been brought down from several hours (4–12 h) under reflux conditions to 10–20 min employing microwave

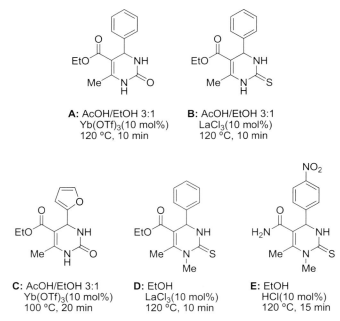

A: AcOH/EtOH 3:1
Yb(OTf)$_3$(10 mol%)
120 °C, 10 min

B: AcOH/EtOH 3:1
LaCl$_3$(10 mol%)
120 °C, 10 min

C: AcOH/EtOH 3:1
Yb(OTf)$_3$(10 mol%)
100 °C, 20 min

D: EtOH
LaCl$_3$(10 mol%)
120 °C, 10 min

E: EtOH
HCl(10 mol%)
120 °C, 15 min

Fig. 5.6 Optimized reaction conditions for selected dihydropyrimidine products.

irradiation in sealed vessels. The optimization cycles described above can be carried out within a few hours, providing an optimized set of conditions (conditions A–E, Fig. 5.6) that can be applied for the ensuing sequential library generation.

5.3.6
Automated Sequential Library Production

With this set of five optimized reaction conditions in hand (Fig. 5.6), the production of a small DHPM library was performed. As a set of structurally diverse representative building blocks, 17 individual CH-acidic carbonyl compounds, 25 aldehydes, and 8 ureas/thioureas were chosen. Combination of all these building blocks would lead to a library of 3400 individual DHPMs. To demonstrate the practicability of the presented concept, a representative subset library of 48 DHPM analogues involving all of the aforementioned building blocks was generated [2].

Utilizing the automation features of the instrument (liquid handler/gripper capabilities, Fig. 3.19), stock solutions of defined concentrations of all the CH-acidic carbonyl components were prepared and stored in designated rack positions of the robot. Similarly, solutions of all the solid aldehydes were prepared, except for those of which the solubility was insufficient to attain the required concentration, which were thus weighed directly into the reaction vials. Liquid reagents were not used as stock solutions, but were dispensed neat. All urea/thiourea derivatives and the catalysts were weighed directly into the process vials before capping. Irradiation using the conditions specified produced the desired DHPMs in isolated yields of 18–92%. A detailed description of the library generation process has been reported elsewhere [3].

The experimental results indicate that considerable variations in all three building blocks are tolerated for the Biginelli reaction. Furthermore, this demonstrates the main benefit of employing a sequential procedure as the general reaction protocol can be individually refined for troublesome building block combinations to maximize the individual yields. In contrast, utilizing parallel rotors in multimode instruments would allow only the same reaction conditions for all substrates (see Section 4.4, Fig. 4.3). For the preparation of relatively small libraries where delicate chemistries are to be performed, a sequential synthetic format may be preferable. Recent evidence shows that these reactions can also be carried out on a significantly larger scale (400 mL) without re-optimization of the reaction conditions [11].

5.4
Limitations and Safety Aspects

In general, most reactions that can be carried out under thermal heating can be performed and accelerated by microwave irradiation. As discussed in Section 2.2, the efficiency of the microwave heating is highly dependent on the dielectric properties of the reaction mixture. Most results suggesting rate enhancements and improved yields can be explained in terms of simple thermal effects. However, for two main reasons, some reactions may not be suitable for performance in microwave reactors:

a) **Low absorbing mixtures:** If the reaction mixture, including the individual reagents therein, is not sufficiently polar to couple efficiently with microwaves, there is a risk of damaging the instrument or parts of the equipment. This is mainly a problem associated with multimode instruments, if low-absorbing materials are irradiated with high microwave output power for extended time periods. The introduced microwave energy has to be absorbed somehow, and if it is not received by the reagents, vessel sensors or other accessories are inadvertently heated. To avoid such processes, highly microwave-absorbing substances should be added to the mixtures. Ionic liquids are some of the best additives that may be incorporated to interact with microwaves; even very small amounts of these substances increase the dielectric loss of the sample and additionally serve as reaction mediators or catalysts. As ionic liquids are often difficult to remove from the reaction mixture, it is more convenient to add passive heating elements such as Weflon™-coated stirring bars or silicon carbide chips.

   Modern microwave instruments are equipped with special safety features to protect the magnetrons from damage. Specially designed reflector systems prevent microwaves from being reflected to the magnetrons or a double-check temperature measurement avoids overheating by promptly switching off the magnetrons if the output power is not efficiently converted.

b) **Overload with high absorbing mixtures:** On the other hand, an extremely polar mixture can be heated up too rapidly and exceed the safety limits of the instrument or the vessels (pressure rate, temperature limit), causing damage to the equipment. This may occur if the vessels are overloaded with too high a volume or too concentrated mixtures. The use of more dilute mixtures or the programming of heating ramps can avoid these problems.

To reduce the risk of container failure, the pressure vessels are equipped with several safety features. These can include an effective self-venting system where unforeseen overpressure is released by a quick open-resealing step, or the use of safety disks which rupture when their pressure limit is reached. The small vials (0.2–20 mL) of some monomode reactors are protected by the pressure limit (20 bar) of the caps used, which is significantly lower than the operating limit of the vials themselves (40–50 bar).

However, in case of any venting activity, it is essential that the exhaust system immediately starts to withdraw the solvent vapors. Especially in large-scale syntheses, the formation of ignitable mixtures can be critical. Organic solvents are often highly flammable and the microwave radiation might act as an ignition source for the system and cause severe explosions. Thus, proper protection of the microwave instrumentation used is absolutely necessary to minimize the potential hazard to the operator. All multimode ovens possess electronic solvent sensors, which measure the concentration of organic solvents in the atmosphere within the cavity. If a critical limit is reached, the magnetrons are turned off and the solvent vapors are exhausted.

Another crucial point is the thermal stability of the solvents used. In general, thermally labile compounds benefit from the short reaction times when employing microwave heating, but on the other hand the applied temperatures are often somewhat higher compared to classical heating. Thus, when exposed to high temperatures, several solvents and reagents may decompose, forming hazardous compounds. Potential risk compounds include those containing nitro or azide groups, as well as ethers, as these are known to cause explosions when heated. For instance, diethyl ether, which forms explosive peroxides and has a very low boiling point, should not be used in microwave-assisted reactions.

Precautions should be taken, especially in a scale-up approach, when dealing with exothermic reactions in the microwave field. Due to the rapid energy transfer of microwaves, any uncontrolled exothermic reaction is potentially hazardous (thermal runaway). Temperature increase and pressure rise may occur too rapidly for the instrument's safety measures and cause vessel rupture.

For accurate temperature monitoring when conducting a temperature-controlled program, a minimum filling volume of the vessels is crucial. In the case of IR temperature measurement from the bottom of a vessel, only a very small amount of reaction mixture (ca. 50 µL) is sufficient to obtain a precise temperature feedback in a monomode instrument (CEM Discover series). On the other hand, a rectangular mounted IR sensor, as used in Biotage instruments (see Section 3.5) requires a certain minimum filling volume (200 µL for the smallest reaction vials; see Fig. 3.21).

Multimode instruments with an IR sensor mounted in the cavity side wall (see Section 3.4) certainly need larger volumes for precise temperature monitoring. Immersion temperature probes require a well-defined minimum volume for accurate measurement, depending on the total vessel volume. It must be ensured that the temperature probe has extensive contact with the reaction mixture, even when the mixture is stirred, in order to obtain reliable, reproducible results.

Care has to be taken if reactions are performed that lead to the evolution of gaseous components, for example those involving ammonium formate, which acts as a hydrogen-transfer agent and additionally releases significant amounts of ammonia and carbon dioxide. When planning an experiment, the volume of gas that will be formed has to be kept in mind, in order not to exceed the operational limits through the build-up of pressure. In general, this is not a safety issue for well-designed systems, as the venting mechanisms release the overpressure and protect the equipment from damage, but it certainly impedes the experimental progress.

A common phenomenon that is encountered when using microwave irradiation to heat chemicals is so-called superheating (see Section 2.5.2), whereby solvents are heated well above their boiling points. This effect is exploited in some synthetic strategies, but it might be hazardous as superheated mixtures can suddenly start boiling. Therefore, stirring is always recommended in organic synthesis, but the size of the stir bar used is crucial. If the metallic core is a quarter of the wavelength of the radiation ($\lambda = 12.25$ cm) it will operate as an antenna, causing spark discharges and destruction of the stirring bar or even explosions of the vessels.

The use of metals or metallic compounds in microwave-assisted reactions can also lead to damage to the reaction vessels. As metals interact intensively with microwaves, the formation of extreme hot spots may occur, which might weaken the vessel surface due to the onset of melting processes. This will destroy the stability of the vessels and may cause explosive demolition of the reaction containers. If catalysts are used which can produce elemental metal precipitates (for example, of palladium or copper), stirring is recommended to avoid the deposition of thin metal layers on the inner surfaces of the reaction vessels.

Operating with chemicals and pressurized containers always carries a certain risk, but the safety features and the precise reaction control of the commercially available microwave reactors protect the users from accidents, perhaps more so than with any classical heating source. The use of domestic microwave ovens in conjunction with flammable organic solvents is hazardous and must be strictly avoided as these instruments are not designed to withstand the resulting conditions when performing chemical transformations.

References

[1] B. L. Hayes, *Microwave Synthesis: Chemistry at the Speed of Light*, CEM Publishing, Matthews, NC, **2002**.

[2] A. Stadler, C. O. Kappe, *J. Comb. Chem.* **2001**, *3*, 624–630.

[3] C. O. Kappe, A. Stadler, in *Combinatorial Chemistry, Part B* (Eds.: B. B. Bunin, G. Morales), Elsevier Sciences, **2003**, pp 197–223.

[4] C. O. Kappe, *Eur. J. Med. Chem.* **2000**, *35*, 1043–1052.

[5] C. O. Kappe, *Acc. Chem. Res.* **2000**, *33*, 879–888.

[6] C. O. Kappe, *QSAR Comb. Sci.* **2003**, *22*, 630–645.

[7] C. O. Kappe, A. Stadler, *Org. React.* **2004**, *63*, 1–117.

[8] K. Lewandowski, P. Murer, F. Svec, J. M. J. Fréchet., *Chem. Commun.* **1998**, 2237–2238; K. Lewandowski, P. Murer, F. Svec, J. M. J. Fréchet, *J. Comb. Chem.* **1999**, 105–112.

[9] H. Tye, *Drug Discovery Today* **2004**, *9*, 485–491.

[10] M. D. Evans, J. Ring, A. Schoen, A. Bell, P. Edwards, D. Berthelot, R. Nicewonger, C. M. Baldino, *Tetrahedron Lett.* **2003**, *44*, 9337–9341; H. Tye, M. Whittaker, *Org. Biomol. Chem.* **2004**, *2*, 813–815; N. J. McLean, H. Tye, M. Whittaker, *Tetrahedron Lett.* **2004**, *45*, 993–995.

[11] A. Stadler, B. H. Yousefi, D. Dallinger, P. Walla, E. Van der Eycken, N. Kaval, C. O. Kappe, *Org. Proc. Res. Dev.* **2003**, *7*, 707–716.

6

Literature Survey Part A: General Organic Synthesis

In this chapter, we summarize recent applications of controlled microwave heating technology in organic synthesis. The term "controlled" here refers to the use of a dedicated microwave reactor for synthetic chemistry purposes (single- or multi-mode). Therefore, the exact reaction temperature *during the irradiation process* has been adequately determined in the original literature source. Although the aim of this chapter is not primarily to speculate about the existence or non-existence of microwave effects (see Section 2.5), the results of adequate control experiments or comparison studies with conventionally heated transformations are sometimes presented. Because of the inherent difficulties in conducting such experiments, the reader should not draw any definitive conclusions about the involvement or non-involvement of "microwave effects" from these experimental results. In terms of processing techniques (see Chapter 4), preference is given to transformations in solution under sealed-vessel conditions, since this reflects the recent trend in the literature and these transformations are, in principle, scalable in batch or continuous-flow mode. Thus, unless specifically noted otherwise, sealed-vessel microwave technology was employed. Most of the examples have been taken from the period 2002–2004. Earlier examples of controlled microwave-assisted organic synthesis (MAOS) are limited and can be found in previous review articles and books [1–13]. Microwave-assisted solid-phase organic synthesis and related transformations involving immobilized reagents or catalysts are described in Chapter 7. Here, in some instances, results obtained with domestic microwave ovens have been included.

6.1
Transition Metal-Catalyzed Carbon–Carbon Bond Formations

Homogeneous transition metal-catalyzed reactions represent one of the most important and most extensively studied reaction types in MAOS. Using traditional heating under reflux conditions, transition metal-catalyzed carbon–carbon and carbon–heteroatom bond-forming reactions typically need hours or days to reach completion and often require an inert atmosphere. Over the past few years, the groups of Hallberg, Larhed, and others have demonstrated that many of these transformations can be significantly enhanced by employing microwave heating under sealed-vessel con-

Microwaves in Organic and Medicinal Chemistry. C. Oliver Kappe, Alexander Stadler
Copyright © 2005 WILEY-VCH Verlag GmbH & Co. KGaA, Weinheim
ISBN: 3-527-31210-2

ditions ("microwave flash heating"), in most cases without requiring an inert atmosphere [7]. The use of metal catalysts in conjunction with microwaves may have significant advantages in comparison with traditional heating methods, since the inverted temperature gradients under microwave conditions (see Fig. 2.6) may lead to an increased lifetime of the catalyst by elimination of wall effects. In fact, the elimination of wall effects and low thermal gradients (bulk heating) in microwave-heated reactions has frequently been invoked to rationalize the outcome of microwave-assisted reactions involving homogeneous transition metal catalysts (see Section 2.5.2).

6.1.1
Heck Reactions

The Heck reaction, a palladium-catalyzed vinylic substitution, is typically conducted with alkenes and organohalides or pseudohalides as the reactants. Numerous elegant synthetic transformations based on C–C bond-forming Heck chemistry have been developed, both in classical organic synthesis and natural product chemistry [14]. Solution-phase Heck chemistry was successfully carried out by MAOS as early as 1996, allowing a reduction of reaction times from several hours under conventional reflux conditions to sometimes less than 5 min [15]. These early examples of microwave-assisted Heck reactions have been extensively reviewed by the Larhed group and will not be further discussed herein [7]. Scheme 6.1 shows a recent example of a standard Heck reaction reported by Kappe and coworkers involving aryl bromides and acrylic acid and furnishing the corresponding cinnamic acid derivatives [16]. Optimization of the reaction conditions under small-scale (2 mmol) single-mode microwave conditions led to a protocol that employed acetonitrile as solvent, 1 mol% palladium acetate/tri(*ortho*-tolyl)phosphine as catalyst/ligand system, triethylamine as base, and a reaction temperature of 180 °C for 15 min. Interestingly, the authors have discovered that here the rather expensive homogeneous catalyst system can be replaced by 5% palladium-on-charcoal (< 0.1 mol% concentration of palladium) without the need to change any of the other reaction parameters [16]. Yields for the Heck reaction providing cinnamic acids (X = H) were very similar using either homogeneous or heterogeneous Pd catalysis. In the same article [16], the authors also demonstrate that it is possible to directly scale-up the 2 mmol Heck chemistries to 80 mmol (ca. 120 mL total reaction volume) by switching from a single-mode to a larger multimode microwave reactor (see Fig. 3.16). Importantly, the

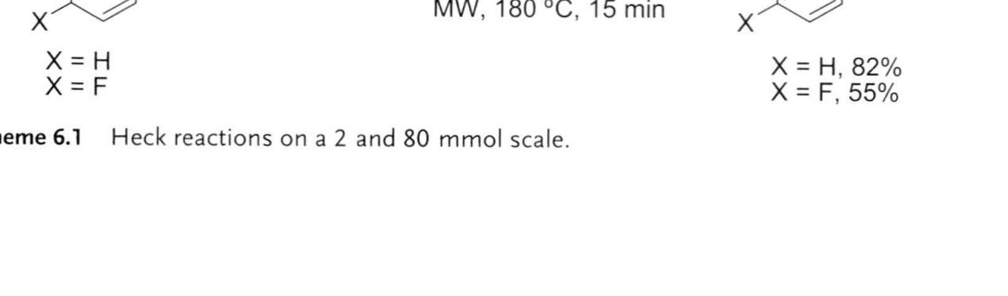

X = H
X = F

X = H, 82%
X = F, 55%

Scheme 6.1   Heck reactions on a 2 and 80 mmol scale.

optimized small-scale reaction conditions could be directly applied for the larger scale run, giving rise to very similar product yields (see Section 4.5).

In 2002, Larhed and coworkers reported microwave-promoted Heck arylations using the ionic liquid 1-butyl-3-methylimidazolium hexafluorophosphate (bmimPF$_6$) as reaction medium [17]. This reaction (see Scheme 4.17) and the unique properties of ionic liquids in the context of microwave synthesis are described in detail in Section 4.3.3.2. More recently, the same group has exploited the combination of bmimPF$_6$ and dioxane in a Heck coupling of both electron-rich and electron-poor aryl chlorides with butyl acrylate (Scheme 6.2) [18]. Transition metal-catalyzed carbon–carbon bond-forming reactions involving unreactive aryl chlorides have represented a synthetic challenge for a long time. Only recently, due to advances in the development of highly active catalyst/ligand systems, have such transformations become accessible [19]. For the Heck coupling shown in Scheme 6.2, the air-stable but highly reactive tri-*tert*-butylphosphonium tetrafluoroborate described by Netherton and Fu [20] was employed as a ligand precursor using Herrmann's palladacycle [*trans*-di(μ-acetato)bis[o-(di-o-tolyl-phosphino)benzyl]dipalladium(II)] [21] as palladium pre-catalyst. Depending on the reactivity of the aryl chloride, 1.5–10 mol% of palladium catalyst (3–20% of ligand), 1.5 equivalents of N,N-dicyclohexylmethylamine as a base, and 1 equivalent of bmimPF$_6$ in dioxane were irradiated at 180 °C under sealed-vessel conditions (no inert gas atmosphere) with the aryl chloride and butyl acrylate for 30–60 min. Under these optimized conditions, the desired cinnamic esters were obtained in moderate to excellent yields (Scheme 6.2) [18].

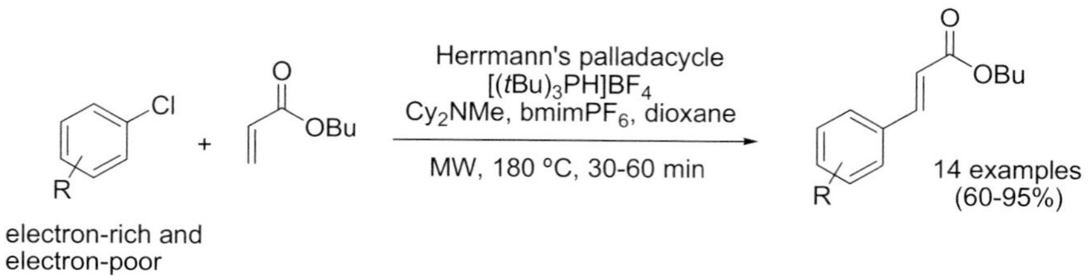

Scheme 6.2 Heck reactions of aryl chlorides involving air-stable phosphonium salts as ligand precursors.

In a recent article by Botella and Nájera, controlled mono- and double-Heck arylations in water catalyzed by an oxime-derived palladacycle were described [22]. When the reaction was carried out under microwave irradiation at 120 °C in the presence of dicyclohexylmethylamine with only 0.01 mol% of palladium catalyst (palladium acetate or palladacycle), monoarylation took place in only 10 min with a very high turnover frequency (TOF) of > 40000 (Scheme 6.3). As regards diarylation, 1 mol% of the palladacycle catalyst and 2 equivalents of iodobenzene had to be utilized to obtain moderate to good yields of diarylated product. Whereas microwave heating at 120 °C provided a 31% yield after 10 min, a 66% isolated yield of product was obtained by heating the reaction mixture under reflux for 13 h at 100 °C. Here, the

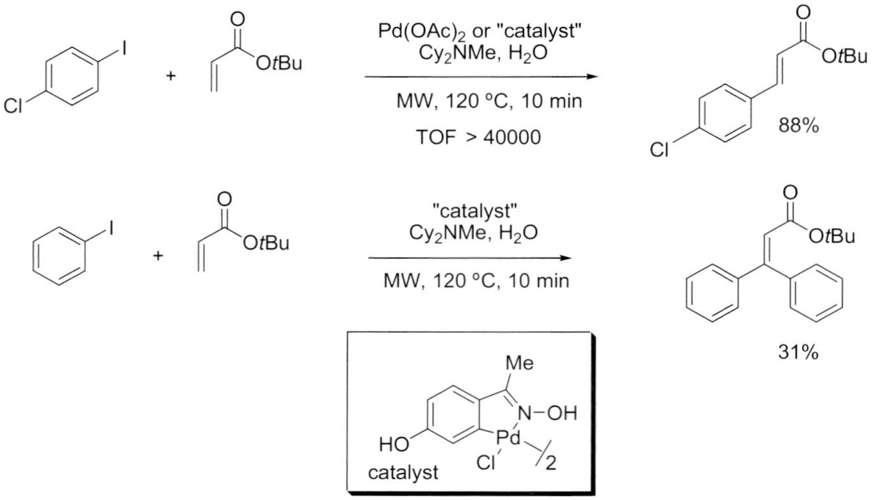

Scheme 6.3 Mono- and double-Heck arylations in water using oxime-derived palladacycles.

use of palladium acetate as catalyst gave very poor conversions under both thermal and microwave conditions.

This protocol could be extended to a range of different α,β-unsaturated carbonyl compounds and either activated or deactivated aryl iodides [22]. An application of related Heck chemistry to the synthesis of methylated resveratrol (3,4′,5-trihydroxy-(*E*)-stilbene) is shown in Scheme 6.4 [23]. The phytoalexin resveratrol exhibits a variety of interesting biological and therapeutic properties, among them activity against several human cancer cell lines. Botella and Nájera have shown that the trimethyl ether of resveratrol (Scheme 6.4) can be rapidly prepared by microwave-assisted Heck reaction of the appropriate aryl iodide and styrene derivatives, using the same oxime-derived palladacycle as indicated in Scheme 6.3.

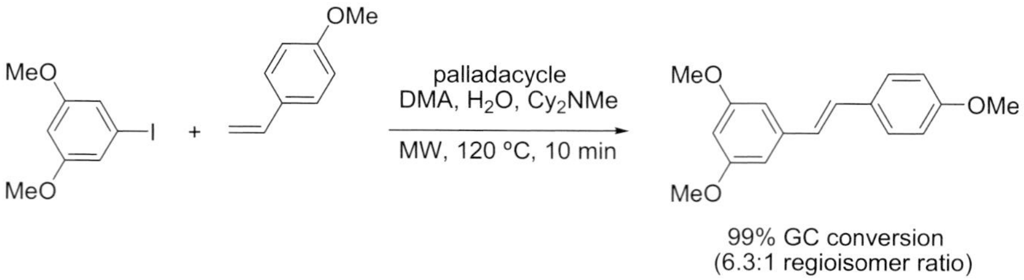

Scheme 6.4 Synthesis of resveratrol trimethyl ether via Heck arylation.

Using a different catalytic system, the Larhed group was able to perform regioselective microwave-promoted chelation-controlled double-β-arylations of terminal alkenes (Scheme 6.5) [24]. In this Heck approach, the authors used vinyl ethers as chelating alkenes and aryl bromides as coupling partners, employing Herrmann's palladacycle as a palladium source. By appropriate selection of the experimental parameters, it was possible to achieve symmetrical and non-symmetrical terminal β,β-diarylations with both electron-rich and electron-poor aryl bromides. One-pot microwave-heated symmetrical bis-arylations were carried out by employing an excess of the aryl bromide (Ar^1Br, 3.0–5.0 equivalents) with a small amount of Herrmann's palladacycle (0.5 mol%) in 10% aqueous DMF. To increase the stability of the under-ligated catalytic palladium(0) system, a highly ionic reaction cocktail was preferred, containing lithium chloride and sodium acetate as additives and potassium carbonate as base. Optimum conditions involved microwave heating at 160–180 °C for 10–55 min. In order to obtain unsymmetrical products, the mono-β-arylated alkenes were reacted with aryl bromides (Ar^2Br) under almost identical reaction conditions to those employed in the symmetrical examples (Scheme 6.5) [24]. For related Heck vinylations of chelating vinyl ethers, see [470].

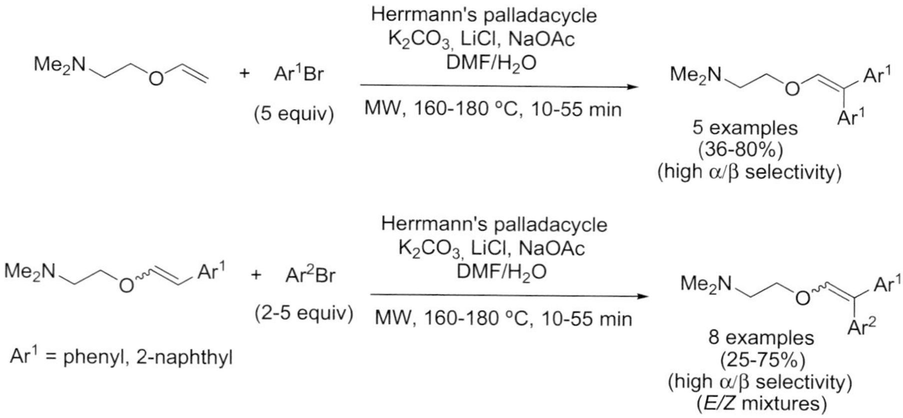

Scheme 6.5   Double-Heck arylations of chelating alkenes with aryl bromides.

A very recent addition to the already powerful spectrum of microwave Heck chemistry has been the development of a general procedure for carrying out oxidative Heck couplings, that is, the palladium(II)-catalyzed carbon–carbon coupling of arylboronic acids with alkenes using copper(II) acetate as a reoxidant [25]. In a 2003 publication (Scheme 6.6), Larhed and coworkers utilized lithium acetate as a base and the polar and aprotic N,N-dimethylformamide as solvent. The coupling

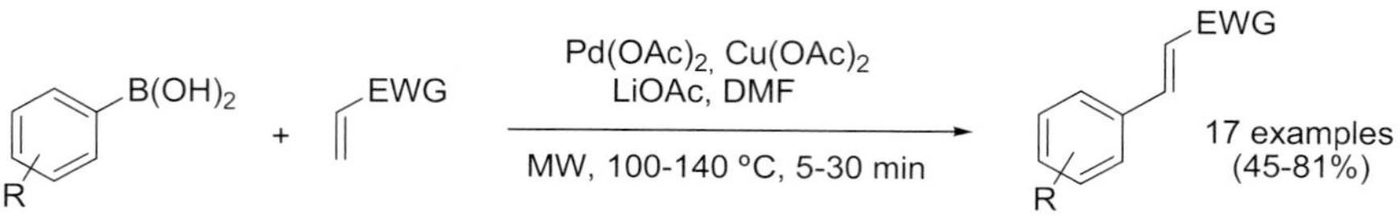

Scheme 6.6   Oxidative Heck coupling of boronic acids and alkenes using copper(II) acetate as a reoxidant.

reaction could be conducted in air without any discernible difference in outcome compared to when it was performed under an inert nitrogen atmosphere. Generally, a reaction temperature of 100–140 °C and an irradiation time of 5–30 min produced moderate to good yields of arylated products with the (*E*)-configuration predominating. Electron-rich arylboronic acids were found to be the superior coupling partners, affording high yields of adducts without any apparent side reactions. The protocol has also been extended to fluorous alkenes (see Section 7.3) [25].

In 2004, the same group presented a modified method for performing oxidative Heck arylations, exploiting molecular oxygen gas for environmentally benign reoxidation and a stable 1,10-phenanthroline bidentate ligand (dmphen) to promote the palladium(II) regeneration and to control the regioselectivity (Scheme 6.7) [26]. While the conventional thermal reaction often required 18 h to reach completion, microwave irradiation in pre-pressurized vessels (3 bar oxygen pressure, see also Section 4.3.2) allowed the oxidative Heck coupling to proceed within 1 h [26].

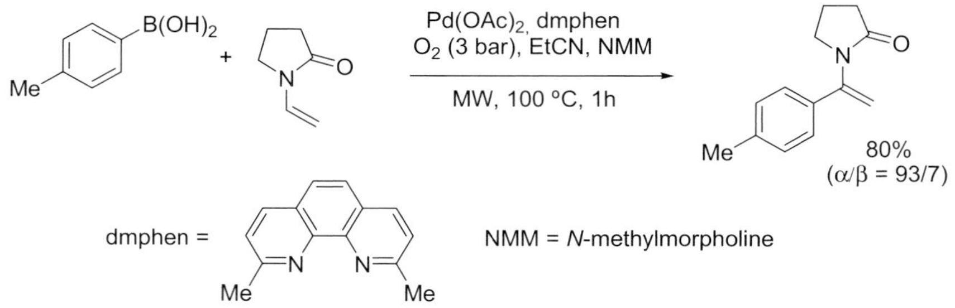

Scheme 6.7 Oxidative Heck coupling of boronic acids and alkenes using dioxygen as a reoxidant.

Microwave-assisted Heck reactions have also been carried out with triflates as coupling partners, involving some very complex molecules. Winterfeld and coworkers have reported a multigram synthesis of a complex non-symmetrical bis-steroidal diene by microwave-promoted coupling of the corresponding alkene and triflate steroidal moieties (Scheme 6.8) [27].

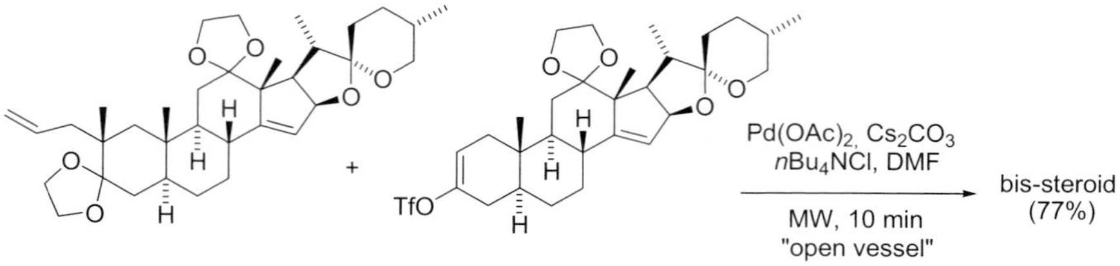

Scheme 6.8 Heck reaction for the synthesis of bis-steroids.

A synthetically useful application of an intramolecular microwave-assisted Heck reaction has been described by Gracias and coworkers (Scheme 6.9) [28]. In their approach toward the synthesis of N-containing seven-membered heterocycles, the

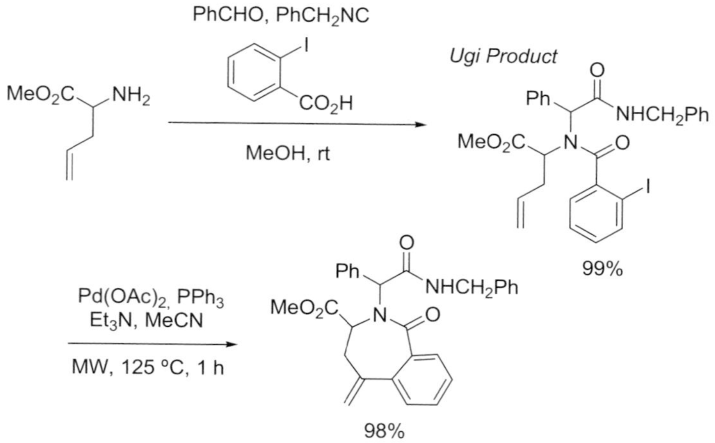

Scheme 6.9 Sequential Ugi/Heck cyclizations for the synthesis of seven-membered N-heterocycles.

initial product of an Ugi four-component reaction was subjected to an intramolecular Heck cyclization using 5 mol% palladium acetate/triphenylphosphine as the catalytic system. For the example shown in Scheme 6.9, microwave irradiation at 125 °C in acetonitrile for 1 h provided a 98% isolated yield of product. A number of related sequential Ugi/Heck cyclizations were reported in the original publication, some involving aryl bromides instead of iodides.

Similarly, the Tietze group has described an intramolecular microwave-promoted Heck reaction for the construction of the B ring in the synthesis of enantiopure B-nor-estradiol analogues (Scheme 6.10 a) [29]. The Heck coupling took place from below, *anti* to the angular methyl group, to form a single diastereoisomer. The best

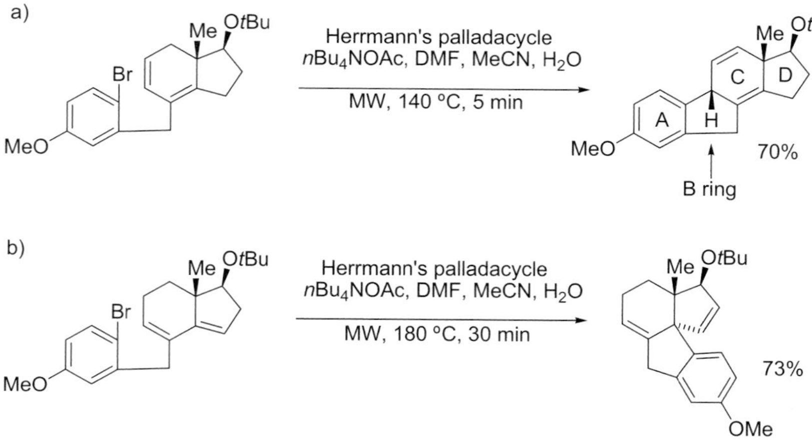

Scheme 6.10 Intramolecular Heck coupling for the construction of B-nor steroids.

results were obtained using 5 mol% of Herrmann's palladacycle as catalyst and tetrabutylammonium acetate as additive in a mixture of DMF, acetonitrile, and water. Under these conditions, the reaction time could be shortened to 5 min, and this also led to suppression of the aromatization of ring C with opening of ring D as an unwanted side reaction, which was observed when the reaction was performed under standard conditions (120 °C, 18 h). Treatment of the isomeric *seco*-B-norsteroid under identical reaction conditions led to a novel spirocyclic ring system as a single diastereomer (Scheme 6.10 b) [30]. The isolated product yield was 73% under both microwave conditions (180 °C, 30 min) and after conventional oil-bath heating (120 °C, 18 h).

Another application of an intramolecular Heck reaction using microwave irradiation was disclosed by Sørensen and Pomb-Villar [31]. In the context of synthesizing analogues of a group of seratonin (5-HT$_3$) receptor antagonists structurally related to the tricyclic carbazolone ring system, the preparation of a cyclopenta[*b*]indol-1-one intermediate via intramolecular Heck reaction of a suitable iodo precursor was envisaged (Scheme 6.11). Optimum conditions were found to involve the use of 5 mol% palladium(II) acetate/tri(*ortho*-tolyl)phosphine as a catalytic system and tetrabutylammonium chloride as an additive. The analogous bromoaniline precursor turned out to be less reactive in the desired cyclization reaction, resulting in only 30% conversion under similar microwave-promoted reaction conditions. Surprisingly, however, a 95% yield of the desired product could be obtained when the Heck cyclization was carried out under conventional heating in an oil bath at 120 °C for 16 h. An enantioselective Heck reaction is depicted in Scheme 6.53 a.

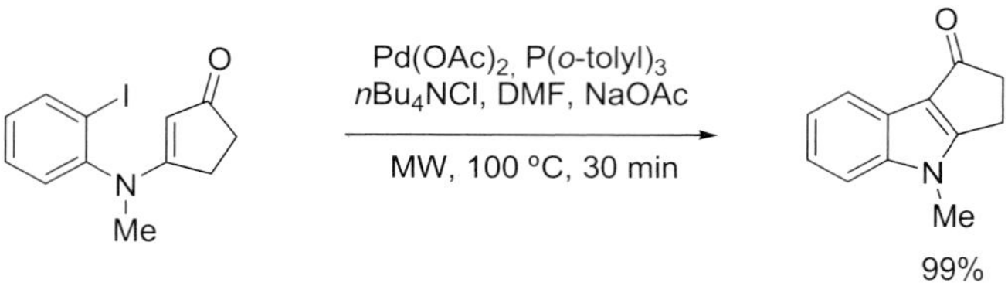

Scheme 6.11 Intramolecular Heck coupling for the preparation of cyclopenta[*b*]indoles.

6.1.2
Suzuki Reactions

The Suzuki reaction (the palladium-catalyzed cross-coupling of aryl halides with boronic acids) is arguably one of the most versatile and at the same time also one of the most often used cross-coupling reactions in modern organic synthesis [32]. Carrying out high-speed Suzuki reactions under controlled microwave conditions can today be considered almost a routine synthetic procedure, given the enormous literature precedent for this transformation [7].

A significant advance in Suzuki chemistry has been the observation that Suzuki couplings can be readily carried out using water as solvent in conjunction with microwave heating [33, 34]. Water, being cheap, readily available, non-toxic, and non-flammable, has clear advantages as a solvent for use in organic synthesis (see Section 4.3.3.1). With its comparatively high loss tangent (tan δ) of 0.123 (see Table 2.3), water is also a potentially very useful solvent for microwave-mediated synthesis, especially in the high-temperature region accessible using sealed-vessel technology. Leadbeater and Marco have described very rapid, ligand-free palladium-catalyzed aqueous Suzuki couplings of aryl halides with arylboronic acids (Scheme 6.12) [33]. Key to the success of this method was the use of 1 equivalent of tetrabutylammonium bromide (TBAB) as a phase-transfer catalyst additive. The role of the ammonium salt is to facilitate the solubility of the organic substrates and to activate the boronic acid by formation of an $[ArB(OH)_3]^-$ $[R_4N]^+$ species. By using controlled microwave heating at 150 °C for 5 min with only 0.4 mol% of palladium acetate as catalyst, a wide variety of aryl bromides and iodides were successfully coupled with arylboronic acids (Scheme 6.12) [33]. Aryl chlorides also reacted but required higher temperatures (175 °C).

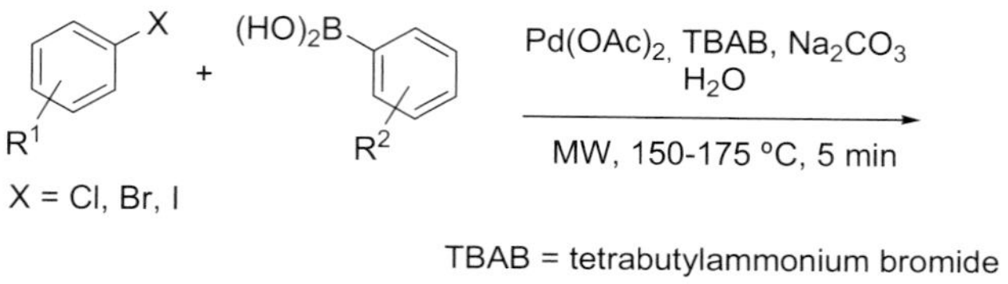

Scheme 6.12 Ligand-free Suzuki reactions using TBAB as an additive.

The same Suzuki couplings could also be performed under microwave-heated open-vessel reflux conditions (110 °C, 10 min) on a ten-fold larger scale, giving nearly identical yields to the closed-vessel runs [33, 35]. Importantly, nearly the same yields were obtained when the Suzuki reactions were carried out in a pre-heated oil bath (150 °C) instead of using microwave heating, clearly indicating the absence of any specific or non-thermal microwave effects [34].

In another modification, the same authors reported that, somewhat surprisingly, it was also possible to carry out the Suzuki reactions depicted in Scheme 6.12 in the absence of any added palladium catalyst [36–38]. These "transition metal-free" aqueous Suzuki-type couplings again utilized 1 equivalent of tetrabutylammonium bromide (TBAB) as an additive, 3.8 equivalents of sodium carbonate as a base, and 1.3 equivalents of the corresponding boronic acid (150 °C, 5 min). High yields were obtainable for aryl bromides and iodides, whereas aryl chlorides proved unreactive under the conditions used. The reaction is also limited to electron-poor and electron-neutral boronic acids. It was subsequently discovered that ultra-low concentrations of palladium (50 ppb) contained in the sodium carbonate base were responsible for these Suzuki reactions to proceed [468]. Comparatively low catalyst loadings of aqueous Suzuki couplings have also been reported by Nájera and coworkers [472].

Cross-coupling reactions of unactivated aryl chlorides represent one of the most challenging problems in transition metal-catalyzed transformations. A series of air- and moisture-stable (N-heterocyclic carbene)palladium(allyl) complexes (NHC) has been shown by Nolan and coworkers to catalyze Suzuki cross-coupling reactions of aryl chlorides with boronic acids (Scheme 6.13) [39]. This catalytic system is compatible with microwave conditions, and rapid couplings were observed within 1.5 min at 120 °C employing 2 mol% of the NHC ligand in the presence of a strong base (sodium *tert*-butoxide, 3 equivalents). Under these high-temperature conditions, the catalyst appeared to be stable as no palladium black was formed. The conventionally heated reactions (60 °C) required several hours to reach completion. In the same article, the authors also reported on microwave-assisted dehalogenations of aryl chlorides using the same catalytic system but switching to 2-propanol as a protic solvent (Scheme 6.13) [39]. By using 1.05 equivalents of sodium *tert*-butoxide, the amount of catalyst required to dehalogenate 4-chlorotoluene was reduced to 0.025 mol% of the NHC system at 120 °C (2 min). Similar Suzuki couplings of aryl chlorides under microwave conditions were disclosed by Bedford and coworkers [471].

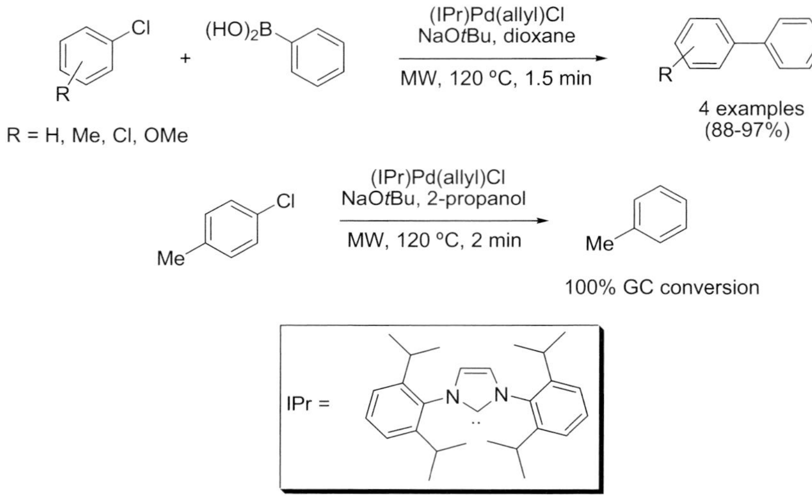

Scheme 6.13 Suzuki couplings and dehalogenations catalyzed by (*N*-heterocyclic carbene)palladium(allyl)Cl complexes.

Many recent examples in the literature demonstrate the versatility of the Suzuki protocol under microwave conditions. In the example shown in Scheme 6.14, Gong and He [40] described the direct synthesis of unprotected 4-aryl phenylalanines by a microwave-assisted Suzuki protocol utilizing standard coupling conditions. In most cases, 5 mol% of palladium(II) bis(triphenylphosphine) dichloride [$Pd(PPh_3)_2Cl_2$] and the use of sodium carbonate as a base allowed the high-yielding coupling of 4-borono-phenylalanine with a variety of (hetero)aryl halides within 10 min of microwave irradiation (150 °C). A control experiment in a pre-heated oil bath under otherwise identical conditions showed a significantly lower conversion. These authors also investigated the effect of switching the positions of the boronic acid and halide functionalities in the coupling partners. This complementary method-

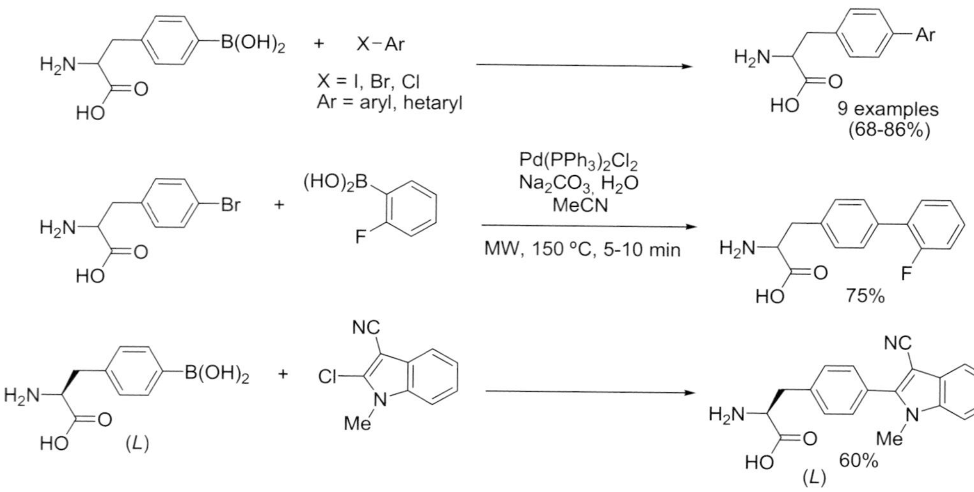

Scheme 6.14 Synthesis of unprotected 4-aryl phenylalanine by Suzuki cross-coupling reactions.

ology led to equally high yields. It was also possible to employ enantiomerically pure borono-phenylalanine in the Suzuki coupling with a heteroaryl chloride. In addition to the excellent conversion, no significant racemization was observed in the coupling product.

The application of a microwave-assisted Suzuki reaction in the synthesis of electron-rich phenethylamines and apogalanthamine analogues has been described by Van der Eycken and coworkers (Scheme 6.15) [41]. Here, the use of sodium hydrogen carbonate and 5 mol% tetrakis(triphenylphosphine)palladium(0) as catalyst in a mixture of N,N-dimethylformamide and water was found to give the optimum medium for the microwave-assisted Suzuki reactions (120–150 °C, 10–15 min). In order to access the required apogalanthamine analogues, this reaction was also attempted with boronic acids bearing highly electron-withdrawing substituents in the sterically unfavorable *ortho* position. In order to prevent proto-deboronation, to which these systems are prone, the reaction conditions were slightly modified (cesium carbonate) in order to provide an 84% yield of the desired biaryl coupling product. An acid-mediated, microwave-assisted deprotection procedure followed by a one-pot reductive amination then furnished the desired apogalanthamine analogue, 10,11-dimethoxy-5,6,7,8-tetrahydrodibenzo[c,e]azocine (Scheme 6.15) [41].

A recent publication by the group of Barbarella has disclosed the rapid preparation of poorly soluble unsubstituted and modified α-quinque- and sexithiophenes by the extensive use of bromination/iodination steps and microwave-assisted Suzuki and Sonogashira cross-couplings (Scheme 6.16) [42]. Suzuki reactions were either carried out under solvent-free conditions on a strongly basic potassium fluoride/ alumina support for the synthesis of soluble oligothiophenes, or in solution phase for the preparation of the rather insoluble α-quinque- and sexithiophenes. In both cases, 5 mol% of [1,1′-bis(diphenylphosphino)ferrocene]dichloropalladium(II)

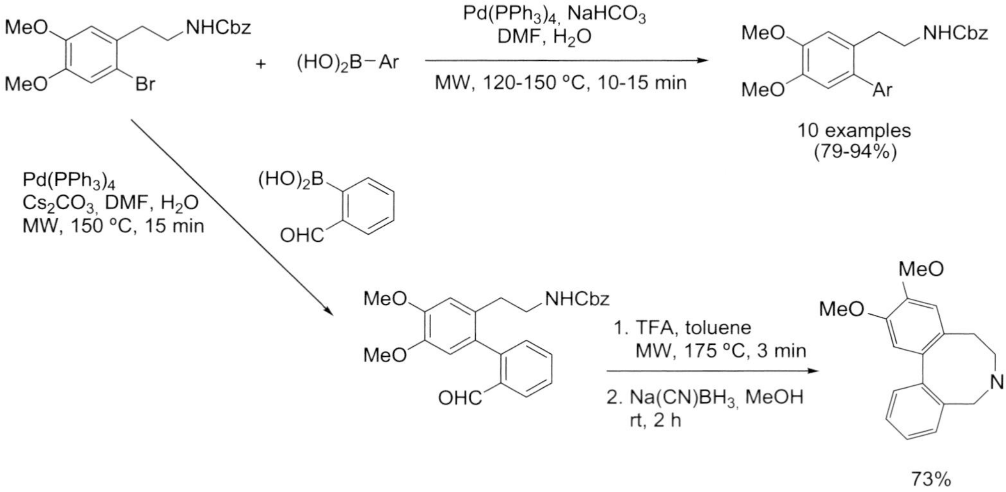

Scheme 6.15 Synthesis of apogalanthamine analogues using Suzuki cross-coupling reactions.

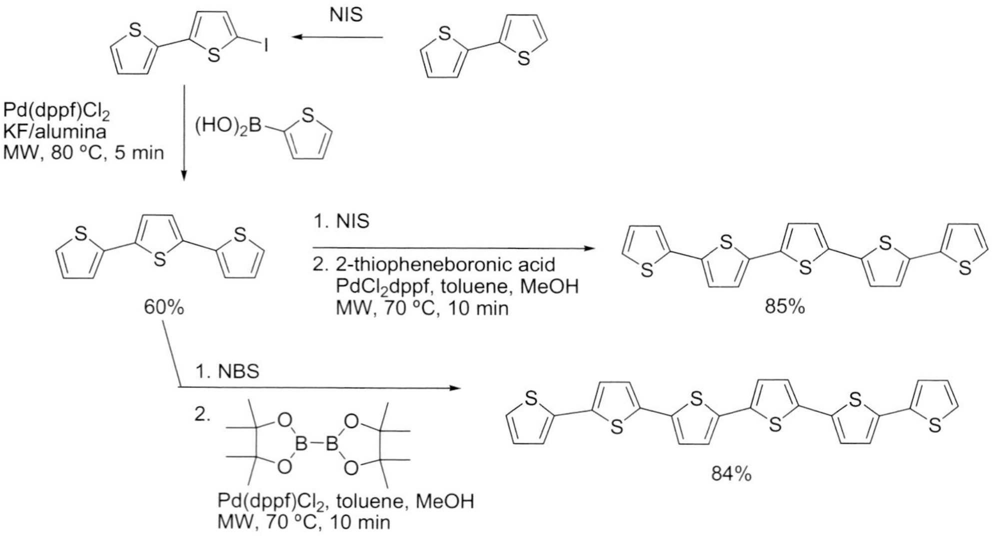

Scheme 6.16 Synthesis of oligothiophenes using Suzuki cross-coupling reactions.

[Pd(dppf)Cl$_2$] was used as the catalyst. A particularly noteworthy aspect of this publication is the introduction of a one-pot borylation/Suzuki reaction of 5-bromoterthiophene with commercially available bis(pinacolato)diboron. The resulting sexithiophene was obtained within 10 min of microwave irradiation in 84% isolated yield. The same concept was also applied for the synthesis of modified oligothiophenes (not shown) [42]. Several of the prepared oligothiophenes show liquid-crystalline properties.

The asymmetric diboration of alkenes provides versatile, reactive 1,2-diboron intermediates in a catalytic enantioselective fashion. Unsymmetrical 1,2-bis(boronate)s, such as those derived from terminal alkenes, engage in selective cross-coupling reactions involving the more accessible, less hindered primary carbon–boron bond [43]. In this context, Morken and coworkers have reported a catalytic asymmetric carbohydroxylation of alkenes by a tandem diboration/Suzuki cross-coupling/oxidation reaction. In the example shown in Scheme 6.17, the sterically encumbered terminal alkene was converted into the desired 1,2-diboron intermediate by a rhodium-catalyzed diboration [5 mol% rhodium(I)-(2,5-norbornadiene)acetylacetonate, (nbd)Rh(acac); 5 mol% (S)-(−)-1-(2-diphenylphosphino-1-naphthyl)isoquinoline, (S)-quinap] at room temperature. After the addition of phenyl triflate as a coupling partner and 10 mol% of [1,1'-bis(diphenylphosphino)ferrocene]dichloropalladium(II) [Pd(dppf)Cl$_2$] as a catalyst, the selective Suzuki coupling was performed under microwave conditions at 80 °C for 1 h, and this was followed by oxidation of the remaining carbon–boron bond with hydrogen peroxide. The anticipated chiral secondary alcohol, the product of a net carbohydroxylation reaction, was isolated in 70% yield with 93% enantiomeric excess (*ee*) [43].

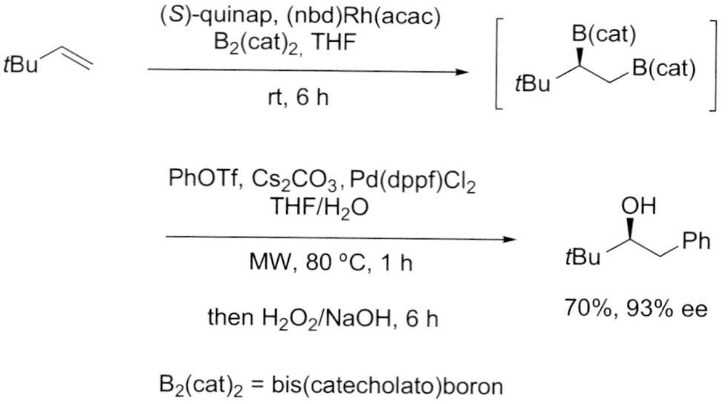

Scheme 6.17 Catalytic asymmetric carbohydroxylation of alkenes.

A recent publication by Turner and coworkers has described the microwave-assisted palladium-catalyzed cross-coupling of β-chloroalkylidene/arylidene malonates ("vinyl chlorides") with boronic acids in a Suzuki-type fashion (Scheme 6.18) [44]. The Suzuki arylation reaction was found to proceed with a wide range of arylboronic acids including electron-rich, electron-deficient, and sterically demanding systems. For most reactions, the presence of 1 mol% of the commercially available air-stable palladium catalyst [(*t*Bu)$_2$P(OH)]$_2$PdCl$_2$ provided a high yield of the desired β-aryl/alkylarylidene malonates within 30 min in tetrahydrofuran as solvent. The Suzuki reaction was found to be compatible with arylboronic acids bearing amino, hydroxy, and chloro functionalities, leading to functionalized products that could be further derivatized (Scheme 6.18). The catalyst was also found to be active in pro-

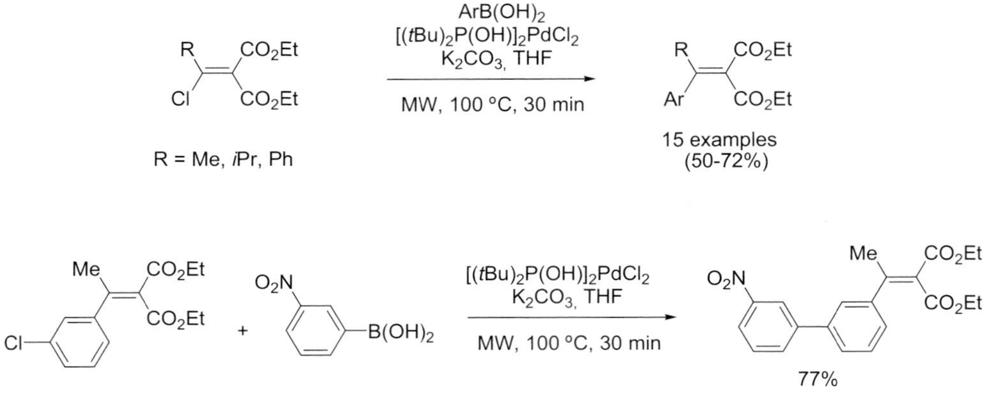

Scheme 6.18 Palladium-catalyzed cross-coupling of β-chloroalkylidene/arylidene malonates.

moting other standard Suzuki reactions of aryl chlorides under microwave conditions [44].

Recent literature examples involve the use of the Suzuki protocol for the high-speed decoration of various heterocyclic scaffolds of pharmacological or biological interest, including pyrimidines [45], pyridazines [46], pyrazines [47], chromanes [48], and pyrazoles [49] (Scheme 6.19).

In this context, a highly convergent three-step microwave-assisted sequence for the preparation of diversely functionalized pyrazolopyrimidines was reported by Schultz and coworkers (Scheme 6.20) [50]. In the first step, 4-chloropyrazolopyrimi-dine was brominated on a multigram scale using *N*-bromosuccinimide in aceto-nitrile under microwave irradiation conditions (100 °C, 10 min). The resulting 4-chloro-5-bromopyrazolopyrimidine was then subjected to microwave-promoted nucleophilic displacement (see also Section 6.12) of the C4 chlorine atom by primary and secondary amines (12 examples) under mild acidic conditions in dioxane as solvent. After adding an appropriate base (potassium phosphate) and palladium catalyst (20 mol% [1,1′-bis(diphenylphosphino)ferrocene]dichloropalladium(II) [Pd(dppf)Cl$_2$]), Suzuki coupling with a variety of boronic acids (12 examples) was performed in the same reaction vessel by microwave heating at 180 °C for 10 min. A diverse set of substituents was utilized in this two-step one-pot sequence, providing the desired 4,5-disubstituted pyrazolopyrimidines in excellent overall yields. Boronic acids and amines bearing hydroxyl, amino, ketone, amide, and chloro groups were are well tolerated in this protocol [50].

An even simpler protocol for performing nucleophilic substitutions (aminations) and Suzuki reactions in one pot was reported by the Organ group for the generation of a 42-member library of styrene-based nicotinic acetylcholine receptor (nAChR) antagonists (Scheme 6.21) [49]. After considerable experimentation, the authors found that simultaneous nucleophilic displacement and Suzuki coupling could be carried out very effectively by charging the microwave process vessel with the palladium catalyst (0.5 mol% palladium-on-charcoal), the boronic acid [R^{1}B(OH)$_2$], the

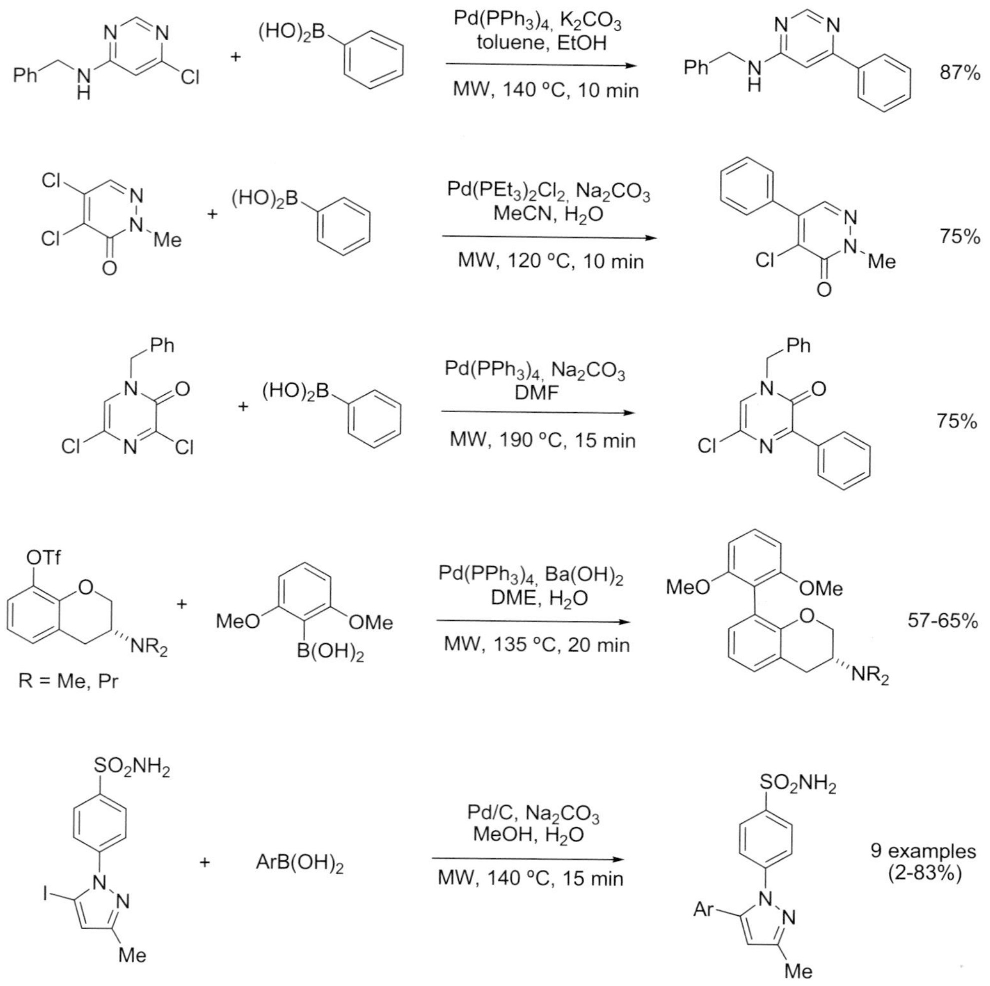

Scheme 6.19 Scaffold decoration of heterocycles using Suzuki cross-coupling chemistry.

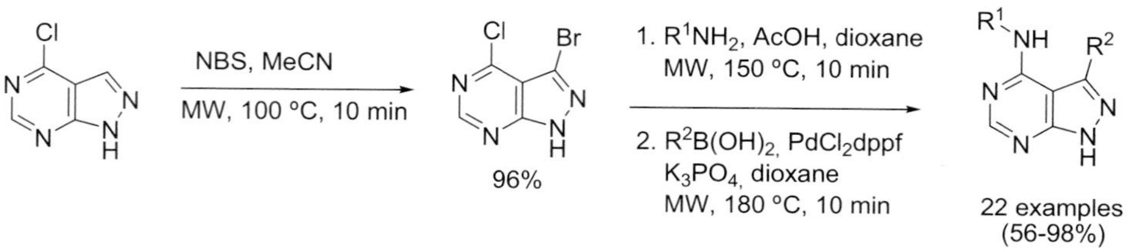

Scheme 6.20 Sequential one-pot nucleophilic aromatic substitution and Suzuki cross-coupling reactions.

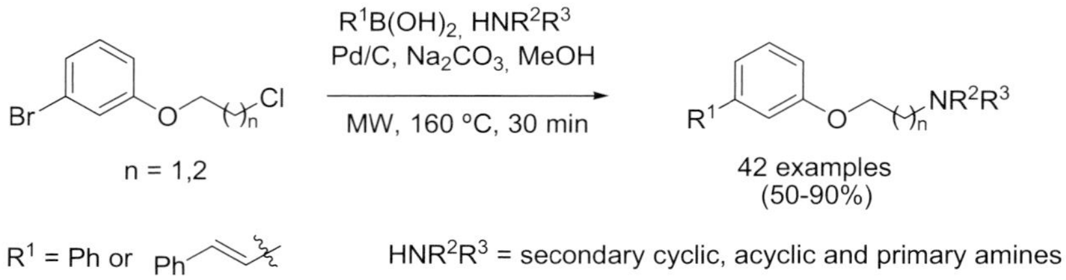

Scheme 6.21 One-pot nucleophilic substitutions and Suzuki cross-coupling reactions for the synthesis of nAChR antagonists.

amine substrate (HNR^2R^3), and an inorganic base required to drive the Suzuki reaction to completion. The desired basic target compounds were isolated as their hydrochloride salts in high yield and purity by utilizing a "catch-and-release" strategy.

The group of Larhed has reported the preparation of focused libraries of inhibitors of the malarial proteases plasmepsin I (Plm I) and plasmepsin II [51]. Using high-speed microwave Suzuki reactions, optimization of the initial lead compound used in the study led to inhibitors with greatly improved activity. A combinatorial optimization protocol utilizing an experimental design technique afforded plasmepsin inhibitors with K_i values in the low nanomolar range and with high selectivity versus the human protease cathepsin D. For the decoration of the P1' subunit, a Suzuki protocol involving 8 mol% of palladium(II)bis(triphenylphosphine) dichloride as catalyst was utilized. Although the yields remained low to moderate, enough material was generated for screening and biological evaluation. One of the most active compounds disclosed in this investigation showed a K_i value (Plm I) of 4.1 nM (Scheme 6.22) [51].

In a somewhat related study by Hallberg and coworkers, C_2-symmetric compounds based on a 1,2-dihydroxyethylene scaffold with elongated P1/P1' side chains

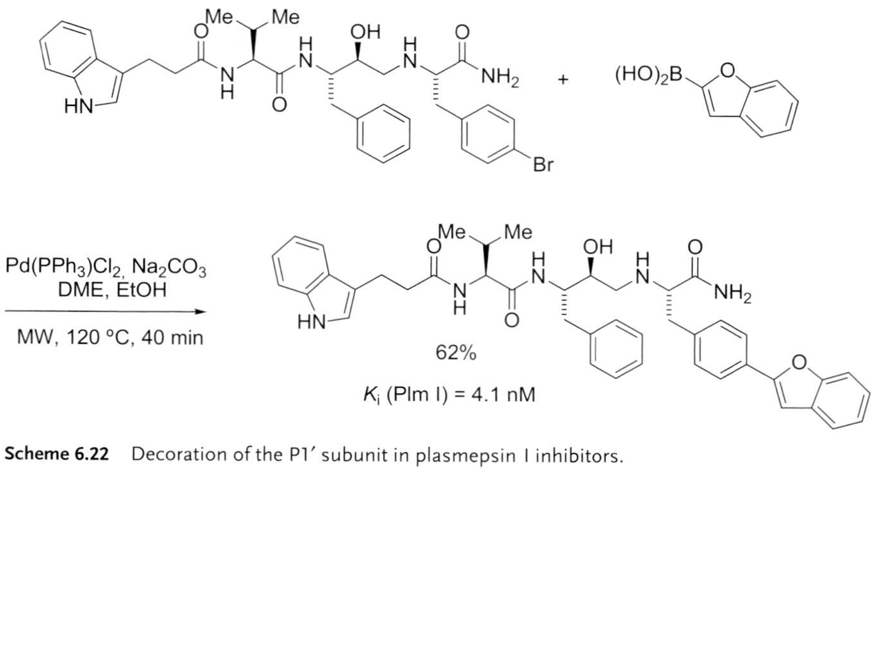

Scheme 6.22 Decoration of the P1' subunit in plasmepsin I inhibitors.

were synthesized using a series of microwave-assisted double-Suzuki, Heck, and Sonogashira coupling reactions (Scheme 6.23) [52]. The synthesized compounds exhibited picomolar to nanomolar inhibition constants for plasmepsins I and II,

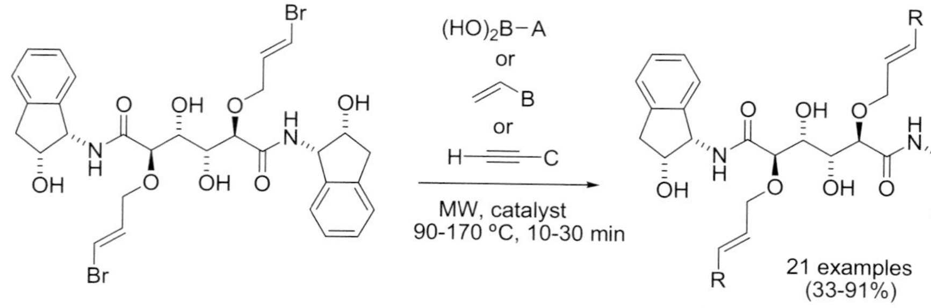

Scheme 6.23 Synthesis of plasmepsin I and II inhibitors with elongated P1/P1′ side chains.

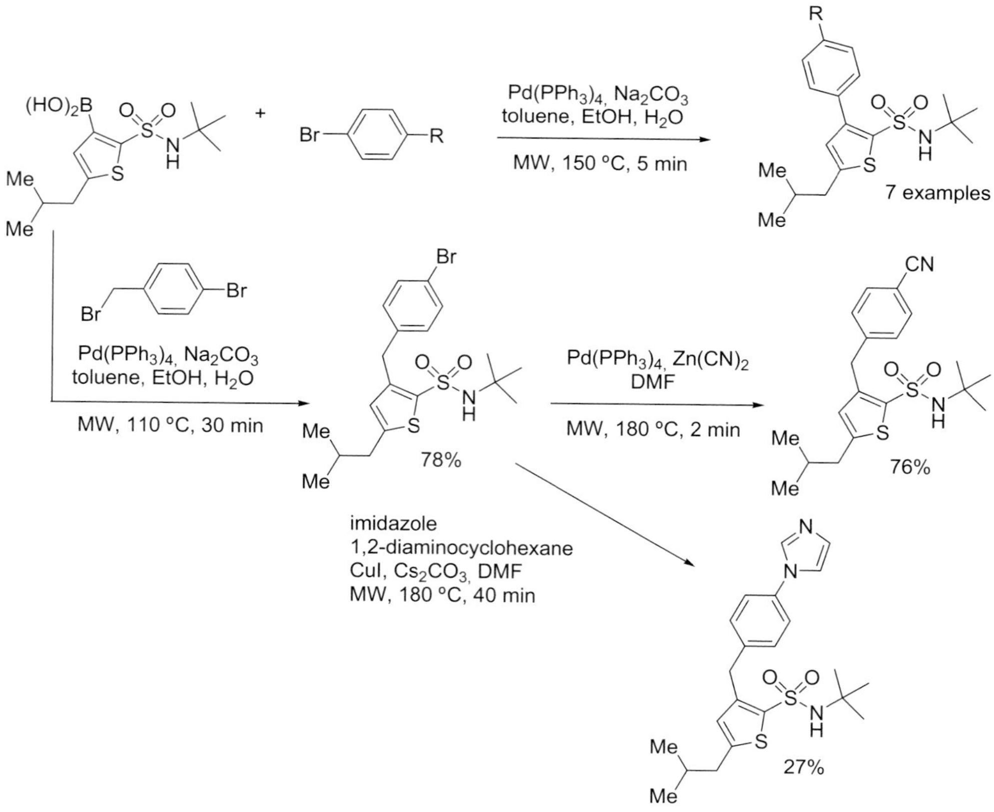

Scheme 6.24 Synthesis of selective non-peptide AT$_2$ receptor antagonists.

with no measurable affinity for the human enzyme cathespsin D. Because of their reliability and robustness, microwave-assisted Suzuki reactions are often used in medicinal chemistry projects [53, 54].

The same group of authors has recently reported a combination of various palladium- and copper-catalyzed Suzuki, cyanation, and Ullmann condensation reactions for the synthesis of thiophene-based selective angiotensin II AT_2 receptor antagonists (Scheme 6.24) [55].

A rather unusual Suzuki-type coupling involving a cyclic boronate was recently discovered by Zhou and coworkers (Scheme 6.25) [56]. Treatment of 2-methoxybiaryls with boron tribromide unexpectedly led to the formation of a 10-hydroxy-10,9-boroxarophenanthrene derivative. This formation most likely proceeds through intramolecular electrophilic aromatization of a reactive dibromoaryloxyborane intermediate. The boroxarene structures are interesting synthetic intermediates. Treatment with an aryl iodide under Suzuki-type reaction conditions (palladium catalyst, base) led to the formation of the expected carbon–carbon coupling product in high yield (Scheme 6.25) [56].

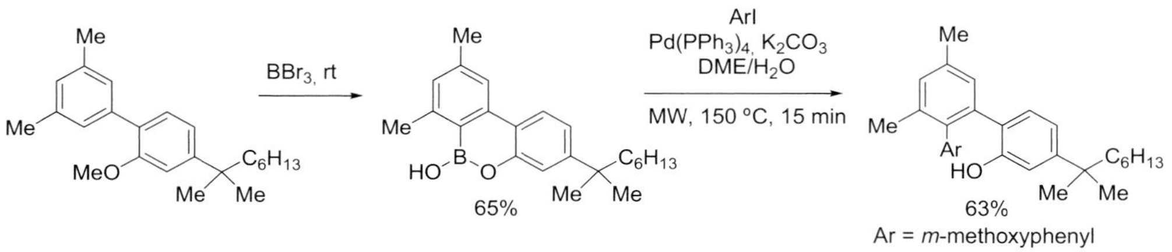

Scheme 6.25 Suzuki-type couplings involving boroxarophenanthrenes.

Apart from projects related to drug discovery, Suzuki reactions are also utilized in the synthesis of advanced functional materials, such as polymers and dyes. The Burgess group has described several organometallic couplings of fluorescein and rhodamine derivatives, including microwave-assisted Suzuki cross-coupling reactions (Scheme 6.26) [57]. The authors found that the key to successful high-yielding Suzuki couplings with this series of molecules was the use of a water-soluble palladium catalyst. A 2 mol% quantity of a sulfonic acid-derived triarylphosphine ligand was used in an acetone/water mixture (100 °C, 15 min) to couple a fluorescein-type boronic acid (prepared by microwave-assisted borylation, see Section 6.2.3) with a rosamine-derived bromide (Scheme 6.26). The product was subsequently converted to a protic form, and this was followed by counterion metathesis with the tetra[3,5-bis(trifluoromethyl)phenyl]boronate (BARF) anion to allow convenient isolation by chromatography.

The synthesis of fully conjugated semiconducting *para*-phenylene ladder polymers by microwave-assisted palladium-mediated "double" Suzuki and Stille-type reactions has been demonstrated by Scherf and coworkers (Scheme 6.27) [58]. The procedure, which yields polymeric material in ca. 10 min, has no adverse effects on the quality of the polymers and displays a high degree of reproducibility. Compared

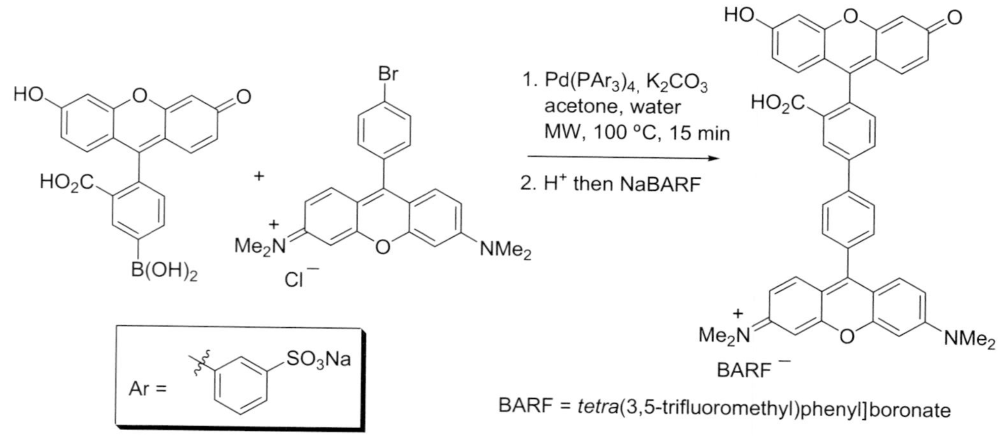

Scheme 6.26 Suzuki-type couplings for the preparation of fluorescent dyes.

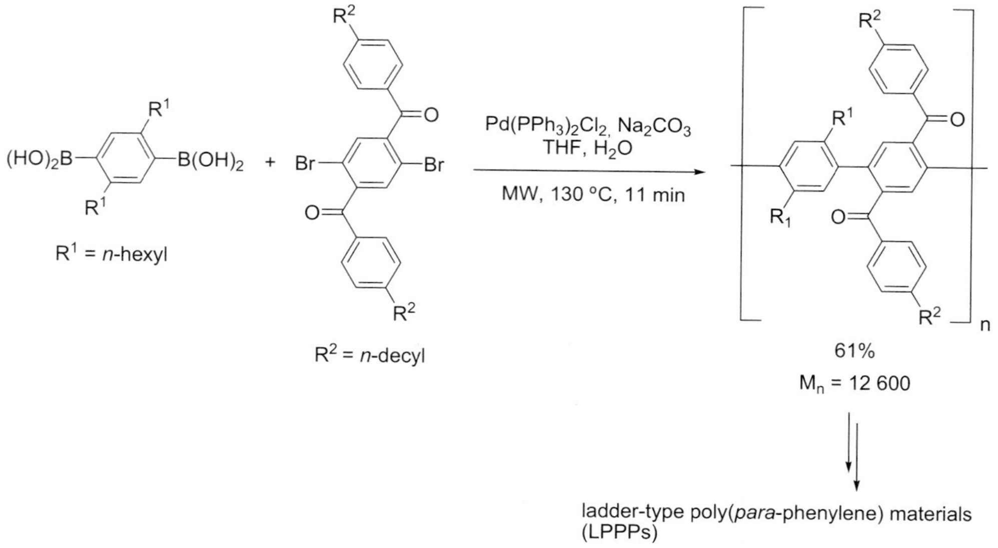

ladder-type poly(*para*-phenylene) materials
(LPPPs)

Scheme 6.27 Semiconducting polymers via Suzuki-type couplings.

to the classical thermal protocols, reaction times were reduced from 1–3 days to less than 1 h. Comparing the results achieved by conventional heating and by microwave-assisted heating, the molecular weights were generally found to be of similar magnitudes, and the ^{1}H and ^{13}C NMR data of the polymers were identical.

Apart from examples involving Suzuki reactions with standard soluble palladium catalysts, there is a growing number of publications reporting the use of immobilized, recyclable palladium catalysts for carrying out Suzuki and other cross-cou-

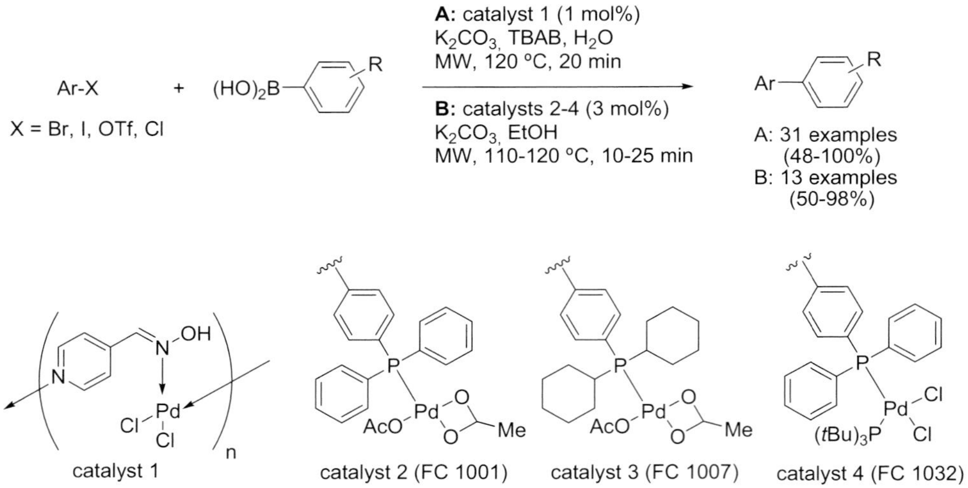

Ar-X + (HO)$_2$B–⟨ ⟩–R

X = Br, I, OTf, Cl

A: catalyst 1 (1 mol%)
K$_2$CO$_3$, TBAB, H$_2$O
MW, 120 °C, 20 min

B: catalysts 2–4 (3 mol%)
K$_2$CO$_3$, EtOH
MW, 110–120 °C, 10–25 min

Ar–⟨ ⟩–R

A: 31 examples
(48–100%)
B: 13 examples
(50–98%)

catalyst 1   catalyst 2 (FC 1001)   catalyst 3 (FC 1007)   catalyst 4 (FC 1032)

Scheme 6.28   Suzuki couplings involving immobilized palladium catalysts.

pling transformations under microwave conditions [59, 60]. Two recent examples are highlighted in Scheme 6.28. Kirschning and coworkers [59] and a team from Abbott Laboratories [60] have reported very efficient microwave-promoted Suzuki coupling reactions involving immobilized palladium catalysts. Both methods can be used to couple a range of aryl halides and triflates with boronic acids in solvents such as water or ethanol. While the Kirschning group used a novel type of solid palladium(II) pre-catalyst that can easily be prepared from 4-pyridine-aldoxime and sodium tetrachloropalladate(II) (catalyst 1), the Abbott researchers employed polyethylene-supported, so-called "FibreCat" palladium catalysts (catalysts 2–4) under similar conditions. The insoluble catalyst 1 can be recycled at least 14 times without any appreciable loss of activity when contained in a so-called IRORI® Kan [59]. Using water as solvent and tetrabutylammonium bromide (TBAB) as an additive, a small library of biaryls was prepared. Coupling was observed with aryl bromides, iodides, chlorides, and triflates. Similar coupling behavior was also observed with the FibreCat palladium catalysts [60]. Compared with the homogeneous control reactions, the supported palladium reactions were much cleaner, usually yielding a colorless solution upon their completion. Here, rapid purification of the reaction mixture was achieved by solid-phase extraction (SPE), passing the crude mixture through an SPE cartridge filled with silica carbonate, which sequestered any excess boronic acid.

Further examples of microwave-assisted Suzuki cross-couplings involving supported substrates/catalysts or fluorous-phase reaction conditions are described in Chapter 7.

Somewhat related to the Suzuki cross-coupling reaction involving organoboron reagents is the Hiyama coupling of organosilanes with halides and triflates to form unsymmetrical biaryl compounds. Seganish and DeShong have described rapid, microwave-promoted Hiyama couplings of bis(catechol) silicates with aryl bromides (Scheme 6.29) [61]. Suitable reaction conditions entailed the use of 5 mol% of tris-(dibenzylideneacetone)dipalladium(0) as a palladium source, 5 mol% of 2-(dicyclo-hexylphosphanyl)biphenyl as ligand, and 1.5 equivalents of tetrabutylammonium fluoride (TBAF) in tetrahydrofuran as the solvent. Exposure of the reaction mixture to microwave irradiation at 120 °C for 10 min typically provided good yields of biaryl coupling products for a wide range of substrates. The only functional group that was found to fail in the coupling studies was the amino group, probably as a result of catalyst poisoning.

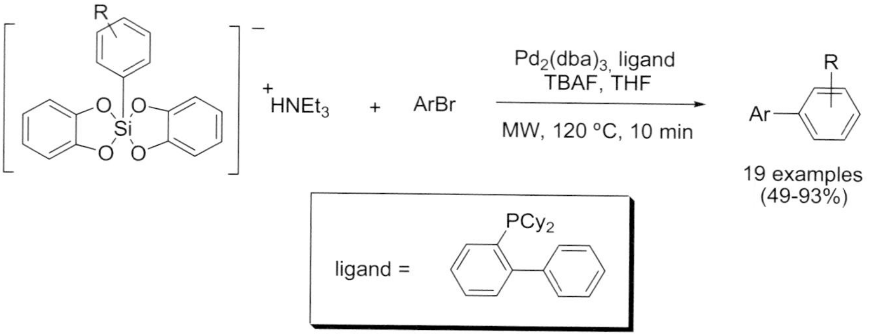

Scheme 6.29   Hiyama couplings of of bis(catechol) silicates with aryl bromides.

6.1.3
Sonogashira Reactions

The Sonogashira reaction (palladium and copper co-catalyzed coupling of terminal alkynes with aryl and vinyl halides) enjoys considerable popularity as a reliable and general method for the preparation of unsymmetrical alkynes [62]. General proto-cols for microwave-assisted Sonogashira reactions under controlled conditions were first reported in 2001 by Erdélyi and Gogoll [63]. Typical reaction conditions for the coupling of aryl iodides, bromides, chlorides, and triflates involve *N,N*-dimethyl-formamide as solvent, diethylamine as base, and palladium(II)bis(triphenylpho-sphine) dichloride (2–5 mol%) as a catalyst, with copper(I) iodide (5 mol%) as an additive [63]. In a recent publication by the group of Gogoll, such a protocol was utilized in a rapid "domino" Sonogashira sequence in order to synthesize amino esters, as outlined in Scheme 6.30 [64].

Essentially the same experimental protocol was employed by Vollhardt and co-workers in order to synthesize *ortho*-dipropynylated arenes, which served as precur-sors to tribenzocyclynes through an alkyne metathesis reaction (Scheme 6.31) [65]. Here, the Sonogashira reaction was carried out in a pre-pressurized (ca. 2.5 bar of propyne) sealed microwave vessel (see Section 4.3.2). Double Sonogashira coupling of the dibromodiiodobenzene was completed within 3.75 min at 110 °C. It is worth mentioning that the authors did not carry out the subsequent tungsten-mediated alkyne metathesis chemistry under microwave conditions to shorten the exceedingly

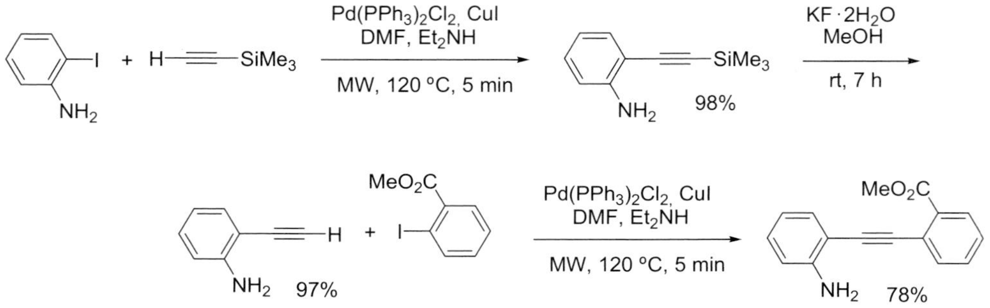

Scheme 6.30 "Domino" Sonogashira sequence for the synthesis of bis(aryl)-alkynes.

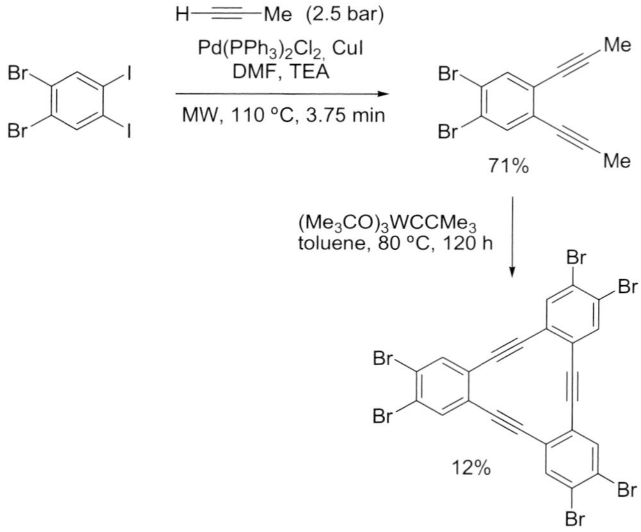

Scheme 6.31 Double Sonogashira reactions in pre-pressurized vessels.

long reaction times and perhaps to improve the low yield (see Scheme 6.72 for a microwave-assisted alkyne metathesis reaction). Related double Sonogashira chemistry has been described by Nielsen and coworkers [473].

Microwave-assisted Sonogashira protocols have also been used for the decoration or functionalization of various heterocyclic core structures. Some recent examples involving pyrazines [47], pyrimidines [66], and thiophenes [42] are shown in Scheme 6.32.

A microwave-driven Sonogashira coupling step is involved in the total synthesis of azaphilones, a structurally diverse family of natural products containing a highly oxygenated bicyclic core and a quaternary center. Porco and his colleagues have described the alkynylation of densely functionalized bromobenzaldehydes with methyl propargyl ether and propargyl cyclohexyl amide under standard Sonogashira coupling conditions (10 mol% palladium(II)bis(triphenylphosphine) dichloride, 20 mol% copper(I) iodide) (Scheme 6.33) [67]. The resulting *ortho*-alkynylbenzaldehydes were obtained in 68% and 65% isolated yield, respectively, and were subsequently converted to the desired bicyclic substrates (non-natural azaphilones) through a gold(III)-catalyzed cycloisomerization–oxidation sequence.

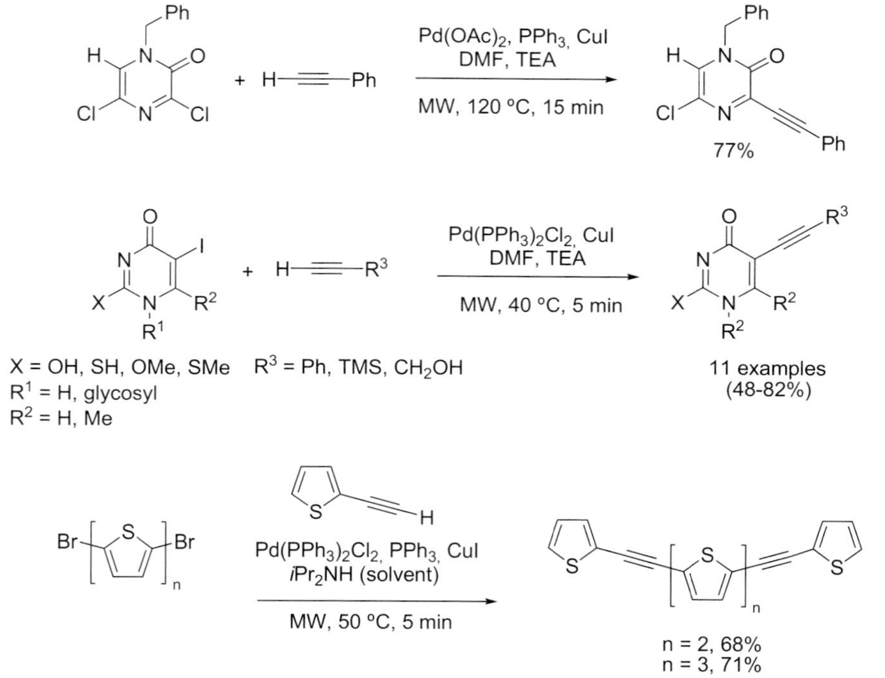

Scheme 6.32 Scaffold decoration of heterocycles using Sonogashira cross-coupling chemistry.

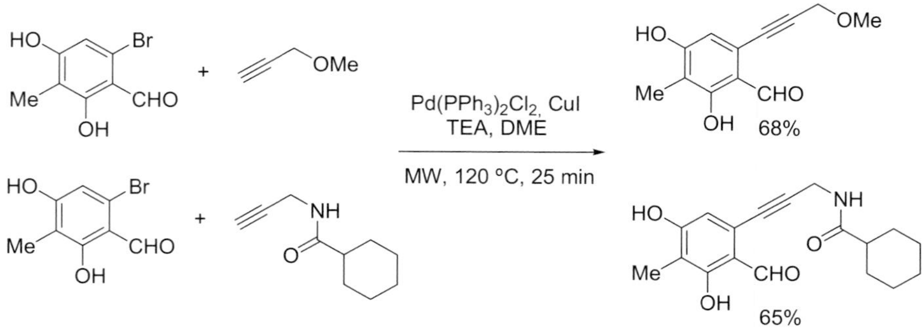

Scheme 6.33 Alkynylation of bromobenzaldehydes.

Hopkins and Collar have reported a one-pot Sonogashira/heteroannulation strategy for the synthesis of 6-substituted-5*H*-pyrrolo[2,3-*b*]pyrazines (Scheme 6.34) [68]. The reaction could either be performed by a two-step protocol, by first performing a classical Sonogashira coupling on 2-amino-3-chloropyrazine, followed by base-induced cyclization (Scheme 6.34 a), or by a one-step method by directly reacting the corresponding sulfonamide, *N*-(3-chloropyrazin-2-yl)methanesulfonamide, with terminal alkynes in the presence of a suitable palladium catalyst (3 mol%) (Scheme 6.34 b). The microwave conditions (150 °C, 20 min) tolerated much functional diversity (both electron-withdrawing and electron-donating substituents). Halogens as well as cyano groups were also tolerated, along with silyl protecting groups.

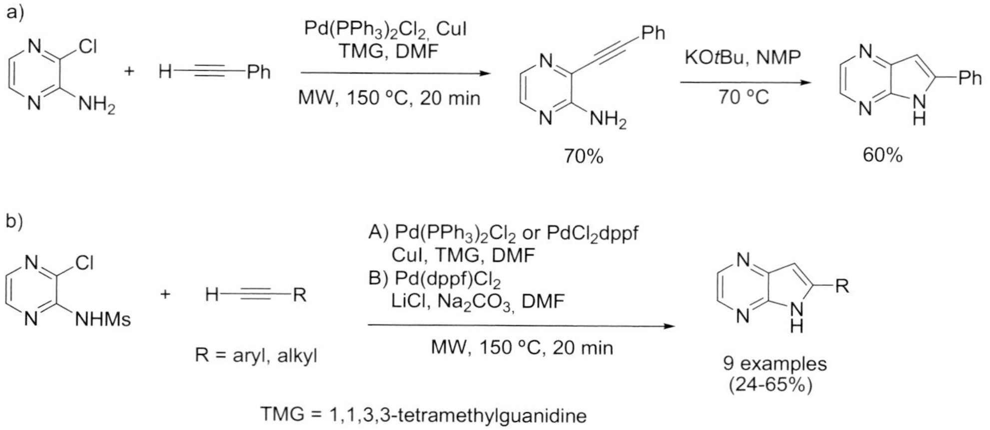

a)

70%   60%

b)

R = aryl, alkyl

9 examples
(24–65%)

TMG = 1,1,3,3-tetramethylguanidine

Scheme 6.34   Palladium-catalyzed heteroannulation.

An application of the Sonogashira reaction in supramolecular chemistry and chemosensor development is highlighted in Scheme 6.35. Swager and coworkers employed a microwave-assisted double Sonogashira–Hagihara coupling for the synthesis of rotaxanated conjugated sensory polymers [69]. Microwave irradiation of a suitable diiodo rotaxane with an aryl diacetylene under standard microwave Sonogashira coupling conditions [tetrakis(triphenylphosphine)palladium(0), copper(I) iodide] at 115 °C for 50 min provided the corresponding poly(*para*-phenylene ethynylene)s, which displayed interesting macromolecular photophysical properties. Using the microwave approach, the reaction time for the synthesis of the rotaxanated conjugated polymer was reduced from 2 days to less than 1 h.

Similarly, functionalized poly(*meta*-phenyleneethynylene)s were obtained by Khan and Hecht through polycondensation routes involving microwave-assisted Sonogashira couplings (Scheme 6.36) [70]. The rate-limiting *in situ* deprotection of the trimethylsilyl (TMS) protecting group (by water) minimized the concentration of free terminal alkyne in the polymerization mixture and therefore limited competing side reactions. Optimum conditions for the one-pot activation/coupling procedure

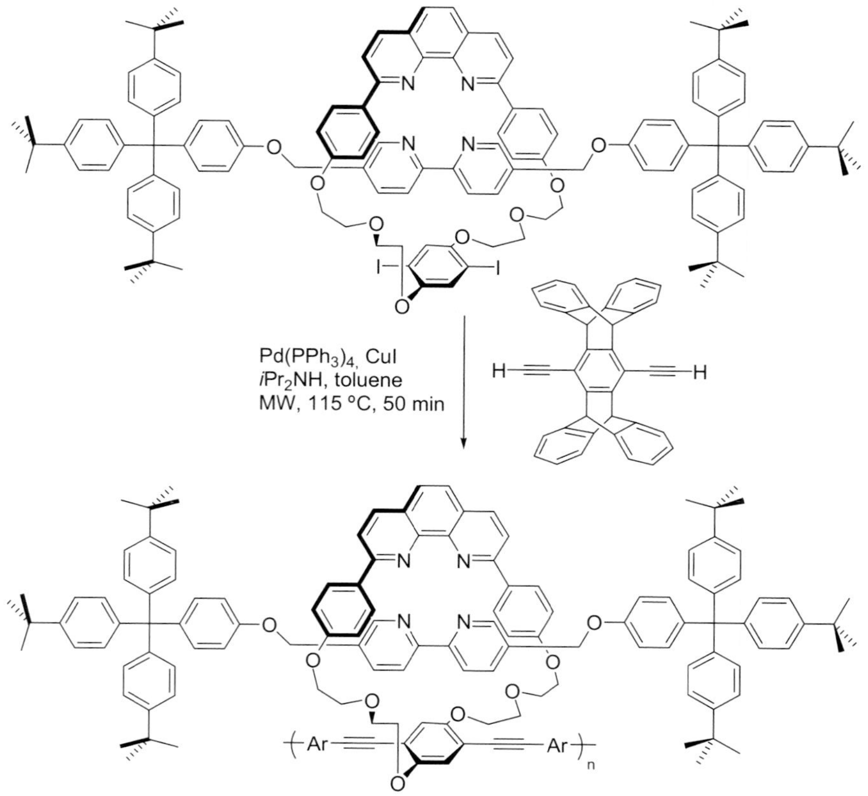

Scheme 6.35 Sonogashira coupling for the synthesis of sensory polymers.

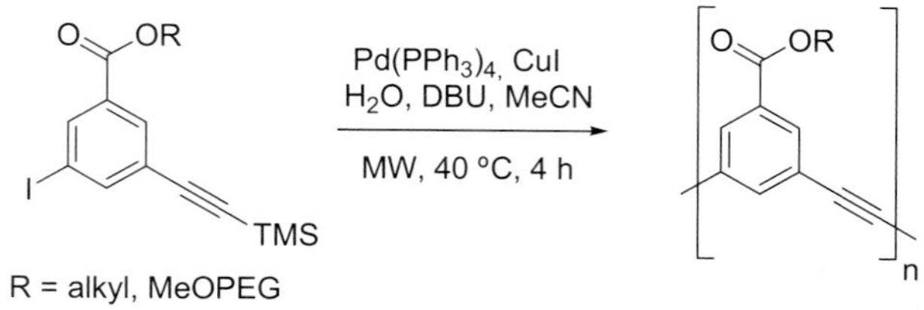

Scheme 6.36 Synthesis of lengthy and defect-free poly(meta-phenyleneethynylene)s via polycondensation.

involved the use of 6 mol% each of tetrakis(triphenylphosphine)palladium(0) and copper(I) iodide, along with 1,8-diazabicyclo[5.4.0]undec-7-ene (DBU) as a strong base. Utilizing microwave irradiation at 40 °C for 4 h, a 94% yield of diyne defect-free poly(meta-phenyleneethynylene) was obtained, showing respectable chain lengths and polydispersity.

As with the Suzuki reaction (see Scheme 6.12), there have been two independent reports by the groups of Leadbeater [71] and Van der Eycken [72] that have shown that it is also possible to perform "transition metal-free" Sonogashira couplings (Scheme 6.37 a) [468]. Again, these methods rely on the use of microwave-heated water as solvent, a phase-transfer catalyst (tetrabutylammonium bromide or polyethylene glycol), and a base (sodium hydroxide or sodium carbonate). To date, these "metal-free" procedures have proved successful for aryl bromides and iodides, and typical reaction conditions involve heating at ca. 170 °C for 5–25 min. Employing an organic solvent (1-methyl-2-pyrrolidone), He and Wu demonstrated the possibility of performing palladium-free Sonogashira couplings using 10 mol% of copper(I) iodide as a catalyst (Scheme 6.37 b) [73]. However, the coupling is only successful with aryl iodides, and relatively harsh conditions (195 °C, 2–6 h) need to be employed, which limits the practicability of this method.

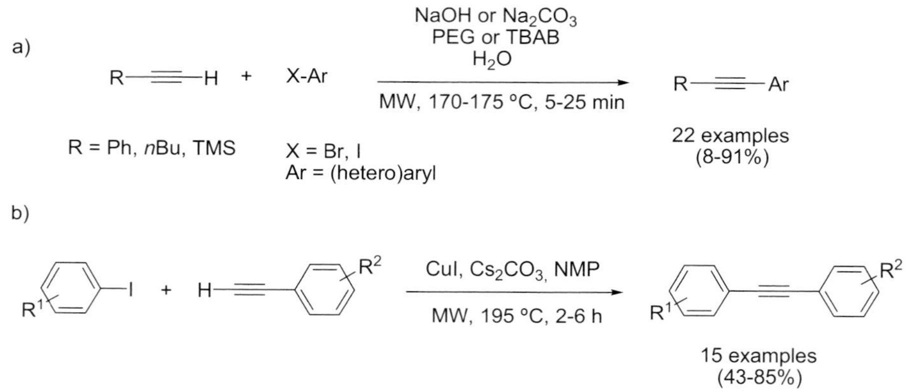

a)

R——≡——H + X-Ar $\xrightarrow[\text{MW, 170-175 °C, 5-25 min}]{\substack{\text{NaOH or Na}_2\text{CO}_3 \\ \text{PEG or TBAB} \\ \text{H}_2\text{O}}}$ R——≡——Ar

R = Ph, *n*Bu, TMS   X = Br, I   22 examples
                        Ar = (hetero)aryl   (8-91%)

b)

R^1—⟨ ⟩—I + H—≡—⟨ ⟩—R^2 $\xrightarrow[\text{MW, 195 °C, 2-6 h}]{\text{CuI, Cs}_2\text{CO}_3, \text{NMP}}$ R^1—⟨ ⟩—≡—⟨ ⟩—R^2

                                                            15 examples
                                                            (43-85%)

Scheme 6.37   (a) Transition metal-free, (b) copper(I)-catalyzed Sonogashira-type couplings.

6.1.4
Stille Reactions

Compared to the previously described transition metal-catalyzed transformations in this chapter, microwave-assisted Stille reactions [74] involving organotin reagents as coupling partners are comparatively rare. A few examples describing both inter- and intramolecular Stille reactions in heterocyclic systems are summarized in Scheme 6.38 [47, 75–77]. Additional examples involving fluorous Stille reactions are described in Section 7.3.

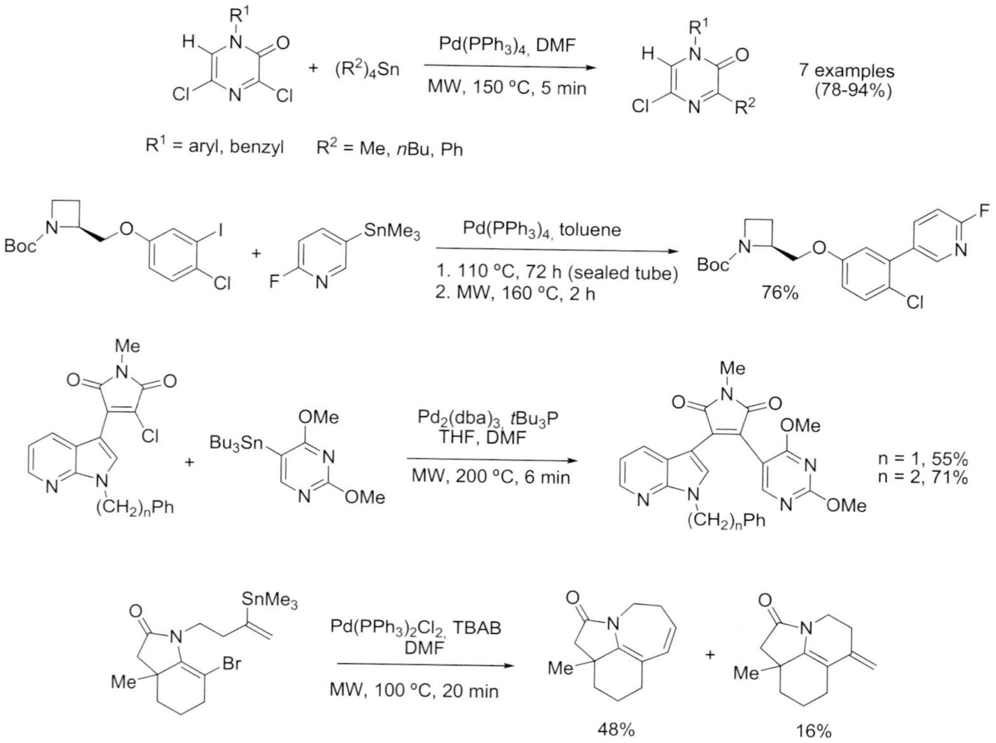

Scheme 6.38 Scaffold decoration and modification of heterocycles using Stille cross-coupling chemistry.

6.1.5
Negishi, Kumada, and Related Reactions

Until recently, very little work had been published on Negishi (organozinc reagents) [78] and Kumada (organomagnesium reagents) [79] cross-coupling chemistry under microwave conditions. There are two examples in the peer-reviewed literature describing Negishi cross-coupling reactions of activated aryl bromides [80] and heteroaryl chlorides [81] with organozinc halides (Scheme 6.39). Öhberg and Westman [80] have reported that microwave-assisted Negishi couplings of activated aryl bromides with arylzinc bromides can be performed within 1 min of microwave heating at 160 °C using 5 mol% of palladium(II)bis(triphenylphosphine) dichloride as the catalyst (Scheme 6.39 a). Importantly, this protocol, initially carried out on a less than 1 mmol scale, was directly scalable to 40 mmol in a larger multimode batch reactor without re-optimization of the reaction conditions (see also Section 4.5) [16]. On the other hand, the group of Stanetty has demonstrated that activated heteroaryl chlorides can be coupled with heteroarylzinc iodides or chlorides under similar con-

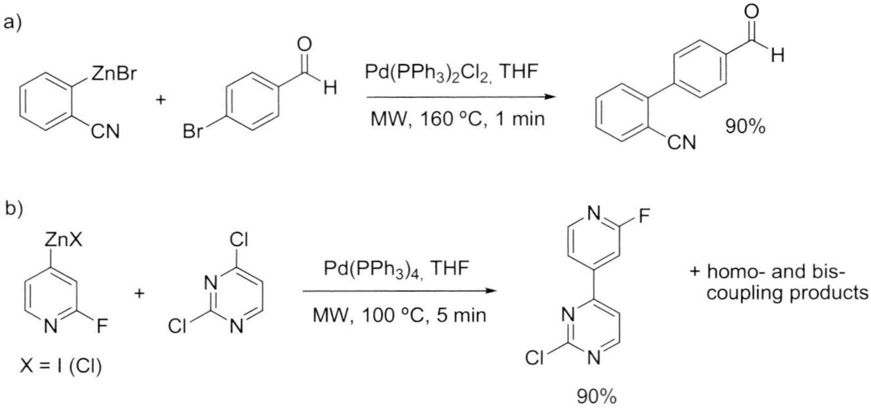

a)

b)

Scheme 6.39 Aryl and heteroaryl cross-coupling using Negishi chemistry.

ditions to furnish pyridinyl-pyrimidines in high yields (Scheme 6.39 b) [81]. Optimized conditions utilized 0.5 mol% tetrakis(triphenylphosphine)palladium(0) as the catalyst and tetrahydrofuran as solvent. The highest selectivity in favor of the desired 4-pyrimidinyl coupling product was obtained at a reaction temperature of 100 °C, with only small amounts of homo- and bis-coupling products being formed as by-products.

A more general procedure for high-speed microwave-assisted Negishi and Kumada couplings of unactivated aryl chlorides has recently been reported by Walla and Kappe (Scheme 6.40) [82]. This procedure uses 0.015–2.5 mol% of tris(dibenzyl-

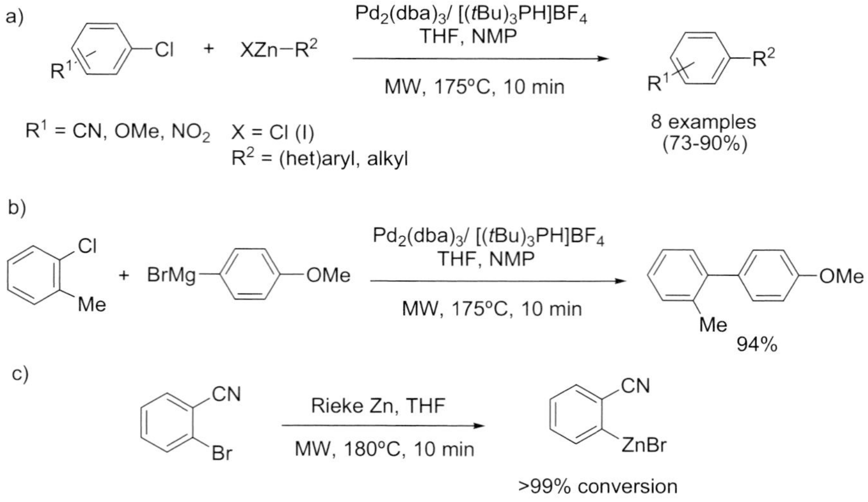

a)

b)

c)

Scheme 6.40 Negishi and Kumada cross-coupling reactions.

ideneacetone)dipalladium(0) as a palladium source and the air-stable tri-*tert*-butyl-phosphonium tetrafluoroborate as ligand precursor. Successful couplings were observed for both arylorganozinc chlorides and iodides (Scheme 6.40 a). Using this methodology, it was also possible to successfully couple aryl chlorides with *alkyl*zinc reagents such as *n*-butylzinc chloride in a very rapid manner without the need for an inert atmosphere. Optimized conditions utilized sealed-vessel microwave irradiation in a THF/1-methyl-2-pyrrolidone mixture at 175 °C for 10 min. Applying the same reaction conditions, Kumada cross-couplings with organomagnesium (Grignard) reagents were also successfully accomplished (Scheme 6.40 b) [82]. In the same article, the authors also described rapid microwave-assisted methods for the preparation of the corresponding organozinc and -magnesium compounds by insertion of Rieke zinc or magnesium metal, respectively, into aryl halides (Scheme 6.40 c) [82].

A similar protocol for high-speed Negishi cross-coupling transformations has been developed by Mutule and Suna (Scheme 6.41) [83]. Here, the required organozinc reagents were readily prepared from aryl iodides using freshly prepared zinc–copper couple. Zinc insertion occurred readily in polar solvents such as *N,N*-dimethylformamide using a six-fold excess of the zinc–copper couple under microwave conditions at 80–125 °C, within 2–15 min depending on the reactivity of the aryl iodide. Subsequent Negishi cross-coupling with 4-bromobenzaldehyde was accomplished by adding the organozinc reagent (supernatant solution after centrifugation) to a solution of the bromide in *N,N*-dimethylformamide in the presence of 3 mol% of palladium(II)bis(triphenylphosphine) dichloride (microwave heating, 100–120 °C, 5–15 min) [83].

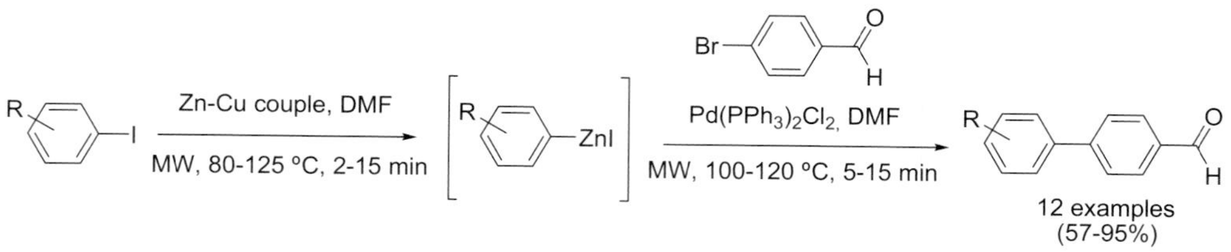

Scheme 6.41 Sequential arylzinc formation and Negishi cross-coupling.

Application of this high-speed Negishi coupling methodology for the preparation of enantiopure 2,2′-diarylated 1,1′-binaphthyls was reported by the Kappe and Putala groups (Scheme 6.42) [84]. Reaction times as short as 40 s in certain cases were sufficient to achieve complete conversions in the stereoconservative Negishi coupling of commercially available 2,2′-diiodo-1,1′-binaphthyl (DIBN) with arylzinc chlorides. The catalyst loading for typical runs was 5 mol% of tetrakis(triphenylphosphine)palladium(0), but could be lowered to 0.5 mol% in some instances without appreciable reduction of the coupling efficiency. The enantiopure 2,2′-diarylated 1,1′-binaphthyls thus formed are of interest in material science applications. In the same article, corresponding Negishi alkynations using zinc phenylacetylide and zinc trimethylsilyl-acetylide were also described [84].

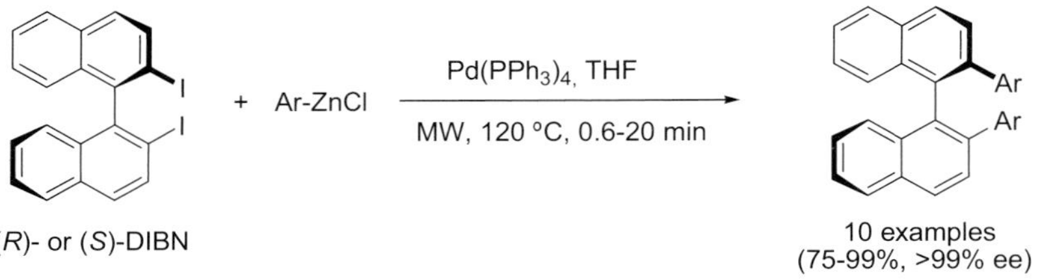

Scheme 6.42 Synthesis of enantiopure 2,2′-diarylated 1,1′-binaphthyls utilizing stereoconservative Negishi cross-coupling reactions.

Through chemistry related to the Negishi reaction, Moloney and coworkers have reported the palladium-catalyzed α-arylation of esters and amides with organozinc reagents in a one-pot fashion (Scheme 6.43) [85]. The required Reformatsky reagents were readily prepared by microwave irradiation of the corresponding bromoacetate or bromoacetamide together with zinc metal (2 equivalents) in THF for 5 min at 100 °C. Addition of this Reformatsky reagent to the coupling partner, an aryl bromide, and the requisite catalyst/ligand in tetrahydrofuran followed by further irradiation for 5–10 min at 120 °C provided the arylacetic esters or amides in good yields. The best results were achieved using 10 mol% of tetrakis(triphenylphosphine)palladium(0) as palladium catalyst.

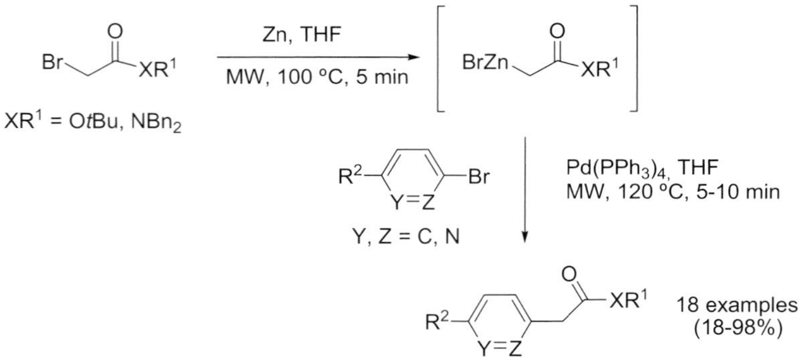

Scheme 6.43 Palladium-catalyzed α-arylation of esters and amides with organozinc reagents.

In addition to the classical Negishi cross-coupling utilizing organozinc reagents, the "zirconium version" involving the coupling of zirconocenes with aryl halides has also been described by the application of sealed-vessel microwave technology. Lipshutz and Frieman have reported the rapid coupling of both vinyl and alkyl zirconocenes (prepared *in situ* by hydrozirconation of alkynes and alkenes, respectively) with aryl iodides, bromides, and chlorides (Scheme 6.44) [86]. While aryl iodides required only 5 mol% of nickel-on-carbon (Ni/C) as a ligand-free heterogeneous catalytic system, the presence of triphenylphosphine as a ligand was necessary for the successful coupling of aryl bromides (10 mol%) and chlorides (20 mol% ligand).

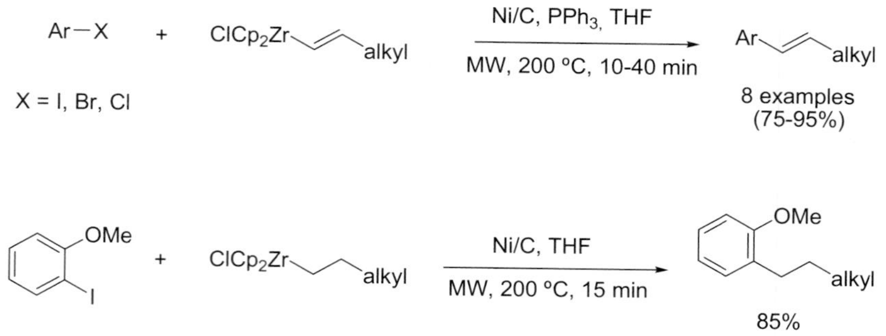

Scheme 6.44 Nickel-catalyzed cross-coupling of zirconocenes and aryl halides.

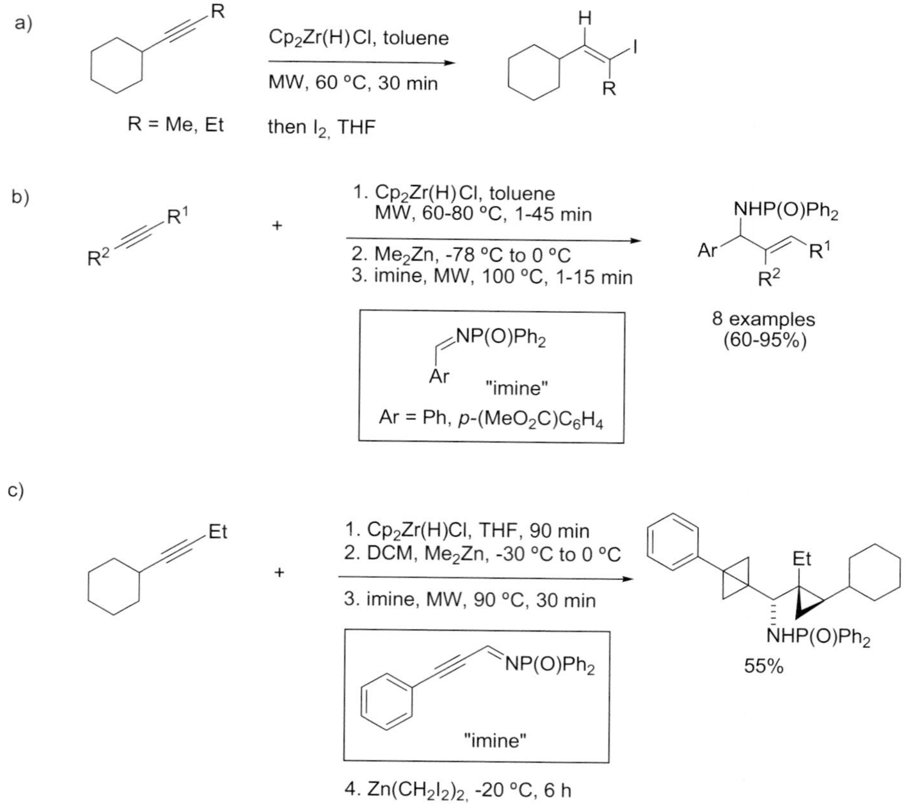

Scheme 6.45 Hydrozirconations and zirconocene-imine additions.

Under these conditions, complete conversion was achieved within 10–40 min at 200 °C using tetrahydrofuran as solvent. The same group subsequently demonstrated that the triphenylphosphine ligands used in the reactions shown in Scheme 6.44 can be conveniently sequestered and subsequently recovered by precipitating them with copper(I) chloride [87]. Independently, a publication from the Fu laboratory has described a similar high-yielding palladium-catalyzed process, in which alkyl bromides are coupled with alkenylzirconium reagents under ligand-free conditions (2.5 mol% palladium(II) acetylacetonate, 2 equivalents of lithium bromide, tetrahydrofuran/1-methyl-2-pyrrolidone, 100 °C, 15 min) [88].

Hydrozirconation is a mild method for the selective preparation of functionalized organometallics and its compatibility with a range of common protecting groups represents a considerable advantage of these species over more traditional organometallic reagents [89]. Wipf and coworkers have recently reported that the hydrozirconation of alkynes with zirconocene hydrochloride can be greatly accelerated by microwave irradiation, as illustrated by the examples depicted in Scheme 6.45 a [90]. Under optimum conditions, hydrozirconation of unsymmetrical internal alkynes was accomplished in 30 min with 2 equivalents of $Cp_2Zr(H)Cl$ at 60 °C. In this context, a synthetically useful one-pot method for the preparation of allylic amides was elaborated, in which an alkyne was first hydrozirconated by microwave irradiation, and this was followed by rapid addition of imines in the presence of dimethylzinc (Scheme 6.45 b) [90]. The synthetically useful allylic amides were isolated in good to excellent overall yields (60–95%). A further extension of this protocol involved the synthesis of *C,C*-dicyclopropylmethylamines through multicomponent condensation of alkynes, imines, and zinc carbenoids [91]. For the example shown in Scheme 6.45 c, the alkynyl imine–zirconocene addition step was accelerated by microwave heating at 90 °C.

6.1.6
Carbonylation Reactions

Taking advantage of the rapid and controlled heating made possible by microwave irradiation of solvents under sealed-vessel conditions, the group of Larhed has reported a number of valuable palladium-catalyzed carbonylation reactions [92–100]. The key feature of many of these protocols is the use of molybdenum hexacarbonyl as a solid precursor of carbon monoxide, required in carbonylation chemistry, utilizing aryl halides (bromides and iodides) as starting materials. At 150 °C, molybdenum hexacarbonyl liberates enough carbon monoxide for rapid *in situ* aminocarbonylation reactions to take place (carbon monoxide is liberated instantaneously at 210 °C). The initially reported conditions involved the use of a combination of Herrmann's palladacycle (7.4 mol% palladium) and 2,2'-bis(diphenylphosphino)-1,1'-binaphthyl (BINAP) as the catalytic system in a diglyme/water mixture, providing the desired secondary and tertiary amides in high yields (Scheme 6.46 a) [92]. As in many other cases, an inert atmosphere was not required. Subsequent experimental improvements of the protocol allowed the use of sterically and electronically more demanding amines (for example, anilines or unprotected amino acids) by using 1,8-

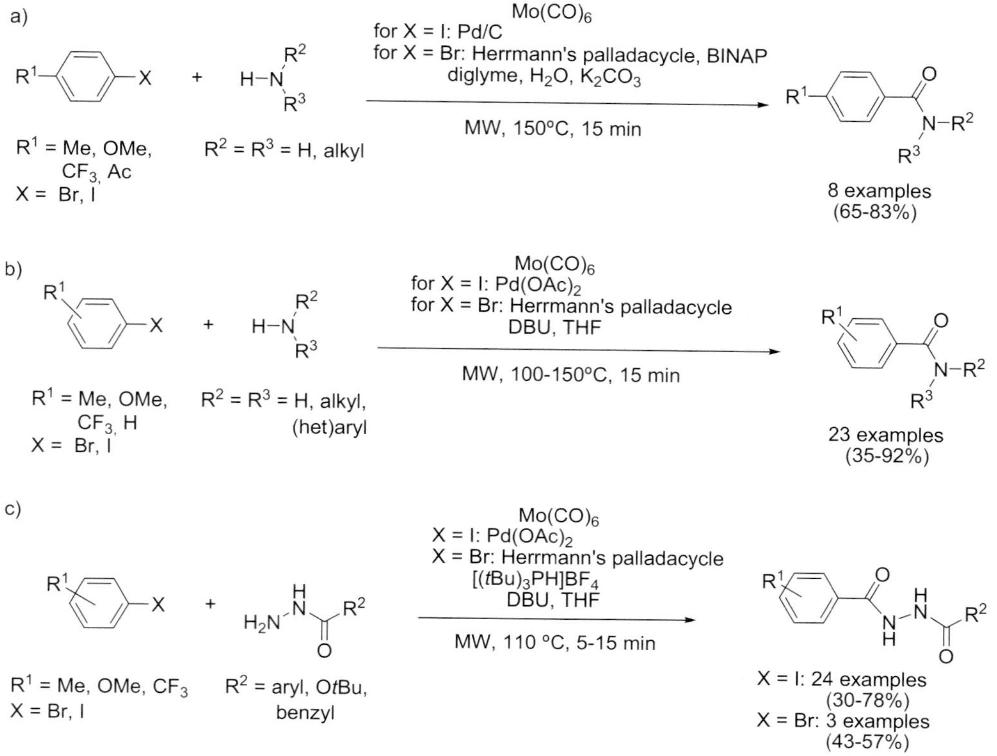

a)

R^1 = Me, OMe,
CF$_3$, Ac
X = Br, I

R^2 = R^3 = H, alkyl

Mo(CO)$_6$
for X = I: Pd/C
for X = Br: Herrmann's palladacycle, BINAP
diglyme, H$_2$O, K$_2$CO$_3$

MW, 150°C, 15 min

8 examples
(65-83%)

b)

R^1 = Me, OMe,
CF$_3$, H
X = Br, I

R^2 = R^3 = H, alkyl,
(het)aryl

Mo(CO)$_6$
for X = I: Pd(OAc)$_2$
for X = Br: Herrmann's palladacycle
DBU, THF

MW, 100-150°C, 15 min

23 examples
(35-92%)

c)

R^1 = Me, OMe, CF$_3$
X = Br, I

R^2 = aryl, OtBu,
benzyl

Mo(CO)$_6$
X = I: Pd(OAc)$_2$
X = Br: Herrmann's palladacycle
[(tBu)$_3$PH]BF$_4$
DBU, THF

MW, 110 °C, 5-15 min

X = I: 24 examples
(30-78%)
X = Br: 3 examples
(43-57%)

Scheme 6.46 Palladium-catalyzed aminocarbonylations using molybdenum hexacarbonyl as a solid source of carbon monoxide.

diazabicyclo[5.4.0]undec-7-ene (DBU) as the base and tetrahydrofuran as solvent for both aryl bromides and iodides (Scheme 6.46 b) [93]. The protocol could also be extended to hydrazidocarbonylations employing similar reaction conditions (Scheme 6.46 c) [94]. A fluorous version of this last method is discussed in relation to Scheme 7.87.

By simple modifications of the general strategy outlined in Scheme 6.46, the corresponding carboxylic acids [92] and esters [95] could be obtained instead of the amides, using water or alcohols as reaction partners instead of amines (Scheme 6.47). The ester-forming carbonylation was also exploited in a tandem carbonylation–lactone formation reaction sequence for the synthesis of phthalides [96]. Here again, optimum conditions involved the use of molybdenum hexacarbonyl as a solid source of carbon monoxide and palladium acetate/1,1′-bis(diphenylphosphino)ferrocene as a catalyst (5 mol%) at 180 °C. This microwave-assisted carbonylation–cyclization method was also applied for the synthesis of other scaffolds, such as dihydroisocoumarins, dihydroisoindones, and phthalimides (not shown) [96], and could also be applied to the preparation of indanones [97]. Further modifications by Alterman and coworkers have resulted in the use of *N,N*-dimethylformamide as a

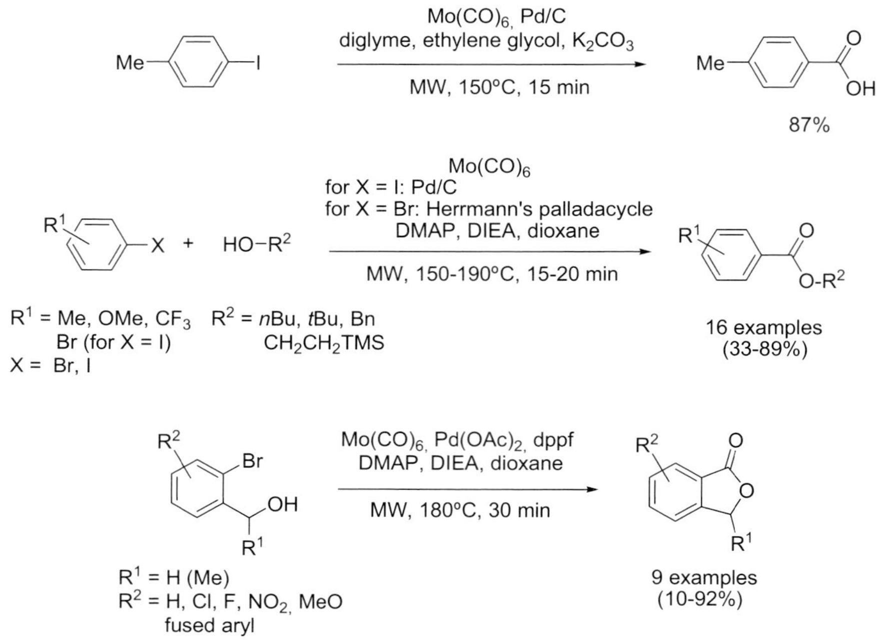

Mo(CO)$_6$, Pd/C
diglyme, ethylene glycol, K$_2$CO$_3$

MW, 150°C, 15 min

Me—⟨C$_6$H$_4$⟩—I → Me—⟨C$_6$H$_4$⟩—C(=O)OH

87%

Mo(CO)$_6$
for X = I: Pd/C
for X = Br: Herrmann's palladacycle
DMAP, DIEA, dioxane

MW, 150-190°C, 15-20 min

R^1—⟨aryl⟩—X + HO–R^2 → R^1—⟨aryl⟩—C(=O)O-R^2

R^1 = Me, OMe, CF$_3$ R^2 = nBu, tBu, Bn
    Br (for X = I)   CH$_2$CH$_2$TMS
X = Br, I

16 examples
(33-89%)

Mo(CO)$_6$, Pd(OAc)$_2$, dppf
DMAP, DIEA, dioxane

MW, 180°C, 30 min

R^1 = H (Me)
R^2 = H, Cl, F, NO$_2$, MeO
    fused aryl

9 examples
(10-92%)

Scheme 6.47 Palladium-catalyzed carbonylation reactions yielding acids, esters, and lactones using molybdenum hexacarbonyl as a solid source of carbon monoxide.

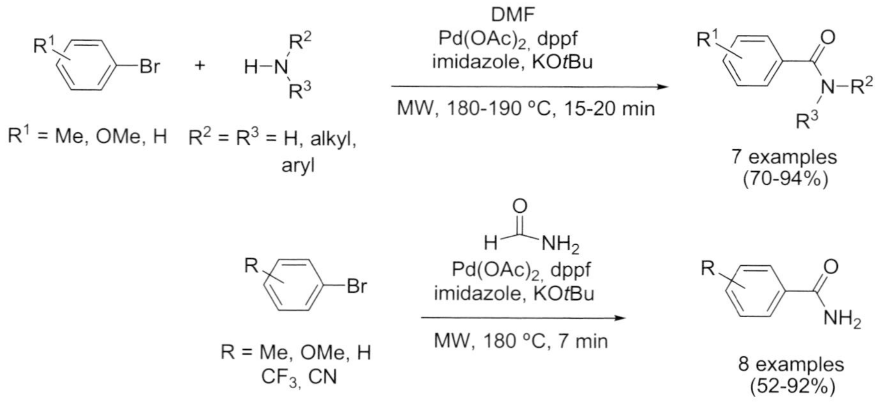

DMF
Pd(OAc)$_2$, dppf
imidazole, KOtBu

MW, 180-190 °C, 15-20 min

R^1—⟨aryl⟩—Br + H–N(R^2)(R^3) → R^1—⟨aryl⟩—C(=O)N(R^3)-R^2

R^1 = Me, OMe, H   R^2 = R^3 = H, alkyl,
                          aryl

7 examples
(70-94%)

H–C(=O)–NH$_2$
Pd(OAc)$_2$, dppf
imidazole, KOtBu

MW, 180 °C, 7 min

R—⟨aryl⟩—Br → R—⟨aryl⟩—C(=O)NH$_2$

R = Me, OMe, H
    CF$_3$, CN

8 examples
(52-92%)

Scheme 6.48 Palladium-catalyzed aminocarbonylations using formamides as sources of carbon monoxide.

source of carbon monoxide [98] and in the use of formamide as a combined source of both ammonia and carbon monoxide (Scheme 6.48) [99]. The latter method is useful for the preparation of primary aromatic amides from aryl bromides. In both cases, strong bases and temperatures of around 180 °C (7–20 min) have to be used to mediate the reaction.

A somewhat related process, the cobalt-mediated synthesis of symmetrical benzo-phenones from aryl iodides and dicobalt octacarbonyl, is shown in Scheme 6.49 [100]. Here, dicobalt octacarbonyl is used as a combined Ar–I bond activator and carbon monoxide source. Employing acetonitrile as solvent, a variety of aryl iodides with different steric and electronic properties underwent the carbonylative coupling in excellent yields. Remarkably, in several cases, microwave irradiation for just 6 s was sufficient to achieve full conversion! An inert atmosphere, a base or other additives were all unnecessary. No conversion occurred in the absence of heat-ing, regardless of the reaction time. However, equally high yields could be achieved by heating the reaction mixture in an oil bath for 2 min.

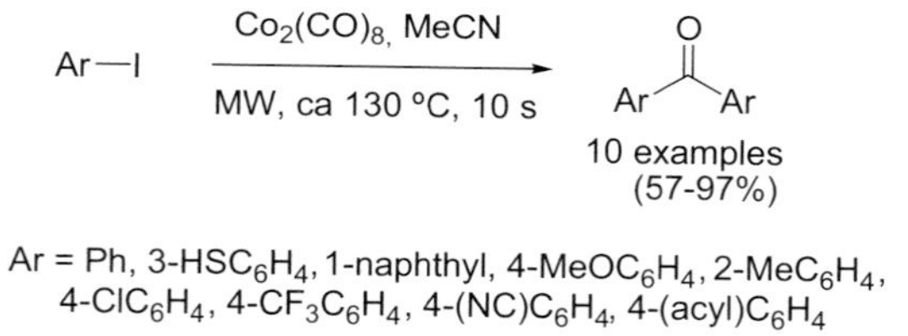

Ar—I $\xrightarrow[\text{MW, ca 130 °C, 10 s}]{\text{Co}_2(\text{CO})_8,\ \text{MeCN}}$ Ar—C(=O)—Ar

10 examples (57-97%)

Ar = Ph, 3-HSC$_6$H$_4$, 1-naphthyl, 4-MeOC$_6$H$_4$, 2-MeC$_6$H$_4$, 4-ClC$_6$H$_4$, 4-CF$_3$C$_6$H$_4$, 4-(NC)C$_6$H$_4$, 4-(acyl)C$_6$H$_4$

Scheme 6.49 Cobalt carbonyl-mediated synthesis of diaryl ketones.

6.1.7
Asymmetric Allylic Alkylations

A frequent criticism of microwave synthesis has been that the typically high reaction temperatures will invariably lead to reduced selectivities. This is perhaps the reason why comparatively few enantioselective processes driven by microwave heating have been reported in the literature. In order for a reaction to occur with high enantios-electivity, there must be a sufficiently large difference in the activation energies for the processes leading to the two enantiomers. The higher the reaction temperature, the larger the difference in energy required to achieve high selectivity. High enan-tioselectivity is observed for processes having large differences in the $\Delta G^{\ddagger}$ values for the reactions that give the two enantiomers. Provided that the selectivity-determin-ing step remains the same, the selectivity decreases with increasing temperature, as is apparent from the relationship $E = \exp(-\Delta\Delta G^{\ddagger}/RT)$ (E is the enantiomeric ratio as originally defined for kinetic resolutions that involve enzymatic systems) [101]. Further, if the selectivity does not change during the catalytic reaction, E correlates with *ee* ($E = (ee - 1)/(ee + 1)$). Plots of *ee* as a function of $\Delta\Delta G^{\ddagger}$ at 0, 100, and 200 °C (Fig. 6.1) show that, for highly selective processes (*ee* > 99% at 0 °C), heating can be

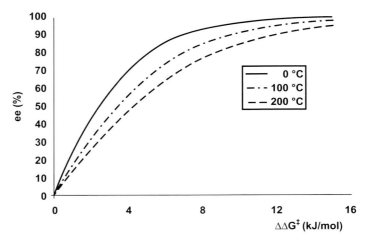

Fig. 6.1 Enantiomeric excess as a function of the difference in activation energy ($\Delta\Delta G^{\ddagger}$) for an enantioselective process at different reaction temperatures. Reproduced with permission from [101].

employed without a deterioration in selectivity, provided that the catalyst is not degraded and the mechanism of the reaction remains unchanged.

Despite these limitations, a number of very impressive enantioselective reactions involving chiral transition metal complexes have been described in the literature. As long ago as 2000, the groups of Moberg, Hallberg, and Larhed reported on microwave-mediated palladium- [102, 103] and molybdenum-catalyzed [104–106] asymmetric allylic alkylation reactions involving neutral carbon, nitrogen, and oxygen nucleophiles. Both processes, carried out under non-inert conditions, yielded the desired products in high chemical yield with typically >98% *ee* (Scheme 6.50).

More recently, Trost and Andersen have applied this concept in their approach to the orally bioavailable HIV inhibitor Tipranavir (Scheme 6.51) [107]. The key C-3α chiral intermediate was synthesized by asymmetric allylic alkylation starting from the corresponding carbonate. Employing 10 mol% of the molybdenum pre-catalyst and 15 mol% of a suitable chiral ligand along with 2 equivalents of dimethyl sodiomalonate as an additive, a 94% product yield was achieved. The reaction was carried out by sealed-vessel microwave heating at 180 °C for 20 min. Thermal heating under reflux conditions (67 °C) required 24 h, although it produced the same chemical yield of the intermediate with slightly higher enantiomeric purity (96% *ee*).

A similar pathway involving a microwave-driven molybdenum-catalyzed asymmetric allylic alkylation as the key step was elaborated by Moberg and coworkers for the preparation of the muscle relaxant (*R*)-baclofen (Scheme 6.52) [108]. The racemic form of baclofen is used as a muscle relaxant (antispasmodic) lipophilic derivative of γ-aminobutyric acid (GABA). Pharmacological studies have shown that the (*R*)-enantiomer is the therapeutically useful agonist of the GABA$_B$ receptor. Asymmetric alkylation of the allylic carbonate precursor with dimethyl malonate afforded

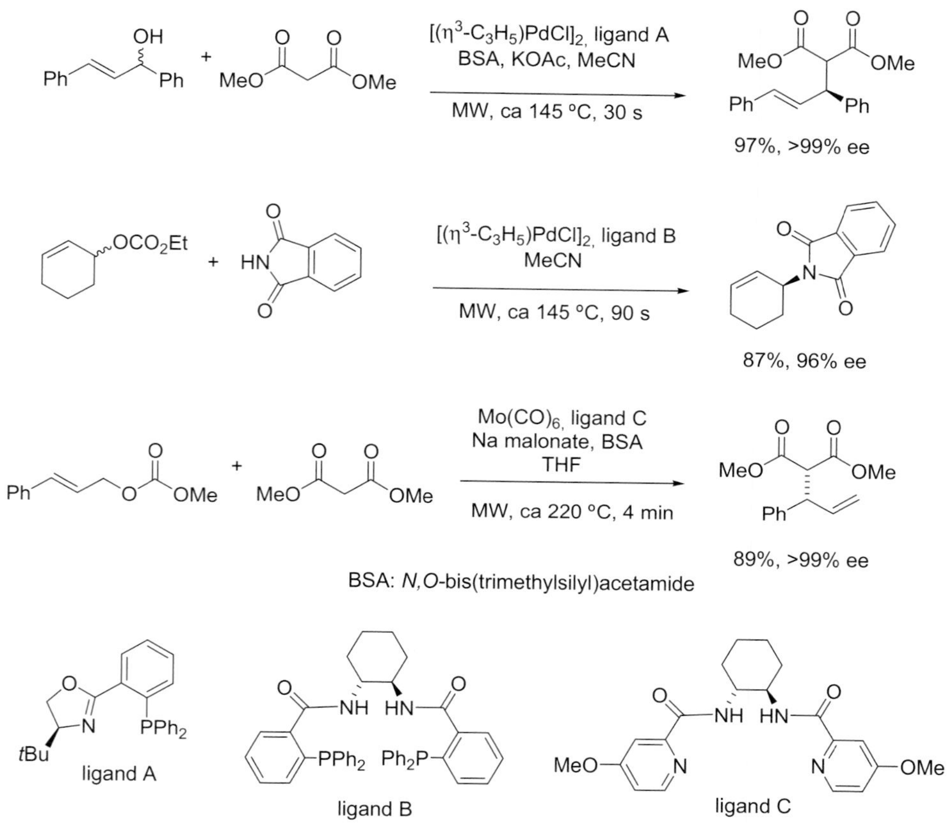

Scheme 6.50 Palladium- and molybdenum-catalyzed asymmetric allylic alkylations.

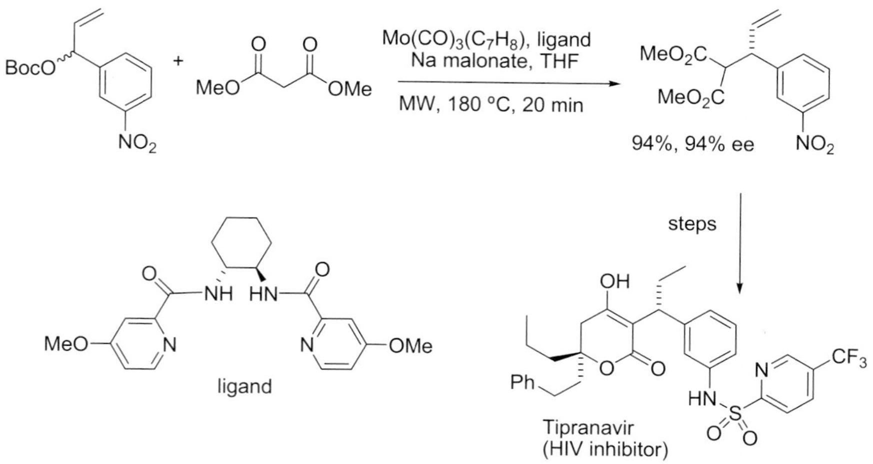

Scheme 6.51 Molybdenum-catalyzed asymmetric allylic alkylation in the synthesis of Tipranavir.

the desired chiral intermediate in high chemical yield and with 96% ee. The reaction was performed in THF as solvent using 4 mol% of molybdenum hexacarbonyl as pre-catalyst at 160 °C. A somewhat lower enantiomeric purity (89% ee) was obtained when an immobilized version of the ligand was employed (see Scheme 7.118).

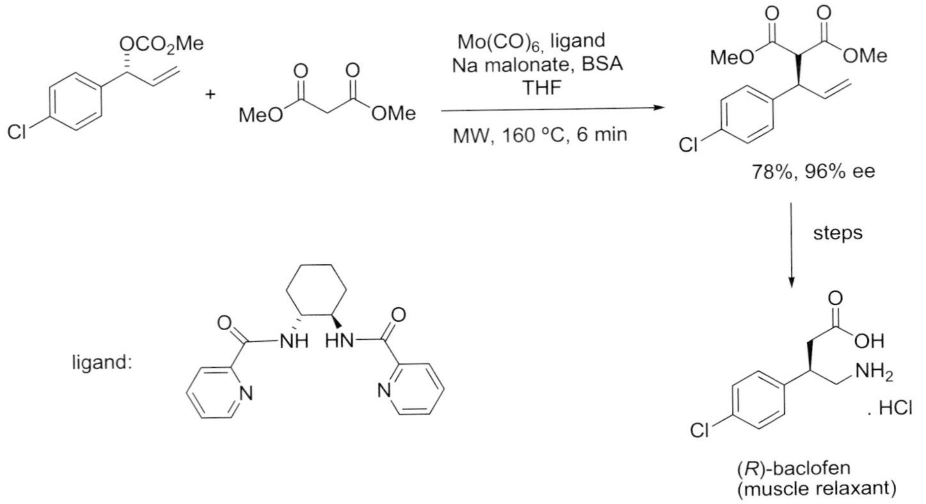

Scheme 6.52 Molybdenum-catalyzed asymmetric allylic alkylation in the synthesis of (R)-baclofen.

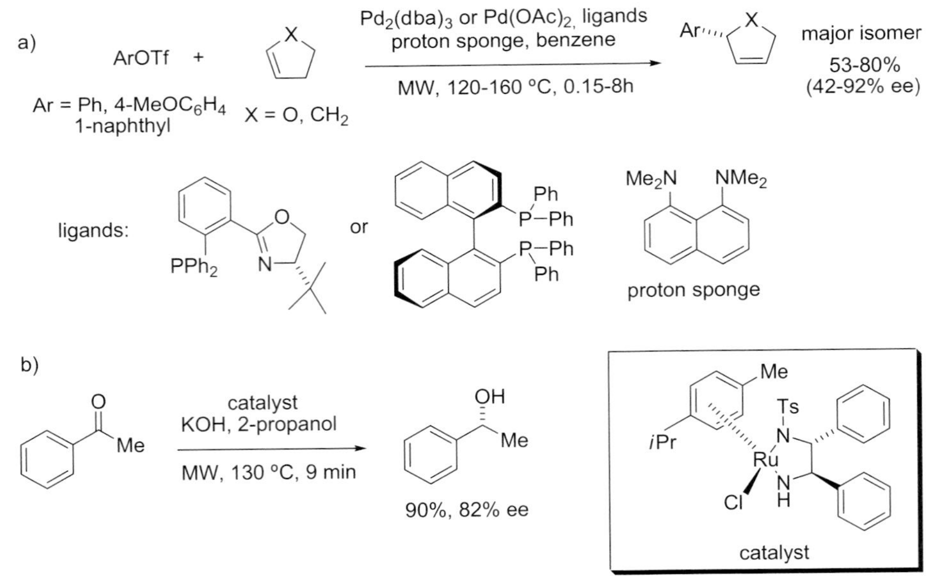

Scheme 6.53 Asymmetric Heck and hydrogen-transfer reactions.

Other enantioselective reactions performed by microwave heating include asymmetric Heck reactions (Scheme 6.53 a) [109] and ruthenium-catalyzed asymmetric hydrogen-transfer processes (Scheme 6.53 b) [110].

6.1.8
Miscellaneous Carbon–Carbon Bond-Forming Reactions

A publication by Nájera and coworkers described the palladium-catalyzed acylation of terminal alkynes with acid chlorides [111]. In general, high yields of the corresponding ynone coupling products were obtained using low loadings of an oxime-derived palladacycle as catalyst at elevated temperatures (Scheme 6.54). Utilizing microwave heating at 80 °C, a 0.2 mol% concentration of the palladacycle catalyst was sufficient to provide 88% conversion within 6 min. Under otherwise identical conditions, a 41% conversion was achieved at room temperature after 25 h (97% conversion at 110 °C for 1 h) [111].

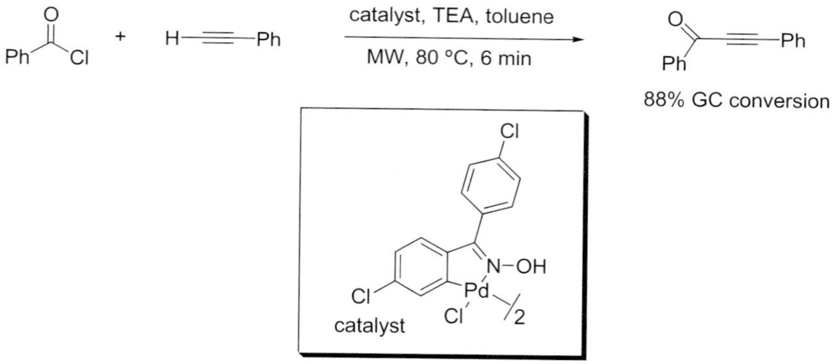

Scheme 6.54 Palladium-catalyzed acylation of terminal alkynes with acid chlorides.

Recently, Liebeskind and Srogl have developed a novel carbon–carbon cross-coupling protocol, involving the palladium(0)-catalyzed, copper(I)-mediated coupling of thioether-type species with boronic acids under neutral conditions [112]. A key feature of these protocols is the requirement of stoichiometric amounts of a copper(I) carboxylate (for example, copper(I) thiophene-2-carboxylate, CuTC) as metal cofactor. Due to the higher thiophilicity of the soft Cu(I) metal, selective sulfide coupling under Liebeskind–Srogl conditions can even be performed in the presence of a Suzuki-active bromide [112]. In this context, Lengar and Kappe have reported a microwave-assisted version of the Liebeskind–Srogl cross-coupling reaction (Scheme 6.55) [113]. Here, rapid carbon–carbon bond forming was achieved employing a 2-methylthio-1,4-dihydropyrimidine derivative as starting material. Coupling with phenylboronic acid was performed under microwave conditions at 130 °C, providing an 84% isolated yield of the desired product within 25 min. Optimum yields were achieved using 3 mol% of tetrakis(triphenylphosphine)palladium(0) as catalyst and 2.5 equivalents of copper(I) thiophene-2-carboxylate as

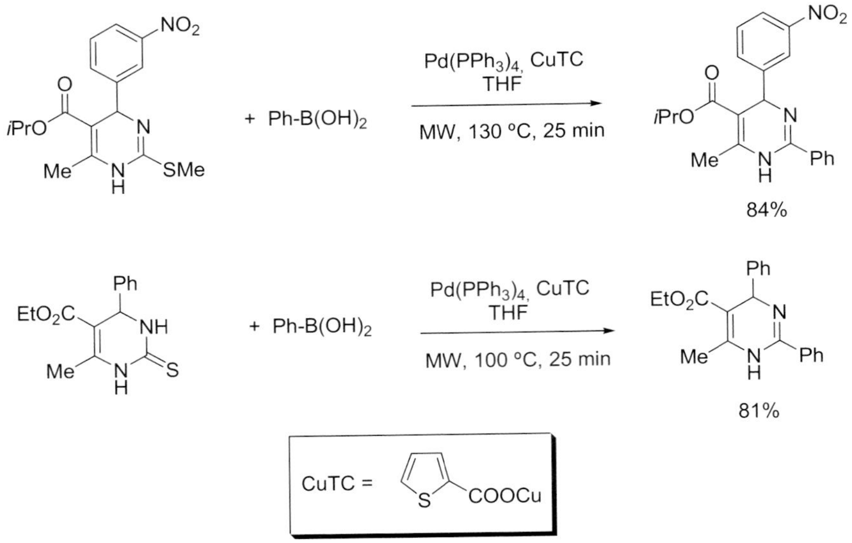

Scheme 6.55 Palladium(0)-catalyzed, copper(I)-mediated Liebeskind–Srogl-type couplings.

an additive. Interestingly, it was also possible to perform carbon–carbon coupling directly with the corresponding cyclic thiourea structures using similar palladium(0)-catalyzed, copper(I)-mediated coupling conditions (Scheme 6.55). The methodology was used to synthesize a small focused library of 2-aryl-1,4-dihydropyrimidines, which are highly potent non-nucleosidic inhibitors of hepatitis B virus replication exhibiting *in vitro* and *in vivo* antiviral activity [113].

A method for the microwave-promoted conversion of aryl triflates to the corresponding nitriles was presented by Zhang and Neumeyer (Scheme 6.56a) [114, 115]. Based on the pioneering work by Alterman and Hallberg [116], these authors have shown that 3-cyano-3-desoxy-10-ketomorphinans – which are key intermediates in the generation of *κ* opioid receptor-selective agonists/antagonists – can be readily prepared by palladium-catalyzed cyanation from the corresponding triflates. Optimum results were achieved by employing 8 mol% of tetrakis(triphenylphosphine)-palladium(0) and 2 equivalents of zinc cyanide in *N,N*-dimethylformamide as solvent. After microwave heating at 200 °C for 15 min, the corresponding nitriles were obtained in excellent yields. The authors found that the presence of a keto group adjacent to the phenyl ring in the substrates is important for the efficient conversion of triflates to nitriles. A related method devised by Arvela and Leadbeater uses aryl bromides or chlorides as starting materials for cyanation reactions (Scheme 6.56b) [117]. The methodology reported for aryl bromides involved the use of either 0.6 equivalents of nickel cyanide or a mixture of sodium cyanide (2 equivalents) and nickel bromide (1 equivalent). With aryl chlorides, a mixture of sodium cyanide and nickel bromide was required and the reaction proceeded via *in situ* formation of the corresponding aryl bromides (see Scheme 6.145). Typically, complete conversions

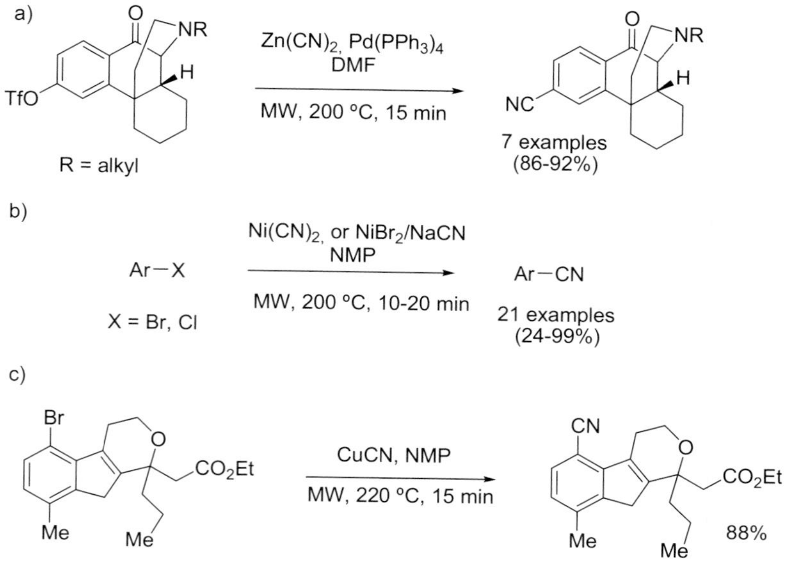
a)
NR
H
O
TfO
R = alkyl
Zn(CN)2, Pd(PPh3)4
DMF
MW, 200 °C, 15 min
NC
7 examples
(86-92%)
b)
Ar–X
X = Br, Cl
Ni(CN)2, or NiBr2/NaCN
NMP
MW, 200 °C, 10-20 min
Ar–CN
21 examples
(24-99%)
c)
Br
CO2Et
Me
CuCN, NMP
MW, 220 °C, 15 min
CN
88%

Scheme 6.56 Transition metal-mediated cyanation reactions.

were achieved after 10 min at 200 °C. Comparable results were reported by Gopalsamy and coworkers, who used copper(I) cyanide under similar reaction conditions (Scheme 6.56 c) [118].

The groups of Loupy and Jun have presented a chelation-assisted rhodium(I)-catalyzed *ortho*-alkylation of aromatic imines with alkenes (Scheme 6.57) [119]. The use of 2 mol% of Wilkinson's catalyst, $RhCl(PPh_3)_3$, and 5 equivalents of the corresponding alkene under solvent-free conditions proved to be optimal, providing the desired *ortho*-alkylated ketones in high yields after acidic hydrolysis. Somewhat lower yields were obtained when the imine preparation and the *ortho*-alkylation were realized in a one-pot procedure.

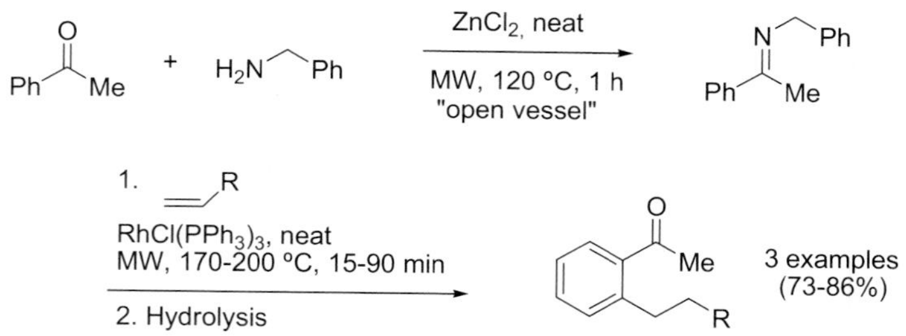
Ph
Me
O
+
H2N
Ph
ZnCl2, neat
MW, 120 °C, 1 h
"open vessel"
N
1.
R
RhCl(PPh3)3, neat
MW, 170-200 °C, 15-90 min
2. Hydrolysis
3 examples
(73-86%)

Scheme 6.57 Rhodium-catalyzed ortho-alkylation of ketimines.

6.2
Transition Metal-Catalyzed Carbon–Heteroatom Bond Formations

6.2.1
Buchwald–Hartwig Reactions

The groups of Buchwald [120] and Hartwig [121] have developed a large variety of useful palladium-mediated methods for C–O and C–N bond formation. These arylations have been enormously popular in recent years and, indeed, the vast amount of published material available describing a wide range of palladium-catalyzed methods, ligands, solvents, temperatures, and substrates has led to a "toolbox" of tunable reaction conditions, the scope of which allows access to most target molecules that incorporate an aryl amine motif. In 2002, Alterman and coworkers described the first high-speed Buchwald–Hartwig aminations using controlled microwave heating (Scheme 6.58) [122, 123]. Using N,N-dimethylformamide as solvent under non-inert conditions, the best results were obtained by employing 5 mol% of palladium acetate as pre-catalyst and 2,2′-bis(diphenylphosphino)-1,1′-binaphthyl (BINAP) as ligand. The procedure proved to be quite general and provided moderate to high yields with both electron-rich and electron-poor aryl bromides within the short time-frame of a microwave-heated reaction.

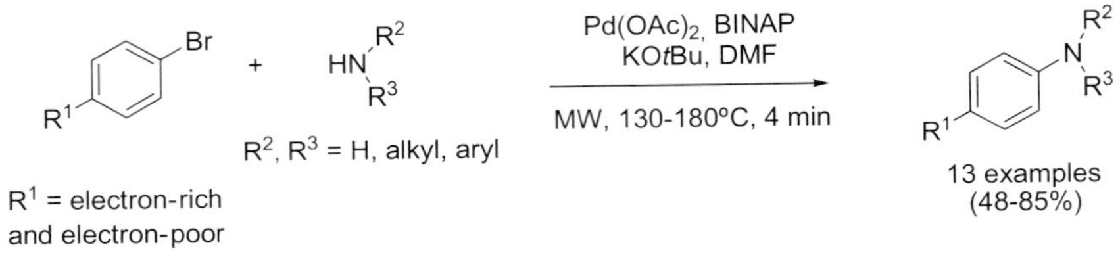

Scheme 6.58 Palladium-catalyzed amination reactions (Buchwald–Hartwig).

Following the pioneering work by Alterman, several microwave-assisted palladium-catalyzed aminations have been reported for a number of different substrates, using different types of palladium sources and ligands. The examples shown in Scheme 6.59 include bromoquinolines [124], aryl triflates [125], and intramolecular aminations in the synthesis of benzimidazoles [126]. In all cases, the use of microwave irradiation dramatically reduced the required reaction times and in many cases also improved the yields. Several authors have also found that the microwave-driven reaction required significantly less catalyst than its conventionally heated counterpart [126].

An extensive and detailed optimization study of microwave-assisted Buchwald–Hartwig reactions was reported by Skjaerbaek and coworkers in the context of elaborating an efficient protocol for the synthesis of aryl aminobenzophenone p38 MAP kinase inhibitors (Scheme 6.60) [127]. Several different strategies involving halide, triflate, and tosylate leaving groups were investigated, in addition to an alternative amination mode (method B). Among the many ligands screened for effi-

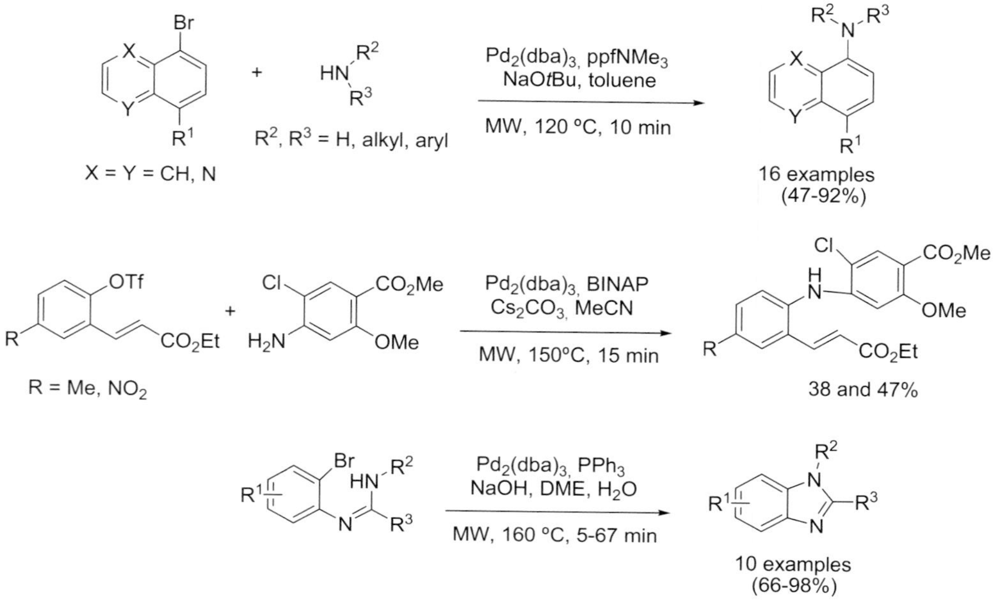

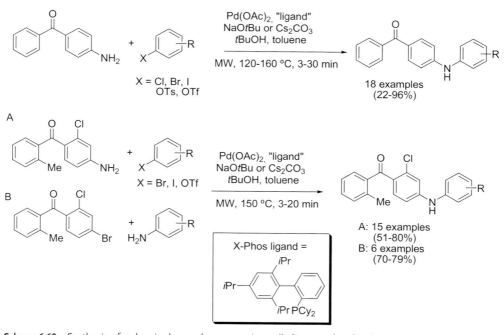

Scheme 6.59 Inter- and intramolecular palladium-catalyzed amination reactions of aryl bromides and triflates.

Scheme 6.60 Synthesis of aryl aminobenzophenones using palladium-catalyzed aminations.

cacy in the palladium-catalyzed amination, the X-Phos ligand system (2 mol% palladium acetate, 4 mol% ligand) provided the highest product yields and cleanest reaction profiles. As a base, both sodium *tert*-butoxide and cesium carbonate worked equally well in a 5:1 toluene/*tert*-butyl alcohol solvent mixture. Amination of an electronically diverse array of aryl halides with a variety of anilines was realized in good to excellent yields in most cases, without the need to work under an inert atmosphere. Depending on the structure and reactivity of the coupling partners, reaction times of 3–30 min at 120–160 °C were required to achieve full conversion.

Utilizing more reactive discrete palladium-*N*-heterocyclic carbene (NHC) complexes (for example, Pd(carb)$_2$) or *in situ* generated palladium/imidazolium salt complexes (1 mol% ligand A), Caddick and coworkers were able to extend the rapid amination protocols described above to electron-rich aryl chlorides (Scheme 6.61) [128].

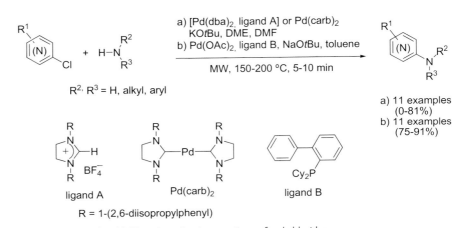

Scheme 6.61   Buchwald–Hartwig amination reactions of aryl chlorides.

Independent investigations by Maes and coworkers have involved the use of commercially available and air-stable 2-(dicyclohexylphosphanyl)biphenyl (ligand B) as a ligand system for the successful and rapid coupling of (hetero)aryl chlorides with amines under microwave Buchwald–Hartwig conditions (0.5–2 mol% palladium catalyst) [129, 130]. Both methods provide very high yields of products within an irradiation time of 10 min.

An application of this methodology toward the synthesis of *N*-arylsulfonamides is highlighted in Scheme 6.62 a. *N*-Arylsulfonamides constitute an important class of therapeutic agents in medicinal chemistry. Over 30 drugs containing this moiety are in clinical use in the areas of antibacterials, non-nucleosidic reverse transcriptase inhibitors, antitumor agents, and HIV-1 protease inhibitors. A group from GlaxoSmithKline has demonstrated that *N*-arylsulfonamides can be readily obtained by palladium-catalyzed intermolecular coupling of heteroaryl chlorides (for example, 4-chloroquinoline) with sulfonamides under microwave conditions [131]. The reactions proceeded at 180 °C with 2–10 mol% of palladium catalyst in the presence of a

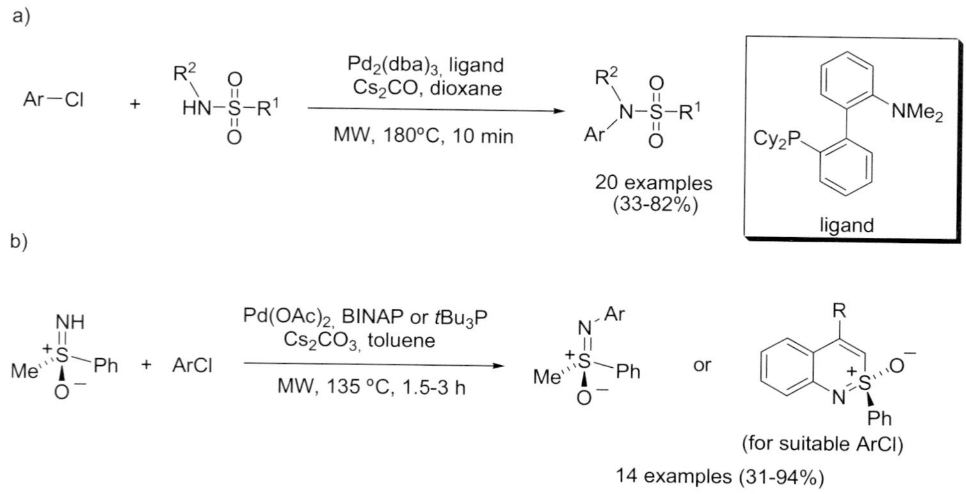

a)

20 examples
(33-82%)

ligand

b)

or

(for suitable ArCl)

14 examples (31-94%)

Scheme 6.62 Palladium-catalyzed *N*-arylations of sulfonamides and sulfoximines with aryl chlorides.

hemilabile *N,P* ligand. Similarly, Harmata and coworkers have disclosed an efficient protocol for the palladium-catalyzed *N*-arylation of enantiopure sulfoximines with aryl chlorides (Scheme 6.62 b) [132]. Optimal results were achieved by using palladium acetate as palladium source and *rac*-2,2′-bis(diphenylphosphino)-1,1′-binaphthyl (BINAP) or tris(*tert*-butyl)phosphine as ligand under microwave irradiation conditions. With aryl chlorides bearing *ortho*-carbonyl substituents, the corresponding benzothiazines were obtained.

6.2.2
Ullmann Condensation Reactions

A recent survey of the literature on the Ullmann and related condensation reactions has highlighted the growing importance and popularity of copper-mediated carbon–nitrogen, carbon–oxygen, and carbon–sulfur bond-forming protocols [133]. In Scheme 6.63, two examples of microwave-assisted Ullmann-type condensations from a group of researchers at Bristol-Myers Squibb are shown. In the first example, (*S*)-1-(3-bromophenyl)ethylamine was coupled with eleven N–H containing heteroarenes in the presence of 10 mol% of copper(I) iodide and 2 equivalents of potassium carbonate base [134]. The comparatively high reaction temperature (1-methyl-2-pyrrolidone, 195 °C) and the long reaction times are noteworthy. For the coupling of 3,5-dimethylpyrazole, for example, microwave heating for 22 h was required to afford a 49% isolated yield of product! The average reaction times were 2–3 h. In the second example, similar conditions were chosen for the reaction of mostly aromatic thiols with aryl bromides and iodides to afford aryl sulfides [135]. The same authors have also described the synthesis of diaryl ethers by copper-catalyzed arylation of phenols with aryl halides [136].

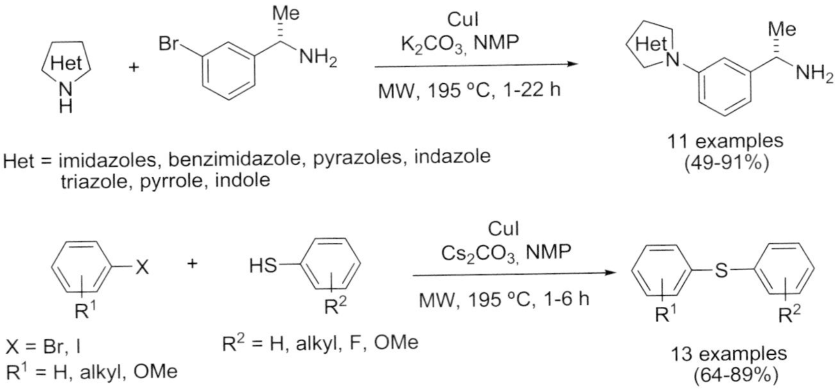
Het
Br
Me
NH2
CuI
K2CO3, NMP
MW, 195 °C, 1-22 h
11 examples
(49-91%)
Het = imidazoles, benzimidazole, pyrazoles, indazole
triazole, pyrrole, indole
X
HS
R1
R2
Cs2CO3, NMP
MW, 195 °C, 1-6 h
X = Br, I
R1 = H, alkyl, OMe
R2 = H, alkyl, F, OMe
13 examples
(64-89%)

Scheme 6.63 Ullmann-type carbon–nitrogen and carbon–sulfur bond formations.

A reaction related to the Ullmann condensation is the Goldberg reaction, that is, the copper-catalyzed amidation of aryl halides. Due to the rather drastic reaction conditions that are usually required, the Goldberg reaction has not been recognized as a powerful synthetic methodology in organic synthesis. A team from Solvay Pharmaceuticals has reported that Goldberg reactions can be efficiently carried out under microwave irradiation conditions, employing 10 mol% of copper(I) iodide as a catalyst and small amounts (2 molar equivalents) of 1-methyl-2-pyrrolidone as a solvent (Scheme 6.64) [137]. Beyond simple acetamides, the reaction could be extended to cyclic amides, such as 3,4-dihydro-2-quinolones and piperazin-2-ones. The reactions were carried out on a comparatively large scale (25 mmol) under open-vessel conditions under a gentle stream of nitrogen. Yields were in the range 48–77%.

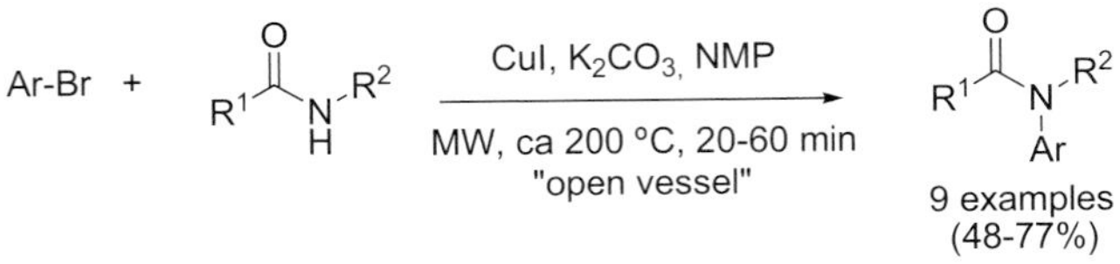
Ar-Br +
R1
R2
CuI, K2CO3, NMP
MW, ca 200 °C, 20-60 min
"open vessel"
Ar
9 examples
(48-77%)

Scheme 6.64 *N*-Arylation of amides via Goldberg reactions.

Recently, interest in copper-catalyzed carbon–heteroatom bond-forming reactions has shifted to the use of boronic acids as reactive coupling partners [133]. One example of carbon–sulfur bond formation is displayed in Scheme 6.65. Lengar and Kappe have reported that, in contrast to the palladium(0)/copper(I)-mediated process described in Scheme 6.55, which leads to carbon–carbon bond formation, reaction of the same starting materials in the presence of 1 equivalent of copper(II) acetate and 2 equivalents of phenanthroline ligand furnishes the corresponding carbon–sulfur cross-coupled product [113]. Whereas the reaction at room temperature needed 4 days to reach completion, microwave irradiation at 85 °C for 45 min in 1,2-dichloroethane provided a 72% isolated yield of the product.

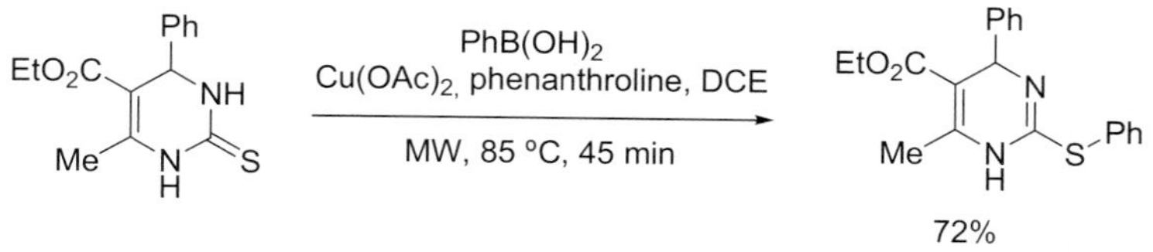

Scheme 6.65 Copper(II)-mediated carbon–sulfur cross-coupling.

A palladium-catalyzed protocol for carbon–sulfur bond formation between an aryl triflate and *para*-methoxybenzylthiol was introduced by Macmillan and Anderson (Scheme 6.66) [138]. Using palladium(II) acetate as a palladium source and 2,2′-bis(diphenylphosphino)-1,1′-binaphthyl (BINAP) as a ligand, microwave heating of the two starting materials in *N,N*-dimethylformamide at 150 °C for 20 min in the presence of triethylamine base led to the formation of the desired sulfide in 85% yield.

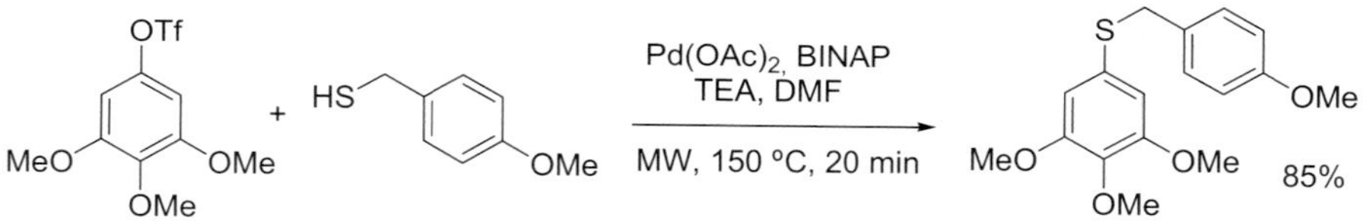

Scheme 6.66 Palladium-catalyzed carbon–sulfur cross-coupling.

6.2.3
Miscellaneous Carbon–Heteroatom Bond-Forming Reactions

Tertiary phosphines are a very important class of ligands in transition metal-catalyzed reactions. Among the various methods for the synthesis of such phosphine ligands, direct carbon–phosphorus (C–P) bond formation by transition metal-catalyzed cross-coupling of unprotected secondary phosphines with aryl halides/triflates can be considered as one of the most valuable procedures. All of these protocols utilize a homogeneous transition metal catalyst and a base, and generally require many hours of reaction time at high temperatures, or otherwise impractical or complex synthetic manipulations. Stadler and Kappe have demonstrated that aryl iodides, bromides, and triflates can be successfully coupled with diphenylphosphine under microwave conditions (Scheme 6.67) [139]. Optimized reaction conditions for aryl iodides involved the utilization of combinations of 1-methyl-2-pyrrolidone, potassium acetate, and 2.5 mol% of palladium(II) acetate as catalyst. A 50% excess of iodobenzene was found to give the best yields and at 180–200 °C a conversion of >90% was typically achieved within 20 min. At higher temperatures, decomposition of the palladium(II) catalyst was observed, resulting in lower isolated product yields and in the deposition of a very thin film of Pd black on the microwave process vial. The reaction was also successfully accomplished using palladium-on-charcoal

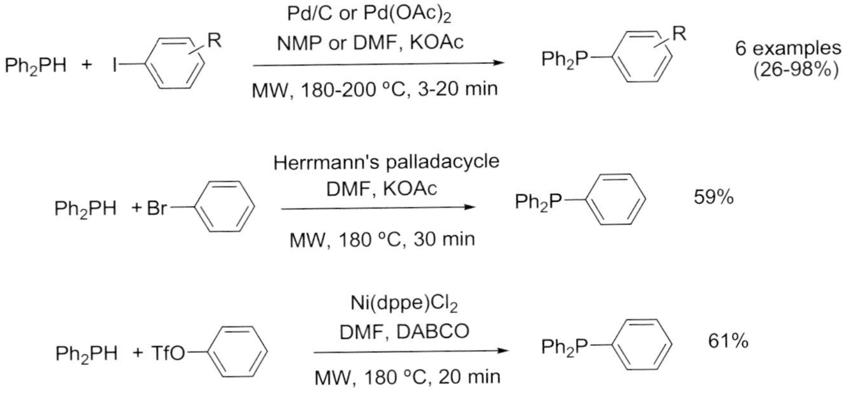

Pd/C or Pd(OAc)$_2$

Ph$_2$PH + I—⟨benzene⟩—R →(NMP or DMF, KOAc / MW, 180-200 °C, 3-20 min)→ Ph$_2$P—⟨benzene⟩—R   6 examples (26-98%)

Herrmann's palladacycle

Ph$_2$PH + Br—⟨benzene⟩ →(DMF, KOAc / MW, 180 °C, 30 min)→ Ph$_2$P—⟨benzene⟩   59%

Ni(dppe)Cl$_2$

Ph$_2$PH + TfO—⟨benzene⟩ →(DMF, DABCO / MW, 180 °C, 20 min)→ Ph$_2$P—⟨benzene⟩   61%

Scheme 6.67 Palladium-catalyzed carbon–phosphorus cross-coupling.

(<0.1 mol% palladium) as a catalyst. For the less reactive aryl bromide and triflate precursors, more active catalytic systems had to be employed (Scheme 6.67) [139].

The large number of boronic acids that are commercially available makes the Suzuki and related types of coupling chemistries highly attractive in the context of high-throughput synthesis and scaffold decoration (see Section 6.1.2). In addition, boronic acids are air- and moisture-stable, of relatively low toxicity, and the boron-derived by-products can easily be removed from the reaction mixtures. Therefore, it is not surprising that efficient and rapid microwave-assisted protocols have been developed for their preparation. In 2002, Fürstner and Seidel outlined the synthesis of pinacol aryl boronates from aryl chlorides bearing electron-withdrawing groups and commercially available bis(pinacol)borane, using a palladium catalyst formed *in situ* from palladium acetate and an appropriate imidazolium chloride (Scheme 6.68, X = Cl) [140]. The very reactive *N*-heterocyclic carbene (NHC) ligand (6–12 mol%) allowed this transformation to proceed to completion within 10–20 min at 110 °C in THF under microwave irradiation in sealed vessels. The conventionally heated process (reflux in THF, ca. 65 °C, argon atmosphere) gave comparable yields, but

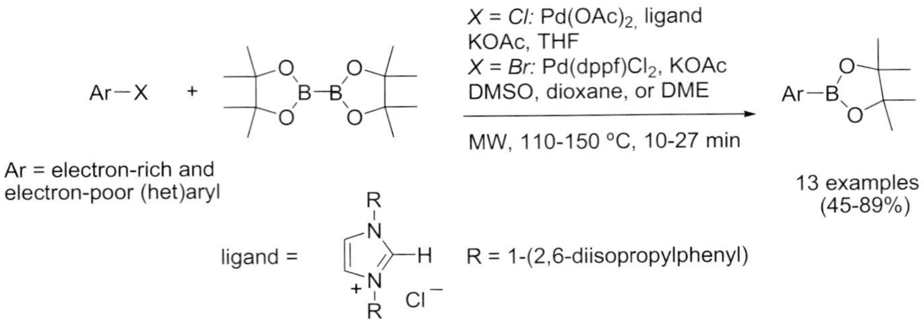

Ar—X + [pinacol diboron] →(*X* = *Cl*: Pd(OAc)$_2$, ligand KOAc, THF / *X* = *Br*: Pd(dppf)Cl$_2$, KOAc DMSO, dioxane, or DME / MW, 110-150 °C, 10-27 min)→ Ar—B⟨pinacol⟩

Ar = electron-rich and electron-poor (het)aryl

13 examples (45-89%)

ligand = [imidazolium structure]—H   R = 1-(2,6-diisopropylphenyl)   Cl$^-$

Scheme 6.68 Palladium-catalyzed formation of aryl boronates.

required 4–6 h to reach completion. A complementary approach subsequently disclosed by Dehaen and coworkers employed electron-rich aryl bromides as substrates (Scheme 6.68, X = Br) and 3 mol% of [1,1′-bis(diphenylphosphino)ferrocene]dichloropalladium(II) [Pd(dppf)Cl$_2$] as a catalyst [141]. Here, a somewhat higher reaction temperature (125–150 °C) was employed, producing a variety of different aryl boronates in good to excellent yields. Burgess and coworkers have employed a similar procedure for the preparation of fluorescein-derived boronic esters (see Scheme 6.26) [57].

6.3
Other Transition Metal-Mediated Processes

6.3.1
Ring-Closing Metathesis

In recent years, the olefin metathesis reaction has attracted widespread attention as a versatile carbon–carbon bond-forming method [142]. Among the numerous different metathesis methods, ruthenium-catalyzed ring-closing metathesis (RCM) has emerged as a very powerful method for the construction of small, medium, and macrocyclic ring systems. In general, metathesis reactions are carried out at room temperature or at slightly elevated temperatures (for example, at 40 °C in refluxing dichloromethane), and sometimes require several hours of reaction time to achieve full conversion. Employing microwave heating, otherwise sluggish RCM protocols have been reported to be completed within minutes or even seconds, as opposed to hours at room temperature [143–146].

One limitation of ring-closing metathesis is that the reaction can be rather sensitive to external substituents. In most cases, only substituents with little or no steric or electronic bias are compatible. The presence of external groups with an electronic bias (ester, nitrile, etc.) often results in little or no yield of the RCM product. Wilson and coworkers have described microwave-enhanced ring-closing metathesis transformations with diolefin substrates bearing external carboxymethyl substituents (Scheme 6.69 a) [147]. The authors reported that most of the studied reactions proceeded only to 40% conversion utilizing 3 mol% of Grubbs II catalyst at room temperature using 1,2-dichloroethane (DCE) as solvent. Heating to 50 °C resulted in about 75% conversion after 5 h, but the reaction proceeded no further. In sharp contrast, microwave irradiation at 50 °C for 5 min gave 85% conversion, and at 150 °C for 5 min gave 97% conversion. Independent investigations by the group of Grigg with a similar set of diolefin substrates (Scheme 6.69 b) yielded similar conclusions [148]. Again, utilizing controlled microwave heating, the RCM reactions could be run much more efficiently than with thermal heating, allowing a significant reduction in catalyst loading and reaction time.

An interesting combination of ring-closing metathesis chemistry with the aza-Baylis–Hillman reaction has recently been described by Balan and Adolfsson and is shown in Scheme 6.70 a [149]. The authors reported that functionalized 2,5-dihydro-

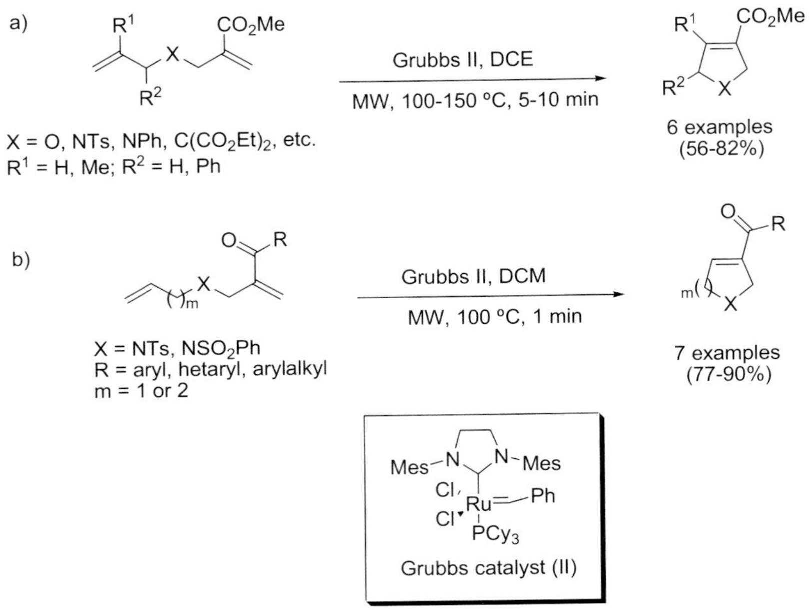

Scheme 6.69 Ring-closing metathesis with external carbonyl substituents.

pyrroles can be obtained by microwave-mediated ruthenium-catalyzed ring-closing metathesis (RCM). The required bis-olefin precursors were conveniently obtained from aza-Baylis–Hillman adducts. Microwave irradiation of a dilute solution of the diene with 5 mol% of Grubbs II catalyst in dichloromethane for 1–2 min at 100 °C produced the desired pyrroles in high yield. The same conditions were used by the group of Lamaty for the ring-closing metathesis of related 2-trimethylsilylethylsulfonyl (SES)-protected substrates (Scheme 6.70 b) [150]. The required starting materials were again prepared by aza-Baylis–Hillman chemistry.

In 2003, Efskind and Undheim reported dienyne and triyne domino RCMs of appropriately functionalized substrates with Grubbs type II or I catalysts (Scheme 6.71, reactions a and b, respectively) [151]. While the thermal processes (toluene, 85 °C) required multiple addition of fresh catalyst (3 × 10 mol%) over a period of 9 h to furnish a 92% yield of product, microwave irradiation for 10 min at 160 °C (5 mol% catalyst, toluene) led to full conversion. The authors ascribe the dramatic rate enhancement to rapid and uniform heating of the reaction mixture and increased catalyst lifetime through the elimination of wall effects. In some instances, use of the Grubbs I catalyst was more efficient than use of the more common Grubbs II equivalent.

An interesting series of ring-closing alkyne metathesis reactions (RCAM) has recently been reported by Fürstner and coworkers (Scheme 6.72) [152]. Treatment of biaryl-derived diynes with 10 mol% of a catalyst prepared *in situ* from molybdenum hexacarbonyl and 4-(trifluoromethyl)phenol at 150 °C for 5 min led to a ca. 70% iso-

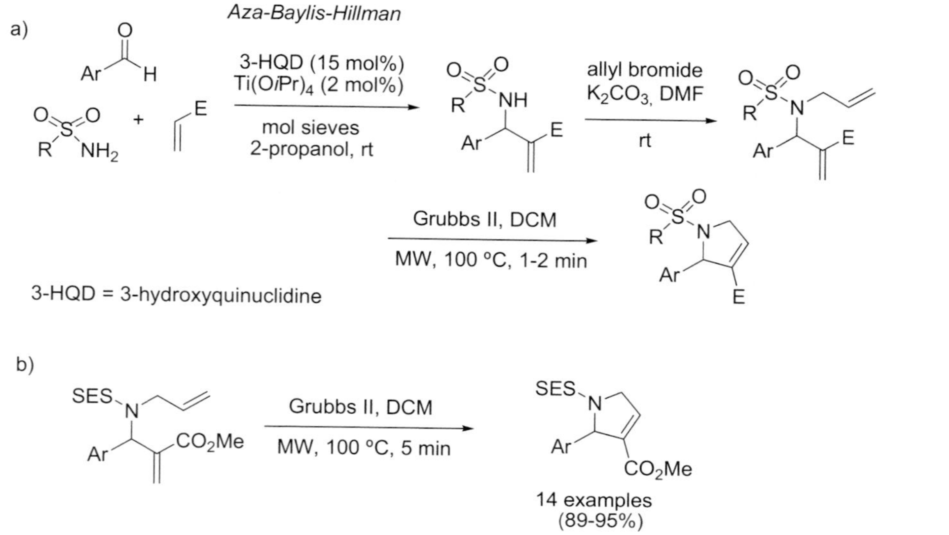

a)

Aza-Baylis-Hillman

3-HQD = 3-hydroxyquinuclidine

b)

14 examples
(89-95%)

Scheme 6.70 Ring-closing metathesis of aza-Baylis–Hillman adducts.

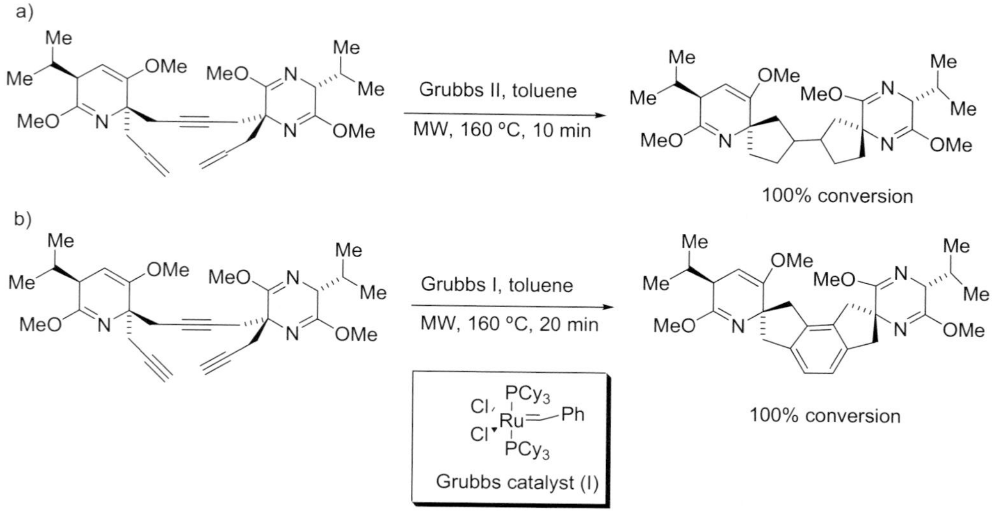

a)

100% conversion

b)

Grubbs catalyst (I)

100% conversion

Scheme 6.71 Dienyne (reaction a) and triyne (reaction b) domino ring-closing metathesis reactions.

lated yield of the desired cycloalkynes, which were further manipulated into a naturally occurring DNA cleaving agent of the turriane family. Conventional heating
under reflux conditions in chlorobenzene for 4 h produced ca. 80% isolated yield of
product under otherwise identical conditions.

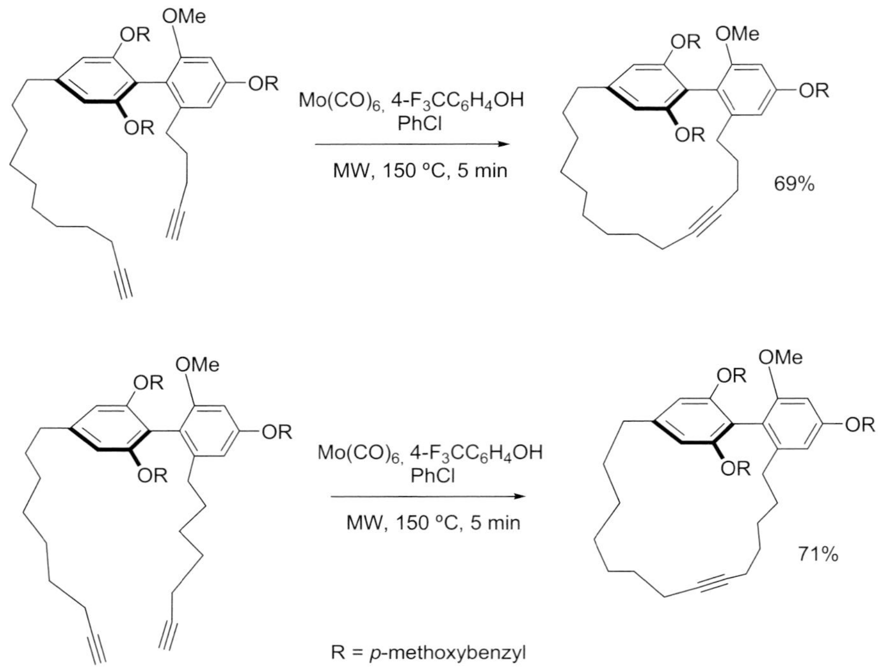

Scheme 6.72 Ring-closing alkyne metathesis reactions.

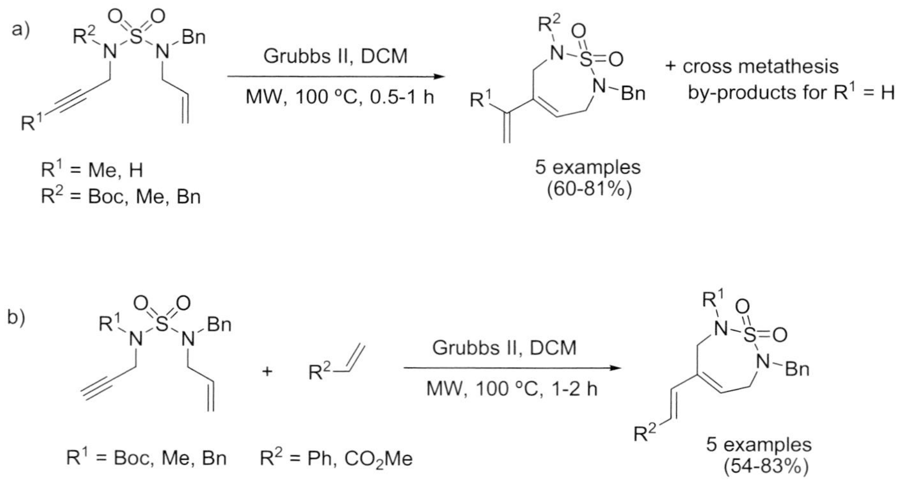

Scheme 6.73 Enyne ring-closing and cross-metathesis reactions.

A series of synthetically valuable ring-closing metathesis (RCM)/alkene cross-metathesis (CM) transformations starting from sulfamide-linked enynes was described by Brown and coworkers in 2004 (Scheme 6.73) [153]. A range of enyne substrates was subjected to ring-closing metathesis using 3–20 mol% of Grubbs II catalyst. While the reactions of internal alkyne substrates were sluggish at room temperature, they proceeded rapidly upon microwave heating at 100 °C, furnishing seven-membered cyclic sulfamides in good yields within 1 h (Scheme 6.73 a). For substrates bearing a terminal alkyne group, the authors developed a one-pot RCM-CM reaction requiring the presence of 2–3 equivalents of alkenes such as styrene and using 6 mol% of Grubbs II catalyst. As anticipated, the desired enyne RCM-CM products were produced selectively in good yields, with the expected (*E*)-isomer predominating.

Related microwave-assisted ring-closing metathesis reactions, leading to seven- [154] and eight-membered aza-heterocyclic products [49], are depicted in Scheme 6.74.

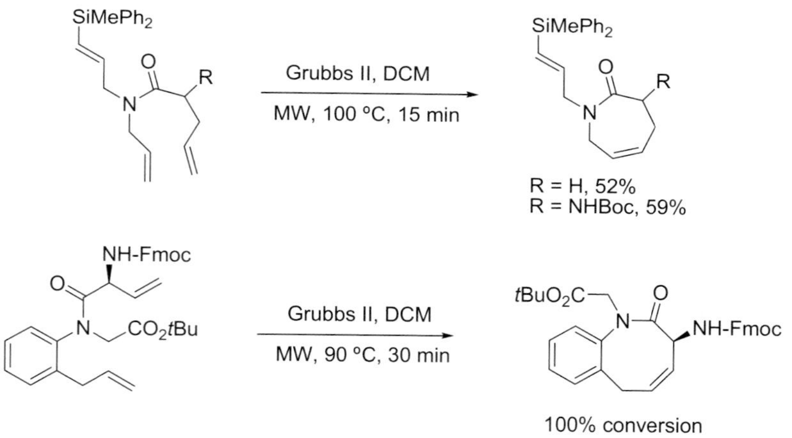

Scheme 6.74 Formation of seven- and eight-membered rings by ring-closing metathesis reactions.

6.3.2
Pauson–Khand Reactions

The [2+2+1] cycloaddition of an alkene, an alkyne, and carbon monoxide is known as the Pauson–Khand reaction and is often the method of choice for the preparation of complex cyclopentenones [155]. Groth and coworkers have demonstrated that Pauson–Khand reactions can be carried out very efficiently under microwave heating conditions (Scheme 6.75 a) [156]. Taking advantage of sealed-vessel technology, 20 mol% of dicobalt octacarbonyl was found to be sufficient to drive all of the studied Pauson–Khand reactions to completion, without the need for additional carbon monoxide. The carefully optimized reaction conditions utilized 1.2 equivalents of

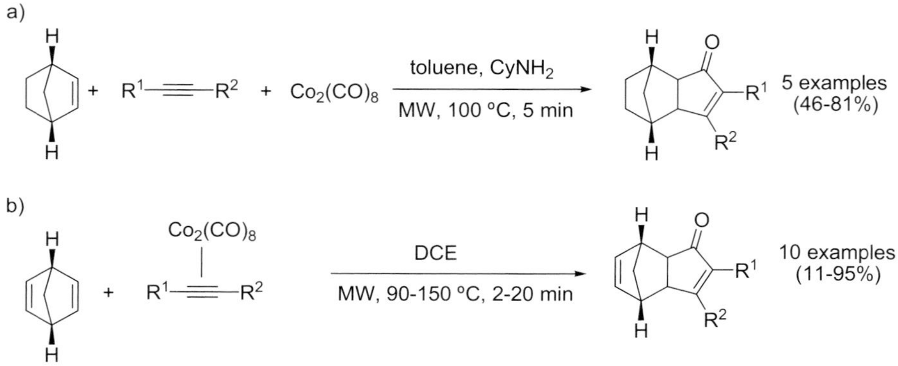

a)

b)

Scheme 6.75 Pauson–Khand [2+2+1] cycloaddition reactions.

cyclohexylamine as an additive in toluene as solvent. Microwave heating at 100 °C for 5 min provided good yields of the desired cycloadducts. Similar results were published independently by Evans and coworkers [157, 158]. Here, however, the preformed alkyne–dicobalt hexacarbonyl complexes were used as substrates (Scheme 6.75 b). These authors were able to perform the microwave-assisted reactions in different mono- and multimode microwave instruments with equal success [157].

6.3.3
Carbon–Hydrogen Bond Activation

Another important reaction principle in modern organic synthesis is carbon–hydrogen bond activation [159]. Bergman, Ellman, and coworkers have introduced a protocol that allows otherwise extremely sluggish inter- and intramolecular rhodium-catalyzed C–H bond activation to occur efficiently under microwave heating conditions. In their investigations, these authors found that heating of alkene-tethered benzimidazoles in a mixture of 1,2-dichlorobenzene and acetone in the presence of di-μ-

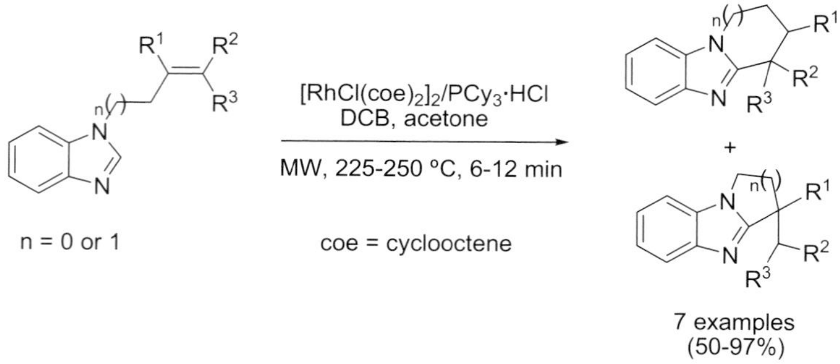

Scheme 6.76 Intramolecular benzimidazole C–H alkene coupling.

chloro-bis-[(cyclooctene)rhodium(I)] (2.5–5 mol%) and tricyclohexylphosphine hydrochloride (5–10 mol%) as the catalyst system provided the desired tricyclic heterocycles in moderate to excellent yields (Scheme 6.76) [160]. Microwave heating at 225–250 °C for 6–12 min proved to be the optimum conditions. The solvents were not degassed or dried before use, but air was excluded by purging the reaction vessel with nitrogen.

6.3.4
Miscellaneous Reactions

In the context of a total synthesis of the C-1–C-28 ABCD unit of the marine macrolide spongistatin 1, the Ley group has reported an efficient methylenation reaction of an advanced ketone intermediate using the titanium-based Petasis reagent (Scheme 6.77 a) [161]. Treatment of the ketone with the Petasis reagent in toluene at 120 °C for 3 h proved to be the optimal conventional conditions, generating the desired alkene in 71% yield. The reaction proved to be much more efficient, however, when carried out under sealed-vessel microwave heating at 160 °C, forming the alkene after 10 min in an improved 82% yield. Again, an ionic liquid (1-ethyl-3-methylimidazolium hexafluorophosphate, emimPF$_6$) was utilized to modify the dielectric properties of the solvent (see Section 4.3.3.2).

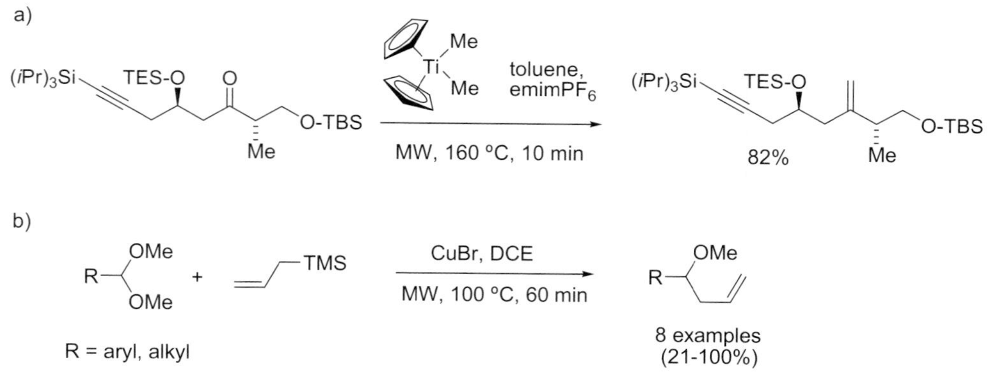

Scheme 6.77 (a) Methylenation of ketones with the Petasis reagent, (b) copper(I)-bromide mediated allylation of acetals.

Jung and Maderna have reported the microwave-assisted allylation of acetals with allyltrimethylsilane in the presence of copper(I) bromide as promoter (Scheme 6.77 b) [162]. Stoichiometric amounts of copper(I) bromide are required and the reaction works best with aromatic acetals free from strongly electron-withdrawing substituents on the aromatic ring.

Mejía-Oneto and Padwa have described the rhodium(II) perfluorobutyrate-catalyzed decomposition of an α-diazo ketoamide precursor (Scheme 6.78) [163]. Microwave heating of a solution of the diazo compound in benzene with a catalytic

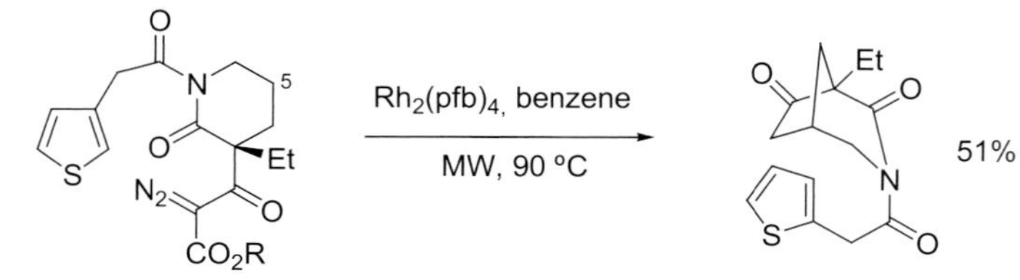

Scheme 6.78 Rhodium(II)-catalyzed carbon–hydrogen insertion.

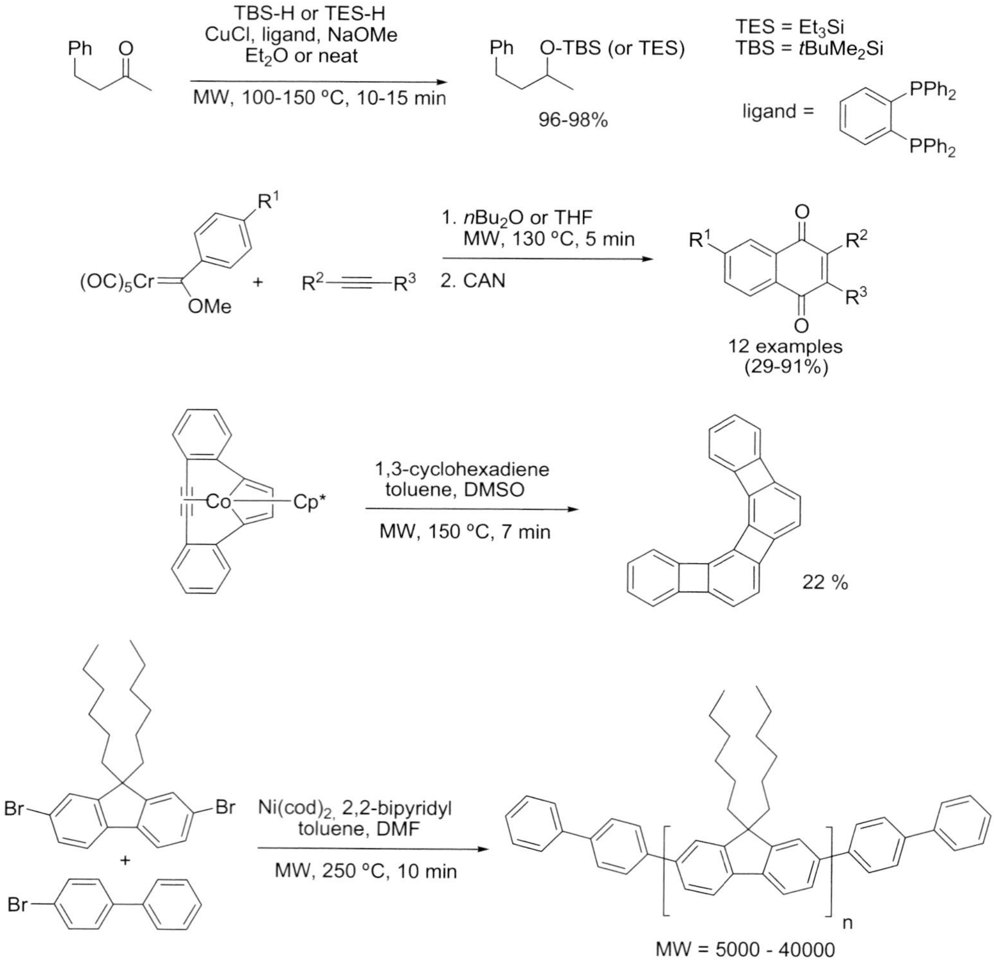

Scheme 6.79 Hydrosilylation of ketones [164], Dötz benzannulation chemistry [165], cobalt-mediated synthesis of angular [4]phenylenes [166], and nickel-mediated coupling polymerizations [167].

amount of the rhodium(II) carboxylate catalyst unexpectedly led to a lactam, formed by a formal insertion of the metal carbene into the carbon–hydrogen bond at the 5-position of the pyridone ring followed by an ethoxydecarboxylation.

Other microwave-assisted reactions involving metal catalysts or metal-based reagents are shown in Scheme 6.79 [164–167].

6.4
Rearrangement Reactions

6.4.1
Claisen Rearrangements

In their synthesis of the natural product carpanone, Ley and coworkers described the microwave-assisted Claisen rearrangement of an allyl ether (Scheme 6.80 a) [168]. A 97% yield of the rearranged product could be obtained by three successive 15 min spells of irradiation at 220 °C, employing toluene doped with the ionic liquid 1-butyl-3-methylimidazolium hexafluorophosphate (bmimPF$_6$) as solvent (see Section 4.3.3.2). Interestingly, a single 45 min irradiation event at the same temperature gave a somewhat lower yield (86%). A related Claisen rearrangement, but on a much more complex substrate, was reported by the same group, again using "pulsed" microwave irradiation conditions. Heating a solution of a propargylic enol ether (Scheme 6.80 b) in 1,2-dichlorobenzene (DCB) at 180 °C for 15 min resulted in a 71% isolated yield of the desired allene as a single diastereomer, which was further elaborated into the skeleton of the triterpenoid natural product azadirachtin [169]. An 88% yield of the product was obtained by applying 15 × 1 min pulses of irradiation. No rationalization for the increased yields obtained in these "pulsed versus continuous irradiation" experiments can be given at present (see also Scheme 2.6).

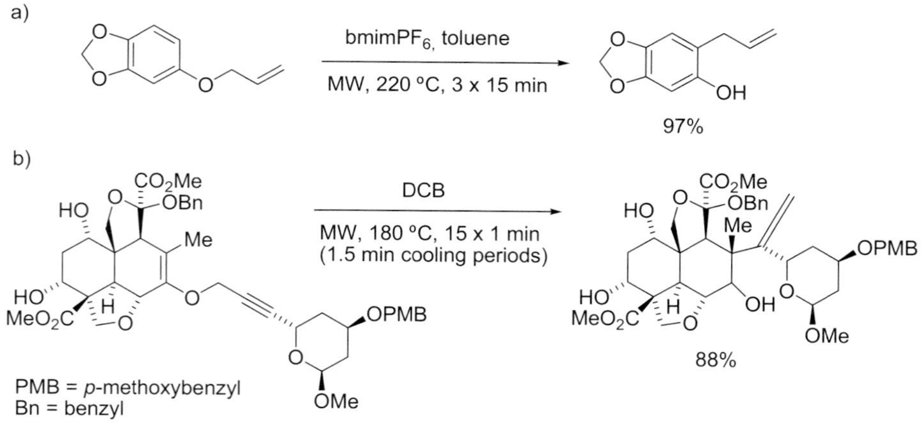

Scheme 6.80 Claisen rearrangements in natural product synthesis.

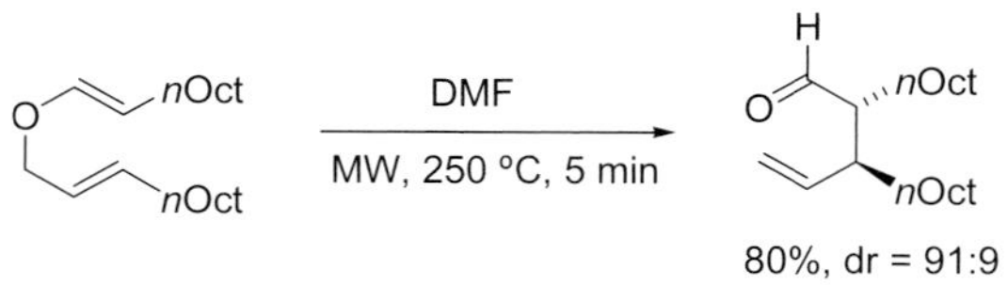

Scheme 6.81 Diastereoselective Claisen rearrangements.

Nordmann and Buchwald have reported a diastereoselective Claisen rearrangement of an allyl vinyl ether to an aldehyde (Scheme 6.81) [170]. Using *N,N*-dimethylformamide as solvent, an 80% yield with a diastereomeric ratio of 91:9 was obtained by microwave heating at 250 °C for 5 min. Conventional heating at 120 °C for 24 h provided somewhat higher yields and selectivities (90% yield, *dr* = 94:6).

A selection of other microwave-assisted Claisen rearrangements is shown in Scheme 6.82. The groups of Wada and Yanagida have discussed the solvent-free double Claisen rearrangement of bis(4-allyloxyphenyl)sulfone into bis(3-allyl-4-hydroxyphenyl)sulfone (Scheme 6.82 a), an important color developer for heat- or pressure-sensitive recording [171]. The process was carried out on a 10 gram scale by first melting the solid starting material by conventional heating and then exposing the melt to microwave irradiation at 180 °C for 5 min, leading to an 87% yield of the desired product. Similarly, 2′-allyloxy-acetophenone was heated by microwave irradiation at 210 °C for 1 h with simultaneous air cooling (see Section 2.5.3) to provide 3′-allyl-2′-hydroxy-acetophenone in quantitative yield (Scheme 6.82 b) [172]. The corresponding thermal process (200 °C) required 44 h to reach completion.

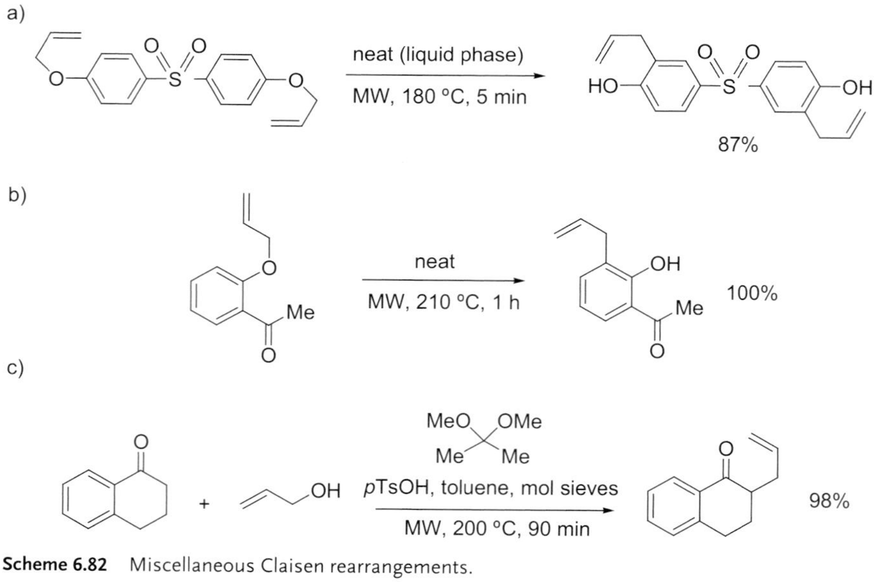

Scheme 6.82 Miscellaneous Claisen rearrangements.

A method for the selective α-monoalkylation of phenyl ketones (for example, α-tetralone) with allyl alcohol, involving the *in situ* formation and acid-catalyzed cracking of the corresponding ketone diallyl ketals, was described by Trabanco and coworkers (Scheme 6.82 c) [173]. The process relied on the use of 2,2-dimethoxypropane and 3 Å molecular sieves as water scavengers, and allowed the preparation of α-allyl-substituted ketones in moderate to good yields. Optimum conditions required the use of 5 equivalents of allyl alcohol, 1.5 equivalents of 2,2-dimethoxypropane, and a catalytic amount of p-toluenesulfonic acid (pTsOH). In the case of α-tetralone, microwave heating at 200 °C for 90 min provided a 98% yield of the desired product, with only minor amounts of diallylated by-product being formed. The optimized protocol was applied to a set of 12 related benzocycloalkanones and phenyl ketones.

6.4.2
Domino/Tandem Claisen Rearrangements

A series of complex sigmatropic rearrangements of allyl tetronates and allyl tetramates to furnish 3-allyltetronic or -tetramic acids, respectively, has been investigated by Schobert and coworkers (Scheme 6.83) [174]. The authors discovered that allyl tetronates (X = O) and allyl tetramates (X = NH) undergo a microwave-accelerated Claisen rearrangement allowing – in contrast to the conventional procedure – the isolation of the Claisen intermediates. Consecutive (homo)sigmatropic [1,5]-hydrogen shifts, such as the oxa-ene (Conia) reaction (Scheme 6.83) leading to 3-(spirocyclopropyl)dihydrofuran-2,4-diones (X = O), are promoted less effectively, which allows the isolation of the Claisen intermediates in these sigmatropic domino sequences. Microwave heating at 110–150 °C for 30–60 min in acetonitrile typically provided mixtures of the desired and readily separable Claisen and Conia products.

In a series of publications over the past few years, the group of Barriault has reported on microwave-assisted tandem oxy-Cope/Claisen/ene and closely related reactions [175–178]. These pericyclic transformations typically proceed in a highly stereoselective fashion and can be exploited for the synthesis of complex natural products possessing decalin skeletons, such as the abietane diterpene wiedamannic

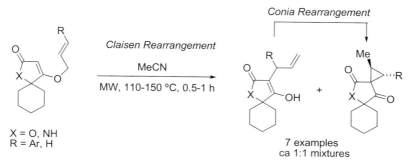

Scheme 6.83 Domino Claisen/Conia rearrangements.

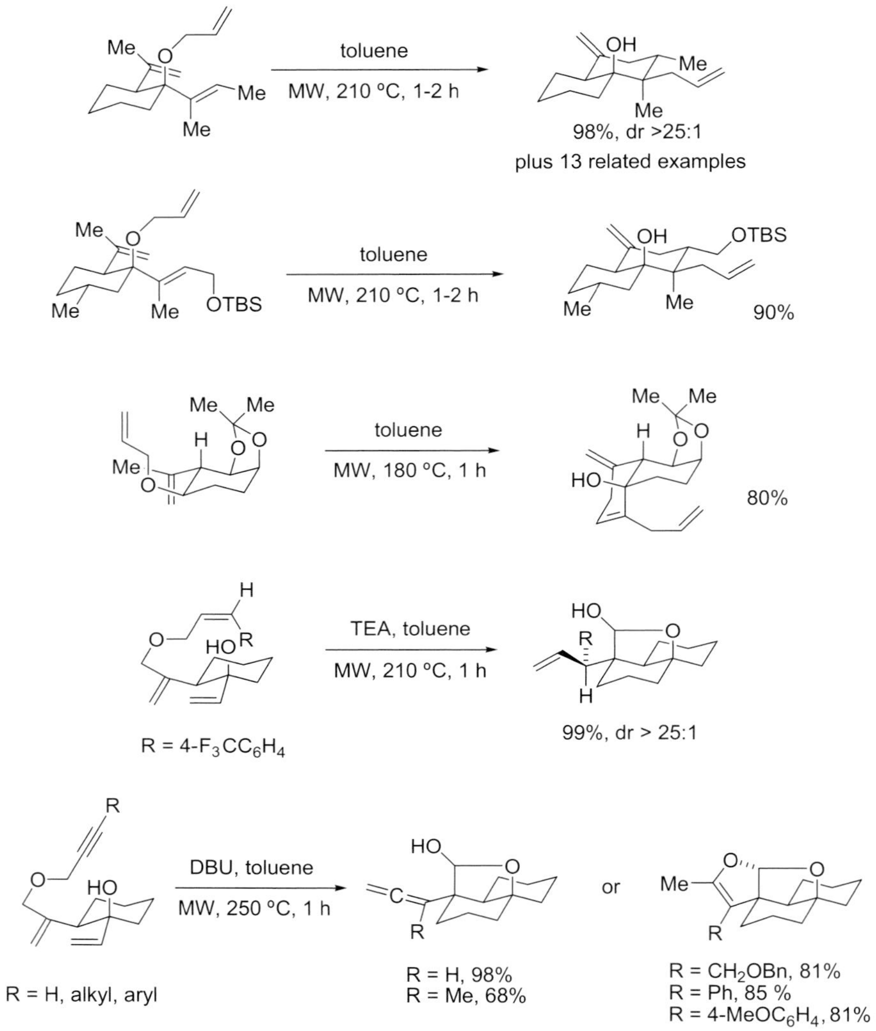

Scheme 6.84 Tandem oxy-Cope/Claisen/ene reactions and related transformations.

acid [177]. Some of the transformations described by Barriault and coworkers are summarized in Scheme 6.84. Typically, microwave heating in an inert solvent such as toluene at temperatures of 180–250 °C is required for these rearrangements to proceed, often in the presence of a strong base.

The base-catalyzed intramolecular cyclization of appropriately substituted 4-alkyn-1-ols, followed by *in situ* Claisen rearrangement, has been investigated by Ovaska and coworkers (Scheme 6.85) [179]. The tandem cyclization-Claisen rearrangements were best carried out in *N,N*-dimethylformamide or phenetole as solvent in the pres-

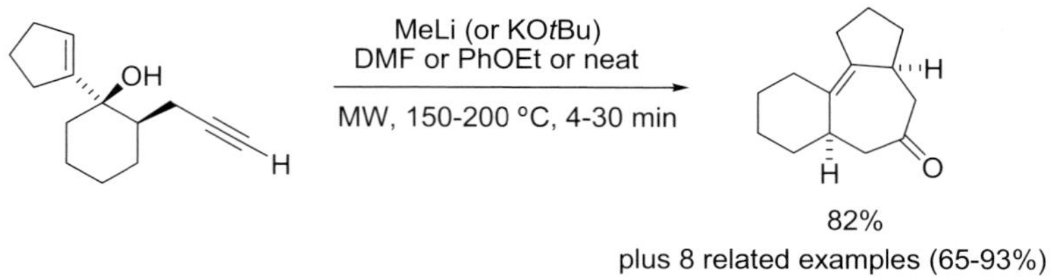

Scheme 6.85 Tandem 5-*exo* cyclization/Claisen rearrangement.

ence of 10 mol% of methyllithium base. In most cases, the resulting cycloheptanoid ring systems were produced in high yields in a matter of minutes upon microwave irradiation at 150–200 °C. Some of the reactions were also performed under solvent-free conditions, providing similar isolated product yields. Several other bicyclo[5.3.0]decane ring systems could also be constructed from relatively simple acetylenic alcohols using this strategy. Microwave-assisted siloxy-cope rearrangements have been described by Davies and Beckwith [474].

6.4.3
Squaric Acid–Vinylketene Rearrangements

Exploring synthetic routes to analogues of the furaquinocin antibiotics, Trost and coworkers utilized a microwave-assisted squaric acid–vinylketene rearrangement to synthesize a dimethoxynaphthoquinone, a protected analogue of furaquinocin E (Scheme 6.86) [180]. Although successfully applied in closely related series of transformations, use of the conventional rearrangement conditions (toluene, 110 °C) in this case led to incomplete conversion. Thus, the reaction was attempted by microwave heating at 180 °C, which afforded an acceptable yield of 58% of the desired product after oxidation to the naphthoquinone.

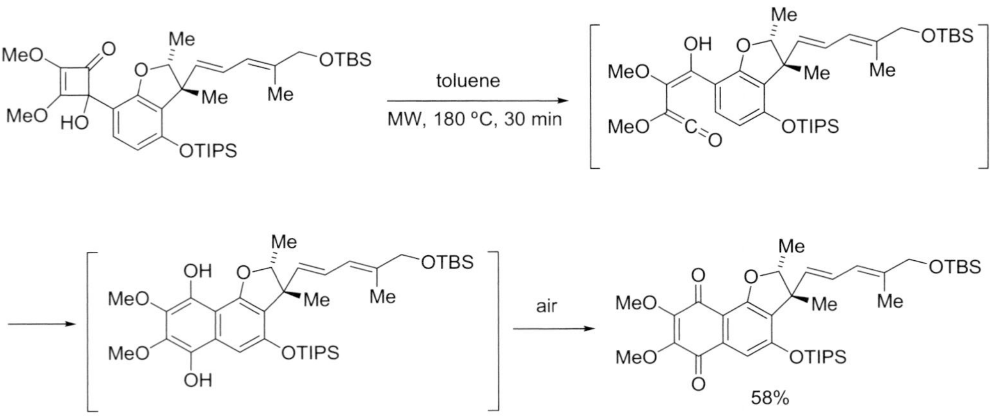

Scheme 6.86 Squaric acid–vinylketene rearrangement.

6.4.4
Vinylcyclobutane–Cyclohexene Rearrangements

A recent publication by the group of Baran has disclosed the total synthesis of ageli-ferin, an antiviral agent with interesting molecular architecture [181]. Microwave irradiation of sceptrin, another natural product, for just 1 min at 195 °C in water as solvent (see Section 4.3.3.1) under sealed-vessel conditions provided ageliferin in 40% yield, along with 52% of recovered starting material (Scheme 6.87). Remark-ably, when the reaction was attempted without microwaves at the same temperature, only the starting material and decomposition products were observed. Microwave heating of sceptrin in deuteriomethanol at 80 °C for 5 min led exclusively to [D$_2$]sceptrin in quantitative yield (see also Scheme 6.173).

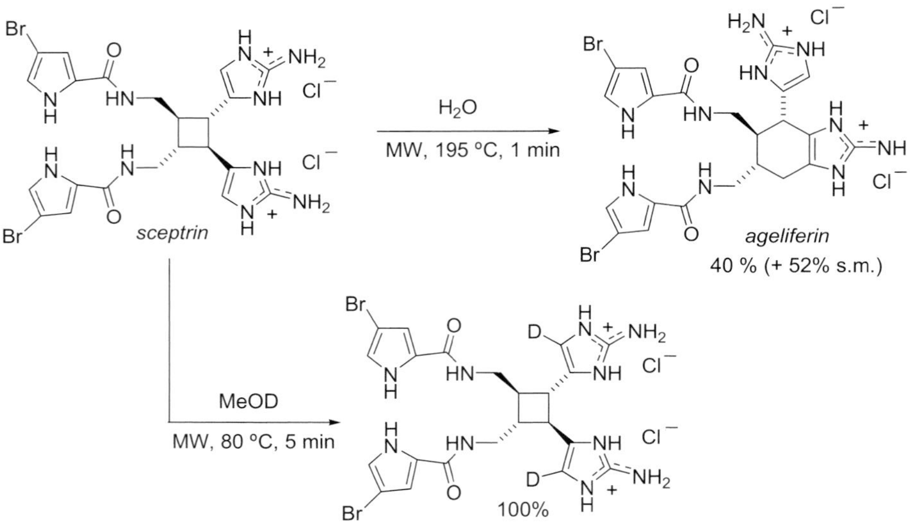

Scheme 6.87 Vinylcyclobutane–cyclohexene rearrangements.

6.4.5
Miscellaneous Rearrangements

The substance 4,12-dibromo[2.2]paracyclophane is the key intermediate en route to several functional C_2-symmetric planar-chiral 4,12-disubstituted[2.2]paracyclo-phanes. Braddock and coworkers have shown that this important intermediate can be obtained by microwave-assisted isomerization of 4,16-dibromo[2.2]paracyclo-phane, itself readily prepared by bromination of [2.2]paracyclophane (Scheme 6.88) [182]. By performing the isomerization in *N,N*-dimethylformamide as solvent (microwave heating at 180 °C for 6 min), in which the pseudo-*para* isomer is insolu-

Scheme 6.88 Isomerization of 4,16-dibromo[2.2]paracyclophane.

ble at room temperature whereas the desired pseudo-*ortho* isomer is soluble, the authors were able to isolate the desired product in 38% yield, with 43% recovery of the starting material.

Microwave-assisted Wolff [183], Curtius [184], and Ferrier rearrangements [185] have also been reported, albeit employing microwave irradiation under uncontrolled conditions.

6.5
Diels–Alder Cycloaddition Reactions

Cycloaddition reactions were among the first transformations to be studied using microwave heating technology and numerous examples have been summarized in previous review articles and book chapters [1–13]. Cycloaddition reactions have been performed with great success with the aid of microwave heating, as conventionally they often require the use of harsh conditions such as high temperatures and long reaction times. Scheme 6.89 shows two examples of Diels–Alder cycloadditions per-

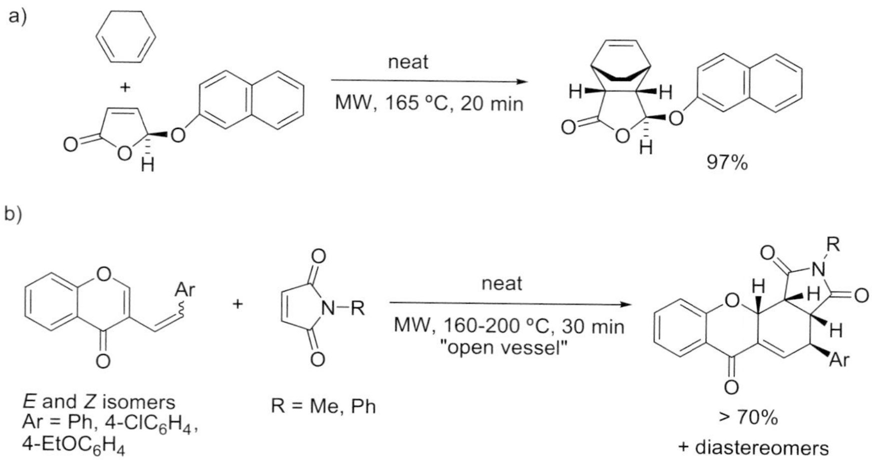

Scheme 6.89 Diels–Alder cycloaddition reactions under solvent-free conditions.

formed by microwave dielectric heating. In both cases, the diene and dienophile were reacted neat without the addition of a solvent. For the transformation shown in Scheme 6.89 a, described by Trost and coworkers, irradiation for 20 min at 165 °C (or for 60 min at 150 °C) gave the cycloadduct in near quantitative yield [186]. In the process reported by the group of de la Hoz (Scheme 6.89 b), open-vessel irradiation of 3-styryl chromones with maleimides at 160–200 °C for 30 min furnished the tetracyclic adducts indicated, along with minor amounts of other diastereoisomers [187].

Diels–Alder cycloaddition of 5-bromo-2-pyrone with the electron-rich *tert*-butyldi-methylsilyl (TBS) enol ether of acetaldehyde, using superheated dichloromethane as solvent, has been investigated by Joullié and coworkers (Scheme 6.90) [188]. While the reaction in a sealed tube at 95 °C required 5 days to reach completion, the antici-pated oxabicyclo[2.2.2]octenone core was obtained within 6 h by microwave irradia-tion at 100 °C. The *endo* adduct was obtained as the main product. Similar results and selectivities were also obtained with a more elaborate bis-olefin, although the desired product was obtained in diminished yield. Related cycloaddition reactions involving 2-pyrones have been discussed in Section 2.5.3 (see Scheme 2.4) [189].

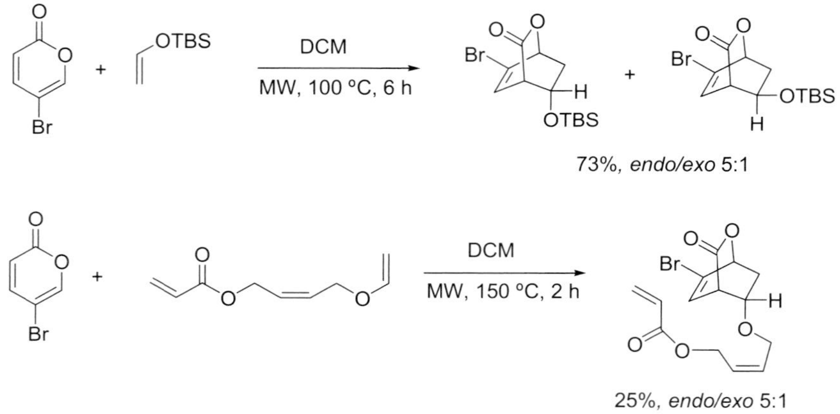

Scheme 6.90 Synthesis of stable oxabicyclo[2.2.2]octenones.

Applying the concept of using solvents doped with ionic liquids in order to allow microwave heating to high temperatures (see Section 4.3.3.2), Leadbeater and Tore-nius studied the Diels–Alder reaction between 2,3-dimethylbutadiene and methyl acrylate (Scheme 6.91) [190]. This reaction is traditionally performed in toluene or

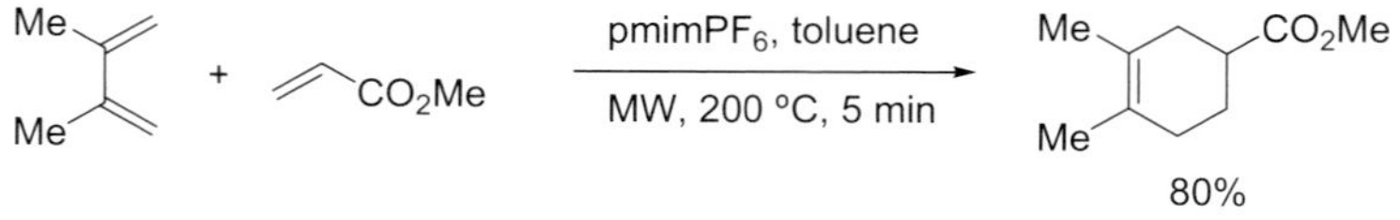

Scheme 6.91 Diels–Alder cycloaddition reactions in ionic liquid-doped solvents.

xylene and takes 18–24 h to reach completion, giving yields of cycloadduct ranging from 9 to 90% depending on the solvent used and the temperature. Using a mixture of toluene and the ionic liquid 1-(2-propyl)-3-methylimidazolium hexafluorophosphate (pmimPF$_6$), the authors were able to perform the cycloaddition within 5 min to obtain the product in 80% yield, this representing a significant rate enhancement over the conventional methods.

Hong and coworkers have investigated the cycloaddition chemistry of fulvenes with a wide variety of alkenes and alkynes in great detail [191]. As one example, the reaction of 6,6-dimethylfulvene with benzoquinone is shown in Scheme 6.92. Under microwave conditions in dimethyl sulfoxide (DMSO) at 120 °C, an unusual hetero-[2+3] adduct was formed in 60% yield, the structure of which was determined by X-ray crystallography. The adduct is a structural analogue of the natural products aplysin and pannellin and differs completely from the reported thermal (benzene, 80 °C) Diels–Alder cycloaddition product of the fulvene and benzoquinone (Scheme 6.92) [191].

Yu and coworkers have reported a recyclable organotungsten Lewis acid as a catalyst for Diels–Alder cycloaddition reactions performed in water or ionic liquids

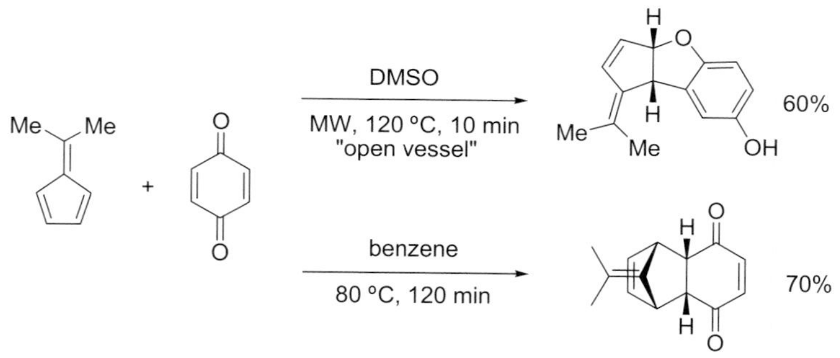

Scheme 6.92 Thermal versus microwave-assisted Diels–Alder cycloaddition reactions of fulvenes with benzoquinones.

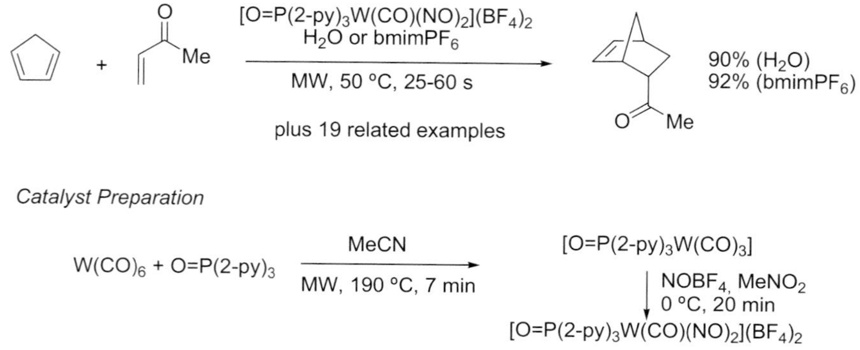

Scheme 6.93 Organotungsten Lewis acid-catalyzed Diels–Alder cycloaddition reactions.

(Scheme 6.93) [192]. Using either of the two solvent systems, all studied cycloaddition reactions were completed in less than 1 min upon microwave irradiation at 50 °C employing 3 mol% of the catalyst. An additional advantage of using the ionic liquid 1-butyl-3-methylimidazolium hexafluorophosphate (bmimPF$_6$) as solvent is that it facilitates catalyst recycling.

Microwave heating has also been employed for performing retro-Diels–Alder cycloaddition reactions, as exemplified in Scheme 6.94. In the context of preparing optically pure cross-conjugated cyclopentadienones as precursors to arachidonic acid derivatives, Evans, Eddolls, and coworkers performed microwave-mediated Lewis acid-catalyzed retro-Diels–Alder reactions of suitable *exo*-cyclic enone building blocks [193, 194]. The microwave-mediated transformations were performed in dichloromethane at 60–100 °C with 0.5 equivalents of methylaluminum dichloride as catalyst and 5 equivalents of maleic anhydride as cyclopentadiene trap. In most cases, the reaction was stopped after 30 min since continued irradiation eroded the product yields. The use of short bursts of microwave irradiation minimized double-bond isomerization.

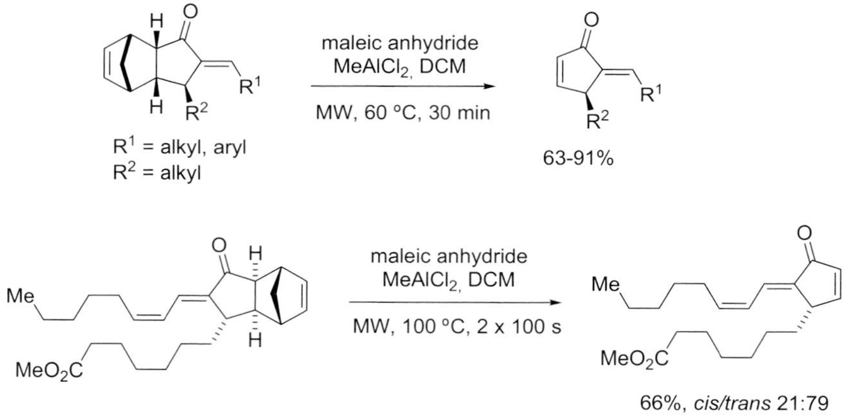

Scheme 6.94   Retro Diels–Alder reactions.

Inter- and intramolecular hetero-Diels–Alder cycloaddition reactions in a series of functionalized 2-(1*H*)-pyrazinones have been studied in detail by the groups of Van der Eycken and Kappe (Scheme 6.95) [195–197]. In the intramolecular series, cycloaddition of alkenyl-tethered 2-(1*H*)-pyrazinones required 1–2 days under conventional thermal conditions involving chlorobenzene as solvent under reflux conditions (132 °C). Switching to 1,2-dichloroethane doped with the ionic liquid 1-butyl-3-methylimidazolium hexafluorophosphate (bmimPF$_6$) and sealed-vessel microwave technology, the same transformations were completed within 8–18 min at a reaction temperature of 190 °C (Scheme 6.95 a) [195]. Without isolating the primary imidoyl chloride cycloadducts, rapid hydrolysis was achieved by the addition of small amounts of water and subjecting the reaction mixture to further microwave irradia-

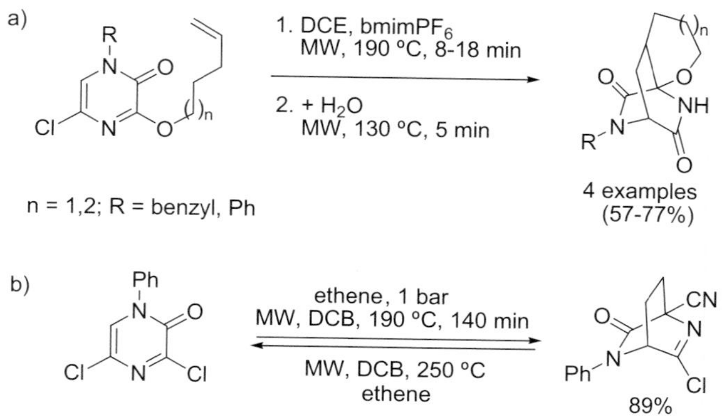

a)

1. DCE, bmimPF$_6$
MW, 190 °C, 8-18 min

2. + H$_2$O
MW, 130 °C, 5 min

n = 1,2; R = benzyl, Ph

4 examples
(57-77%)

b)

ethene, 1 bar
MW, DCB, 190 °C, 140 min

MW, DCB, 250 °C
ethene

89%

Scheme 6.95 Hetero-Diels–Alder cycloaddition reactions of 2(1H)-pyrazinones.

tion (130 °C, 5 min). The isolated overall yields of the desired cycloadducts were in the same range as reported for the conventional thermal protocols.

In the intermolecular series, Diels–Alder cycloaddition of ethene to the pyrazinone heterodiene led to the expected bicyclic cycloadduct (Scheme 6.95 b) [195]. The details of this transformation, performed in pre-pressurized reaction vessels, are described in Section 4.3.2 [196]. Similar cycloaddition reactions have also been studied on a solid phase (Scheme 7.58) [197].

Microwave-assisted Diels–Alder cycloaddition reactions using water-soluble aquo-cobaloxime complexes have been reported by Welker and coworkers [198]. Many more examples of microwave-assisted cycloaddition processes leading to heterocycles are described in Section 6.24.

6.6
Oxidations

The osmium-catalyzed dihydroxylation reaction, that is, the addition of osmium tetroxide to alkenes producing a vicinal diol, is one of the most selective and reliable of organic transformations. Work by Sharpless, Fokin, and coworkers has revealed that electron-deficient alkenes can be converted to the corresponding diols much more efficiently when the pH of the reaction medium is maintained on the acidic side [199]. One of the most useful additives in this context has proved to be citric acid (2 equivalents), which, in combination with 4-methylmorpholine N-oxide (NMO) as a reoxidant for osmium(VI) and potassium osmate [K$_2$OsO$_2$(OH)$_4$] (0.2 mol%) as a stable, non-volatile substitute for osmium tetroxide, allows the conversion of many olefinic substrates to their corresponding diols at ambient temperatures. In specific cases, such as with extremely electron-deficient alkenes (Scheme 6.96), the reaction has to be carried out under microwave irradiation at 120 °C, to produce in the illustrated case an 81% isolated yield of the pure diol [199].

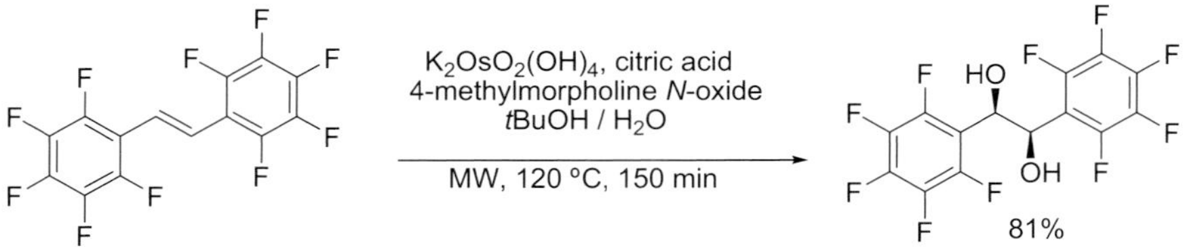

Scheme 6.96 Osmium-catalyzed dihydroxylation of electron-deficient alkenes.

The asymmetric allylic oxidation of bridged bicyclic alkenes using a copper-catalyzed symmetrizing-desymmetrizing Kharasch–Sosnovsky reaction has been reported by the Clark group (Scheme 6.97) [200]. Here, a racemic bridged bicyclic alkene was used as a starting material in the presence of 5 mol% of a copper catalyst and 6 mol% of a chiral pyridyl-bisoxazoline ligand. Microwave heating in acetonitrile with a suitable carbonate substrate provided the allylic oxidation product in 84% yield with 70% ee. A control experiment using conventional heating at the same temperature required 2 days to provide the same product in 80% yield with 66% ee.

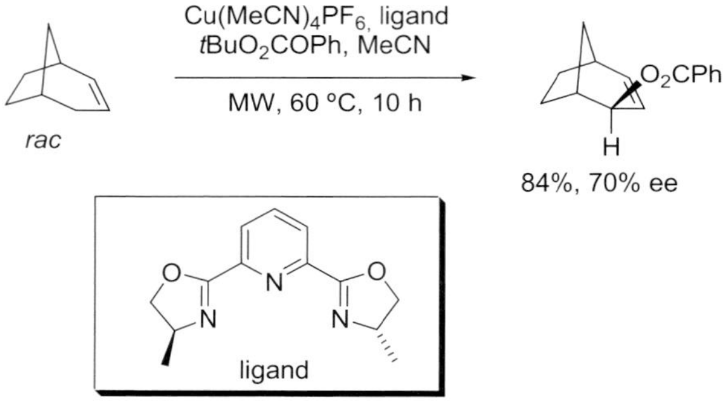

Scheme 6.97 Copper-catalyzed asymmetric allylic oxidation of bridged bicyclic alkenes.

Another industrially important oxidation reaction is the conversion of cyclohexene to adipic acid. The well-known Noyori method uses hydrogen peroxide, catalytic tungstate, and a phase-transfer catalyst to afford the clean oxidation of cyclohexene to adipic acid. Ondruschka and coworkers have demonstrated that a modified protocol employing microwave heating without solvent gave comparable yields of the desired product, but at a much faster rate (Scheme 6.98) [201]. Optimum results were obtained using excess hydrogen peroxide as the oxidant, 1 mol% of methyltrioctylammonium hydrogensulfate as phase-transfer catalyst, and 1 mol% of sodium tungstate as a catalyst for 90 min under microwave reflux conditions (ca. 100 °C).

Rhodium- and ruthenium-catalyzed hydrogen-transfer type oxidations of primary and secondary alcohols have recently been reported by Matsubara and coworkers (Scheme 6.99) [202]. Thus, secondary alcohols were converted into the correspond-

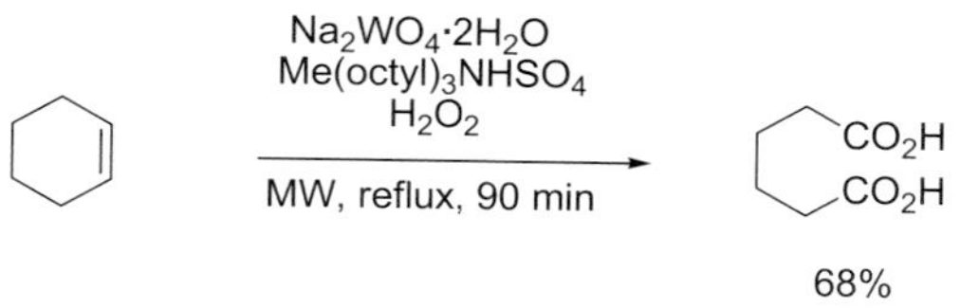

Scheme 6.98 Tungsten-catalyzed oxidation of cyclohexene.

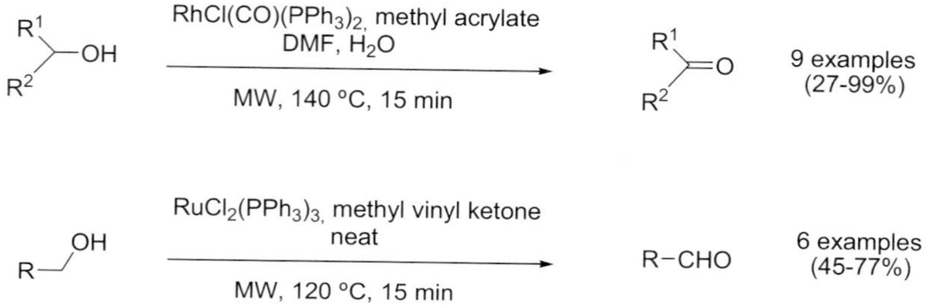

Scheme 6.99 Hydrogen-transfer-type oxidations.

ing ketones using 2 equivalents of methyl acrylate as hydrogen acceptor and 5 mol% of bis(triphenylphosphine)rhodium(I) carbonyl chloride [RhCl(CO)(PPh₃)₂] as transition metal catalyst in a water/N,N-dimethylformamide solvent mixture. Microwave irradiation at 140 °C for 15 min provided the desired ketones in moderate to excellent yields, whereas without microwave irradiation only the starting materials were recovered. Primary alcohols were not oxidized under these conditions but required a ruthenium catalyst. Optimum conditions for the oxidation of primary alcohols involved the utilization of 2.5 mol% of tris(triphenylphosphine)ruthenium(II) dichloride [RuCl₂(PPh₃)₃] and 2 equivalents of methyl vinyl ketone under solvent-free conditions (120 °C, 15 min) [202].

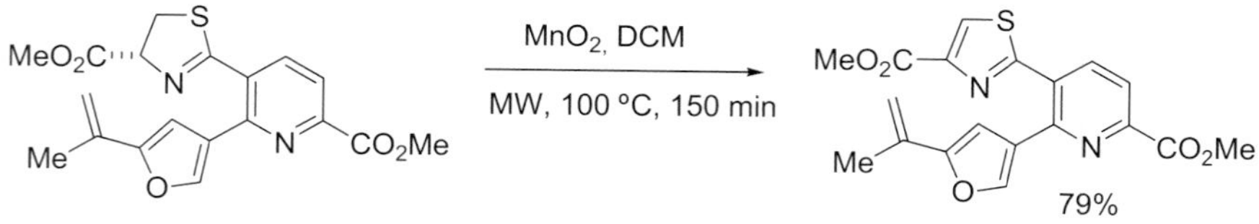

Scheme 6.100 Dehydrogenation of thiazolidines.

The oxidation of a thiazolidine derivative to the corresponding thiazole using activated manganese dioxide in dichloromethane at 100 °C is shown in Scheme 6.100. Further manipulation of this molecule led to dimethyl sulfomycinamate, a methanolysis product of the thiopeptide antibiotic sulfomycin I [203].

The tetrahydrofuranylation of alcohols with hypervalent iodine compounds is described in [479].

6.7
Catalytic Transfer Hydrogenations

A popular reduction process carried out under microwave conditions is catalytic transfer hydrogenation using hydrogen donors such as ammonium formate and a palladium catalyst such as palladium-on-charcoal. Two transformations involving the reduction of aromatic nitro groups are shown in Scheme 6.101. The reaction can be carried out either under sealed-vessel conditions (Scheme 6.101 a) [204] or at atmospheric pressure under reflux (Scheme 6.101 b) [205]. In the latter case, the reaction takes somewhat longer as the temperature is limited by the boiling point of the solvent. Under sealed-vessel conditions, reactions are generally faster, but care must be taken not to generate too much pressure in the microwave reaction vessel due to the formation of ammonia, carbon dioxide, and hydrogen.

Catalytic transfer hydrogenations for the reduction of carbon–carbon double bonds are illustrated in Scheme 4.18. Reductions of azide functionalities to amines with lipases suspended in organic media under microwave conditions have also been reported [206].

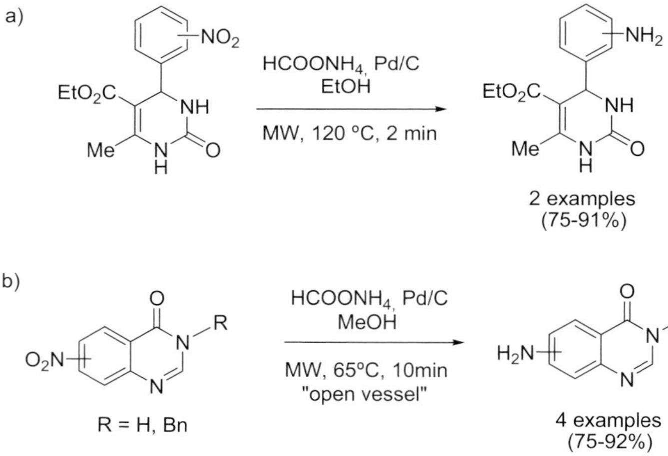

Scheme 6.101   Catalytic transfer hydrogenations.

6.8
Mitsunobu Reactions

The Mitsunobu reaction offers a powerful stereochemical transformation. This reaction is very efficient for inverting the configuration of chiral secondary alcohols since a clean S_N2 process is generally observed ("Mitsunobu inversion"). Considering the fact that Mitsunobu chemistry is typically carried out at or below room temperature, high-temperature Mitsunobu reactions performed under microwave conditions would appear to have little chance of success. In 2001, however, it was established by the Kappe group that Mitsunobu chemistry can indeed be carried out at high temperatures, in the context of an enantio-convergent approach to the aggregation pheromones (R)- and (S)-sulcatol (Scheme 6.102 a) [207]. While the conventional Mitsunobu protocol carried out at room temperature proved to be extremely sluggish, complete conversion of (S)-sulcatol to the (R)-acetate (S_N2 inversion) using essentially the standard Mitsunobu conditions (1.9 equivalents of diisopropylazodicarboxylate, 2.3 equivalents of triphenylphosphine) was achieved within 5 min at 180 °C under sealed-vessel microwave conditions. Despite the high reaction temperatures, no by-products could be identified in these Mitsunobu experiments, with enantiomeric purities of sulcatol (R)-acetate being >98% ee (80% isolated yield).

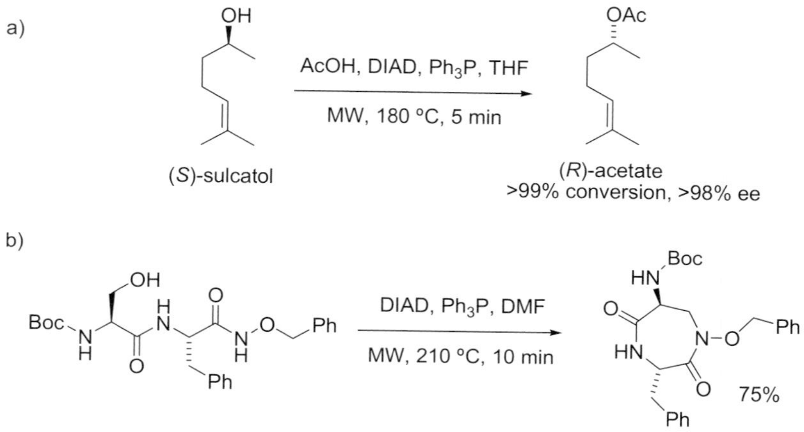

Scheme 6.102 Mitsunobu reactions.

An application of these rather unusual high-temperature Mitsunobu conditions in the preparation of conformationally constrained peptidomimetics based on the 1,4-diazepan-2,5-dione core has recently been disclosed by the group of Taddei (Scheme 6.102 b) [208]. Cyclization of a hydroxy hydroxamate dipeptide using the DIAD/Ph$_3$P microwave conditions (210 °C, 10 min) provided the desired 1,4-diaze-

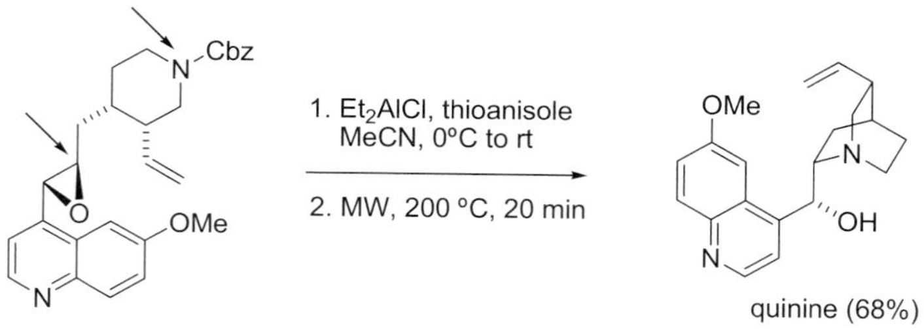

Scheme 6.103 Intramolecular S_N2 reaction in the total synthesis of quinine.

pan-2,5-dione in 75% isolated yield. Standard room temperature conditions (*N,N*-dimethylformamide, 12 h) were significantly less efficient and gave only 46% of the desired compound. These transformations could also be carried out on a polystyrene resin [208].

Another microwave-mediated intramolecular S_N2 reaction forms one of the key steps in a recent catalytic asymmetric synthesis of the cinchona alkaloid quinine by Jacobsen and coworkers [209]. The strategy to construct the crucial quinuclidine core of the natural product relies on an intramolecular S_N2 reaction/epoxide ring-opening (Scheme 6.103). After removal of the benzyl carbamate (Cbz) protecting group with diethylaluminum chloride/thioanisole, microwave heating of the acetonitrile solution at 200 °C for 2 min provided a 68% isolated yield of the natural product as the final transformation in a 16-step total synthesis.

6.9
Glycosylation Reactions and Related Carbohydrate-Based Transformations

Glycosylation reactions involving oxazoline donors are generally rather slow and require prolonged reaction times due to the low reactivity of the donors. Oscarson and coworkers have reported the preparation of spacer-linked dimers of *N*-acetyl-lactosamine using microwave-assisted pyridinium triflate-promoted glycosylations with oxazoline donors (Scheme 6.104 a) [469]. Using 2.2 equivalents each of the oxazoline donor and pyridinium triflate promoter, rapid and efficient coupling was

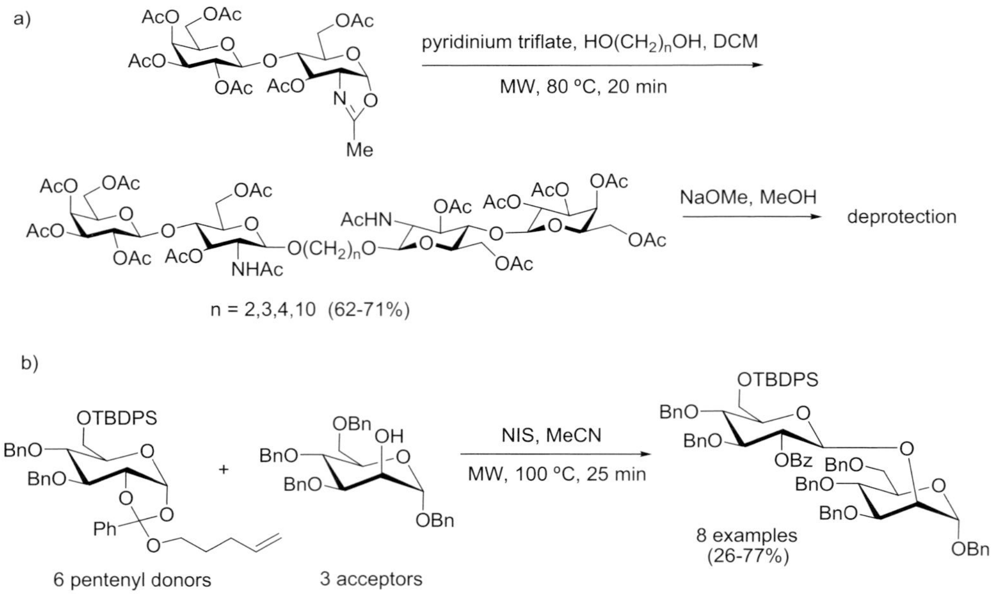

Scheme 6.104 Glycosylation reactions.

achieved in dichloromethane with four different diols. Employing 20 min of microwave irradiation at 80 °C, moderate to high yields of the dimers were obtained, with yields increased by 12–15% over the conventional process. Fraser-Reid and coworkers have recently described related saccharide couplings employing n-pentenyl glycosyl donors and N-iodosuccinimide (NIS) as promoter in acetonitrile (Scheme 6.104 b) [211].

The glucosamine residues in heparin-like glycosaminoglycans have been found to exist as amines, acetamides, and N-sulfonates. To develop a completely general, modular synthesis of heparin, three degrees of orthogonal nitrogen protection are required. The synthesis of fully N-differentiated heparin oligosaccharides has been demonstrated by Lohman and Seeberger (Scheme 6.105) [212]. One of the many synthetic steps involves the simultaneous introduction of N-diacetate and O-acetyl functionalities in a trisaccharide building block. Microwave irradiation in isopropenyl acetate as solvent in the presence of p-toluenesulfonic acid (pTsOH) at 90 °C for 5 h led to the desired product in 86% yield. This transformation could not be accomplished under a variety of thermal conditions, with only poor yields being achieved even after several days.

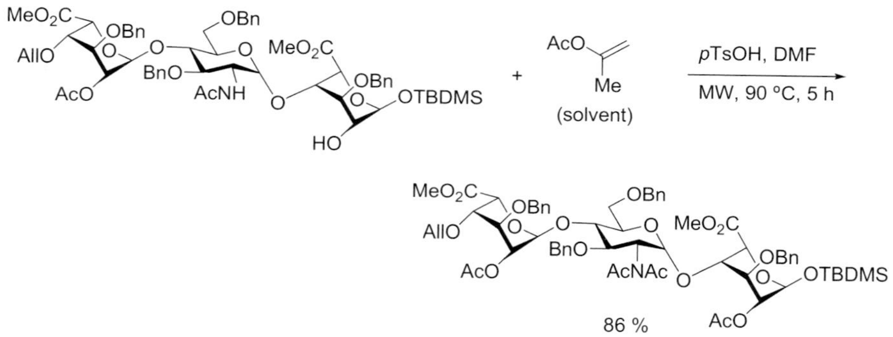

Scheme 6.105 Heparin oligosaccharide synthesis.

The reaction of modified carba-sugars with heterocyclic bases in the synthesis of constrained carbanucleosides has been studied in detail by the group of Sega (Scheme 6.106) [213]. Under conventional conditions employing sodium hydride or cesium carbonate as a base in N,N-dimethylformamide (120 °C, 48 h), the reaction between the two substrates shown in Scheme 6.106 was extremely sluggish and only provided modest yields of the desired carbanucleosides. Performing the reaction under microwave irradiation conditions (205 °C, 4 min) proved to be ineffective; only the starting materials were recovered. Switching to the ionic liquid 1-butyl-3-methylimidazolium tetrafluoroborate (bmimBF₄) as reaction medium (cesium carbonate base, 120 °C, 48 h) did not provide any product either. Only by applying the combination of microwave heating and the use of the ionic liquid as reaction medium could the transformation shown in Scheme 6.106 be successfully accomplished, providing a 64% isolated yield of the desired enantiomerically pure carbanucleoside after 4 min. Similar results were obtained with other heterocyclic bases [213].

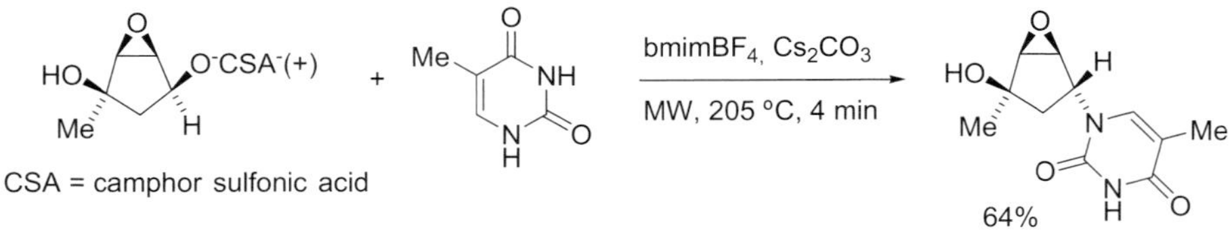

CSA = camphor sulfonic acid

Scheme 6.106 Carbanucleoside synthesis in ionic liquids.

In view of the need for new carbohydrate functionalization methods that mini-mize the employment of protecting groups, tin acetal-mediated regioselective func-tionalization methods are of great importance. A microwave-assisted tin-mediated, regioselective 3-O-alkylation of galactosides has recently been presented by Pieters and coworkers (Scheme 6.107) [214]. Here, dibutylstannylene acetals were generated *in situ* by heating the unprotected carbohydrates with 1.1 equivalents of dibutyltin oxide under microwave conditions at 150 °C for 5 min. After the addition of the alkylating reagent (5 equivalents) and tetrabutylammonium iodide (TBAI) (2.5 equivalents), the mixture was further irradiated at 110–170 °C for 6–20 min. Good yields of products were obtained with short overall reaction times.

A somewhat related microwave-promoted 5′-O-allylation of thymidine has been described by the Zerrouki group (Scheme 6.108) [215]. While the classical method for the preparation of 5′-O-allylthymidine required various protection steps (four synthetic steps in total), the authors attempted the direct allylation of thymidine under basic conditions. Employing sodium hydride as a base at room temperature in N,N-dimethylformamide resulted in the formation of per-allylated compounds along with the desired monoallylated product (75% yield). The best result was achieved when both the deprotonation with sodium hydride (1.15 equivalents) and the subsequent allylation (1.2 equivalents of allyl bromide) were conducted under

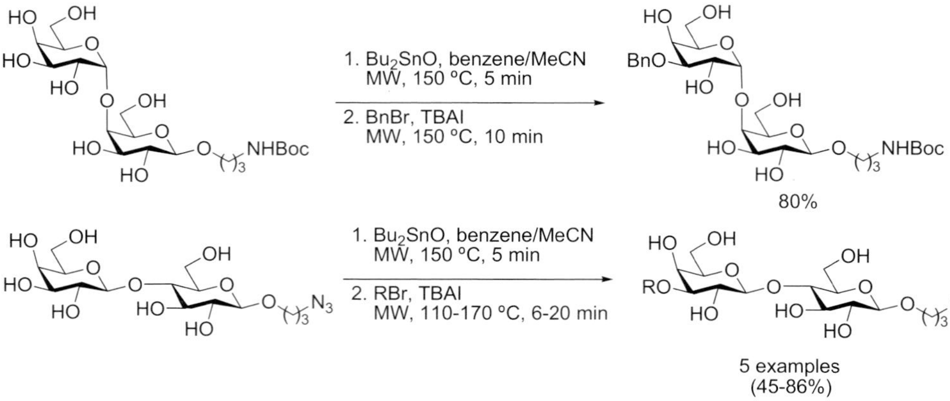

Scheme 6.107 Tin-mediated 3-O-alkylation of galactosides.

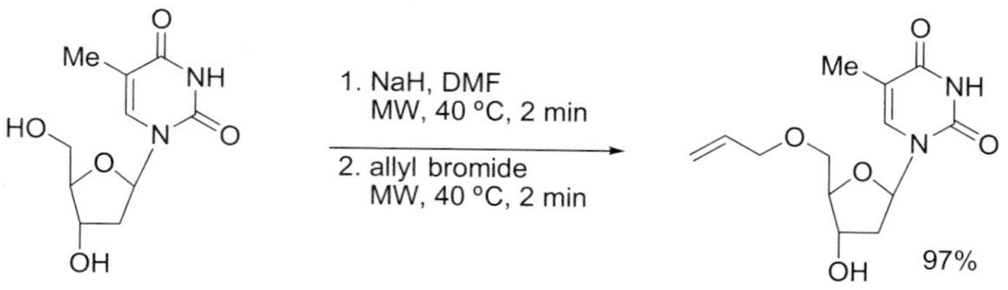

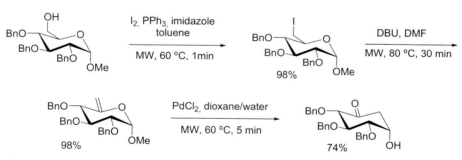

Scheme 6.108 Selective allylation of thymidine.

microwave irradiation for 2 min each. Under these conditions, a 97% yield of 5′-*O*-allylthymidine was isolated [215].

A rapid synthesis of carba-sugars has been reported by Pohl and coworkers starting from a protected D-glucose (Scheme 6.109) [216]. Microwave-enhanced iodination of the selectively protected glucose precursor was accomplished in 1 min in toluene at 60 °C in the presence of iodine and triphenylphosphine. Subsequent base-promoted elimination of hydrogen iodide in the presence of 1,8-diazabicyclo[5.4.0]undec-7-ene (DBU) required 30 min (*N,N*-dimethylformamide, 80 °C), and the key Ferrier rearrangement induced by palladium dichloride was also completed within 5 min of microwave irradiation. The development of this series of microwave-assisted transformations significantly shortened the time required to form the core carbocyclic structure from a protected glucose.

Scheme 6.109 Multistep synthesis of carbaglucose derivatives.

Bajugam and Flitsch [217] have described the synthesis of glycosylamines from mono-, di-, and trisaccharides by direct microwave-assisted Kochetkov amination (Scheme 6.110). The reaction was found to be effective with just a fivefold excess (*w/w*) of ammonium carbonate with respect to the sugar, as compared to the 40- or 50-fold excess needed under thermal conditions. All transformations were completed within 90 min in dimethyl sulfoxide as solvent, maintaining the vessel temperature at an apparent 40 °C using the "heating-while-cooling" technique (see Section 2.5.3).

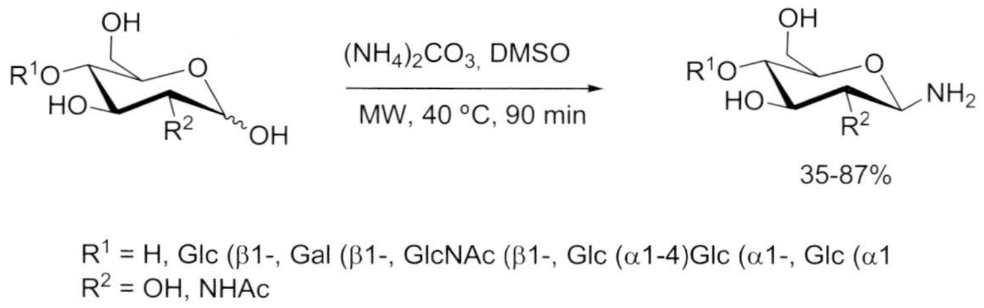

R^1 = H, Glc (β1-, Gal (β1-, GlcNAc (β1-, Glc (α1-4)Glc (α1-, Glc (α1
R^2 = OH, NHAc

Scheme 6.110 Synthesis of β-glycosylamines by Kochetkov amination.

Microwave-assisted epoxide ring-openings of 1,5:2,3-dianhydro-4,6-O-benzyl-idene-D-allitol with nucleobases have been reported [218]. Various rapid microwave-assisted protection and deprotection methods in the area of carbohydrate chemistry are known [210], and two general review articles on microwave-assisted carbohydrate chemistry were published in 2004 [219, 220].

6.10
Multicomponent Reactions

The Mannich reaction has been known since the early 1900s and since then has been one of the most important transformations leading to β-amino ketones. Al-though the reaction is powerful, it suffers from some disadvantages, such as the need for drastic reaction conditions, long reaction times, and sometimes low yields of products. Luthman and coworkers have reported microwave-assisted Mannich reactions that employed paraformaldehyde as a source of formaldehyde, a secondary amine in the form of its hydrochloride salt, and a substituted acetophenone (Scheme 6.111) [221]. Optimized reaction conditions involved the utilization of equi-molar amounts of reactants, dioxane as solvent, and microwave irradiation at 180 °C for 8–10 min to produce the desired β-amino ketones in moderate to good yields. Importantly, in several cases, the reaction was not only performed on a 2 mmol scale using a single-mode microwave reactor, but also was run on a 40 mmol scale using a dedicated multimode instrument (see Scheme 4.27).

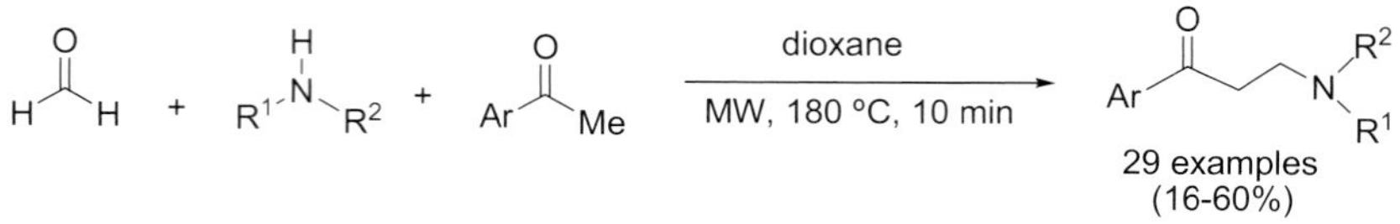

Scheme 6.111 Mannich reactions.

The group of Leadbeater reported a different type of Mannich reaction, which involved condensation of an aldehyde (1.5 equivalents) with a secondary amine and a terminal alkyne, in the presence of copper(I) chloride (10 mol%) to activate the

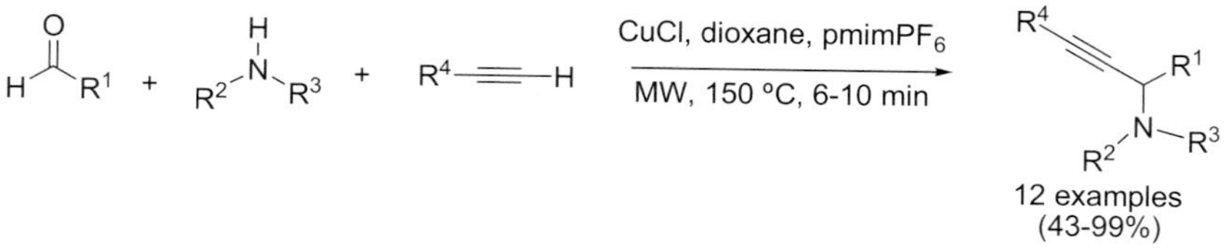

Scheme 6.112 Propargylamines by Mannich-type condensations.

latter (Scheme 6.112) [222]. Optimum yields of the target propargylamines were obtained by microwave irradiation of the three building blocks with the catalyst in dioxane doped with 1-(2-propyl)-3-methylimidazolium hexafluorophosphate (pmimPF$_6$) at 150 °C for 6–10 min. This strategy could also be applied in a solid-phase protocol utilizing a resin-bound amine (see Scheme 7.56) [222]. Tu and coworkers subsequently reported a diastereoselective modification of this method employing chiral amines such as (S)-proline methyl ester and a change of solvent to microwave-heated water [223]. A further variation of this three-component coupling has been described by Varma and coworkers utilizing a solvent-free approach [224].

Related to the Mannich three-component reaction is the Petasis or boronic-Mannich reaction, which involves the reaction between an aldehyde, an amine, and a boronic acid.

A high-speed microwave approach for the Petasis multicomponent reaction has been published by Tye [225]. Using a design of experiments (DoE) approach for reaction optimization (see Section 5.3.4), the best conditions for this transformation involved microwave heating of the components in dichloromethane (1 M concentration) at 120 °C for 10 min (Scheme 6.113). These conditions were successfully applied to a range of Petasis reactions employing either glyoxalin acid or salicylaldehyde as the carbonyl component along with a number of aryl/heteroaryl boronic acids and amine components. The crude products were converted to the corresponding methyl esters in order to simplify purification.

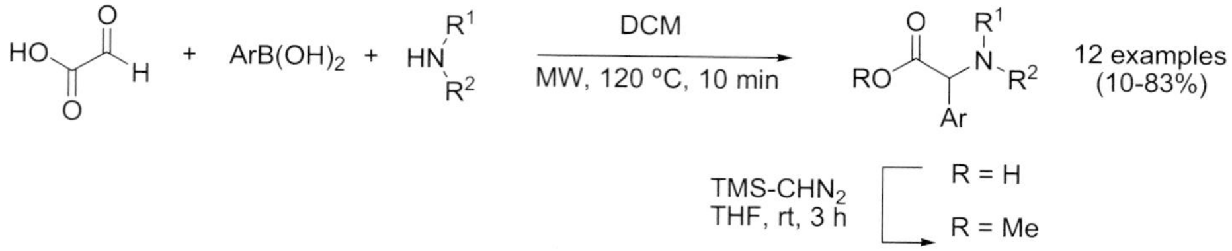

Scheme 6.113 The Petasis (boronic-Mannich) reaction.

Another frequently used multicomponent reaction is the Kindler thioamide synthesis (the condensation of an aldehyde, an amine, and sulfur). The Kappe group has described a microwave-assisted protocol utilizing a diverse selection of 13 aldehyde and 12 amine precursors in the construction of a representative 34-member library of substituted thioamides (Scheme 6.114) [226]. The three-component con-

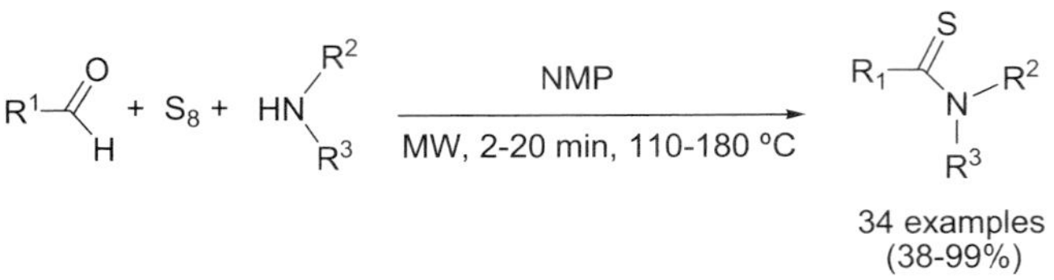

Scheme 6.114 Kindler thioamide synthesis.

densations of aldehydes, amines, and elemental sulfur were carried out using 1-methyl-2-pyrrolidone (NMP) as solvent, employing microwave flash heating at 110–180 °C for 2–20 min. A simple work-up protocol allowed the isolation of synthetically valuable primary, secondary, and tertiary thioamide building blocks in 83% average yield and with >90% purity. Other multicomponent reactions involving the formation of heterocycles are covered in Section 6.24.

6.11
Alkylation Reactions

A simple and atom-economical synthesis of hydrogen halide salts of primary amines directly from the corresponding halides, which avoids the production of significant amounts of secondary amine side products, has been described by researchers from Bristol-Myers Squibb [227]. Microwave irradiation of a variety of alkyl halides or tosylates in a commercially available 7 M solution of ammonia in methanol at 100–130 °C for 15 min to 2.5 h followed by evaporation of the solvent provided the corre-

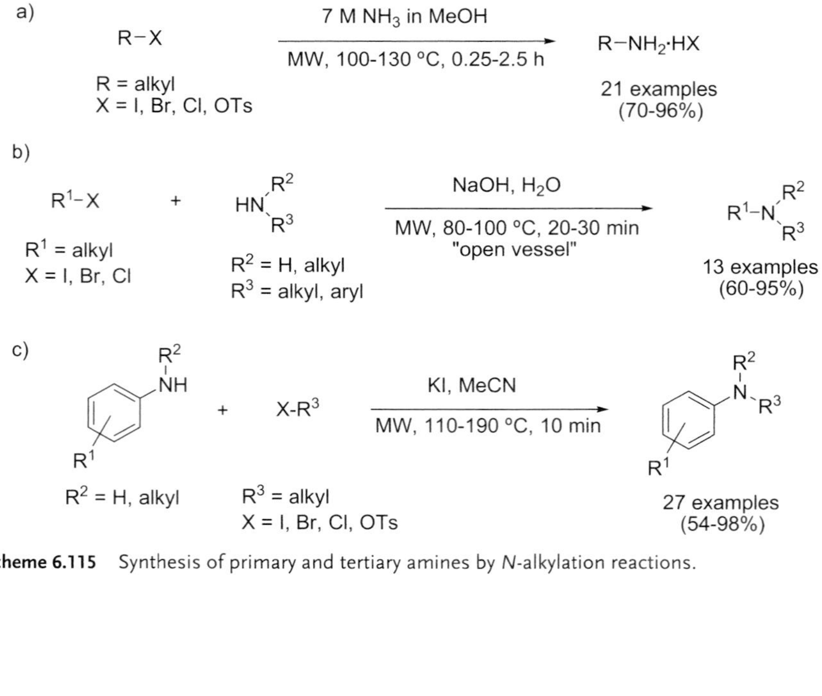

Scheme 6.115 Synthesis of primary and tertiary amines by *N*-alkylation reactions.

sponding primary amines directly as halide or sulfonate salts (Scheme 6.115 a). All reactions gave 2–4% or less of the symmetrical secondary amine side product. A distinct advantage of using the high-temperature microwave approach (100 °C, 15 min) was evident for benzyl chlorides, with which the corresponding reaction at room temperature (23 °C, 2 h) in some cases provided 20–25% of the bis-alkylated secondary amine side product. The microwave method under otherwise identical conditions led to less than 4% of this by-product [227].

Similar results were also obtained by Ju and Varma in the synthesis of tertiary amines from the corresponding alkyl halides and primary or secondary amines (Scheme 6.115 b) [228]. Here, water was used as a solvent and 1.1 equivalents of sodium hydroxide as a base.

The selective *N*-monoalkylation of anilines with alkyl halides and alkyl tosylates under microwave irradiation has been described by Romera and coworkers (Scheme

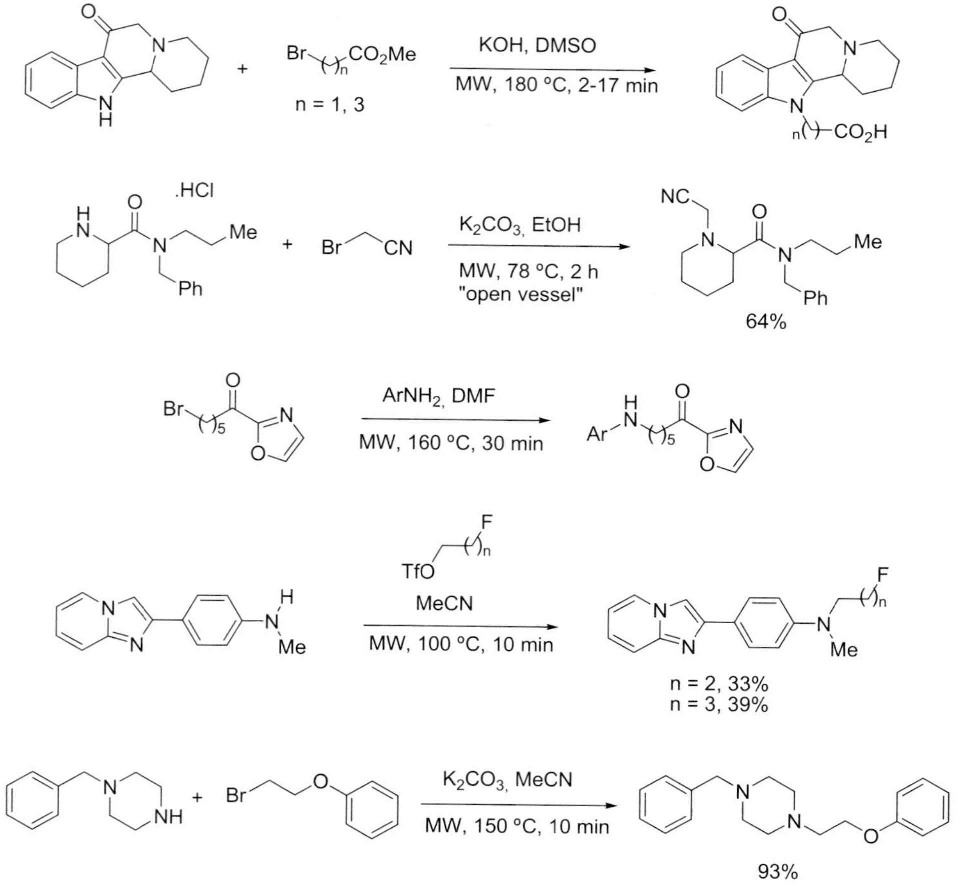

Scheme 6.116 Miscellaneous *N*-alkylation reactions.

6.115 c) [229]. Using 10 mol% of potassium iodide as a catalyst and acetonitrile as solvent, many different *N*-alkylanilines were obtained in good yields with only minor quantities of dialkylation products being formed.

A variety of related microwave-promoted *N*-alkylations involving more elaborate heterocyclic scaffolds are summarized in Scheme 6.116 [230–234]. Additional examples concerning the synthesis of ionic liquids can be found in Section 4.3.3.2.

Besides *N*-alkylation reactions, there are also reports in the literature concerning microwave-promoted *O*-alkylations. A mild method for the *O*-alkylation of phenols with alkyl bromides and chlorides has been developed by Wagner and coworkers (Scheme 6.117 a) [235]. The protocol is applicable to substrates that are sensitive to strong bases or to hydrolysis, or are difficult to extract from an aqueous phase. The procedure uses methanol as a solvent and a stoichiometric amount of potassium carbonate as a weak base. Optimum yields were obtained by heating the phenol with 1.2 equivalents of the alkyl bromide (or 3 equivalents of the less reactive chloride) at 100–140 °C for 15–30 min.

The Organ group has used a similar protocol for the preparation of a key building block required in the synthesis of styrene-based nicotinic acetylcholine receptor (nAChR) antagonists (Scheme 6.117 b) [49]. The authors employed 4 equivalents each of the alkyl halide and potassium carbonate in a water/ethanol mixture and thereby obtained a 97% yield of the alkylated phenol. Subsequent one-pot *N*-alkylation and Suzuki coupling led to the desired nAChR antagonists (see Scheme 6.21) [49]. A procedure for the *O*-alkylation of phenols utilizing polymer-supported bases has been described by Vasudevan and coworkers [232].

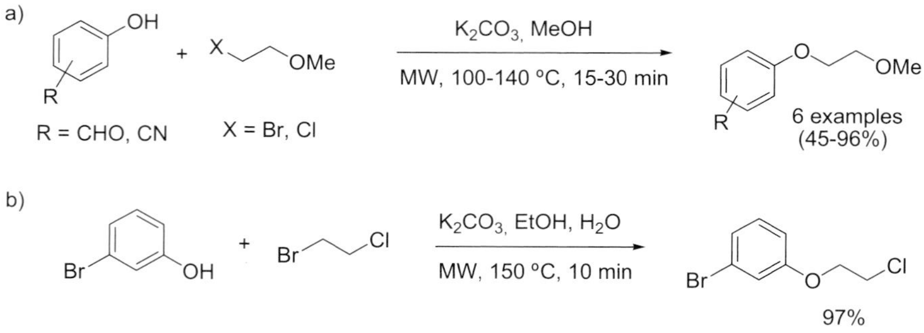

Scheme 6.117 *O*-Alkylations of phenols.

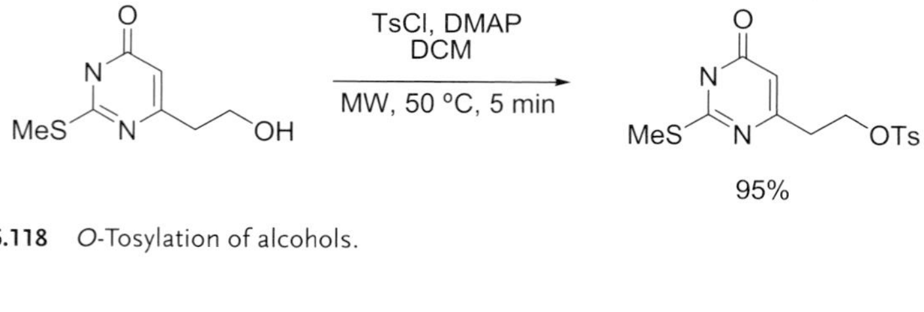

Scheme 6.118 *O*-Tosylation of alcohols.

A rapid *O*-tosylation of a primary alcohol with tosyl chloride (1 equivalent) in the presence of 4-(*N*,*N*-dimethylamino)pyridine (1 equivalent) has been reported by Botta and coworkers (Scheme 6.118) [236]. Microwave heating of the reaction mixture at 50 °C for 5 min provided the desired tosyl ester in 95% yield.

6.12
Nucleophilic Aromatic Substitutions

An alternative to the palladium-catalyzed Buchwald–Hartwig method and the related copper-catalyzed methods for C(aryl)–N, C(aryl)–O, and C(aryl)–S bond formations (see Section 6.2) is offered by nucleophilic aromatic substitution (*S*$_N$Ar) reactions. Thus, a benzene derivative bearing a leaving group may be reacted with an amine, although the arene must generally also bear an electron-withdrawing group. Such nucleophilic aromatic substitution reactions are notoriously difficult to perform and often require high temperatures and long reaction times. A number of publications report efficient nucleophilic aromatic substitutions driven by microwave heating involving either halogen-substituted aromatic or heteroaromatic systems. Scheme 6.119 summarizes some heteroaromatic systems and nucleophiles, along with the reaction conditions devised by Cherng in 2002 for microwave-assisted nucleophilic substitution reactions [237–239]. In general, the microwave-driven processes provide significantly higher yields of the desired products in much shorter reaction times. Typically, the substitutions are carried out neat or in a high-boiling, strongly microwave-absorbing solvent such as 1-methyl-2-pyrrolidone (NMP), hexamethyl phosphoramide (HMPA), or dimethyl sulfoxide (DMSO).

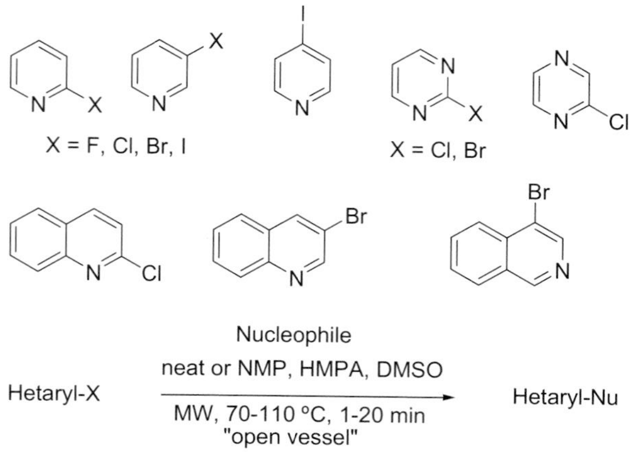

X = F, Cl, Br, I   X = Cl, Br

Hetaryl-X → Hetaryl-Nu

Nucleophile
neat or NMP, HMPA, DMSO
MW, 70-110 °C, 1-20 min
"open vessel"

Nucleophiles: PhNH$_2$, PhCH$_2$NH$_2$, piperidine, pyrazole, benzotriazole
PhSNa, MeSNa, PhONa, EtONa, PhCH$_2$OH, PhCH$_2$CN

Scheme 6.119 Nucleophilic aromatic substitution reactions involving halo-substituted *N*-heteroaromatic ring systems.

In more recent work by other researchers, sealed-vessel microwave technology has been utilized to access valuable medicinally relevant heterocyclic scaffolds or intermediates (Scheme 6.120) [240–245]. Additional examples not shown in Scheme 6.120 can be found in the most recent literature (see also Scheme 6.20) [246–249]. Examples of nucleophilic aromatic substitutions in the preparation of chiral ligands for transition metal-catalyzed transformations are displayed in Scheme 6.121 [106, 108].

A pyridyl bis-*N*-heterocyclic carbene (NHC) ligand has been prepared by Steel and Teasdale based on nucleophilic aromatic substitution of dichloroisonicotinic amides with *N*-methylimidazole (Scheme 6.122) [250]. Microwave heating of the neat reagents at 140 °C for 10 min provided a 91% yield of the corresponding bis-

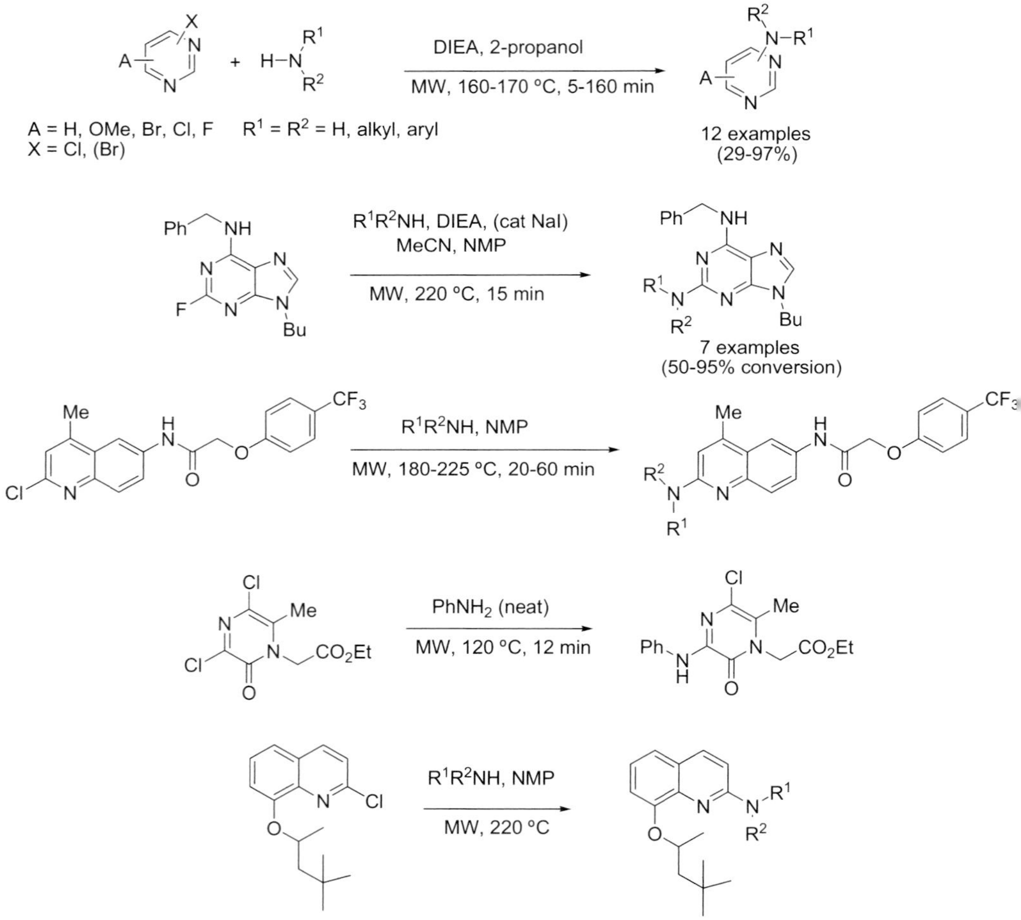

A = H, OMe, Br, Cl, F $R^1 = R^2 = H$, alkyl, aryl
X = Cl, (Br)

12 examples
(29–97%)

7 examples
(50–95% conversion)

Scheme 6.120 Nucleophilic aromatic substitution reactions on medicinally relevant azaheterocyclic cores.

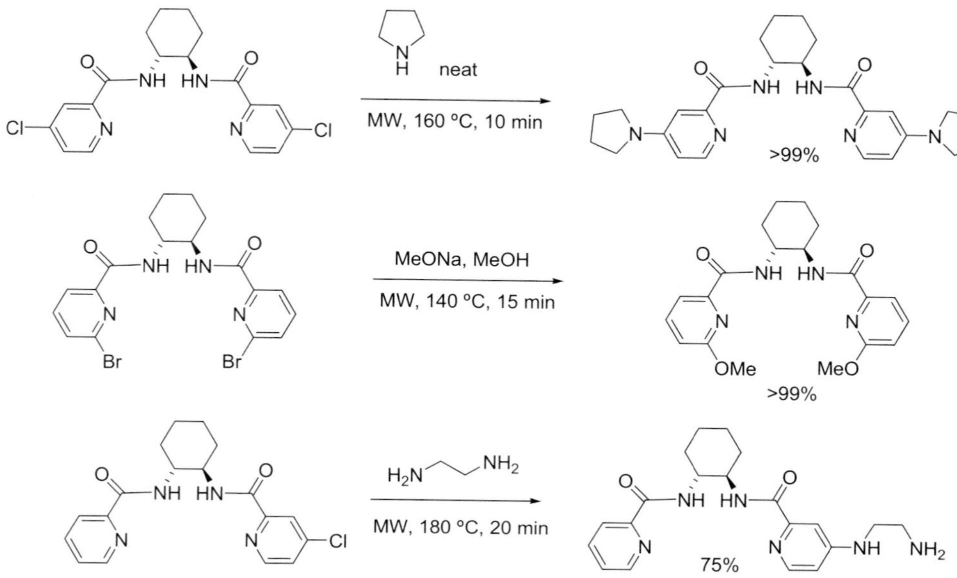

Scheme 6.121 Preparation of chiral ligands through nucleophilic aromatic substitution reactions.

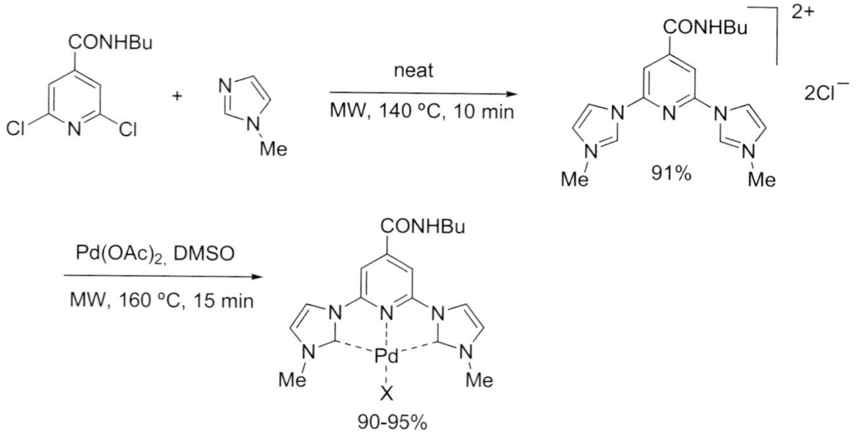

Scheme 6.122 Preparation of a pyridyl bis-*N*-heterocyclic carbene palladium complex.

hydrochloride salt, which was subsequently subjected to complexation by simple ligand exchange with palladium(II) acetate in degassed dimethyl sulfoxide at 160 °C (microwave irradiation, 15 min). The resulting palladium complex proved useful in Suzuki and Heck reactions, and furthermore was immobilized on Tentagel resin to provide a long-lived and recyclable heterogeneous palladium catalyst [250].

In 2001, researchers at Wyeth-Ayerst described the preparation of RFI-641, a potent and selective inhibitor of the respiratory syncytial virus (RSV) [251]. The final key step in the synthesis involved the coupling of a diaminobiphenyl with two equivalents of a chlorotriazine derivative under microwave irradiation (Scheme 6.123). The reaction was carried out in dimethyl sulfoxide/phosphate buffer/sodium hydroxide in an open vessel at 105 °C. This protocol provided RFI-641 in 10% isolated yield.

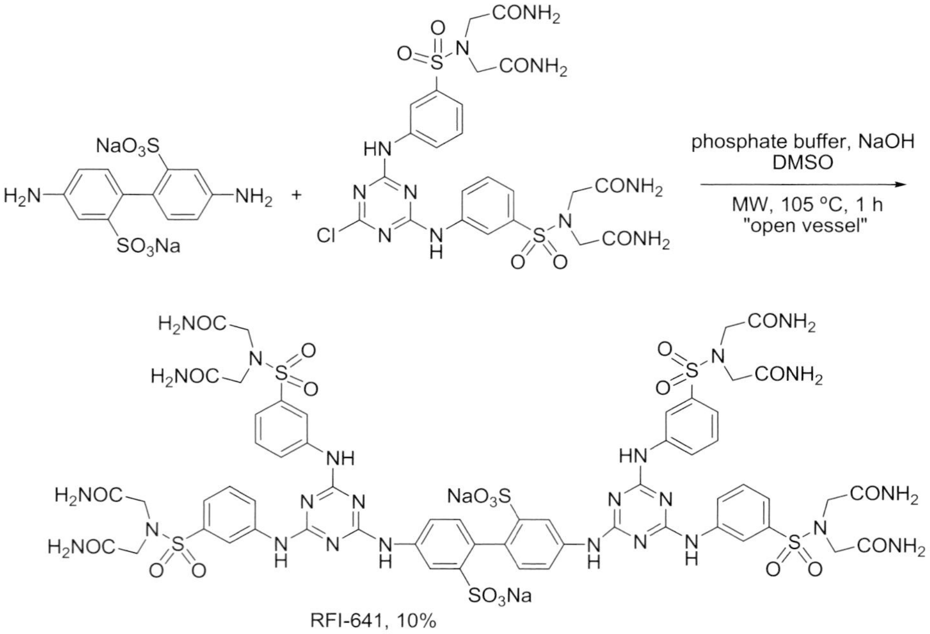

Scheme 6.123 Synthesis of the respiratory syncytial virus inhibitor RFI-641.

Nucleophilic aromatic substitution reactions are also possible on benzene rings when the aryl halide bears strongly electron-withdrawing substituents, such as nitro groups, in the *ortho* or *para* position with respect to the halide atom. Rapid substitution reactions can therefore be performed with nitro-substituted aryl fluorides such as 2,4-dinitrofluorobenzene (Sanger's reagent). In the example shown in Scheme 6.124, described by Cherng [252], amino acids are used as coupling partners with water as solvent. The presence of 2 equivalents of sodium hydrogen carbonate was found to be necessary in order to increase the nucleophilicity of the amino group. Typical reaction times under microwave irradiation were of the order of seconds, leading to *N*-arylated amino acids in excellent yields. Under conventional heating conditions (95 °C oil-bath temperature) no product was detected within 1 min. The influence of the halogen atom on the efficiency of the substitution reaction was also investigated. As expected, 2,4-dinitrobromobenzene and 2,4-dinitrochlorobenzene reacted at a much slower rate and gave inferior yields as com-

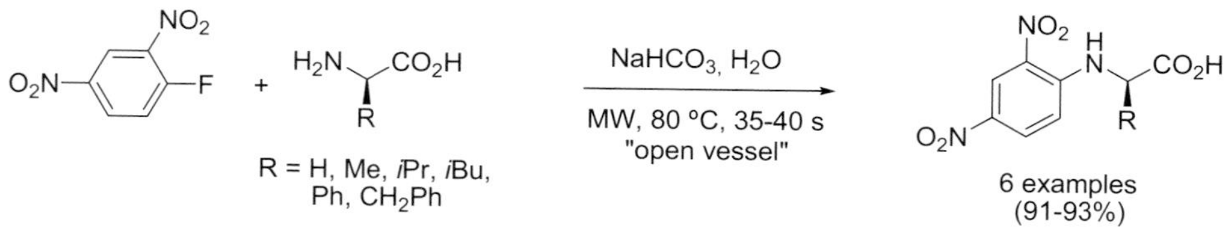

Scheme 6.124 Nucleophilic aromatic substitutions involving amino acids.

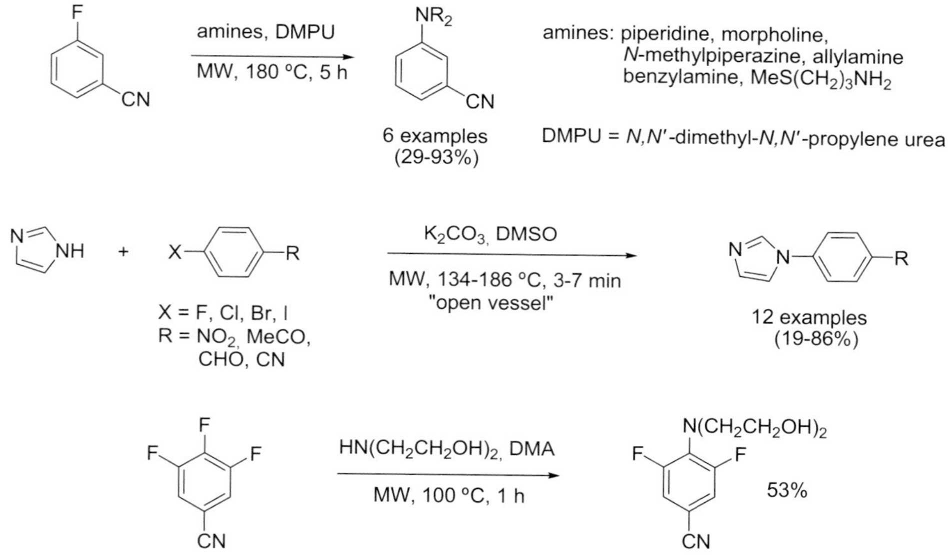

Scheme 6.125 Nucleophilic aromatic substitutions of activated aryl halides.

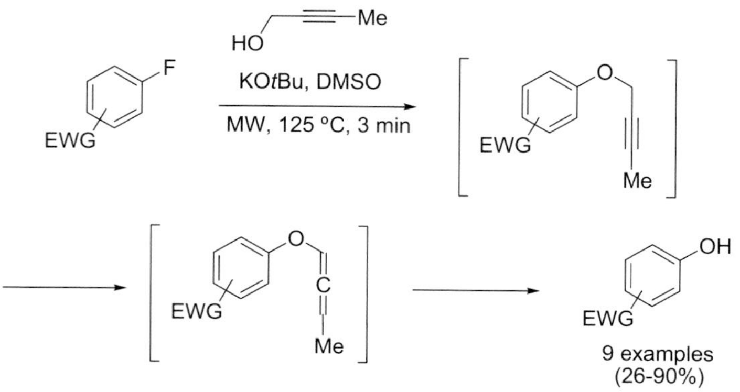

Scheme 6.126 One-pot conversion of aryl fluorides into phenols.

pared to the fluoro analogue (6–15 min, 48–64% yield) [252]. Similar nucleophilic aromatic substitution reactions involving activated aryl halides are shown in Scheme 6.125 [253–255].

A one-pot three-step conversion of aryl fluorides to phenols based on a consecutive nucleophilic aromatic substitution/isomerization/hydrolysis sequence has been reported by Levin and Du (Scheme 6.126) [256]. The authors discovered that 2-butyn-1-ol can function as a hydroxyl synthon through consecutive S_NAr displacement, *in situ* isomerization to the allenyl ether, and subsequent hydrolysis, to afford phenols rapidly and in good yields. In most cases, excesses of 2-butyn-1-ol (1–2 equivalents) and potassium *tert*-butoxide (2–4 equivalents) were required in order to achieve optimum yields.

Direct high-temperature microwave-assisted aminations of a variety of aryl halides [257] or aryl triflates [258] under transition metal-free conditions have also been reported, probably involving a benzyne mechanism.

6.13
Ring-Opening Reactions

6.13.1
Cyclopropane Ring-Openings

The fused cyclopropane 1-acetyl-4-phenyl-3-oxabicyclo[3.1.0]hexane shown in Scheme 6.127 is a key precursor for the synthesis of *endo,exo*-furofuranone derivatives, one of the largest subclasses of lignans, which are known to exhibit varied biological activities. Brown and coworkers found that while Lewis acid-assisted ring-opening of the cyclopropane ring in methanol at reflux temperature was extremely sluggish, the cyclopropane ring could be effectively opened at 120 °C by microwave irradiation (Scheme 6.127) [259, 260]. After screening a variety of conditions, the best results were achieved by employing a stoichiometric amount of zinc triflate in methanol. After 30 min at 120 °C, complete consumption of the starting bicycle was achieved, resulting in the formation of the desired methyl ether product and the corresponding enol ether.

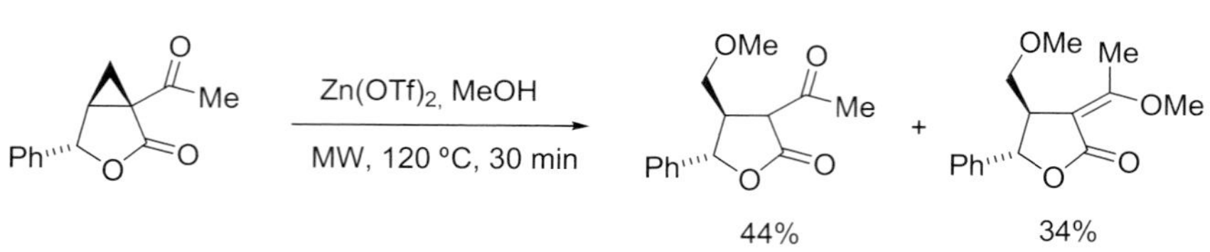

Scheme 6.127 Cyclopropane ring-opening.

6.13.2
Aziridine Ring-Openings

Aziridines with electron-withdrawing groups on the nitrogen atom readily undergo ring-opening when treated with nucleophiles. The nucleophilic attack generally occurs at the least hindered carbon atom. Lake and Moberg have utilized the ring-opening of chiral *N*-tosyl-aziridines for the generation of C_3-symmetric tripodal tris(sulfonamide) ligands (Scheme 6.128) [261]. Exposing aziridine and ammonia in a 4.5:1 molar ratio in methanol to microwave irradiation at 160 °C for 45 min led to the clean formation of the tris(sulfonamide) ligand without any formation of unwanted mono(tosyl)amine by-products. The isolated product yield was 88%, with 12% recovery of the aziridine starting material. In comparison, ring-opening at 50 °C required 4 days to reach completion and furnished only a 70% yield of the product.

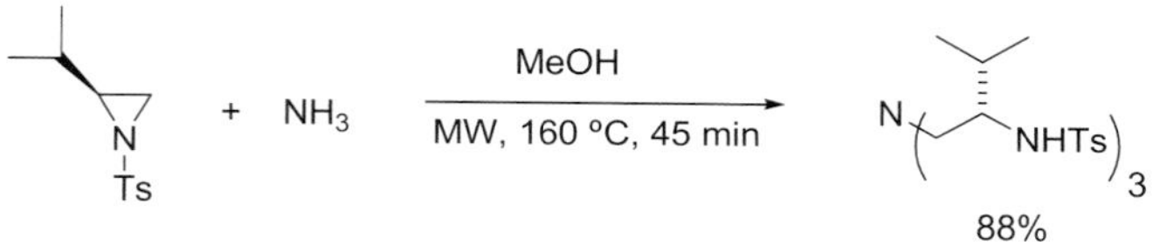

Scheme 6.128   Aziridine ring-opening

6.13.3
Epoxide Ring-Openings

Several recent articles describe the ring-opening of chiral epoxides under microwave irradiation conditions (see also Scheme 6.103). In the context of the preparation of novel β_2-adrenoceptor agonists related to formoterol and salmeterol, Fairhurst and a team from Novartis have described the synthesis of chiral ethanolamines by solvent-free microwave-assisted ring-opening of a suitable chiral epoxide precursor with secondary benzylated amines (Scheme 6.129) [262]. At 110 °C, the reaction occurred

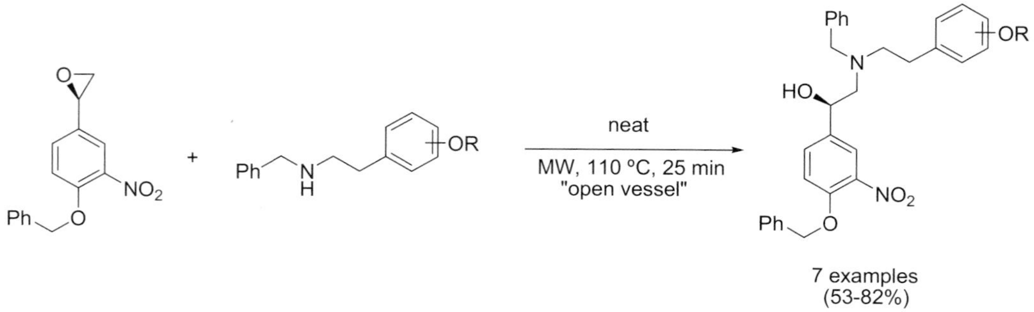

Scheme 6.129   Preparation of β_2-adrenoceptor agonists through epoxide ring-opening.

selectively at the least hindered position, yielding the desired ethanolamine derivatives in 52–82% isolated yield.

Lindsay and Pyne have described related ring-opening reactions of a chiral vinyl epoxide with ammonia or allylamine, as highlighted in Scheme 6.130 [263, 264]. In the case of ammonia, the reaction was simply carried out in concentrated aqueous ammonia (28%) under sealed-vessel microwave conditions. After irradiation for 30 min at 110 °C, the amino alcohol was obtained in 98% yield [263]. In the case of allylamine, acetonitrile was used as a solvent in the presence of 1 equivalent of lithium triflate as a Lewis acid. Again, a high yield of the expected amino alcohol was obtained (120 °C, 1 h, 97% yield) [264]. In both cases, clean S_N2 ring-opening occurred, with no evidence of other regio- or stereoisomers.

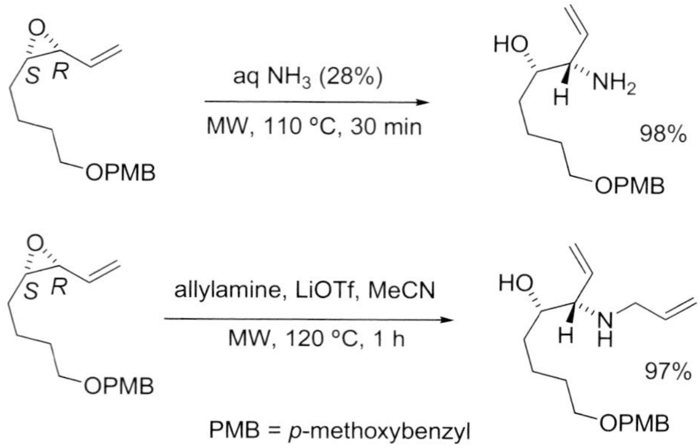

Scheme 6.130 Epoxide ring-opening with amines.

A rather complex microwave-assisted ring-opening of chiral difluorinated epoxycyclooctenones has been studied by Percy and coworkers (Scheme 6.131) [265]. The epoxide resisted conventional hydrolysis, but reacted smoothly in basic aqueous media (ammonia or N-methylimidazole) under microwave irradiation at 100 °C for 10 min to afford unique hemiacetals and hemiaminals in good yields. Other nitrogen nucleophiles, such as sodium azide or imidazole, failed to trigger the reaction. The reaction with sodium hydroxide led to much poorer conversion of the starting material.

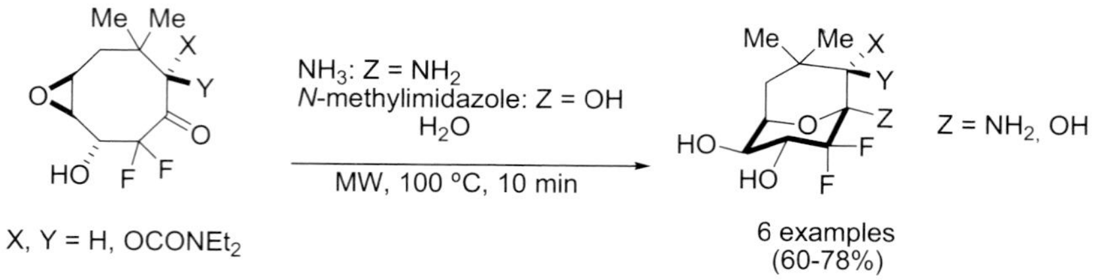

Scheme 6.131 Ring-opening of chiral difluorinated epoxycyclooctenones.

Finally, several authors have reported on ring-opening transformations of epoxides with azides. Van Delft and coworkers have investigated the chelation-controlled ytterbium triflate-mediated azidolysis of a highly functionalized epoxycyclohexene derivative (Scheme 6.132 a) [266]. The authors established that ring-opening to the desired 1,3-diazidocyclitol compound was best achieved by exposing a mixture of the starting epoxide, sodium azide, ytterbium triflate catalyst, triethylamine base, and 3 Å molecular sieves to microwave irradiation at 135 °C. Under these conditions, the desired bis-azide was obtained in 79% yield, as compared to a 49% product yield achieved by conventional heating at 80 °C for 4 days.

Related ring-openings of levoglucosan-derived epoxides with 4 equivalents of lithium azide in the presence of alumina were reported by Cleophax and coworkers in 2004 (Scheme 6.132 b) [267].

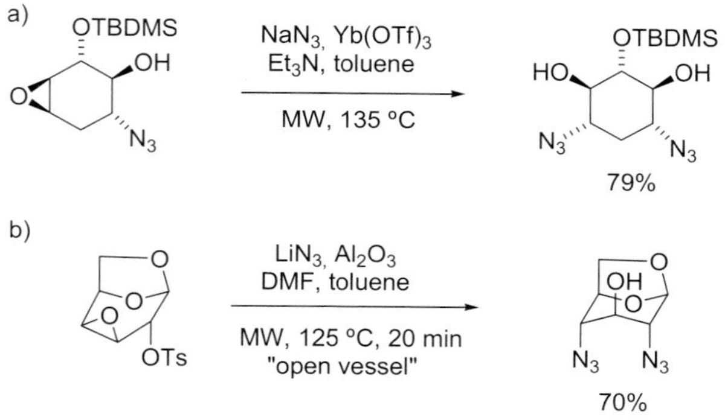

Scheme 6.132 Ring-opening of epoxides with azide ions.

6.14
Addition and Elimination Reactions

6.14.1
Michael Additions

In 2002, Leadbeater and Torenius reported the base-catalyzed Michael addition of methyl acrylate to imidazole using ionic liquid-doped toluene as a reaction medium (Scheme 6.133 a) [190]. A 75% product yield was obtained after 5 min of microwave irradiation at 200 °C employing equimolar amounts of Michael acceptor/donor and triethylamine base. As for the Diels–Alder reaction studied by the same group (see Scheme 6.91), 1-(2-propyl)-3-methylimidazolium hexafluorophosphate (pmimPF$_6$) was the ionic liquid utilized (see Table 4.3). Related microwave-promoted Michael additions studied by Jennings and coworkers involving indoles as heterocyclic amines are shown in Schemes 6.133 b [230] and 6.133 c [268]. Here, either lithium bis(trimethylsilyl)amide (LiHMDS) or potassium *tert*-butoxide (KO*t*Bu) was em-

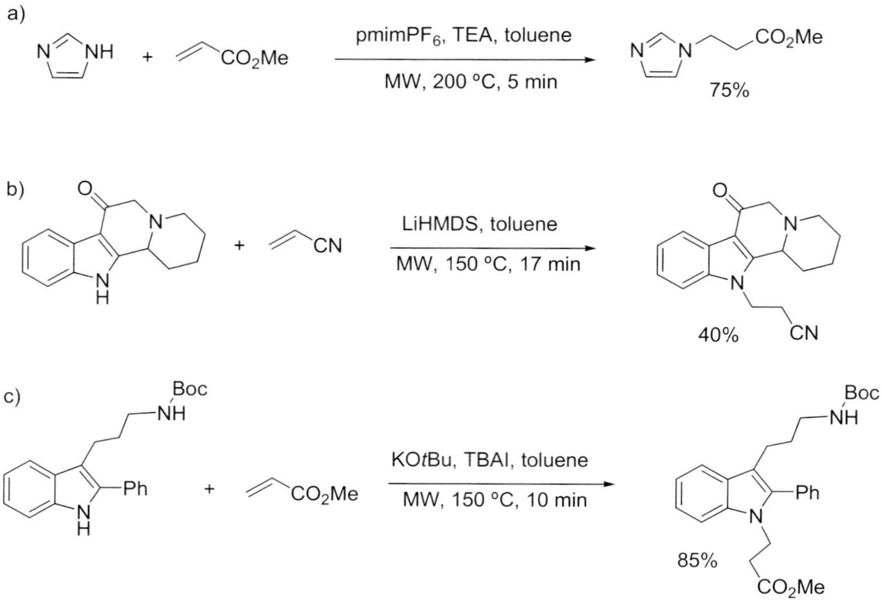

a)

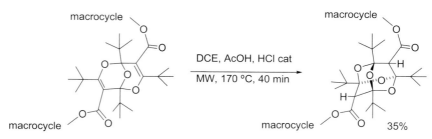

Scheme 6.133   Michael additions involving heterocyclic amines.

ployed as a strongly basic reaction mediator, in the presence of tetrabutylammonium iodide (TBAI).

An acid-catalyzed double-Michael addition of water to the bridged bis-dioxine moiety in a larger macrocyclic framework has been described by the Kollenz group (Scheme 6.134) [269]. While conventional reaction conditions failed to provide any of the desired functionalized 2,4,6,8-tetraoxaadamantane product, microwave heating of the hydrophobic macrocyclic bisdioxine in a 1:1 mixture of 1,2-dichloroethane and acetic acid containing excess concentrated hydrochloric acid at 170 °C for 40 min provided a 35% isolated yield of the desired oxaadamantane compound.

A retro-Michael addition process leading to an aminomethyl-dihydrodipyridopyrazine analogue was described by Guillaumet and coworkers in the context of prepar-

Scheme 6.134   Formation of 2,4,6,8-tetraoxaadamantanes by double Michael addition.

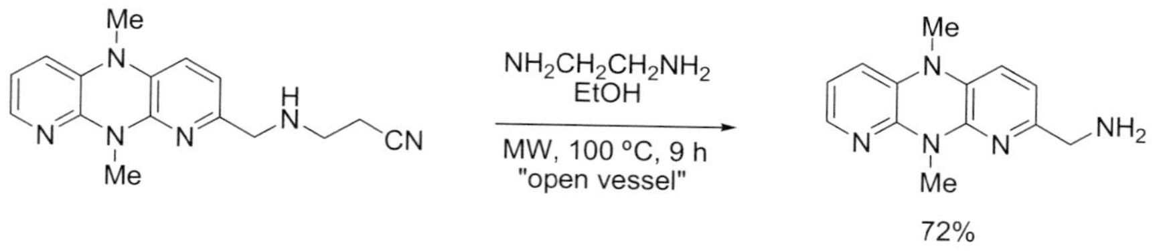

Scheme 6.135 Retro-Michael additions.

ing DNA bis-intercalators as antitumor agents (Scheme 6.135) [270]. Compared to 6 days of heating under conventional reflux conditions, the microwave-assisted reaction was completed in 9 h.

6.14.2
Addition to Alkynes

A direct addition of cycloethers to terminal alkynes has been discovered by Zhang and Li (Scheme 6.136) [271]. The best results were obtained when the reactions were run without additional solvent and in the absence of additives such as transition metal catalysts, Lewis acids, or radical initiators. Typically, the cycloether was used in large excess (200 molar equivalents) as solvent under sealed-vessel conditions. At a reaction temperature of 200 °C, moderate to good yields of the vinyl cycloether products (as mixtures of *cis* and *trans* isomers) were obtained. The reaction is proposed to follow a radical pathway.

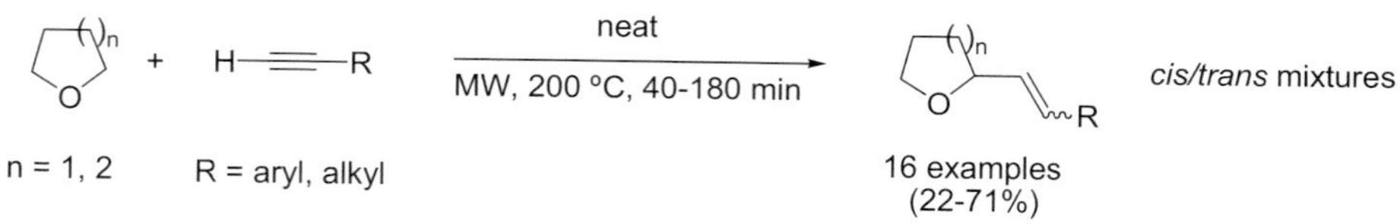

$n = 1, 2$ R = aryl, alkyl

16 examples
(22–71%)

Scheme 6.136 Addition of cycloethers to terminal alkynes.

A hydrophosphination reaction of terminal alkynes with secondary phosphine-borane complexes has been developed by Mimeau and Gaumont as a general synthetic pathway to stereodefined vinylphosphine derivatives [272]. The regioselectivity of the reaction is controlled by the choice of activation method. While in the presence of a palladium catalyst the corresponding α-adducts are obtained (Markovnikov addition), thermal activation leads exclusively to the corresponding β-adducts (anti-Markovnikov addition). For the thermal activation process shown in Scheme 6.137, microwave heating was found to be the best method to achieve fast and efficient reactions. Typically, the secondary phosphine-borane was added to an excess of the alkyne and the mixture was heated under open-vessel conditions with microwaves at 50–80 °C for 30–45 min to allow full conversion to the vinylphosphine products [272]. As regards the stereochemistry, the (Z)-isomer was usually obtained as

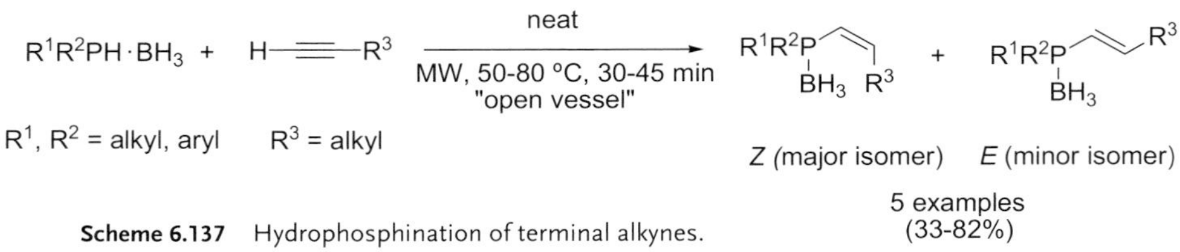

Scheme 6.137 Hydrophosphination of terminal alkynes.

the major product, typically with a high degree of stereoselectivity (>95:5 for $R^1 = R^2 = $ phenyl).

The addition of water to alkynes at high temperatures has already been discussed in Section 4.3.3.1 (see Scheme 4.15) [273].

6.14.3
Addition to Alkenes

Similar to the addition of secondary phosphine-borane complexes to alkynes described in Scheme 6.137, the same hydrophosphination agents can also be added to alkenes under broadly similar reaction conditions, leading to alkylarylphosphines (Scheme 6.138) [274]. Again, the expected anti-Markovnikov addition products were obtained exclusively. In some cases, the additions also proceeded at room temperature, but required much longer reaction times (2 days). Treatment of the phosphine-borane complexes with a chiral alkene such as (–)-β-pinene led to chiral cyclohexene derivatives through a radical-initiated ring-opening mechanism. In related work, Ackerman and coworkers described microwave-assisted Lewis acid-mediated intermolecular hydroamination reactions of norbornene [275].

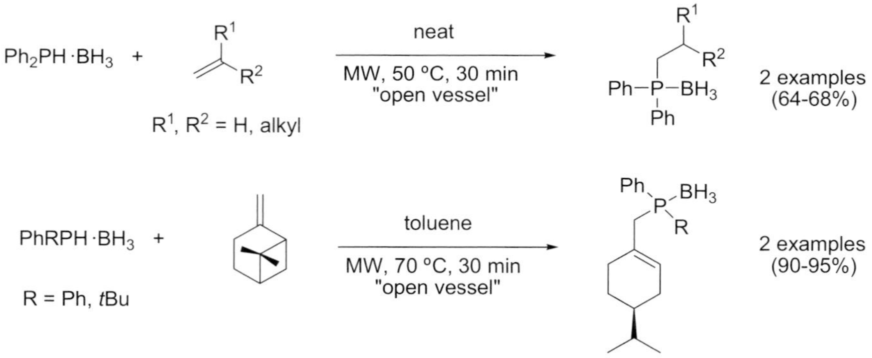

Scheme 6.138 Hydrophosphination of terminal alkenes.

6.14.4
Addition to Nitriles

Bagley and coworkers have described the preparation of primary thioamides by treatment of nitriles with ammonium sulfide in methanol solution (Scheme 6.139) [276]. While the reactions with electron-deficient aromatic nitriles proceeded at room temperature, other aromatic and aliphatic nitriles required microwave heating at 80–130 °C for 15–30 min to furnish the thioamides in moderate to high yields. This protocol avoids the use of hydrogen sulfide gas under high pressure, proceeds in the absence of base, and usually provides thioamides without the need for chromatographic purification.

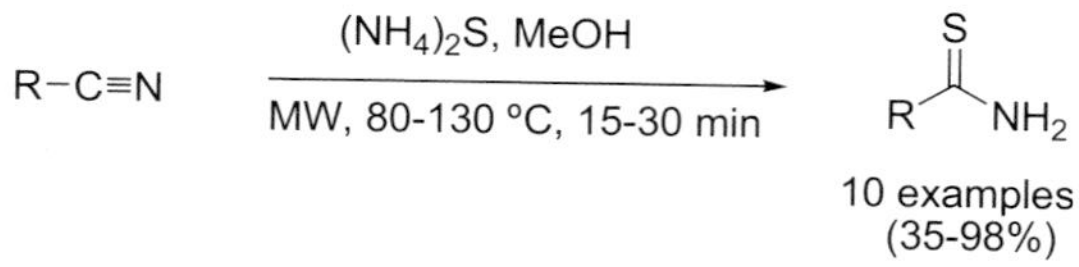

Scheme 6.139 Preparation of primary thioamides.

6.14.5
Elimination Reactions

A high yielding concerted elimination process involving the conversion of N-sulfinyl aldimines into nitriles has been disclosed by Schenkel and Ellman (Scheme 6.140) [277]. These authors found that the self-condensation products derived from N-tert-butanesulfinyl aldimines readily undergo a concerted elimination of tert-butanesulfenic acid to provide the corresponding nitriles. The highest yields were obtained by heating a solution of the starting materials in acetonitrile at 150 °C for 15 min.

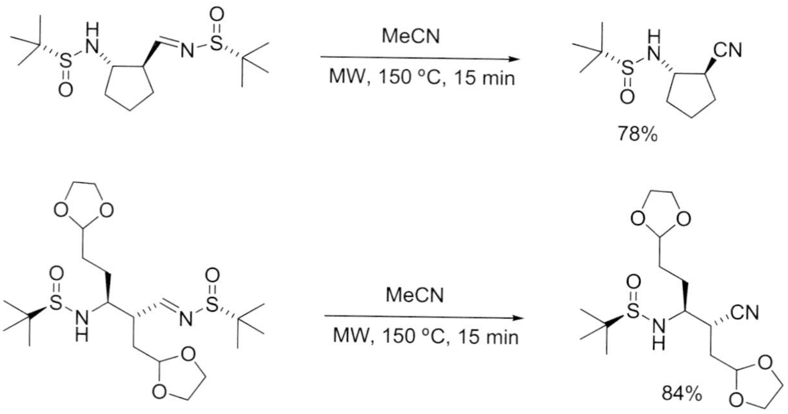

Scheme 6.140 Elimination of tert-butanesulfenic acid.

Incomplete conversion was observed at lower temperatures and significant amounts of undesired decomposition products were obtained at temperatures in excess of 150 °C. These processes were applied to the synthesis of biologically important amine-containing compounds such as the seratonin 5-HT$_4$ agonist SC-53116.

In the context of preparing analogues of chiral 1,2-dimethyl-3-(2-naphthyl)-3-hydroxy-pyrrolidines, which are known non-peptide antinociceptive agents, Collina and coworkers have reported the solvent-free dehydration of hydroxypyrrolidines to pyrrolines under microwave conditions (Scheme 6.141) [278]. In a typical experiment, the substrate was adsorbed onto a large excess of anhydrous ferric(III) chloride on silica gel and then irradiated as a powder under microwave conditions for 30 min at 150 °C. The microwave method leads to dehydration without racemization and provides higher yields in considerably shorter times than the conventionally heated process.

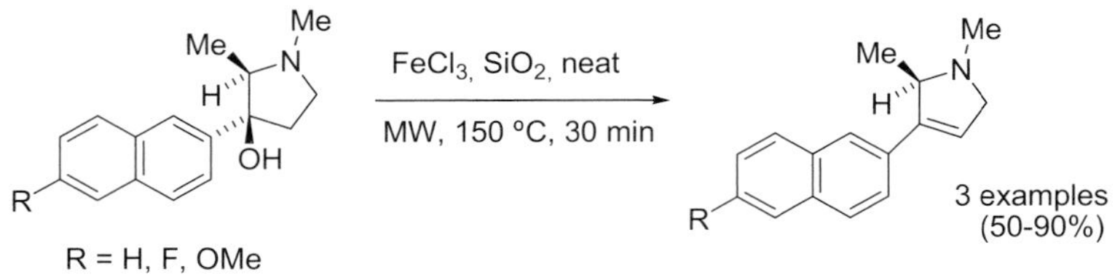

Scheme 6.141 Elimination of water from chiral pyrrolidines.

6.15
Substitution Reactions

A wide range of both electrophilic and nucleophilic substitution reactions performed by controlled microwave heating have been reported in the recent literature. Bose and coworkers have described electrophilic nitrations of electron-rich aromatic systems using dilute nitric acid as a nitrating agent (Scheme 6.142 a) [279]. In their study, 4-hydroxycinnamic acid was treated with an excess of ca. 15% nitric acid. On a 5 gram scale, complete nitration to the corresponding dinitrostyrene derivative was accomplished within 5 min at 80 °C. This nitro compound is a natural product with antifungal activity.

A more elaborate electrophilic nitration system, involving the use of a mixture of tetramethylammonium nitrate and trifluoromethanesulfonic (triflic) anhydride to generate nitronium triflate *in situ*, was reported by Shackelford and a team from Pfizer as a general and mild nitration method (Scheme 6.142 b) [280]. With this reagent, even unreactive heteroaromatics such as 2,6-difluoropyridine could be nitrated. Reactions were conducted under an inert gas in sealed vessels. The nitrating agent, nitronium triflate, was preformed by reacting tetramethylammonium nitrate and triflic anhydride at room temperature for at least 1.5 h before the addition of 2,6-difluoropyridine. The optimum microwave nitration conditions for full conversion involved the utilization of 1.5 equivalents of the nitrating agent and irra-

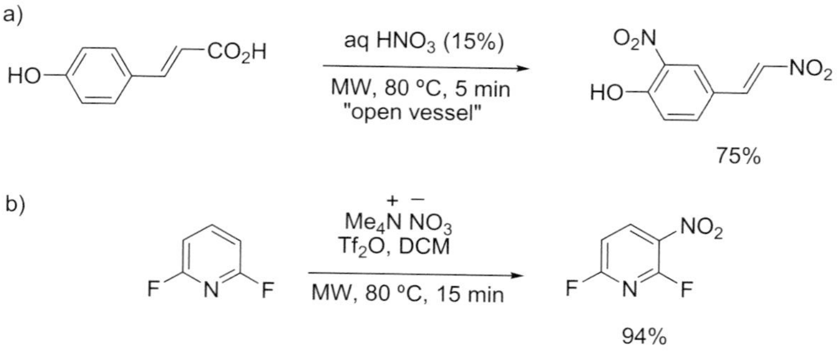

a)

b)

Scheme 6.142 Electrophilic aromatic nitrations.

diation of the solution in dichloromethane at 80 °C for 15 min. This resulted in the formation of analytically pure product in 94% isolated yield. The reaction could be scaled-up from 3.44 mmol using a single-mode microwave reactor to 54.47 mmol in a multimode instrument, providing ca. 9 grams of material in nearly identical yield and purity.

Other electrophilic substitution reactions on aromatic and heteroaromatic systems are summarized in Scheme 6.143. Friedel–Crafts alkylation of N,N-dimethylaniline with squaric acid dichloride was accomplished by heating the two components in dichloromethane at 120 °C in the absence of a Lewis acid catalyst to provide a 23% yield of the 2-aryl-1-chlorocyclobut-1-ene-3,4-dione product (Scheme 6.143 a) [281]. Hydrolysis of the monochloride provided a 2-aryl-1-hydroxycyclobut-1-ene-3,4-dione, an inhibitor of protein tyrosine phosphatases [281]. Formylation of 4-chloro-3-nitrophenol with hexamethylenetetramine and trifluoroacetic acid (TFA) at 115 °C for 5 h furnished the corresponding benzaldehyde in 43% yield, which was further manipulated into a benzofuran derivative (Scheme 6.143 b) [282]. 4-Chloro-5-bromopyrazolopyrimidine is an important intermediate in the synthesis of pyrazolopyrimidine derivatives showing activity against multiple kinase subfamilies (see also Scheme 6.20) and can be rapidly prepared from 4-chloropyrazolopyrimidine and N-bromosuccinimide (NBS) by microwave irradiation in acetonitrile (Scheme 6.143 c) [50]. Similarly, substituted pyrimidinones can be iodinated very effectively at the C5 position with N-iodosuccinimide (NIS) in N,N-dimethylformamide (Scheme 6.143 d) [264]. The last transformation can also be applied in the solid phase, using a resin-bound pyrimidone derivative (see Scheme 7.57) [283].

In the context of developing a catalytic asymmetric total synthesis of quinine (see Scheme 6.103), Jacobsen and coworkers described the bromination of a 4-quinolinone analogue with 1.5 equivalents of dibromotriphenylphosphorane (Ph$_3$PBr$_2$) in acetonitrile at 170 °C (Scheme 6.144 a) [209]. The desired 4-bromo-6-methoxyquinoline was obtained in 86% yield. In the context of synthesizing [18F]-labeled radioligands as positron emission tomography (PET) imaging agents, Neumeyer and colleagues have reported the conversion of phenyltropane-derived tosylates to the corresponding fluoride by using tetrabutylammonium fluoride (TBAF) as fluoride source (Scheme 6.144 b) [284]. Similar transformations have also been reported utilizing

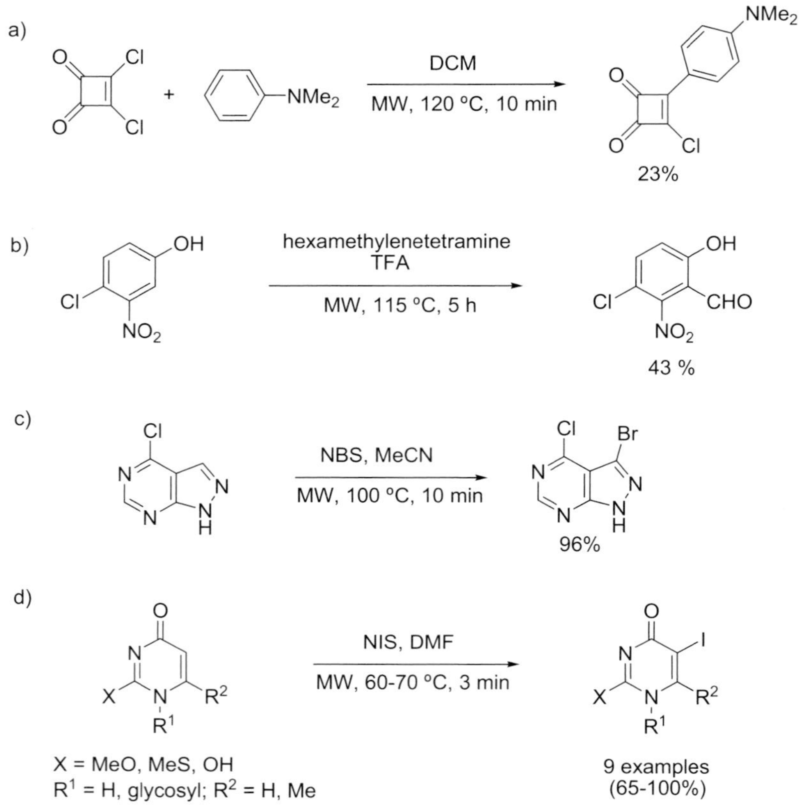

a)

23%

b)

43 %

c)

96%

d)

X = MeO, MeS, OH
R¹ = H, glycosyl; R² = H, Me

9 examples
(65-100%)

Scheme 6.143 Miscellaneous electrophilic aromatic substitutions.

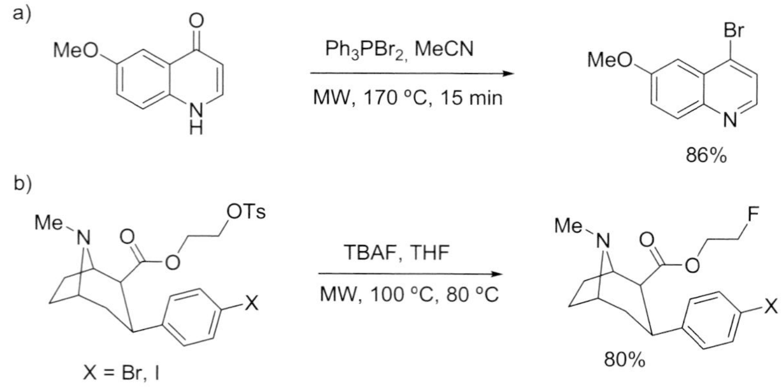

a)

86%

b)

X = Br, I

80%

Scheme 6.144 Bromination and fluorination reactions.

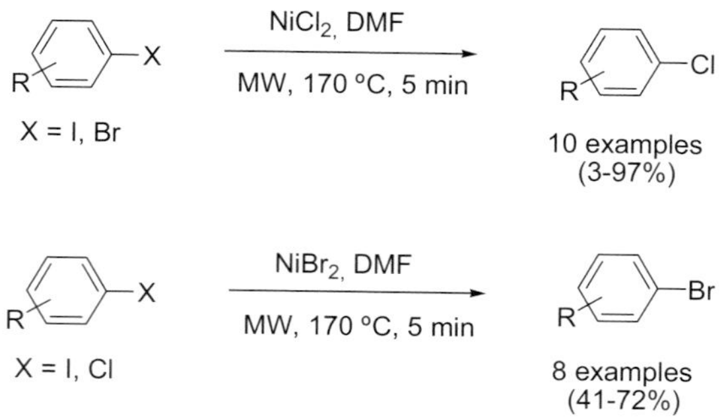

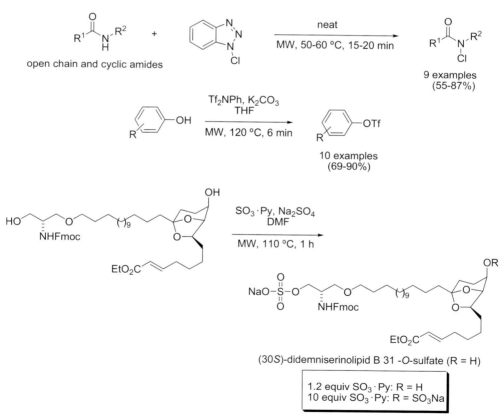

Scheme 6.145 Halide-exchange reactions in aryl halides.

Scheme 6.146 N-Chlorinations, O-triflations, and O-sulfation reactions.

Et₃N·3HF as a fluorinating agent [285, 286] or with other reagents [475, 476]. Benzylic brominations with *N*-bromosuccinimide in the presence of a radical initiator have been investigated by van Koten [287].

Rapid halide-exchange reactions in aryl halides have been investigated by Arvela and Leadbeater using nickel(II) halides as reagents (Scheme 6.145) [288]. The methodology can be used for the conversion of aryl chlorides to bromides, aryl iodides to bromides and chlorides, and aryl bromides to chlorides. The exchange reactions are fast under microwave heating (5 min) and can be performed without the need for exclusion of air and water in *N,N*-dimethylformamide as solvent. Typically, 2 equivalents of the nickel(II) halide are used in these transformations. Additional microwave-assisted substitution reactions, involving *N*-chlorinations [289], *O*-triflations [290], and *O*-sulfation [291] are shown in Scheme 6.146.

6.16
Enamine and Imine Formations

The formation of enamines from the corresponding enols in two series of complex thiazole-containing heterocyclic molecules has been reported by Bagley (Scheme 6.147) [203, 292]. Treatment of the enols with excess ammonium acetate in toluene at 120 °C for 30 min provided the expected products in good yields as single enamine tautomers. The enamines were then manipulated further into thiopeptide antibiotics, again utilizing microwave heating in some of the reaction steps (see also Scheme 6.243).

Similarly, imines can be formed by the condensation of ketones and suitable primary amines under microwave conditions (Scheme 6.148). Since the formation of imines is an equilibrium process, these transformations are best carried out under

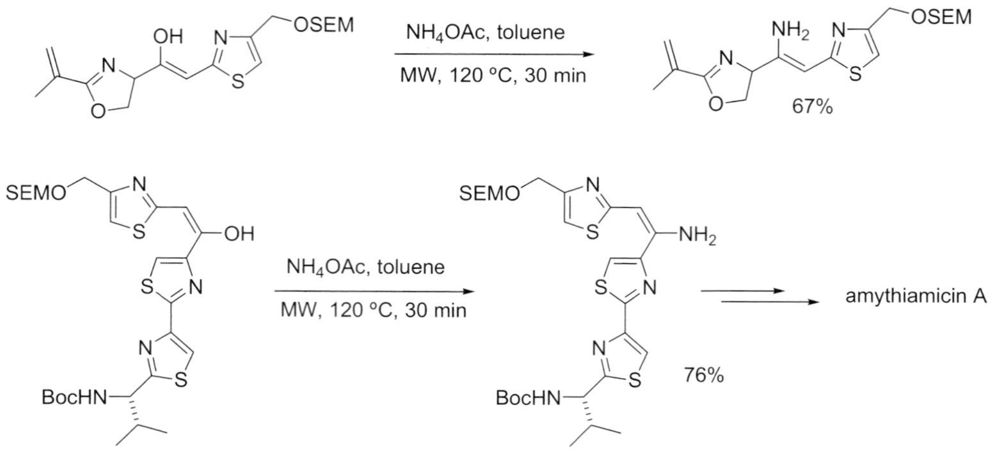

Scheme 6.147 Enamine formations.

open-vessel conditions, allowing the water formed to be removed from the reaction mixture and therefore from the equilibrium (see Section 4.3.1). The groups of Loupy (Scheme 6.148 a) [119] and Langlois (Scheme 6.148 b) [293] have independently described the efficient formation of ketimines from ketones and primary amines using anhydrous zinc(II) chloride (10–50 mol%) as catalyst. While in the example presented in Scheme 6.148 a the reaction was performed in the absence of a solvent, p-xylene was used as a solvent in the second example (Scheme 6.148 b) [293]. The ketimine formed in this study was subsequently used in the synthesis of the marine alkaloid bengacarboline.

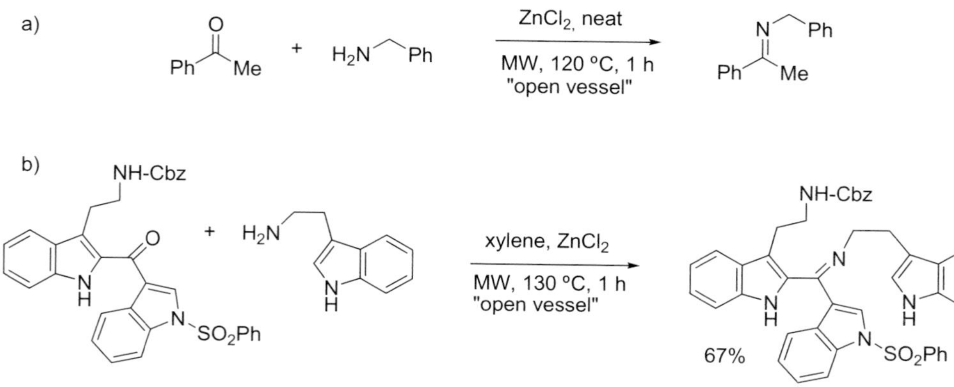

Scheme 6.148 Ketimines from ketones and primary amines.

6.17
Reductive Aminations

Öhberg and Westman were the first to report on reductive aminations under controlled microwave irradiation conditions. In a 2001 publication [80], these authors demonstrated a one-pot, three-step synthesis of thiohydantoins (Scheme 6.149). The first two steps consisted of a reductive amination involving 4-bromobenzaldehyde and an amino acid ester. The aldehyde was treated with the requisite amino acid ester hydrochloride together with 1.1 equivalents of triethylamine base in 1,2-dichloroethene (DCE) as solvent. After 5 min of irradiation at 140 °C, formation of the imine intermediate was complete. After the addition of 1.4 equivalents of sodium triacetoxy borohydride, the reaction mixture was subjected to microwave heating for a further 9 min at 170 °C to give the desired N-benzylated amino acid ester. Addition of an isothiocyanate building block and triethylamine base (2 equivalents of each) followed by microwave heating at 170 °C for a further 5 min ultimately provided the target thiohydantoin products in acceptable overall yield after chromatographic purification [80].

Further applications of microwave-assisted reductive aminations are shown in Schemes 6.150 and 6.151. In the example highlighted in Scheme 6.150, Baran and

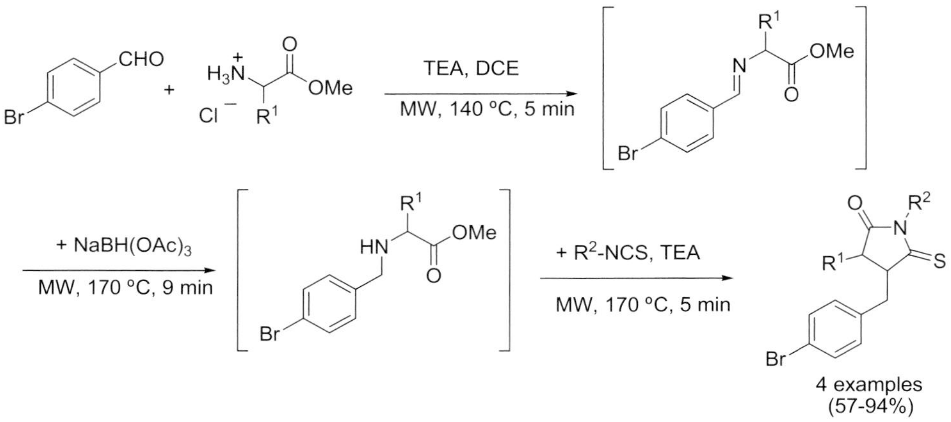

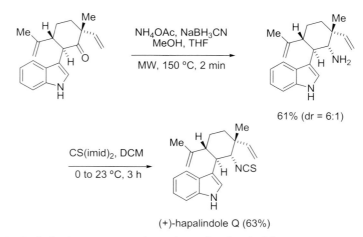

Scheme 6.149 Reductive amination and thiohydantoin synthesis.

Richter utilized a reductive amination in their synthesis of the natural product (+)-hapalindole Q [294]. Employing 10 equivalents of sodium cyanoborohydride and 40 equivalents of ammonium acetate in a methanol/THF mixture (150 °C, 2 min), the primary amine was obtained as a 6:1 mixture of diastereomers. Transformation to the isothiocyanate completed the total synthesis of (+)-hapalindole Q (Scheme 6.150).

In the chemistry described in Scheme 6.151, Coats and a group of researchers from Johnson and Johnson utilized successive reductive aminations and Suzuki cross-coupling reactions to prepare a 192-member library of tropanylidene benzamides [295]. This series of tropanylidene opioid agonists proved to be extremely tolerant with regard to structural variation while maintaining excellent opioid activity.

Scheme 6.150 Reductive amination in the synthesis of (+)-hapalindole Q.

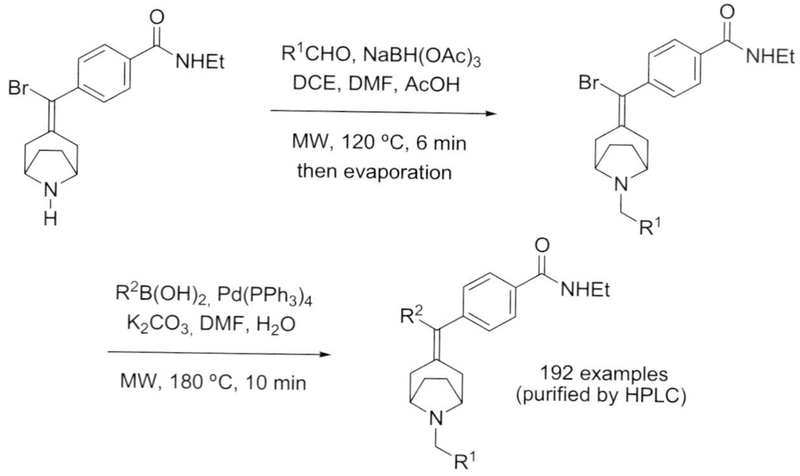

Scheme 6.151 Reductive amination and Suzuki couplings in the synthesis of tropanylidene opioid receptors.

For the solution-phase preparation of functionalized tropanylidenes, the authors simply dispensed solutions of the bromo N–H precursor in 1,2-dichloroethane (DCE) into a set of microwave vials, added the aldehydes (3 equivalents) and a solution of sodium triacetoxy borohydride in dimethylformamide (2 equivalents), and subjected the mixtures to microwave irradiation for 6 min at 120 °C. Quenching the reductive amination with water and subsequent concentration allowed a microwave-assisted Suzuki reaction (Section 6.1.2) to be performed directly on the crude products [295].

Santagada and coworkers have disclosed a reductive amination method for the generation of a reduced peptide bond by reaction of a protected amino acid aldehyde with an *N*-deprotected amino ester using sodium cyanoborohydride as reducing agent [296].

The reduction of an azide group with triphenylphosphine in tetrahydrofuran by microwave heating at 130 °C for 5 min has been described by Kihlberg and colleagues (Scheme 6.152) [297]. The use of diethyl 4-(hydrazinosulfonyl)benzyl phosphonate as an *in situ* diazene precursor for the reduction of trisubstituted *gem*-diiodoalkenes to terminal geminal diodides under microwave conditions has also been reported [298].

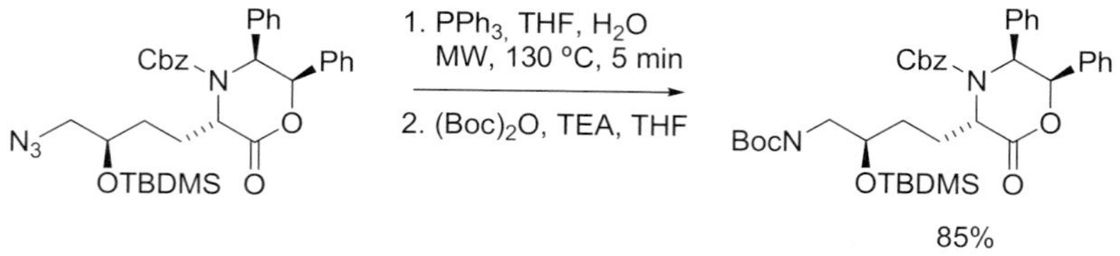

Scheme 6.152 Reduction of an azide group with triphenylphosphine.

6.18
Ester and Amide Formation

Raghavan and coworkers have reported on the preparation of 4-hydroxybenzoic acid esters (parabans) possessing antimicrobial activity by esterification of 4-hydroxy-benzoic acid (Scheme 6.153) [299]. Optimum results were obtained using the alcohol (1-butanol) as solvent in the presence of catalytic amounts of zinc(II) chloride or *p*-toluenesulfonic acid (*p*TsOH) under atmospheric conditions. After 5 min of microwave irradiation at 120 °C, ca. 40% conversion to the ester was observed. Related studies on the synthesis of long-chain aliphatic esters have been described by Mariani and coworkers [300].

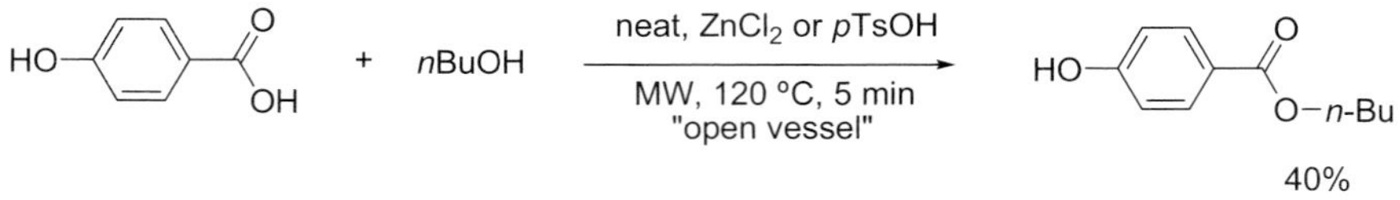

Scheme 6.153 Esterification of benzoic acid.

An unusual class of heterocycles are polyketide-derived macrodiolide natural products. The groups of Porco and Panek have recently shown that stereochemically well-defined macrodiolides of this type can be obtained by cyclodimerization (trans-esterification) of non-racemic chiral hydroxy esters (Scheme 6.154) [301]. Preliminary experiments involving microwave irradiation demonstrated that exposing dilute solutions of the hydroxy ester (0.02 M) in chlorobenzene to sealed-vessel microwave irradiation conditions (200 °C, 7 min) in the presence of 10 mol% of a distannoxane transesterification catalyst led to a 60% isolated yield of the 16-membered macrodiolide heterocycle. Conventional reflux conditions in the same solvent (0.01 M in the hydroxy ester) provided a 75% yield after 48 h at ca. 135 °C.

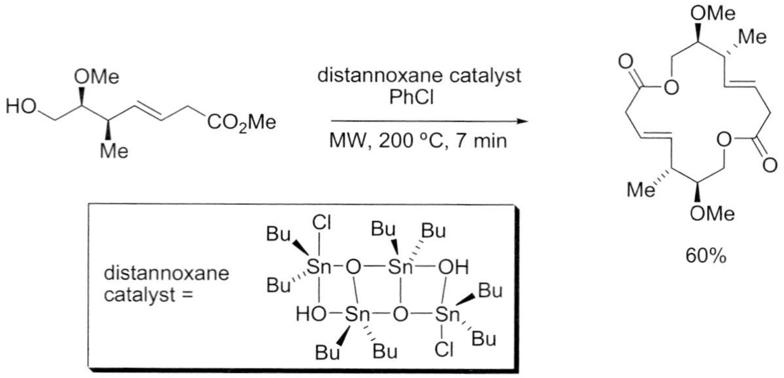

Scheme 6.154 Macrodiolide formation by cyclodimerization.

Toma and coworkers have described the solvent-free synthesis of salicylanilides from phenyl salicylate or phenyl 4-methoxysalicylate and substituted anilines (Scheme 6.155) [302]. By exposing an equimolar mixture of the ester and the amine to microwave irradiation at 150–220 °C for 4–8 min under open-vessel conditions, good yields of the corresponding salicylanilides were obtained. This synthesis was carried out on a multigram scale (0.1 mol).

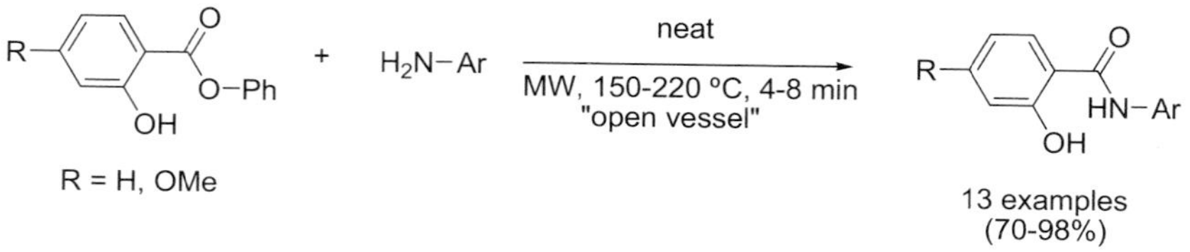

Scheme 6.155 Synthesis of salicylanilides.

A double-acylation reaction involving phosgene was used by Holzgrabe and Heller in their synthesis of diazepinone analogues of the muscarinic receptor antagonist AFDX-384 (Scheme 6.156) [231]. Treatment of a solution of 6-oxo-5,11-dihydrobenzo[*e*]pyrido[3,2-*b*][1,4]diazepine (see Scheme 6.271) and 2 equivalents of *N,N*-diisopropylethylamine (DIEA) in dioxane with 1.75 equivalents of phosgene at room temperature, followed by heating at 85 °C for 2 h, led to the carbonyl chloride, which was transformed to the desired target structure by reaction with a suitable piperidine fragment (1.15 equivalents) and *N,N*-diisopropylethylamine (DIEA) base, again using microwave conditions (110 °C, 10 min). Subsequent *N*-debenzylation provided the desired AFDX-384 analogue [231].

A high-throughput method for the monoacylation of 7-amino-5-aryl-6-cyanopyrido[2,3-*d*]pyrimidines with acid chlorides has been reported by Nicewonger and his team at ArQule (Scheme 6.157) [303]. Since incomplete conversions were achieved

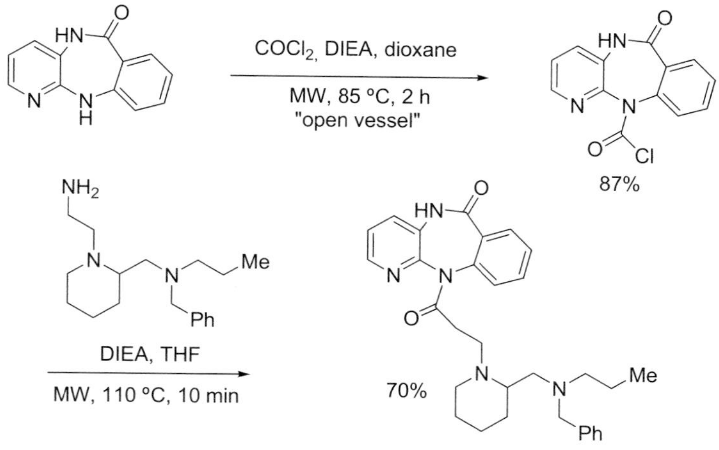

Scheme 6.156 Synthesis of muscarinic receptor antagonist analogues.

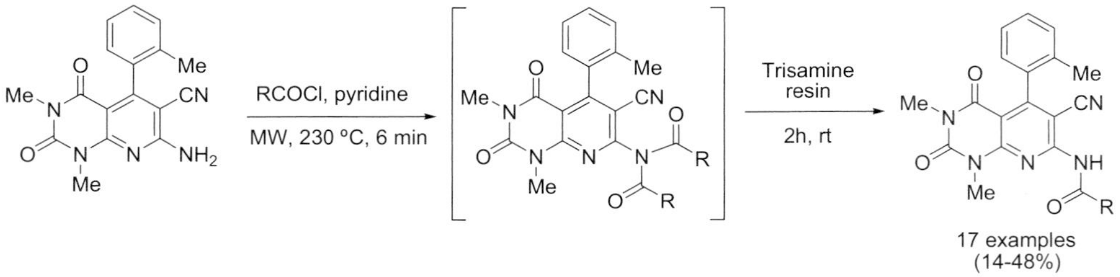

Scheme 6.157 Acylation of 7-amino-5-aryl-6-cyanopyrido[2,3-*d*]pyrimidines.

with either 1.5 or 3 equivalents of the acid chloride, an optimized microwave protocol was elaborated that utilized 9 equivalents of acid chloride at 230 °C using pyridine as solvent. Under these conditions, however, mixtures of mono- and diacylated
products were obtained. Treatment of the crude reaction mixture with an excess of
macroporous Trisamine resin for 2 h allowed complete conversion of the diacylated
product to the monoacylated product while scavenging the excess acid chloride (see
also Scheme 7.111).

Miriyala and Williamson have described the synthesis of β-ketocarboxamides
from primary and secondary amines and 2,2-dimethyl-2H,4H-1,3-dioxin-4-ones as
reactive α-oxoketene precursors (Scheme 6.158) [304]. The experimental procedure
involved heating a mixture of the dioxinone with 2–3 equivalents of the amine at ca.
180 °C for 1–3 min under solvent-free conditions in a sealed vessel by microwave
irradiation. A small collection of 18 β-ketocarboxamides was prepared in very high
yields using this protocol.

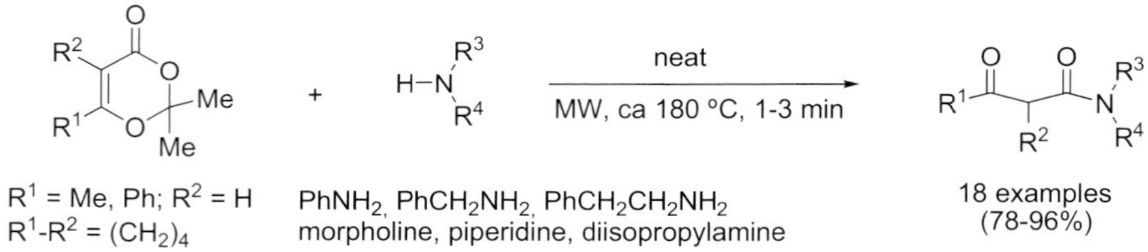

$R^1 = Me, Ph; R^2 = H$ PhNH$_2$, PhCH$_2$NH$_2$, PhCH$_2$CH$_2$NH$_2$
R^1-$R^2 = (CH_2)_4$ morpholine, piperidine, diisopropylamine

18 examples
(78-96%)

Scheme 6.158 Preparation of β-ketocarboxamides.

The preparation of α-ketoamides by a microwave-assisted acyl chloride–isonitrile
condensation has been described in Section 2.5.3 (see Scheme 2.5) [305].

The group of Caddick has described the synthesis of functionalized sulfonamides
by microwave-assisted displacement of pentafluorophenyl (PFP) sulfonate esters
with amines (Scheme 6.159) [306]. Their ease of handling due to their higher crys-
tallinity, along with their long shelf-life and their ability to react under aqueous reaction conditions, makes these an attractive alternative to sulfonyl chlorides. The
microwave-assisted reaction of alkyl PFP esters with amines is a facile process which
proceeds cleanly and in good yields with a number of different amines, including
primary, secondary, and sterically hindered amines, and anilines. Optimum condi-

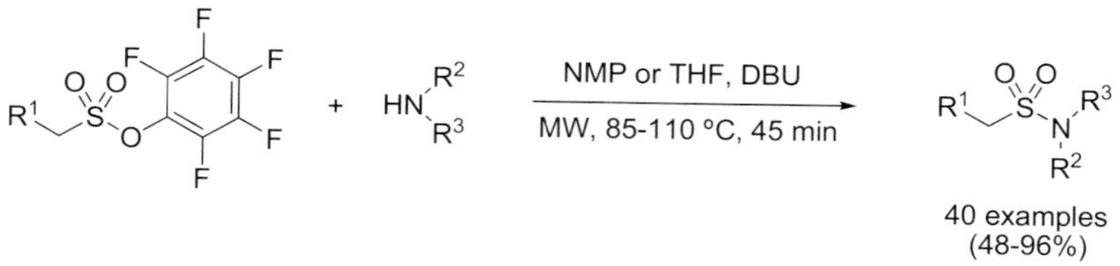

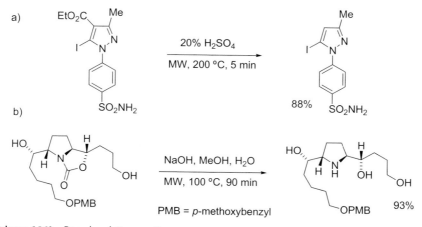

Scheme 6.159 Preparation of sulfonamides.

tions involved a reaction time of 45 min, the use of two equivalents of an amine base such as 1,8-diazabicyclo[5.4.0]undec-7-ene (DBU), and a temperature of 85–110 °C, with either 1-methyl-2-pyrrolidone or tetrahydrofuran as solvent.

6.19
Decarboxylation Reactions

In the context of the preparation of a library of pyrazole-based cyclooxygenase II (COX-II) inhibitors, the Organ group has described the microwave-assisted decarboxylation of a pyrazole carboxylic ester with 20% sulfuric acid (Scheme 6.160 a) [49]. While the conventional protocol (reflux, 100 °C) required 96 h to provide a yield of 86%, full conversion could be achieved within 5 min at 200 °C under microwave heating, leading to an 88% isolated product yield.

Lindsay and Pyne utilized microwave heating for a base-catalyzed cleavage of the oxazolodinone group during the total synthesis of the tricyclic core structure of the cromine alkaloids (Scheme 6.160 b) [263].

Scheme 6.160 Decarboxylation reactions.

Dealkoxycarbonylations of malonic esters have been described by the Curran and Moberg groups (Scheme 6.161). Zhang and Curran found that a variety of malonates and β-ketoesters could be rapidly dealkoxycarbonylated when heated under microwave conditions in wet N,N-dimethylformamide (2.4 equivalents of water) at temperatures of 160–200 °C for 3–30 min (Scheme 6.161 a) [307]. This novel transformation shows good generality for unsubstituted and monosubstituted malonates and ketoesters, while being ineffective for dialkylated analogues. The required time and temperature depended significantly on the substitution pattern, ranging from 3 min for unsubstituted to 30 min for substituted derivatives.

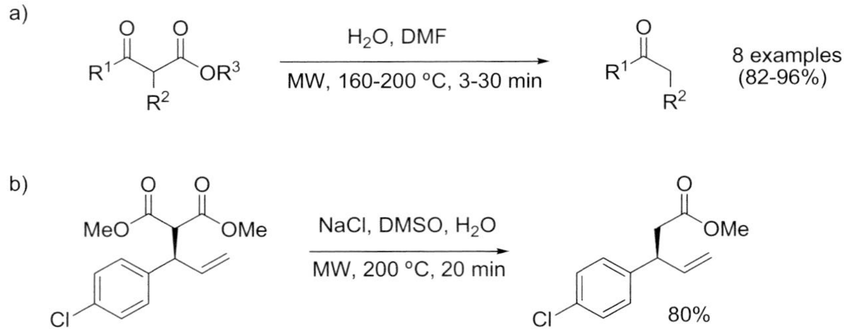

Scheme 6.161 Dealkoxycarboxylation reactions.

A similar dealkoxycarbonylation reaction utilizing the Krapcho conditions was used by Moberg and coworkers in the synthesis of (R)-baclofen (Scheme 6.161 b) from a chiral malonate precursor (see Scheme 6.52) [108].

Related to the transformations described in Scheme 6.161 are the decomposition reactions of mono- and dialkylated Meldrum's acids highlighted in Scheme 6.162 [308].

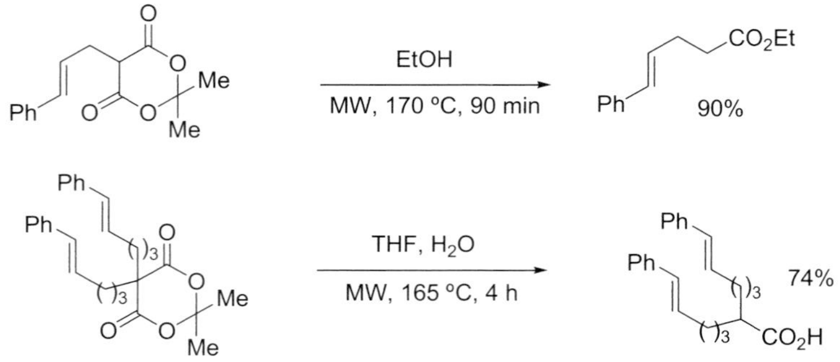

Scheme 6.162 Transesterification and decarboxylation reactions.

6.20
Free Radical Reactions

There are only a limited number of examples in the literature that involve radical chemistry under controlled microwave heating conditions [309]. Wetter and Studer have described radical carboaminoxylations of various non-activated alkenes and difficult radical cyclizations (Scheme 6.163) [310]. The thermally reversible homolysis of alkoxyamines generates the persistent radical 2,2,6,6-tetramethylpiperidinyl-1-ol (TEMPO) and a stabilized transient malonyl radical, which subsequently reacts with an alkene to afford the corresponding carboaminoxylation product. Under conventional conditions (DMF, sealed tube, 135 °C) these radical addition processes take up to 3 days. Using sealed-vessel microwave heating at 180 °C, higher yields were obtained for all but one example when comparing microwave heating at 180 °C for 10 min with thermal heating at 135 °C for 3 days.

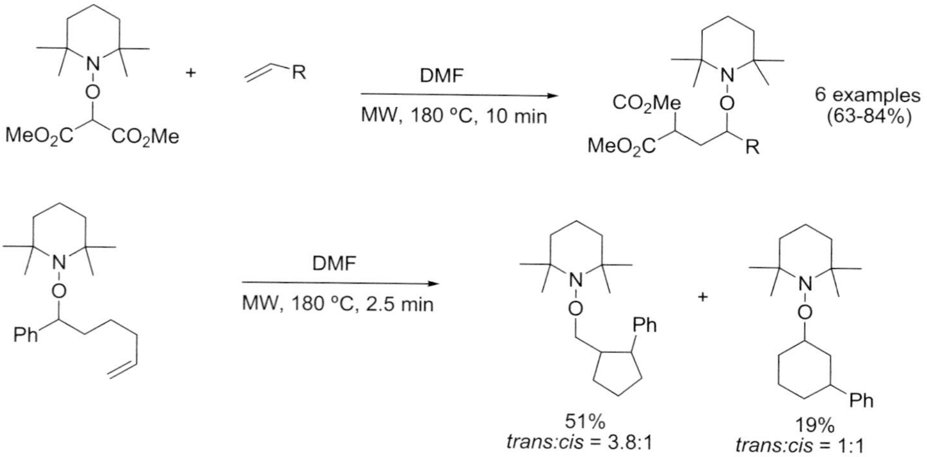

Scheme 6.163 Radical carboxaminations with malonyl radicals.

The same group recently disclosed a related free radical process, namely an efficient one-pot sequence comprising a homolytic aromatic substitution followed by an ionic Horner–Wadsworth–Emmons olefination, for the production of a small library of α,β-unsaturated oxindoles (Scheme 6.164) [311]. Suitable TEMPO-derived alkoxyamine precursors were exposed to microwave irradiation in N,N-dimethylformamide for 2 min to generate an oxindole intermediate via a radical reaction pathway (intramolecular homolytic aromatic substitution). After the addition of potassium tert-butoxide base (1.2 equivalents) and a suitable aromatic aldehyde (10–20 equivalents), the mixture was further exposed to microwave irradiation at 180 °C for 6 min to provide the α,β-unsaturated oxindoles in moderate to high overall yields. A number of related oxindoles were also prepared via the same one-pot radical/ionic pathway (Scheme 6.164).

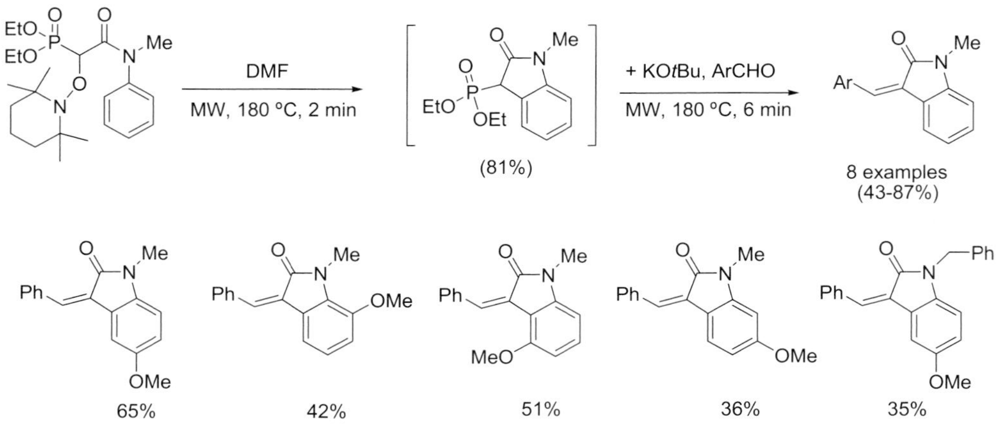

Scheme 6.164 One-pot homolytic aromatic substitution/Horner–Wadsworth–Emmons olefinations.

In 2004, Ericsson and Engman reported on rapid radical group-transfer cycliza-tions of organotellurium compounds. They found that primary and secondary alkyl aryl tellurides, prepared by arenetellurolate ring-opening of epoxides/O-allylation, underwent rapid (3–10 min) group-transfer cyclization to afford tetrahydrofuran de-rivatives in good yields when heated in a microwave cavity at 250 °C in ethylene gly-col or at 180 °C in water (Scheme 6.165) [312]. To proceed to completion, similar transformations had previously required extended photolysis in refluxing benzene.

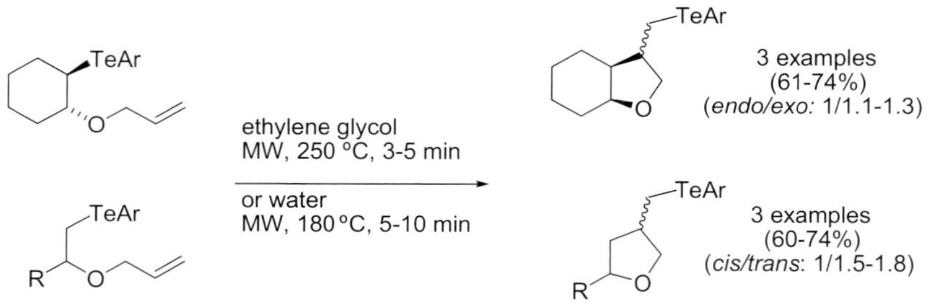

Scheme 6.165 Radical group transfer cyclizations.

Radical processes are, of course, of great importance in the field of polymer syn-thesis, and the applications of controlled microwave heating in this area are rapidly growing. The group of Ritter has described the direct preparation of (meth)acryl-amide monomers from (meth)acrylic acid and an amine using microwave irradia-tion in a solvent-free environment. Irradiation of equimolar mixtures of the reagents in a sealed monomode reactor (no temperature control) for 30 min provided the desired amides in moderate to high yields and with good purities (Scheme 6.166) [313]. The addition of a radical initiator such as 2,2′-azoisobutyronitrile (AIBN) to the starting mixture led directly to the corresponding poly(meth)acrylamides in a

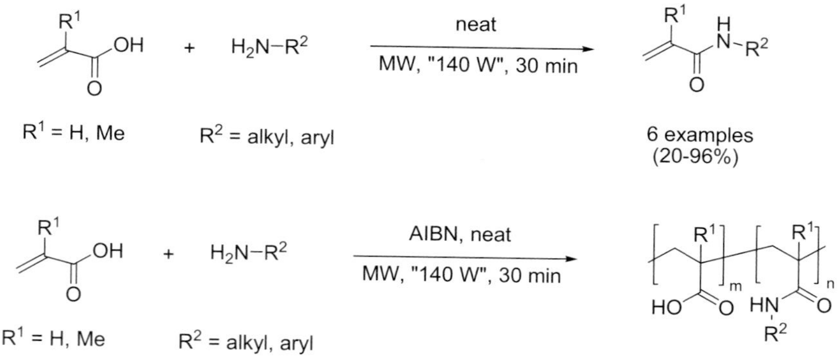

Scheme 6.166 Radical polymerizations.

single step. Additional examples of microwave-assisted radical polymerization reactions have been described [314] and reviewed [315] by Schubert and coworkers.

6.21
Protection/Deprotection Chemistry

In carbohydrate chemistry, protection and deprotection reactions play a significant role. Apart from the recent examples described in Section 6.9, Oscarson and coworkers summarized carbohydrate-based microwave-assisted protecting group manipulations in a 2001 review article [210]. Simultaneous O-debenzylation and carbon–carbon double-bond reduction in a series of alkenoic acids by catalytic transfer hydrogenation has been reported by the group of Pohl [316]. In the example shown in Scheme 6.167, treatment of the starting material in ethylene glycol with 10 equivalents of ammonium formate as hydrogen donor and a catalytic amount of palladium-on-charcoal led to complete debenzylation and C=C double-bond reduction within 5 min at 120 °C (see also Scheme 6.101 for a similar reduction of aromatic nitro groups). Further deprotection of the N-methoxy-N-methyl- (Weinreb) amide to

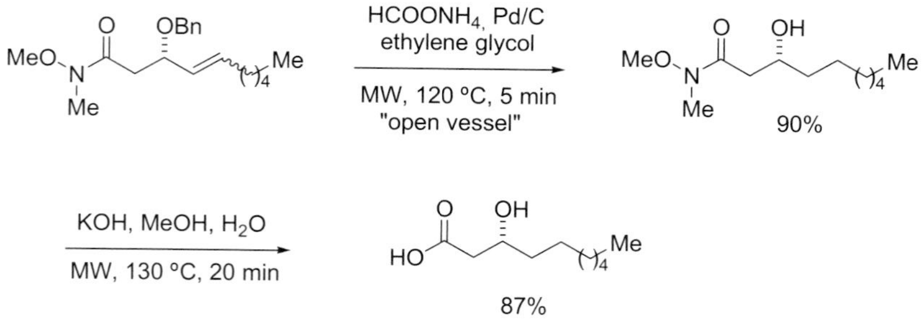

Scheme 6.167 Catalytic transfer hydrogenations.

the corresponding acid (without elimination of the β-hydroxyl moiety) was accomplished by irradiation of the former in a dilute potassium hydroxide/methanol/water mixture at 130 °C. After 20 min, the desired acid was obtained in 87% isolated yield. The same transformation at room temperature required 4 days to reach completion.

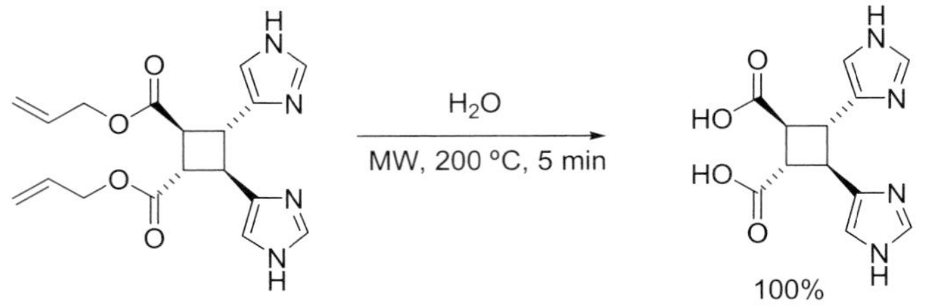

Scheme 6.168 Deallylation reactions.

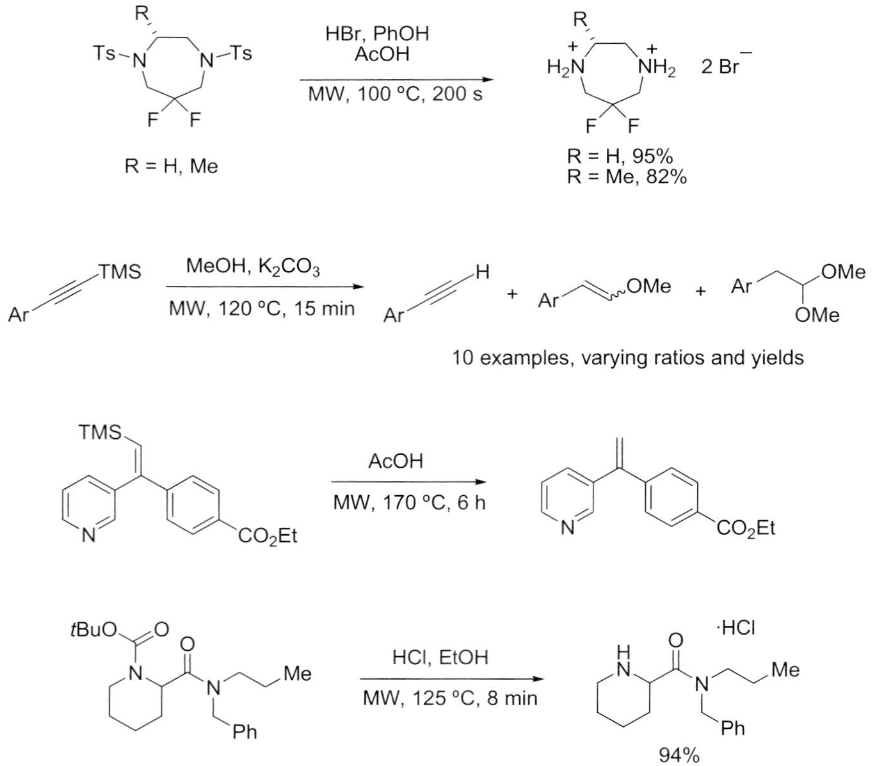

Scheme 6.169 Miscellaneous deprotection reactions.

The Baran group has reported an unusual deprotection of allyl esters in microwave-superheated water. A diallyl ester structurally related to the sceptrin natural products (see Scheme 6.87) was cleanly deprotected at 200 °C within 5 min (Scheme 6.168) [181]. Other standard deprotection transformations carried out under microwave conditions, specifically N-detosylations [317], trimethylsilyl (TMS) removal [318, 319], and N-*tert*-butoxycarbonyl (Boc) deprotection [231], are summarized in Scheme 6.169.

6.22
Preparation of Isotopically Labeled Compounds

The rapid synthesis of short-lived radiolabeled (for example ^{11}C, ^{18}F) substances used in positron emission tomography (PET) was one of the first applications of single-mode microwave-assisted synthesis, and this area has been extensively reviewed [6]. A typical application of this technique is shown in Scheme 6.170. In recent years, considerable effort has been devoted to the design, synthesis, and pharmacological characterization of radiofluorinated derivatives of the 5-HT$_{1A}$ receptor antagonist WAY-100635 for the *in vivo* study of these receptors in the human brain by PET. 6-[^{18}F]Fluoro-WAY-100635 can be efficiently synthesized in one step from the corresponding 6-nitro precursor by nucleophilic heteroaromatic fluorination [320]. As radiofluorinating agent, the activated K[^{18}F]F-Kryptofix® 222 complex was employed (Kryptofix 222 = 4,7,13,16,21,24-hexaoxa-1,10-diazabicyclo[8.8.8]hexacosane). High incorporation yields were observed after 1 min of single-mode microwave irradiation of the nitro precursor in dimethyl sulfoxide solution (no reaction temperature given). The same group has also described related nucleophilic aromatic substitutions on other substituted pyridine cores [321]. For other recent applications of microwave heating in the synthesis of fluorine-18 labeled materials, see ref. [322].

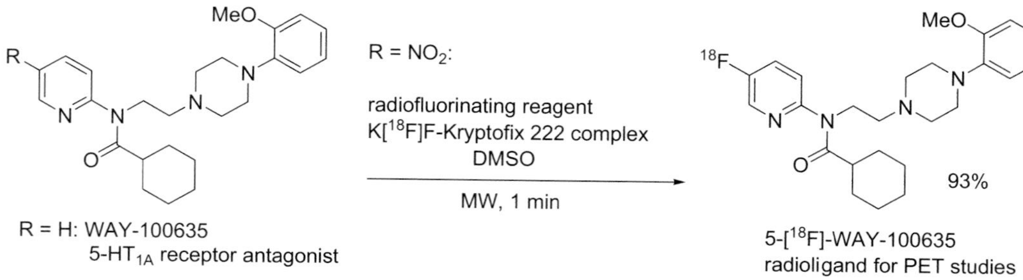

R = H: WAY-100635
5-HT$_{1A}$ receptor antagonist

R = NO$_2$:

radiofluorinating reagent
K[^{18}F]F-Kryptofix 222 complex
DMSO
———————————
MW, 1 min

5-[^{18}F]-WAY-100635
radioligand for PET studies

93%

Scheme 6.170 Incorporation of ^{18}F by nucleophilic heteroaromatic substitution.

The incorporation of a fluorine-18 label can also be achieved by standard aliphatic nucleophilic substitution chemistry, as exemplified in Scheme 6.171. Here, the widely used reagent [^{18}F]-β-fluoroethyl tosylate was utilized to prepare several important ^{18}F-labeled compounds [323].

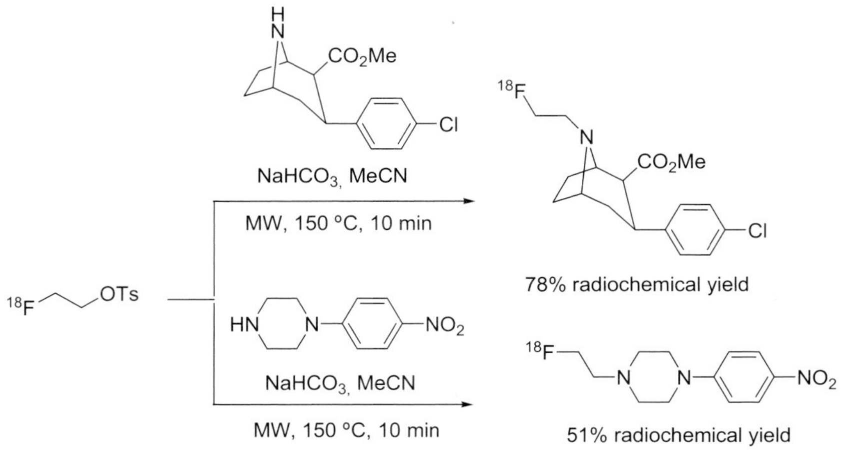

Scheme 6.171 Incorporation of ^{18}F by nucleophilic substitution.

A rapid synthesis of carbon-14 labeled [1-^{14}C]levulinic acid from simple building blocks has been demonstrated by Johansen and coworkers (Scheme 6.172) [324]. In all three of the synthetic steps, starting from bromo[1-^{14}C]acetic acid, microwave heating was used to accelerate the reactions, allowing a total preparation time of less than 1 h. The labeled levulinic acid was subsequently transformed into (5Z)-4-bromo-5-(bromomethylene)-2(5H)-furanone in a bromination/oxidation sequence (not shown), a potent quorum sensing inhibitor.

The synthesis of ^{188}Re [325] and ^{99m}Tc [326] complexes as radiochemical labeling agents using microwave irradiation has been investigated by the group of Park.

Apart from the preparation of radiotracers, microwave-assisted transformations have also been utilized to carry out simple hydrogen–deuterium exchange reactions. In the case of acetophenone, for example, simple treatment with deuterium oxide as solvent in the presence of molecular sieves at 180 °C for 30 min led to complete

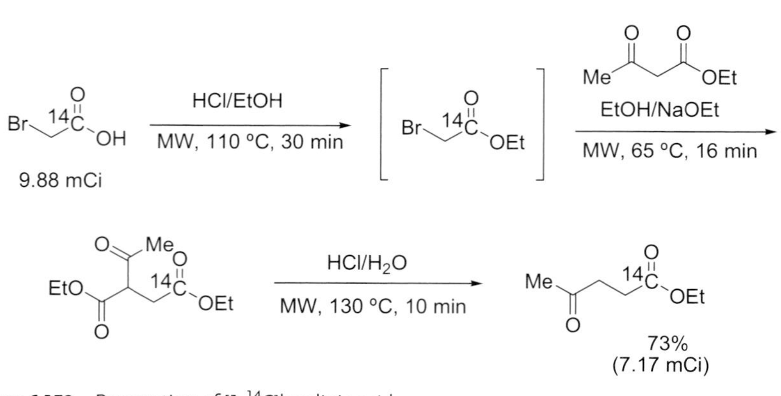

Scheme 6.172 Preparation of [1-^{14}C]levulinic acid.

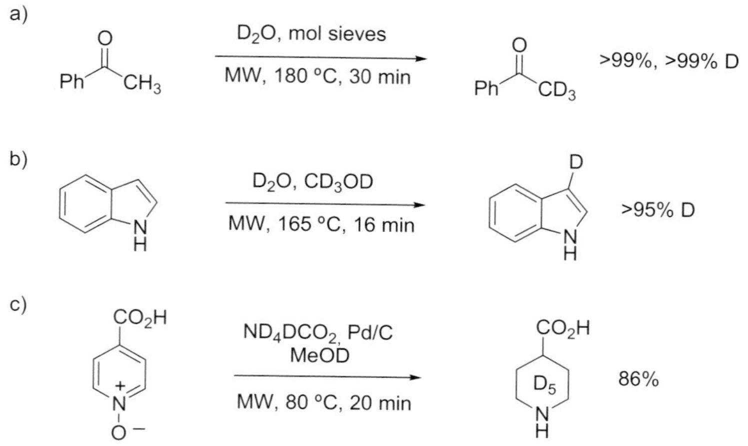

a)

b)

c)

Scheme 6.173 Preparation of deuterium-labeled compounds.

incorporation of deuterium into the methyl group. The presence of the molecular sieves was found to be essential for the exchange of the acidic protons (Scheme 6.173 a) [327]. In the same article, the authors also reported the preparation of deuterium-labeled organophosphonium salts [327].

Masjedizadeh and coworkers have recently described similar microwave-promoted hydrogen–deuterium exchange reactions in a series of heterocycles using mixtures of deuterium oxide and deuteriomethanol (Scheme 6.173 b) [328]. The rapid exchange method was applied to the deuteration of the anti-tumor antibiotic bleomycin A under catalyst-free conditions [328].

The application of deuterated ammonium formate as a deuterium source in transfer deuteration reactions of aromatic heterocycles has been reported by Derdau (Scheme 6.173 c) [329]. It was found that the reaction time could be reduced from 12–18 h (50 °C, oil bath) to 20 min at 80 °C by employing microwave irradiation, using deuteriomethanol as solvent and 10% palladium-on-charcoal as a catalyst.

6.23
Miscellaneous Transformations

In the context of preparing potential inhibitors of dihydrofolate reductase (DHFR), the group of Organ has developed a rapid microwave-assisted method for the preparation of biguanide libraries (Scheme 6.174) [330]. Initial optimization work was centered around the acid-catalyzed addition of amines to dicyandiamide. It was discovered that 150 °C was the optimum temperature for reaction rate and product recovery, as heating beyond this point led to decomposition. While the use of hydrochloric acid as catalyst led to varying yields of product, evaluation of trimethylsilyl chloride in acetonitrile as solvent led to improved results. As compared to the protic

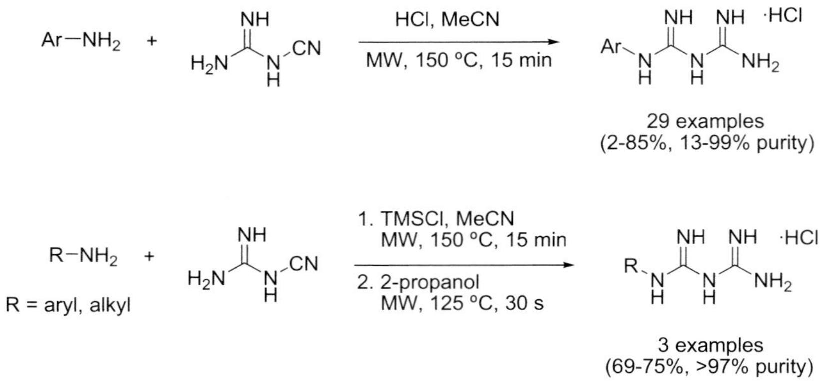

Scheme 6.174 Preparation of biguanide libraries.

conditions, the reaction rate was greatly enhanced under trimethylsilyl chloride (TMSCl) catalysis. The reaction yielded significant product after just 1 min, and the best results were typically obtained by exposing equimolar amounts of amine and dicyandiamide in acetonitrile (150 °C, 15 min) to 1.1 equivalents of trimethylsilyl chloride. After the addition of 3 equivalents of 2-propanol and further microwave heating at 125 °C for 30 s, the desired biguanide products precipitated as their hydrochloride salts. New, improved lead structures were discovered by screening of the 60-member compound library prepared in this fashion.

An important transformation in organic synthesis is the Wittig olefination. Dai and coworkers have described highly regioselective Wittig olefinations of cyclohexa-

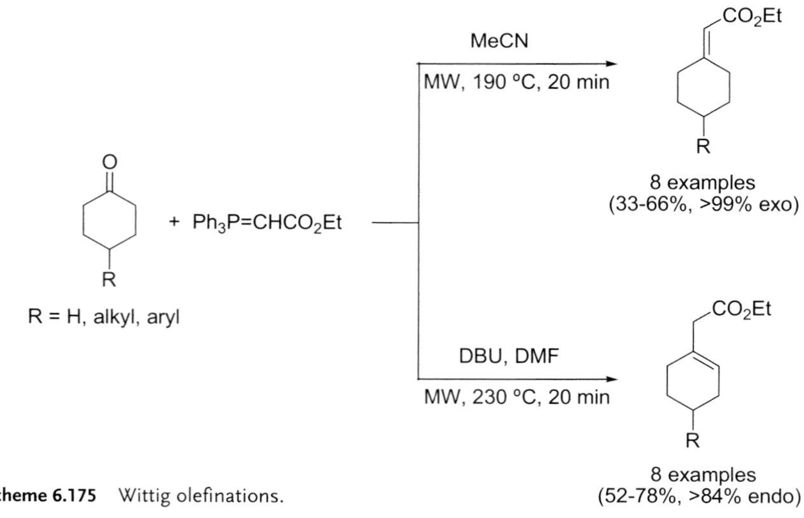

Scheme 6.175 Wittig olefinations.

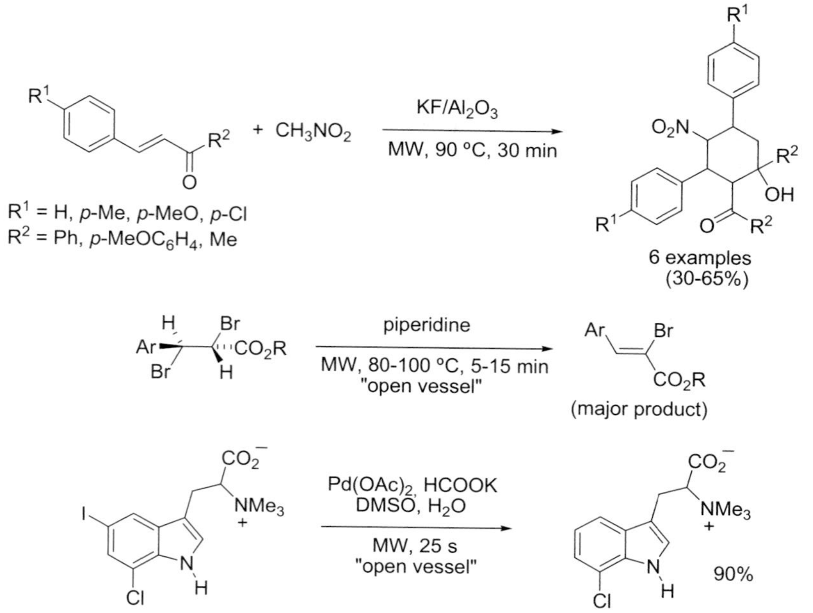

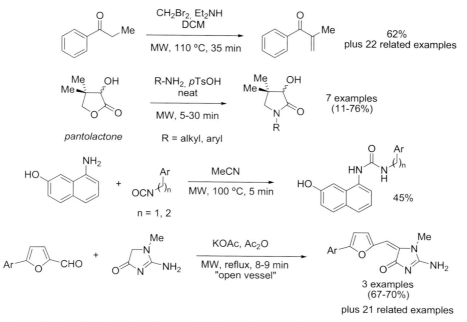

Scheme 6.176 Nitrocyclohexanol synthesis [332], an α,β-dibromoester transformation [333], and a dehalogenation reaction [334].

Scheme 6.177 α-Methylenation of ketones [335], lactam formation from lactones [336], urea formation [337], and Knoevenagel condensation [338, 478].

nones with (carbethoxymethylene)triphenylphosphorane, a stabilized phosphorus ylide, under controlled microwave heating (Scheme 6.175) [331]. In these studies, significant base and temperature effects were noted. When the Wittig reaction was carried out in acetonitrile at 190 °C in the absence of a base, a very high selectivity in favor of the *exo* product was obtained. On the other hand, when the same olefination was carried out at 230 °C in *N,N*-dimethylformamide in the presence of 20 mol% of 1,8-diazabicyclo[5.4.0]undec-7-ene (DBU) as a strong base, the thermodynamically more stable *endo* products were obtained with a >84:16 isomer ratio. In both cases, a threefold excess of the ketone was employed. For a microwave-assisted Wittig olefination utilizing polymer-supported triphenylphosphine, see Scheme 7.96.

Some other microwave-assisted transformations carried out under controlled conditions are summarized in Scheme 6.176 [332–334] and Scheme 6.177 [335–338].

6.24
Heterocycle Synthesis

Heterocyclic scaffolds form the cores of many pharmaceutically relevant substances. Not surprisingly, therefore, many publications in the area of microwave-assisted organic synthesis, both from academia and industry, deal with the preparation of heterocycles [5]. In this final section of this chapter, the description of heterocycle synthesis is structured according to ring-size and the number of heteroatoms in the ring.

6.24.1
Three-Membered Heterocycles with One Heteroatom

The vanadium-catalyzed epoxidation of hindered homoallylic alcohols has been described by Prieto and coworkers [339]. Reaction times for the epoxidation in a series of *cis*- and *trans*-2-methyl-alkenols were significantly reduced from 6–10 days to

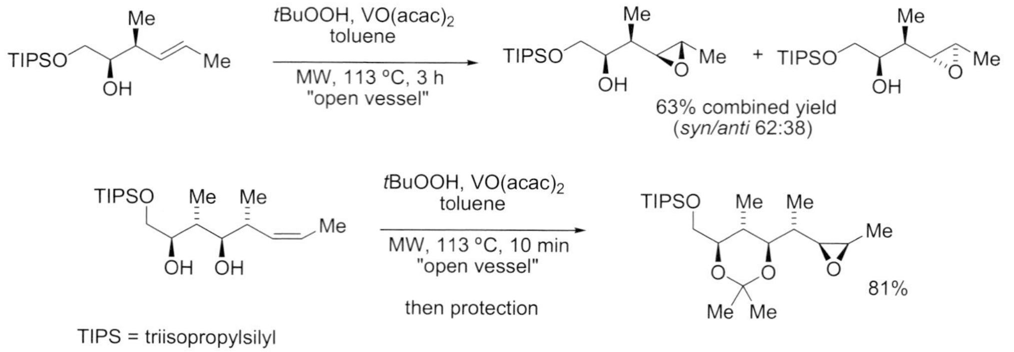

Scheme 6.178 Epoxidation of homoallylic alcohols.

less than 3 h using open-vessel microwave irradiation. Standard conditions for the one-pot oxidation of the alkenes involved the utilization of a bis(acetylacetonate)-oxovanadium(IV)/*tert*-butyl hydroperoxide reagent mixture (1.4 mol% of catalyst, 1.1 equivalents of peroxide oxidant) in toluene solution. For the examples shown in Scheme 6.178, moderate to high chemical yields and diastereoselectivities were achieved. This process, being carried out under open-vessel conditions (see Section 4.3.1), was fully scalable and could be performed on a 30 gram scale [339].

6.24.2
Four-Membered Heterocycles with One Heteroatom

In 2001, Linder and Podlech studied the microwave-assisted decomposition of diazo-ketones derived from α-amino acids [340]. In the presence of imines, the initially formed ketene intermediates reacted spontaneously by [2+2] cycloaddition to form β-lactams with a *trans* substitution pattern at positions C-3 and C-4 (Scheme 6.179) [340]. In order to avoid the use of the high-boiling solvent 1,2-dichlorobenzene, most transformations were carried out in 1,2-dimethoxyethane under sealed-vessel conditions. Solvent-free protocols, in which the substrates were adsorbed onto an inorganic alumina support, led only to the corresponding homologated β-amino acids. Evidently, traces of water present on the support trapped the intermediate ketene.

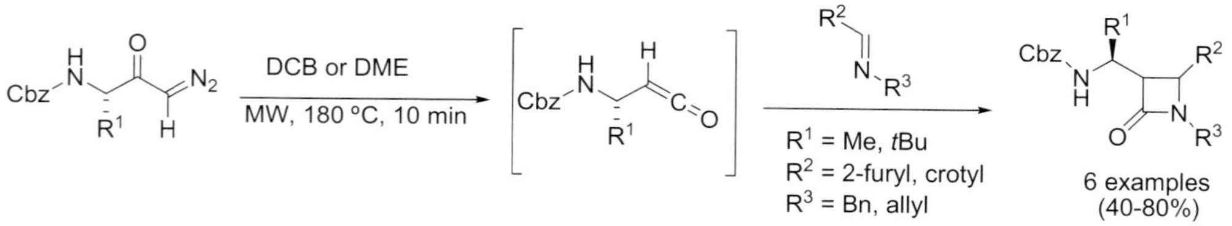

Scheme 6.179 Formation of β-lactams.

6.24.3
Five-Membered Heterocycles with One Heteroatom

6.24.3.1   Pyrroles
Pyrrole is one of the most prominent heterocycles, having been known for more than 150 years, and it is the structural skeleton of several natural products, synthetic pharmaceuticals, and electrically conducting materials. A simple access to the pyrrole ring system involves the conversion of cyclic anhydrides into five-membered imides. Mortoni and coworkers have described the conversion of 2-methylquinoline-3,4-dicarboxylic acid anhydride to a quinoline-3,4-dicarboximide library by treatment of the anhydride with a diverse set of primary amines under microwave conditions (Scheme 6.180) [341]. The authors studied a range of different conditions, including "dry media" protocols (see Section 4.1) whereby the starting materials were adsorbed onto an inorganic support and then irradiated with microwaves. For the transforma-

tions presented in Scheme 6.180, the best results were achieved with wet montmorillonite K10 clay, which afforded complete conversions even with amines that showed low reactivity on other supports such as silica or alumina. Optimum yields were achieved when a mixture of the anhydride and the support was ground in a mortar until a homogeneous powder was obtained. After the addition of 1 equivalent of the primary amine, the material was irradiated at 150 °C for 10–75 min, leading to high product yields. Alternatively, similarly high yields could also be achieved using toluene as solvent, although longer reaction times were typically required.

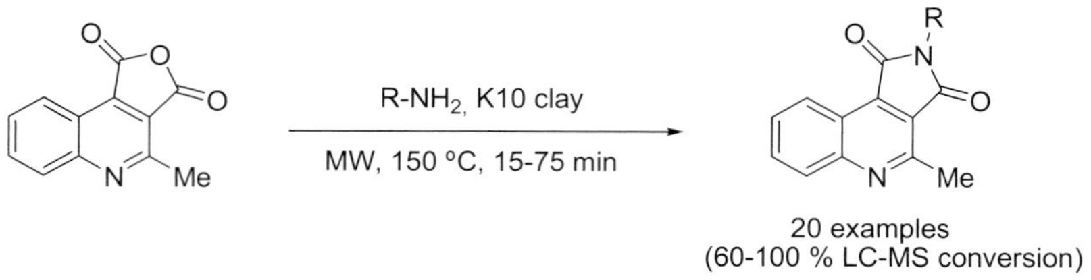

$R-NH_2$, K10 clay

MW, 150 °C, 15-75 min

20 examples
(60-100 % LC-MS conversion)

Scheme 6.180 Formation of imides from anhydrides.

One of the most common approaches to pyrrole synthesis is the Paal–Knorr reaction, in which 1,4-dicarbonyl compounds are converted to pyrroles by acid-mediated dehydrative cyclization in the presence of a primary amine. The group of Taddei has reported a microwave-assisted variation of the Paal–Knorr procedure, whereby a small array of tetrasubstituted pyrroles was obtained (Scheme 6.181) [342]. The pyrroles were effectively synthesized by heating a solution of the appropriate 1,4-dicarbonyl compound in the presence of 5 equivalents of the primary amine in acetic acid at 180 °C for 3 min. The same result was obtained by heating an identical mixture under open-vessel microwave conditions (reflux) for 5 min. Interestingly, the authors were unable to achieve meaningful product yields when attempting to carry out the same transformation by oil-bath heating.

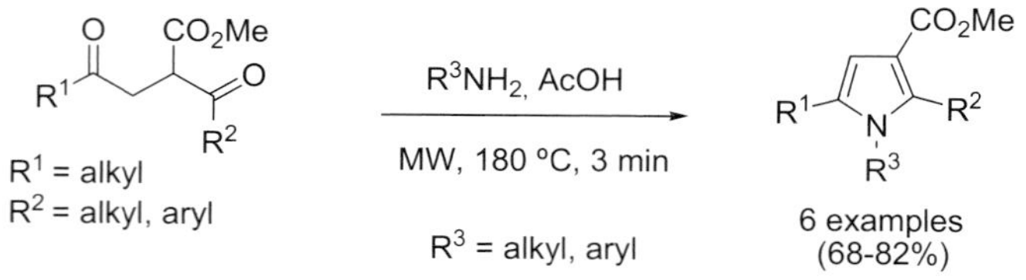

R^1 = alkyl
R^2 = alkyl, aryl

R^3NH_2, AcOH

MW, 180 °C, 3 min

R^3 = alkyl, aryl

6 examples
(68-82%)

Scheme 6.181 Paal–Knorr pyrrole synthesis.

A different approach toward highly substituted pyrroles involving a one-pot sila-Stetter/Paal–Knorr strategy was realized by Bharadwaj and Scheidt (Scheme 6.182) [343]. In this multicomponent synthesis, catalyzed by a thiazolium salt, an acyl anion conjugate addition reaction of an acylsilane (sila-Stetter) was coupled *in situ* with the conventional Paal–Knorr approach. Employing microwave conditions at 160 °C for 15 min, the acylsilane was combined with the α,β-unsaturated ketone in

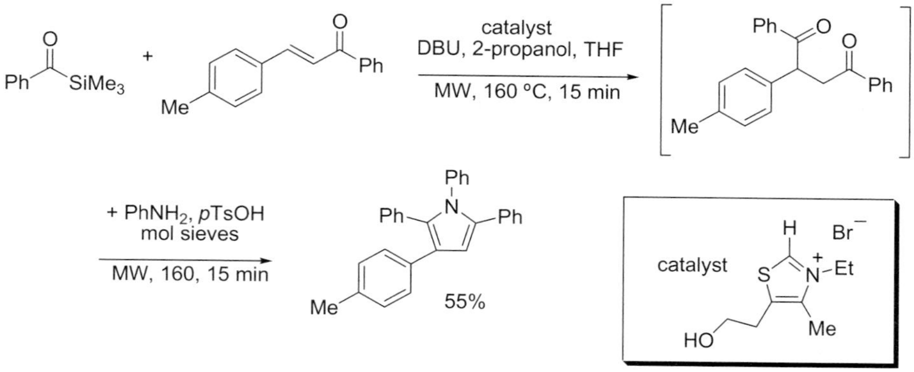

Scheme 6.182 Sila-Stetter/Paal–Knorr pyrrole synthesis.

the presence of 20 mol% of the thiazolium salt catalyst and 30 mol% of 1,8-diazabi-cyclo[5.4.0]undec-7-ene (DBU) base in addition to 4 equivalents of 2-propanol. This sequence was followed by the addition of aniline and p-toluenesulfonic acid (pTsOH), and a second 15 min heating cycle at 160 °C smoothly provided the desired pyrrole in 55% yield. This streamlined approach generated the target hetero-cycle in 3% of the time required using conventional heating (30 min versus 16 h) [343].

Tejedor and coworkers have utilized a combination of two domino processes for a microwave-promoted synthesis of tetrasubstituted pyrroles [344]. The protocol com-bines two coupled domino processes: the triethylamine-catalyzed synthesis of enol-protected propargylic alcohols and their sequential transformation into pyrroles through a spontaneous rearrangement from 1,3-oxazolidines (Scheme 6.183). Over-all, these two linked and coupled domino processes build up two carbon–carbon bonds, two carbon–nitrogen bonds, and an aromatic ring in a regioselective and effi-cient manner. The tetrasubstituted pyrroles could be directly synthesized from the enol-protected propargylic alcohols and the primary amines by microwave irradia-

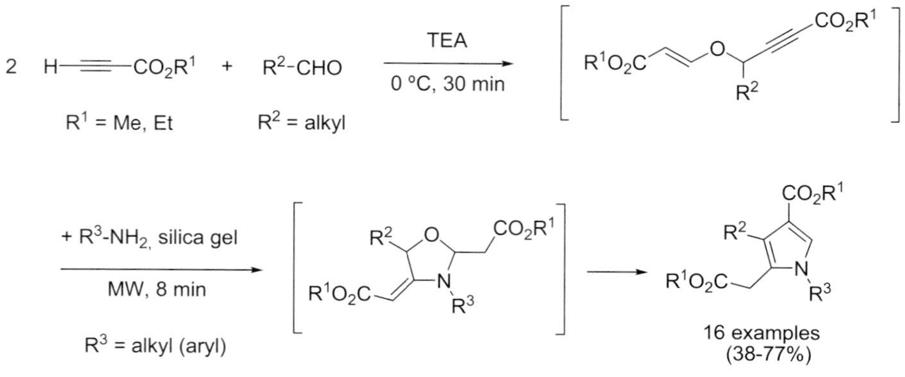

Scheme 6.183 Pyrroles from coupled domino processes.

tion (domestic oven) of a mixture of the two components adsorbed onto silica gel. The process is general for the amine and tolerates a range of functionalities in the aldehyde, leading to the desired diverse pyrrole structures in moderate to good overall yields.

Another method for preparing pyrrole rings is by Ugi-type three-component condensation (Scheme 6.184). In the protocol published by Tye and Whittaker [345], levulinic acid was reacted with two different isonitriles and four amine building blocks (1.5 equivalents) to provide a set of eight pyrrole derivatives. While the previously published protocol at room temperature required a reaction time of up to 48 h and provided only moderate product yields, the microwave method (100 °C, 30 min) optimized by a Design of Experiments (DoE) approach (see Section 5.3.4), led to high yields of the desired lactams for most of the examples studied.

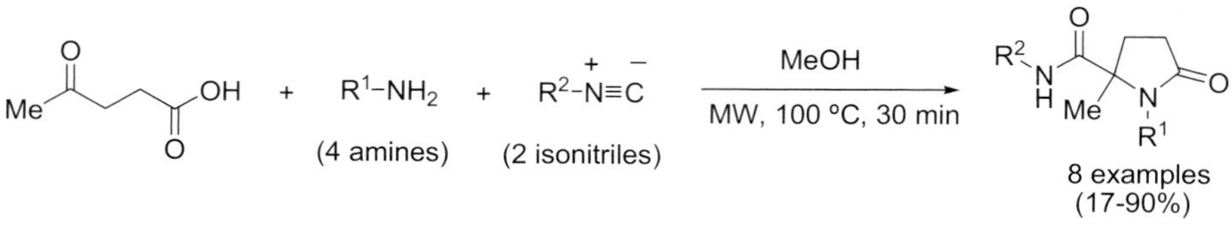

Scheme 6.184 Ugi-type three-component condensation reactions.

In addition to cyclocondensation reactions of the Paal–Knorr type, cycloaddition processes play a prominent role in the construction of pyrrole rings. Thus, 1,3-dipolar cycloadditions of azomethine ylides with alkene dipolarophiles are very important in the preparation of pyrroles. The group of de la Hoz has studied the microwave-induced thermal isomerization of imines, derived from α-aminoesters, to azomethine ylides (Scheme 6.185) [346]. In the presence of equimolar amounts of β-nitrostyrenes, three isomeric pyrrolidines (nitroproline esters) were obtained under solvent-free conditions in 81–86% yield within 10–15 min at 110–120 °C through a [3+2] cycloaddition process. Interestingly, using classical heating in an oil bath (toluene reflux, 24 h), only two of the three isomers were observed.

An intramolecular variation of the same reaction principle has been used by Bashiardes and coworkers to synthesize fused pyrrolidine and pyrrole derivatives

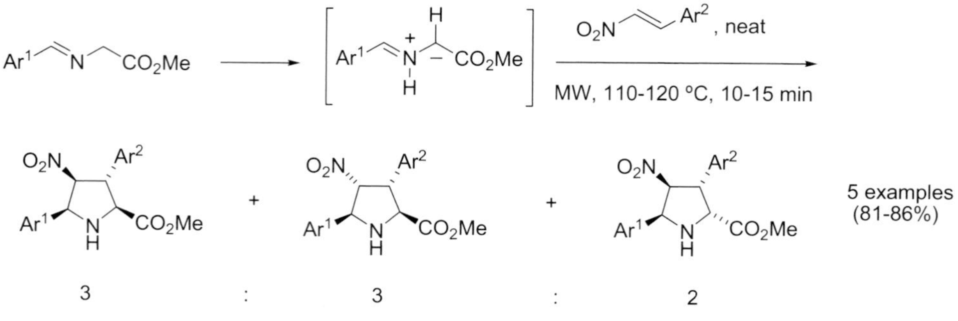

Scheme 6.185 Azomethine ylide–alkene [3+2] cycloadditions.

(Scheme 6.186) [347]. The condensation of *O*-allylic and *O*-propargylic salicylalde-hydes with α-amino esters was carried out either in the absence of a solvent or – if both components were solids – in a minimal volume of xylene. All reactions performed under microwave conditions rapidly proceeded to completion within a few minutes and typically provided higher yields compared to the corresponding thermal protocols. In the case of intramolecular alkene cycloadditions, mixtures of hexa-hydrochromeno[4,3-*b*]pyrrole diastereoisomers were obtained, whereas transformations involving alkyne tethers provided chromeno[4,3-*b*]pyrroles directly after *in situ* oxidation with elemental sulfur (Scheme 6.186). Independent work by Pospíšil and Potáček involved very similar transformations under strictly solvent-free conditions [348].

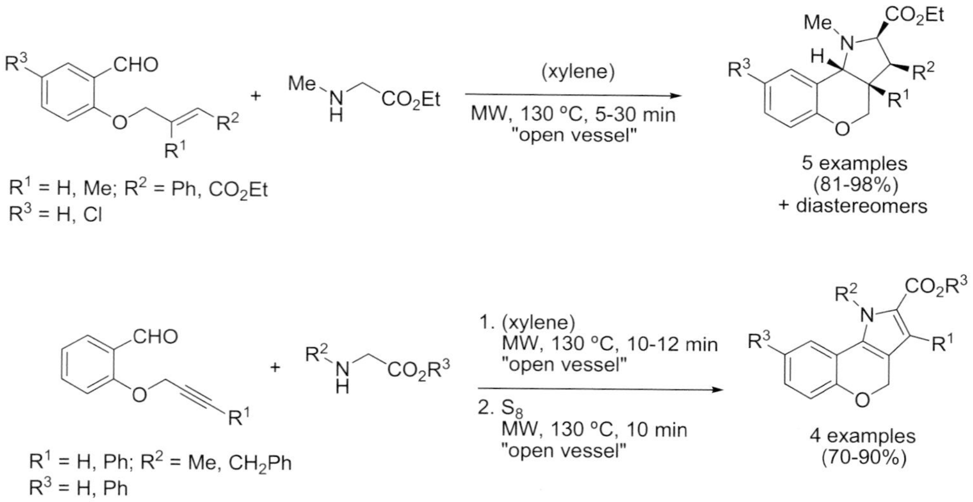

R¹ = H, Me; R² = Ph, CO₂Et
R³ = H, Cl

5 examples
(81-98%)
+ diastereomers

R¹ = H, Ph; R² = Me, CH₂Ph
R³ = H, Ph

4 examples
(70-90%)

Scheme 6.186 Intramolecular azomethine ylide–alkene/alkyne [3+2] cycloadditions.

In 2001, Sarko and coworkers disclosed the synthesis of an 800-membered solution-phase library of substituted prolines based on multicomponent chemistry (Scheme 6.187) [349]. The process involved microwave irradiation of an α-amino ester with 1.1 equivalents of an aldehyde in 1,2-dichloroethane or *N*,*N*-dimethyl-formamide at 180 °C for 2 min. After cooling, 0.8 equivalents of a maleimide dipo-larophile was added to the solution of the imine, and the mixture was subjected to microwave irradiation at 180 °C for a further 5 min. This produced the desired products in good yields and purities, as determined by HPLC, after scavenging excess aldehyde with polymer-supported sulfonylhydrazide resin. Analysis of each compound by LC-MS verified its purity and identity, thus indicating that a high quality library had been produced.

A classical method for synthesizing indoles is by the Fischer indolization, which involves the cyclization of arylhydrazones in the presence of strong acids. Lipińska

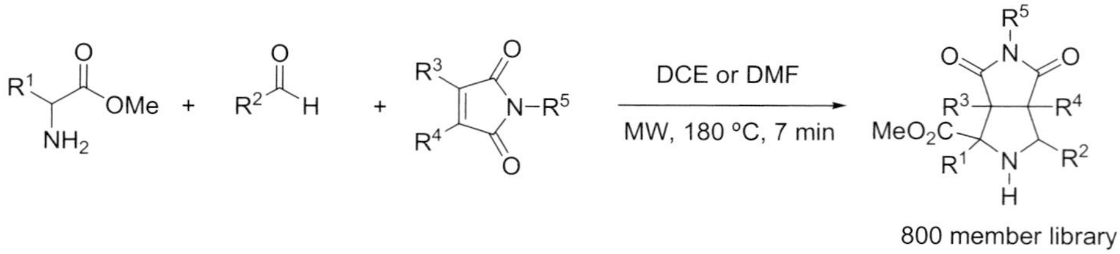

Scheme 6.187 Fused prolines from azomethine ylide–maleimide [3+2] cycloadditions.

has reported the generation of 2-heteroaryl-5-methoxyindoles using a solvent-free, clay-catalyzed protocol, starting from the corresponding *in situ* generated arylhydrazones (Scheme 6.188) [350]. The microwave-induced Fischer indole synthesis was performed on montmorillonite K 10 clay modified with zinc chloride, providing the desired indoles within a relatively short timeframe. The 2-substituted indole products were subsequently transformed into 9-methoxyindolo[2,3-*a*]quinolizine alkaloids of the sempervirine type.

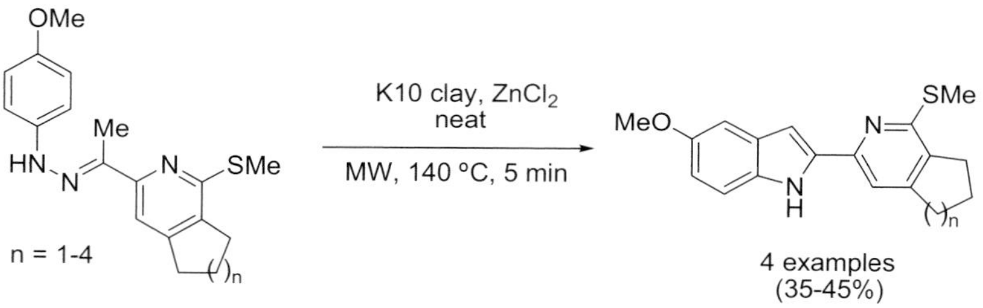

Scheme 6.188 Fischer indole synthesis.

The Leimgruber–Batcho synthesis is another widely used method for preparing indole-containing structures. The reaction depends on the acidity of a methyl group positioned adjacent to an aromatic nitro group. Direct condensation with *N,N*-dimethylformamide dimethyl acetal (DMFDMA) under acid catalysis facilitates the introduction of the future indole α-carbon as the enamine. Subsequent catalytic reduction of the nitro group leads to spontaneous cyclization with the formation of an indole derivative. Under conventional conditions, the first condensation reaction usually requires overnight heating in *N,N*-dimethylformamide, which often leads to less than optimal yields. Ley and coworkers have described microwave-assisted Leimgruber–Batcho reactions in the presence of 2 mol% of anhydrous copper(I) iodide in *N,N*-dimethylformamide/DMFDMA mixtures as solvent (Scheme 6.189) [351]. Typically, irradiation times ranging from 10 min to 10 h (pulsed irradiation, see Section 2.5.3) at 180 °C led to acceptable yields of the (hetero)aromatic enamines. The nitro intermediates were then subjected to catalytic hydrogenation using 10 mol% palladium-on-charcoal (Pd/C) in methanol, or to transfer hydrogenation using an encapsulated nanoparticulate palladium catalyst (10 mol%), formic acid,

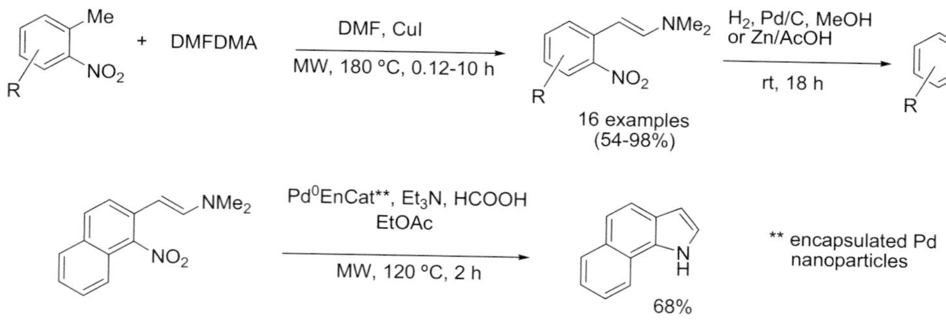

Scheme 6.189 Leimgruber–Batcho indole synthesis.

and triethylamine (5 equivalents each) as the reducing agents. For example, *trans*-2-[β-(dimethylamino)vinyl]nitronaphthalene was smoothly converted to the corresponding 1*H*-benz[g]indole within 2 h at 120 °C by microwave irradiation in ethyl acetate solution (Scheme 6.189) [351]. Other microwave-assisted transformations involving immobilized palladium catalysts are described in Section 7.6.

6.24.3.2 Furans

In analogy to the Paal–Knorr pyrrole synthesis described by Taddei and coworkers [342] (Scheme 6.181), similar reaction conditions were used by these authors to cyclize 1,4-dicarbonyl compounds to give furans (Scheme 6.190). Thus, heating a solution of a 1,4-dicarbonyl compound in ethanol/water in the presence of a catalytic amount of hydrochloric acid at 140 °C for 3 min provided an excellent yield of the corresponding trisubstituted furan derivative.

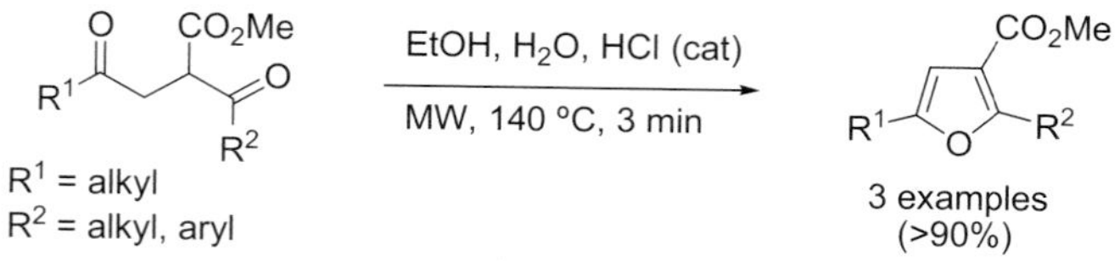

Scheme 6.190 Paal–Knorr synthesis of furans.

A more complex two-step furan synthesis has been described by Jen and coworkers in the context of preparing 2,5-dihydrofuran derivatives as electron acceptors for highly effective nonlinear optical (NLO) chromophores (Scheme 6.191) [352]. In the first step, α-hydroxy ketones are condensed with equimolar quantities of CH-acidic cyano precursors in the presence of 10 mol% of sodium ethoxide in ethanol to furnish reactive 2-imino-2,5-dihydrofuran derivatives. These intermediates can then be further condensed with a second CH-acidic carbonyl or cyano compound (1.2 equivalents) under similar reaction conditions to provide the desired 2-methylene-2,5-dihydrofurans bearing strongly electron-withdrawing groups at the *exo*-methylene functionality. By again using microwave heating, these 2,5-dihydrofuran

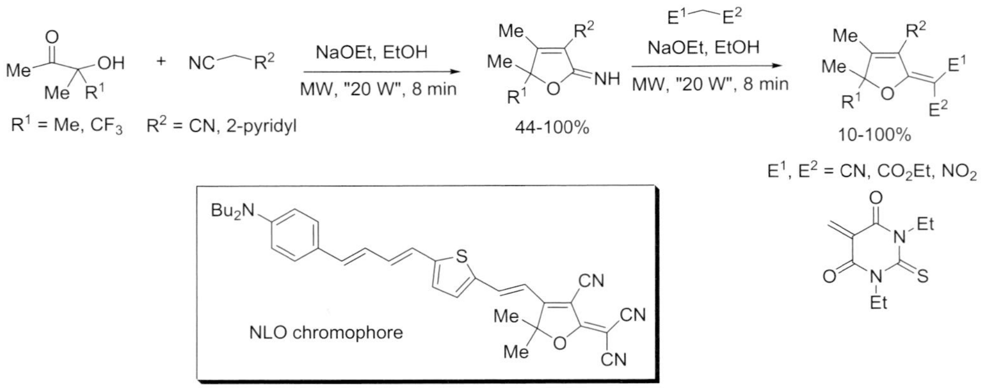

Scheme 6.191 Preparation of 2,5-dihydrofurans.

derivatives can be easily coupled with aromatic, heteroaromatic, or polyene conjugated bridges through their acidic methyl termini at C4 [352]. The properties of the resulting NLO chromophores can therefore be tuned very rapidly utilizing high-speed microwave synthesis.

Furo[3,4-*c*]pyrrolediones are important intermediates in the synthesis of diketo-pyrrolopyrrole (DPP) pigments. Smith and coworkers have described the preparation of several different 3,6-diaryl-substituted furo[3,4-*c*]pyrrole-1,4-diones by microwave-assisted cyclization of readily available 4-aroyl-4,5-dihydro-5-oxo-2-arylpyrrole-3-carboxylates (Scheme 6.192) [353]. While conventional heating in Dowtherm® A at 230–240 °C for 64 h provided only moderate product yields, microwave irradiation of the neat starting material at 250 °C for 10 min provided significantly increased yields.

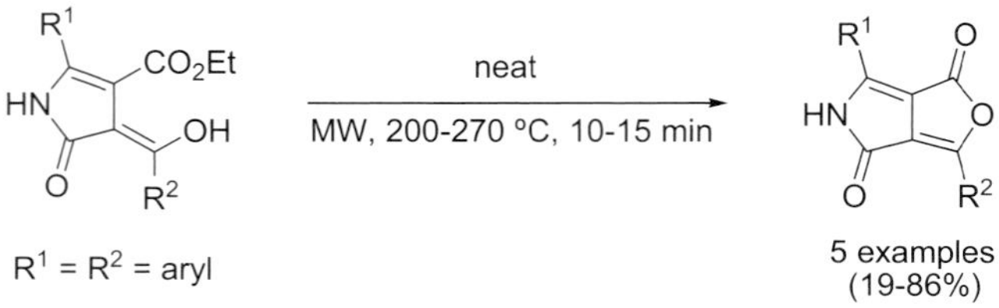

Scheme 6.192 Preparation of furo[3,4-*c*]pyrrolediones.

6.24.3.3 **Thiophenes**

In the context of preparing benzothienyloxy phenylpropanamines as inhibitors of serotonin and norepinephrine uptake, a group from Eli Lilly and Company has developed a two-step synthesis of benzo[*b*]thiophenes (Scheme 6.193) [354]. Thus, a 2-mercapto-3-phenylpropenoic acid derivative was cyclized with iodine in 1,2-dimethoxyethane at 120 °C to give 5-fluoro-4-methoxybenzothiophene-2-carboxylic acid in 67% yield. Decarboxylation under strongly basic conditions involving 1,8-di-azabicyclo[5.4.0]undec-7-ene (DBU) as base in *N,N*-dimethylacetamide (DMA) as

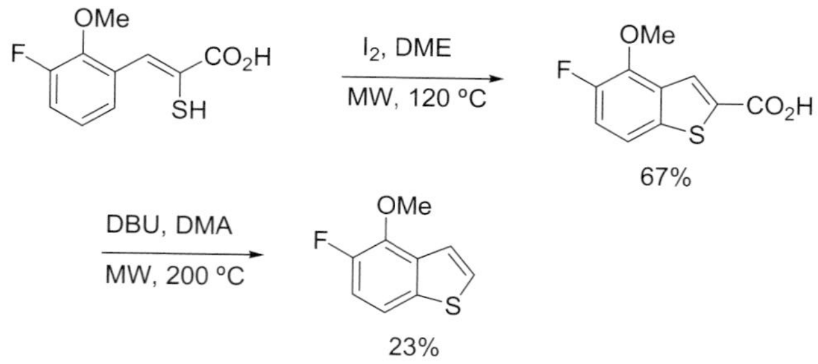

Scheme 6.193 Preparation of benzo[b]thiophenes.

solvent at 200 °C led to the desired benzo[b]thiophene intermediate in moderate yield, which was subsequently further manipulated into the benzothienyloxy phenylpropanamine target structures [354]. A subsequent publication by Eli Lilly and Company disclosed an improved protocol for the decarboxylation of a related benzo[b]thiophene-2-carboxylic acid that provided the final product in 93% yield on a 170 mmol scale using either the DBU/DMA conditions or the traditional quinoline base as solvent [355].

6.24.4
Five-Membered Heterocycles with Two Heteroatoms

6.24.4.1 Pyrazoles
The groups of Giacomelli and Taddei have developed a rapid solution-phase protocol for the synthesis of 1,4,5-trisubstituted pyrazole libraries (Scheme 6.194) [356]. The transformations involved the cyclization of a monosubstituted hydrazine with an enamino-β-ketoester derived from a β-ketoester and N,N-dimethylformamide dimethyl acetal (DMFDMA). The sites for molecular diversity in this approach are the substituents on the hydrazine (R^3) and on the starting β-keto ester (R^1, R^2). Subjecting a solution of the β-keto ester in DMFDMA as solvent to 5 min of microwave irradiation (domestic oven) led to full and clean conversion to the corresponding enamine. After evaporation of the excess DMFDMA, ethanol was added to the crude reaction mixture followed by 1 equivalent of the hydrazine hydrochloride and 1.5 equivalents of triethylamine base. Further microwave irradiation for 8 min provided – after purification by filtration through a short silica gel column – the desired pyrazoles in >90% purity.

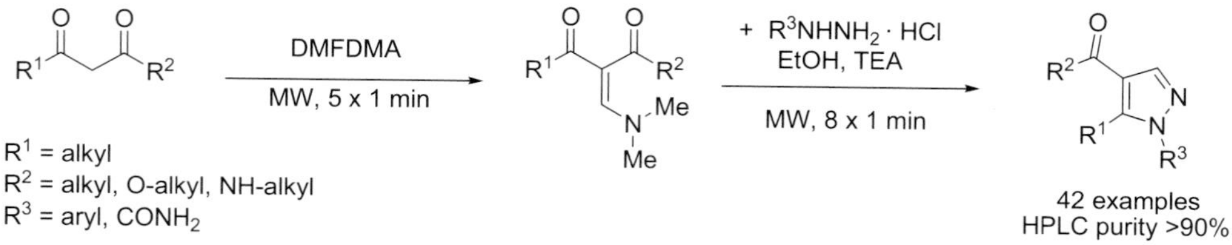

R^1 = alkyl
R^2 = alkyl, O-alkyl, NH-alkyl
R^3 = aryl, CONH$_2$

42 examples
HPLC purity >90%

Scheme 6.194 Synthesis of 1,4,5-trisubstituted pyrazoles.

A somewhat related approach was followed by Molteni and coworkers, who have described the three-component, one-pot synthesis of fused pyrazoles by reacting cyclic 1,3-diketones with DMFDMA and a suitable bidentate nucleophile, such as a hydrazine derivative (Scheme 6.195) [357]. Again, the reaction proceeds by initial formation of an enamino ketone as the key intermediate from the 1,3-diketone and DMFDMA precursors, followed by a tandem addition–elimination/cyclodehydration step. The details of this reaction, carried out in superheated water as solvent, have been described in Section 4.3.3.1.

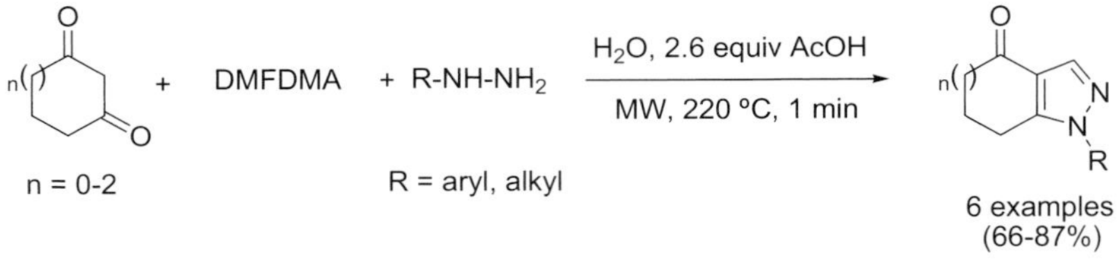

$$n = 0\text{-}2 \qquad\qquad R = \text{aryl, alkyl}$$

H₂O, 2.6 equiv AcOH

MW, 220 °C, 1 min

6 examples
(66-87%)

Scheme 6.195 Three-component condensation leading to fused pyrazoles.

As shown in Scheme 6.196, hydrazines also serve as building blocks for the preparation of medicinally interesting 3,5-diaryl-5-alkyl-4,5-dihydropyrazoles. As reported by Cox and coworkers in 2004, β-alkyl chalcones rapidly add hydrazine hydrate (2 equivalents) within 30 min under microwave conditions at 150 °C to form N1-unsubstituted 4,5-dihydropyrazoles [358]. These unstable intermediates react efficiently with a number of electrophiles to form stable N1-acyl dihydropyrazoles. The current methodology allows the incorporation of many substitution patterns not available from the previously published approaches. Substituted hydrazines provide similar dihydropyrazole products, although the yields using arylhydrazines are low. Microwave heating is not a requirement for these condensation reactions, as refluxing overnight in ethanol provided similar results; however, the shorter reaction times and the ability to easily perform multiple reactions in parallel with an automated handling system makes the microwave route a more attractive option [358].

An alternative strategy for generating 4,5-dihydropyrazoles is to perform 1,3-dipolar cycloaddition reactions of nitrile imines and alkenes. Langa and coworkers have

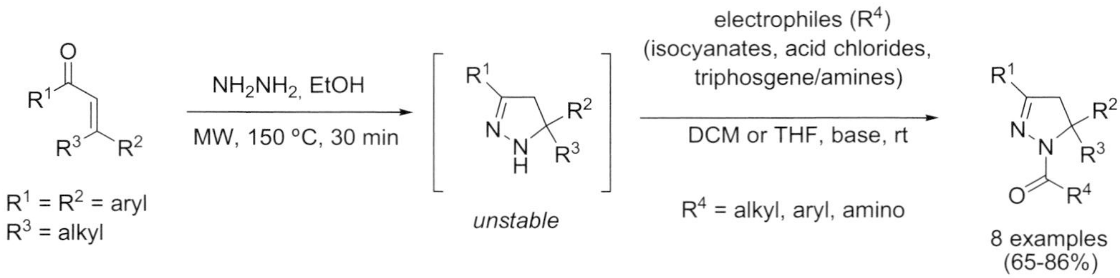

$$R^1 = R^2 = \text{aryl}$$
$$R^3 = \text{alkyl}$$

NH₂NH₂, EtOH

MW, 150 °C, 30 min

unstable

electrophiles (R⁴)
(isocyanates, acid chlorides,
triphosgene/amines)

DCM or THF, base, rt

$$R^4 = \text{alkyl, aryl, amino}$$

8 examples
(65-86%)

Scheme 6.196 4,5-Dihydropyrazole synthesis.

reported the preparation of pyrazolino[60]fullerenes by nitrile imine–fullerene [3+2] cycloadditions (Scheme 6.197 a) [359]. The required nitrile imines were generated *in situ* from the corresponding hydrazones in chloroform. After evaporation of the solvent and addition of C_{60}, triethylamine, and toluene, subsequent irradiation with microwaves under open-vessel conditions for 25 min provided the desired isoindazolylpyrazolino[60]fullerene dyads in moderate yield. Similarly, microwave-assisted sidewall functionalization of single-wall carbon nanotubes (SWNT) through Diels–Alder cycloaddition with *ortho*-quinodimethane has been demonstrated by the same group (Scheme 6.197 b) [360]. The required *ortho*-quinodimethane was generated *in situ* from 4,5-benzo-1,2-oxathiin-2-oxide (sultin) by refluxing in 1,2-dichlorobenzene under open-vessel microwave irradiation conditions. In the presence of ester-functionalized SWNTs, cycloaddition takes place within 45 min. Conventional refluxing in 1,2-dichlorobenzene for 3 days leads only to a low degree of conversion.

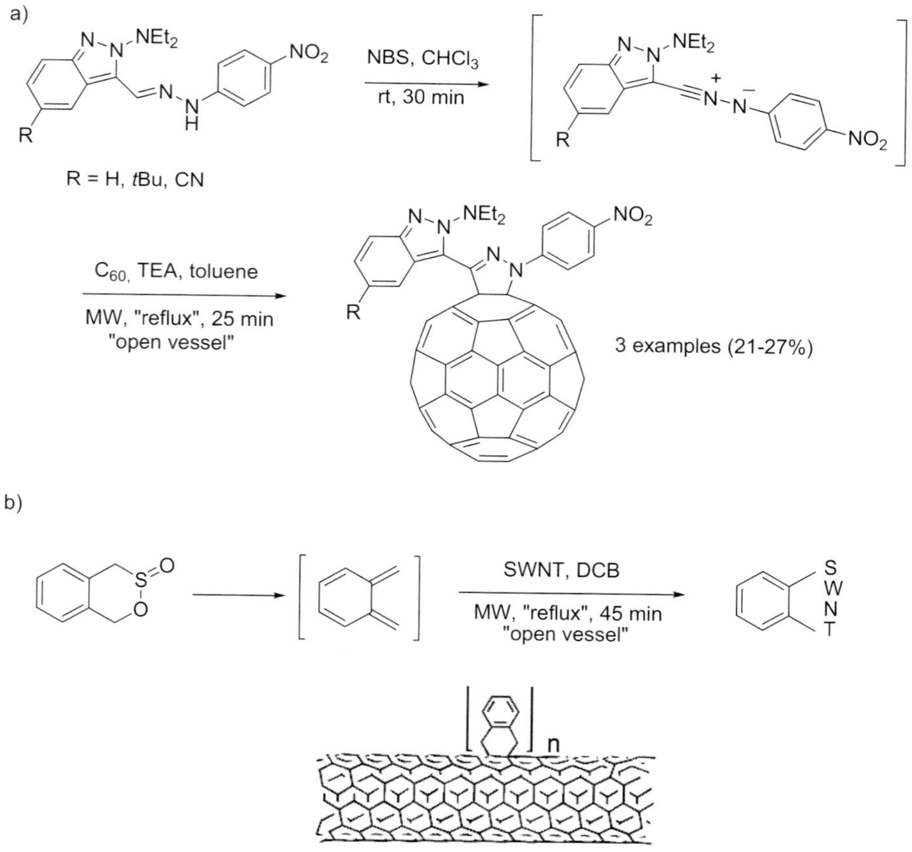

Scheme 6.197 Functionalization of fullerene and single-wall carbon nanotubes through cycloaddition chemistry.

6.24.4.2 **Imidazoles**

A simple, high-yielding synthesis of 2,4,5-trisubstituted imidazoles from 1,2-dike-tones and aldehydes in the presence of ammonium acetate has been reported by Wolkenberg and coworkers (Scheme 6.198) [361]. Utilizing microwave irradiation (180 °C), alkyl-, aryl-, and heteroaryl-substituted imidazoles were formed in very high yields ranging from 76–99% within 5 min by condensing 1,2-diketones with aldehydes in the presence of 10 equivalents of ammonium acetate in acetic acid. Further microwave-assisted alkylation of 2,4,5-trimethylimidazole with benzyl chloride in the presence of a base led to the alkaloid lepidiline B in 43% overall yield. Lepidiline B, with its symmetrical imidazolium structure, exhibits micromolar cytotoxicity against several human cancer cell lines. The microwave-assisted two-step synthesis was evaluated, optimized, and completed within 2 h. Notably, the preparation of the intermediate 2,4,5-trimethylimidazole was technically simpler, faster, and higher yielding than by previous routes [361].

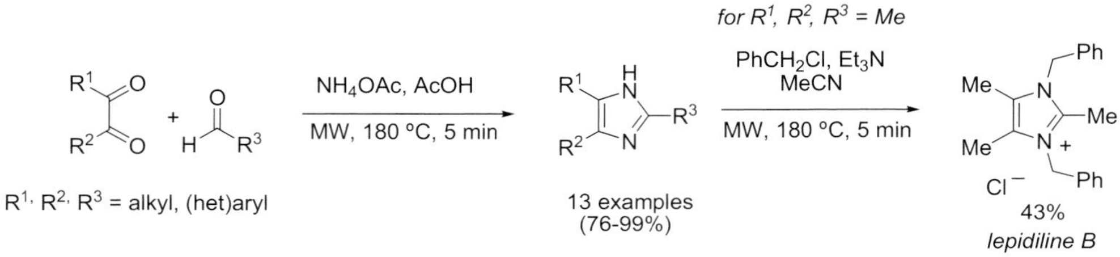

Scheme 6.198   Preparation of imidazoles from 1,2-diketones.

A closely related protocol for the synthesis of imidazoles was independently investigated by Sparks and Combs (Scheme 6.199) [362]. Here, the authors employed readily available unsymmetrical keto-oximes as building blocks, initially leading to *N*-hydroxyimidazoles. Diaryl keto-oximes were condensed with various aldehydes (1.1 equivalents) in the presence of 4 equivalents of ammonium acetate under microwave conditions at 160 °C. In this way, the *N*-hydroxyimidazoles were formed

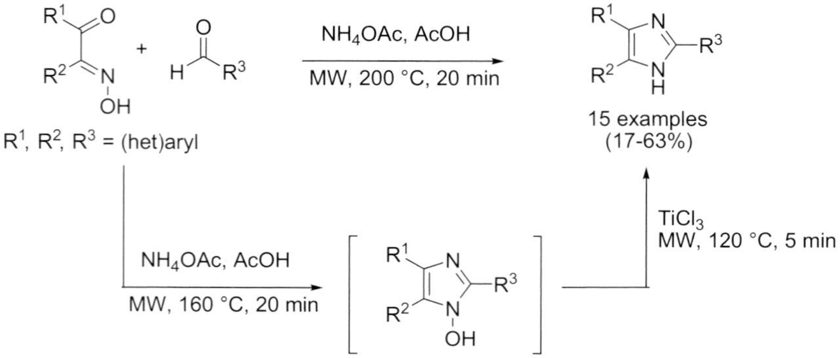

Scheme 6.199   Preparation of imidazoles from keto-oximes.

in high yields and were subsequently quantitatively reduced with titanium trichloride (120 °C, 5 min) to imidazoles. The authors discovered that at higher reaction temperatures (200 °C) treatment of the keto-oximes with aldehydes and ammonium acetate led directly to the desired imidazoles by *in situ* cleavage of the thermolabile N–O bond. Utilizing these optimized conditions, a diverse set of 2,4,5-tri(hetero)-arylimidazoles was prepared.

Multicomponent reactions (MCRs) are of increasing importance in organic and medicinal chemistry. The Ugi four-component condensation, in which an amine, an aldehyde or ketone, a carboxylic acid, and an isocyanide combine to yield an α-acyla-mino amide, is particularly interesting due to the wide range of products obtainable through variation of the starting materials. The reaction of heterocyclic amidines with aldehydes and isocyanides using 5 mol% of scandium triflate as a catalyst in an Ugi-type three-component condensation (Scheme 6.200) generally requires extended reaction times of up to 72 h at room temperature for the generation of the desired fused 3-aminoimidazoles. Tye and coworkers have demonstrated that this process can be speeded up significantly be performing the reaction under microwave conditions [363]. A reaction time of 10 min at 160 °C using methanol as solvent (in some cases ethanol was employed) produced similar yields of products as the same process at room temperature, but in a fraction of the time.

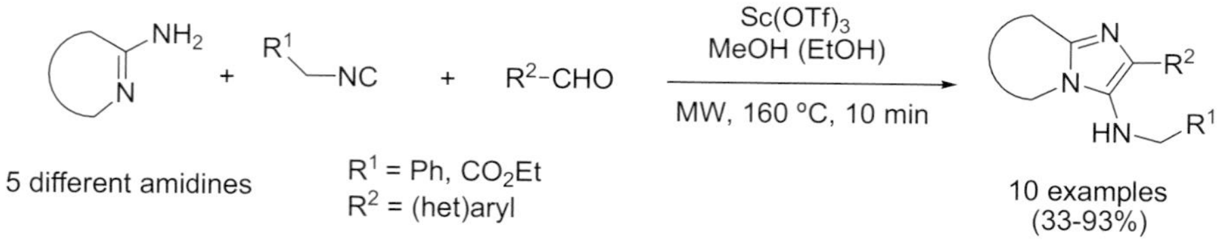

Scheme 6.200 Ugi-type three-component condensation reactions.

A different multicomponent route to imidazoles has been described by the group of O'Shea, involving the diversity-tolerant three-component condensation of an aldehyde, a 2-oxo-thioacetamide, and an alkyl bromide (5 equivalents) in the presence of ammonium acetate (Scheme 6.201) [364]. This allowed the preparation of a 24-membered 4(5)-alkylthio-1*H*-imidazole demonstration library from 21 different aldehydes, 12 alkyl bromides, and two 2-oxo-thioacetamides. The library was synthesized in a parallel format using a custom-built reaction vessel. Alkylthioimidazoles

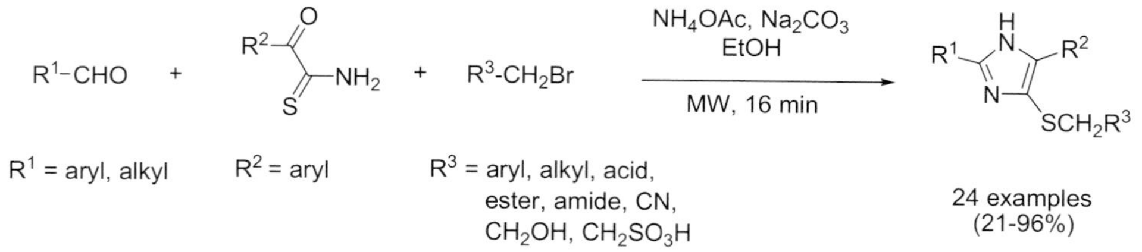

Scheme 6.201 Three-component condensation reactions leading to 4-alkylthioimidazoles.

have been shown to be potential acyl-CoA/cholesterol acyltransferase inhibitors, analgesic agents, and angiotensin II receptor antagonists.

The solvent-free preparation of 1,2,3-trisubstituted imidazolidin-4-ones from aldehydes and N-substituted α-amino acid amides has been reported by Pospíšil and Potáček (Scheme 6.202) [365]. The general procedure simply involved heating equimolar mixtures of the aldehyde and amine building blocks under open-vessel microwave irradiation for 5 min at 200 °C. After cooling to room temperature, the imidazolidin-4-one products were purified by flash chromatography.

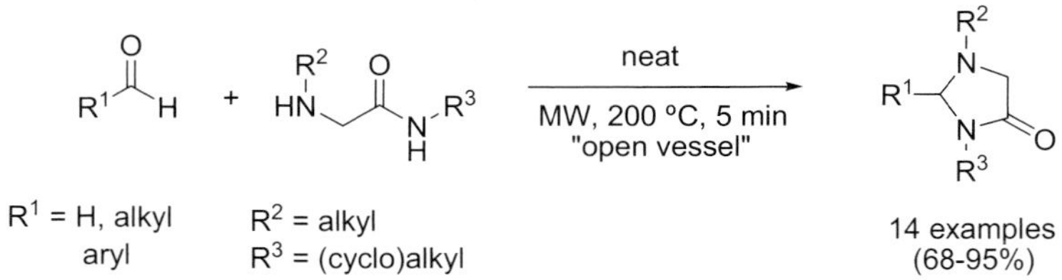

R^1 = H, alkyl R^2 = alkyl
     aryl R^3 = (cyclo)alkyl

14 examples
(68-95%)

Scheme 6.202 Synthesis of imidazolidin-4-ones.

Kim and Varma have described the preparation of a range of cyclic ureas from diamines and urea [366]. In the example highlighted in Scheme 6.203, ethylenediamine and urea were condensed in the presence of 7.3 mol% of zinc(II) oxide in N,N-dimethylformamide as solvent at 120 °C to furnish imidazolidin-2-one in 95% isolated yield. Key to the success of this method is that the reaction needs to be performed under reduced pressure in order to remove the ammonia formed from the reaction mixture. This method was extended to a variety of diamines and amino alcohols [366].

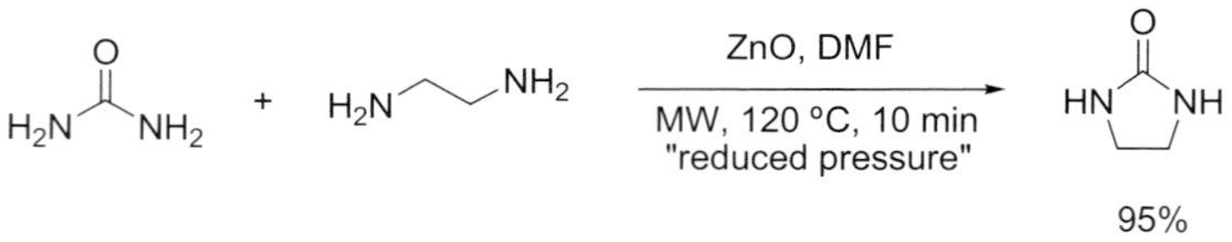

95%

Scheme 6.203 Synthesis of imidazolidinones.

Merriman and colleagues have reported the cyclization of N-acyl-1,2-diaryl-1,2-ethanediamine derivatives, obtained by way of a solid-phase approach, to 4,5-diarylimidazolines by treatment with trimethylsilyl polyphosphate (TMS-PP) in dichloromethane solution (Scheme 6.204) [367]. The best results were obtained by micro-

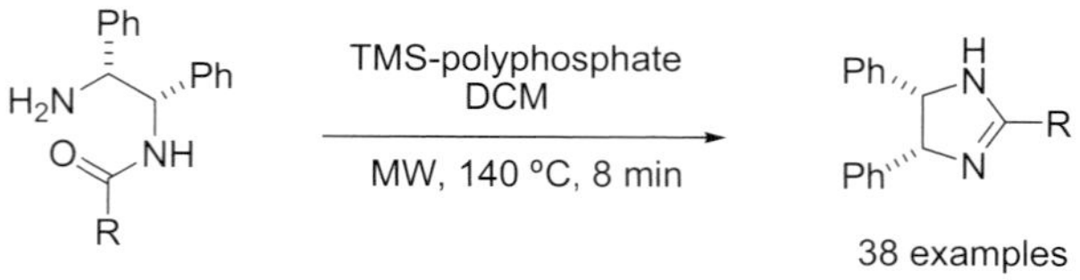

38 examples

Scheme 6.204 Synthesis of 4,5-diarylimidazolines.

wave heating at 140 °C for 8 min. A library of 38 compounds was prepared by this method, leading to a novel and potent family of P2X$_7$ receptor antagonists.

Three different microwave-assisted synthetic routes to benzimidazole derivatives are summarized in Scheme 6.205, involving the condensation of 1,2-phenylenedi-amines with either carboxylic acids (Scheme 6.205 a and b) [368, 369] or two equivalents of aldehydes (Scheme 6.205 c) [370], or by cyclization of N-acylated-diamino-pyrimidines mediated by a strong base (Scheme 6.205 d and e) [371, 372].

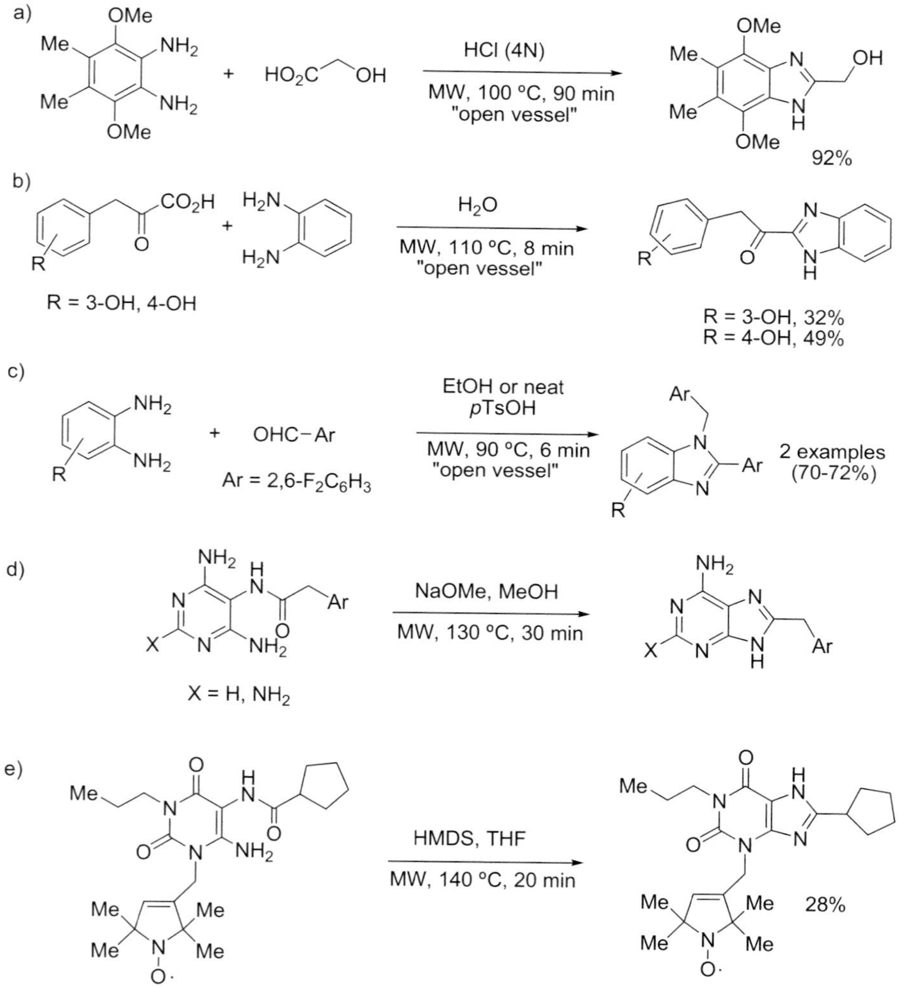

Scheme 6.205 Synthesis of benzimidazole derivatives.

6.24.4.3 Isoxazoles

One obvious synthetic route to isoxazoles and dihydroisoxazoles is by [3+2] cycloadditions of nitrile oxides with alkynes and alkenes, respectively. In the example elaborated by Giacomelli and coworkers shown in Scheme 6.206, nitroalkanes were converted *in situ* to nitrile oxides with 1.25 equivalents of the reagent 4-(4,6-dimethoxy[1,3,5]triazin-2-yl)-4-methylmorpholinium chloride (DMTMM) and 10 mol% of *N*,*N*-dimethylaminopyridine (DMAP) as catalyst [373]. In the presence of an alkene or alkyne dipolarophile (5.0 equivalents), the generated nitrile oxide 1,3-dipoles undergo cycloaddition with the double or triple bond, respectively, thereby furnishing 4,5-dihydroisoxazoles or isoxazoles. For these reactions, open-vessel microwave conditions were chosen and full conversion with very high isolated yields of products was achieved within 3 min at 80 °C. The reactions could also be carried out utilizing a resin-bound alkyne [373]. For a related example, see [477].

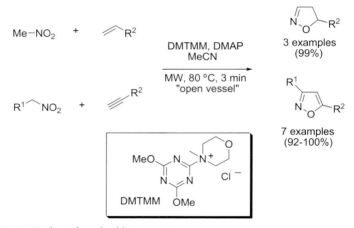

Scheme 6.206 Nitrile oxide cycloaddition reactions.

Related to the nitrile oxide cycloadditions presented in Scheme 6.206 are 1,3-dipolar cycloaddition reactions of nitrones with alkenes leading to isoxazolidines. The group of Comes-Franchini has described cycloadditions of (*Z*)-α-phenyl-*N*-methylnitrone with allylic fluorides leading to enantiopure fluorine-containing isoxazolidines, and ultimately to amino polyols (Scheme 6.207) [374]. The reactions were carried out under solvent-free conditions in the presence of 5 mol% of either scandium(III) or indium(III) triflate. In the racemic series, an optimized 74% yield of an *exo/endo* mixture of cycloadducts was obtained within 15 min at 100 °C. In the case of the enantiopure allyl fluoride, a similar product distribution was achieved after 25 min at 100 °C. Reduction of the isoxazolidine cycloadducts with lithium aluminum hydride provided fluorinated enantiopure polyols of pharmaceutical interest possessing four stereocenters.

A remarkable switch in selectivity was observed by Wagner and coworkers in the cycloaddition of nitrones to free and coordinated (*E*)-cinnamonitrile (Scheme 6.208)

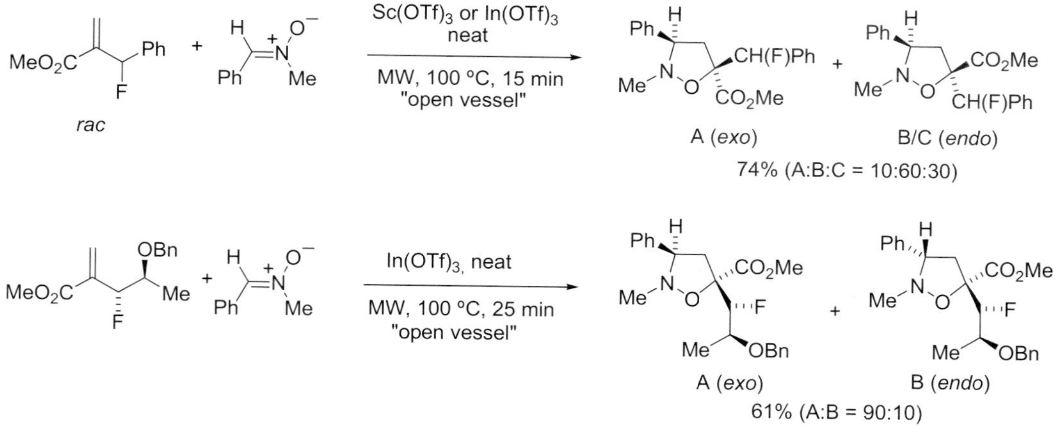

Scheme 6.207 Nitrone–allyl fluoride cycloaddition reactions.

[375]. While the reaction of the nitrone with free cinnamonitrile occurs exclusively at the C=C bond furnishing isoxazolidine-4-carbonitriles (Scheme 6.208 a), cycloaddition of the same nitrone to transition metal-coordinated cinnamonitrile occurs at the nitrile C=N bond, leading to 1,2,4-oxadiazoline complexes, from which the heterocyclic ligand could be released and isolated in high yield (Scheme 6.208 b). Microwave irradiation in dichloromethane solution (100 °C) enhances the rates of both transformations considerably, without changing their regioselectivity with respect to the thermal reactions. The two nitrile ligands in complexes of the type [MCl$_2$(cinnamonitrile)$_2$] (M = platinum or palladium) are significantly different in reactivity. Thus, short-term microwave irradiation allows the selective synthesis of

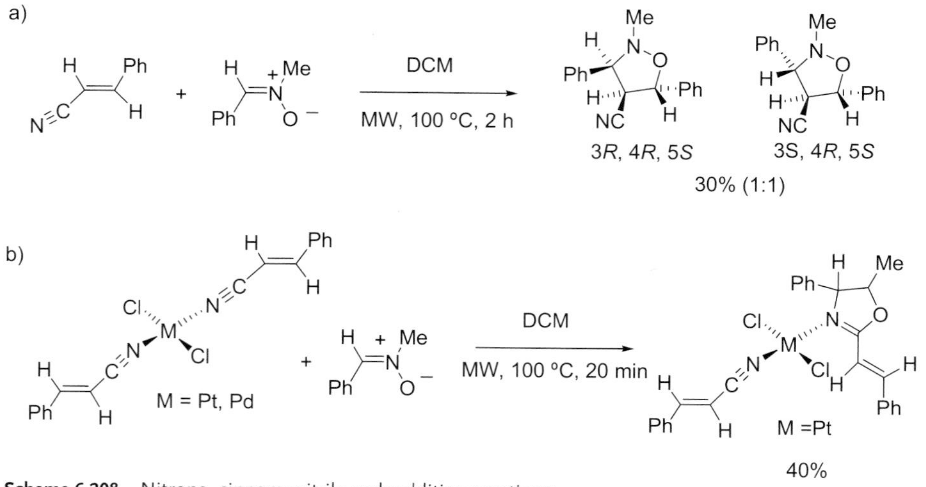

Scheme 6.208 Nitrone–cinnamonitrile cycloaddition reactions.

the mono-cycloaddition product (M = platinum), even in the presence of an excess of nitrone. Using longer reaction times, this complex can be further transformed into the bis-cycloaddition product. Related work was published by the same authors in 2004 [376].

6.24.4.4 Oxazoles

A simple two-step synthesis of 5*H*-alkyl-2-phenyloxazol-4-ones has been reported by Trost and coworkers (Scheme 6.209) [377]. α-Bromo acid halides were condensed with benzamide in the presence of pyridine base at 60 °C to form the corresponding imides. Microwave irradiation of the imide intermediates in *N,N*-dimethylacetamide (DMA) containing sodium fluoride at 180 °C for 10 min provided the desired 5*H*-alkyl-2-phenyloxazol-4-ones (oxalactims) in yields of 44–82%. This class of heterocycles served as excellent precursors for the asymmetric synthesis of α-hydroxycarboxylic acid derivatives [377].

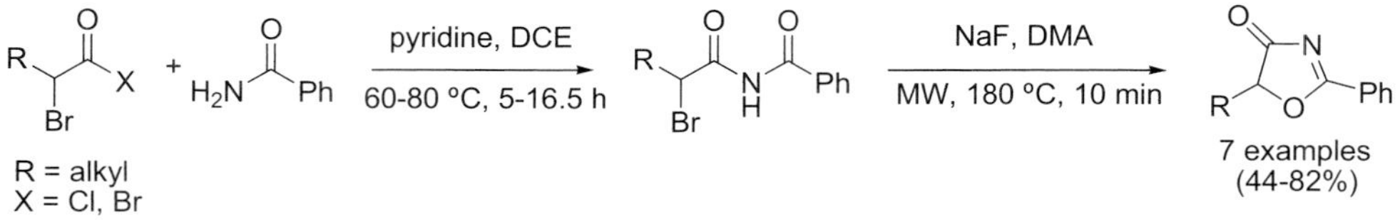

Scheme 6.209 Preparation of 5*H*-alkyl-2-phenyloxazol-4-ones.

An alternative procedure for the synthesis of aliphatic 2-substituted oxazoline hydroxamates was described by Pirrung and colleagues in the context of preparing inhibitors of *E. coli* LpxC zinc amidase [378]. As shown in Scheme 6.210 a, the protocol involved the cyclization of suitable amides, formed *in situ* by acylation of a serine-derived *O*-2,4-dimethoxybenzyl (DMB)-protected hydroxamate. The cyclization

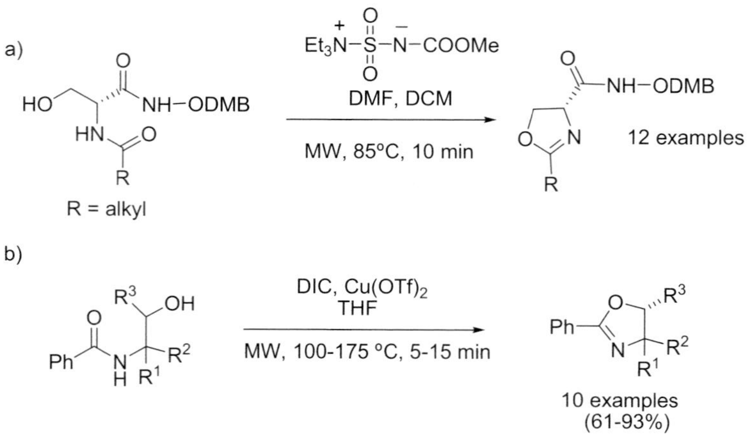

Scheme 6.210 Preparation of 2-substituted oxazolines.

was best performed by employing 1.25 equivalents of Burgess' reagent at 85 °C for 10 min. Owing to the use of Burgess' reagent and the formation of N-acylated by-products, chromatographic purification of the intermediate O-protected oxazoline was required. The purified oxazoline hydroxamates were immediately deprotected with 2% trifluoroacetic acid in hexafluoroisopropanol. Linclau and coworkers have reported a related strategy whereby (chiral) N-(β-hydroxy)amides were cyclized with 1 equivalent of diisopropylcarbodiimide (DIC) in the presence of 5 mol% of copper(II) triflate in tetrahydrofuran (Scheme 6.210 b) [379]. Microwave heating for 5 min at 150 °C provided good to excellent yields of the desired 2-oxazolines.

As demonstrated by the Katritzky group, simple oxazolines can be obtained by treatment of β-amino alcohols with readily available N-acylbenzotriazoles (Scheme 6.211 a) [380]. Thus, microwave irradiation of a solution of 2-amino-2-methyl-1-propanol with 0.5 equivalents of an N-acylbenzotriazole in chloroform at 80 °C for 10 min produced the desired oxazolines along with the uncyclized intermediates, N-(2-hydroxy-1,1-dimethylethyl)amides. Addition of 3 equivalents of thionyl chloride to the reaction mixture and subsequent further microwave irradiation for 2 min (80 °C) resulted in complete conversion of the uncyclized intermediates to the oxazoline product. An analogous protocol was used to generate the corresponding thiazolines (Scheme 6.211 b) [380]. Marrero-Terrero and coworkers have reported a similar preparation of 4,4-disubstituted 2-oxazolines by solvent-free condensation of β-amino alcohols with carboxylic acids using zinc oxide as an inorganic support [381].

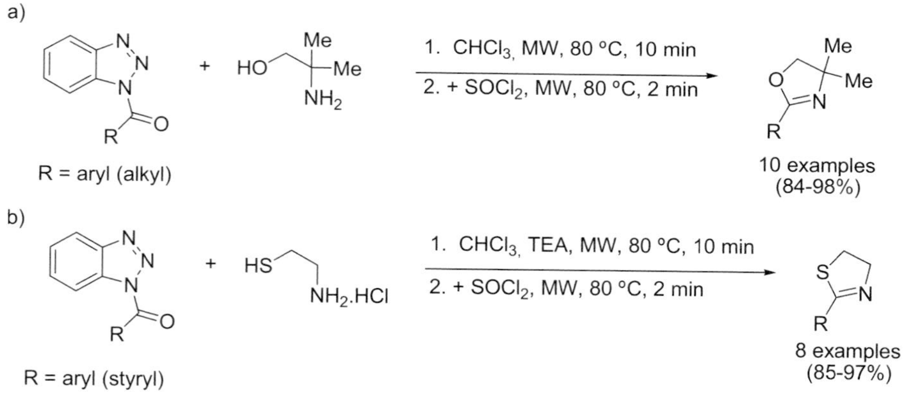

a)

R = aryl (alkyl)

1. CHCl₃, MW, 80 °C, 10 min
2. + SOCl₂, MW, 80 °C, 2 min

10 examples
(84-98%)

b)

R = aryl (styryl)

1. CHCl₃, TEA, MW, 80 °C, 10 min
2. + SOCl₂, MW, 80 °C, 2 min

8 examples
(85-97%)

Scheme 6.211 Preparation of oxazolines.

The condensation of enantiomerically pure amino alcohols (derived from amino acids) with aldehydes to furnish 1,3-oxazolidines was studied by Kuhnert and Danks in 2001 (Scheme 6.212) [382]. Under solvent-free conditions, microwave irradiation of equimolar mixtures of the amino alcohol and the aldehyde for less than 3 min provided high isolated yields of 1,3-oxazolidines with excellent diastereoselectivity. In the case of (−)-ephedrine, prolonged microwave irradiation (3 min) produced quantitative conversions and high diastereoselectivities. For shorter irradiation times (80 s) mixtures of the two diastereomers were obtained with moderate conversions.

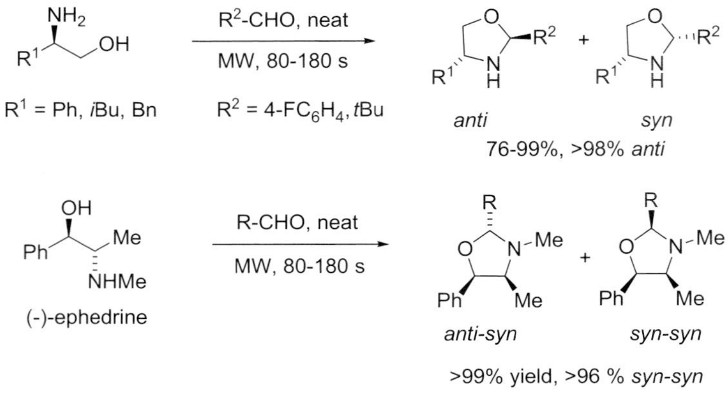

Scheme 6.212 Preparation of 1,3-oxazolidines.

Apparently, the microwave conditions are suitable to drive the equilibrium between the two diastereomers toward the thermodynamically more stable *syn-syn* isomer, which was confirmed in a separate control experiment [382]. Similar results were obtained by Holzgrabe and coworkers for related systems using chloroform as solvent under microwave conditions [383].

For the synthesis of benzoxazoles, Player and coworkers have developed a simple method that involves microwave heating of a 2-aminophenol with 1.1 equivalents of an acid chloride in dioxane or xylene (Scheme 6.213) [384]. The best results were

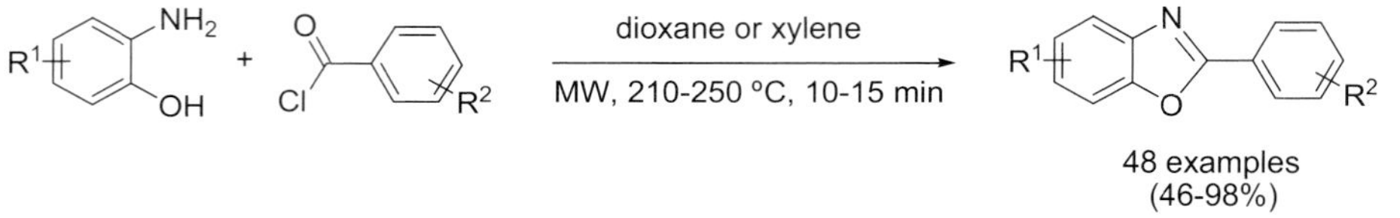

Scheme 6.213 Preparation of benzoxazoles.

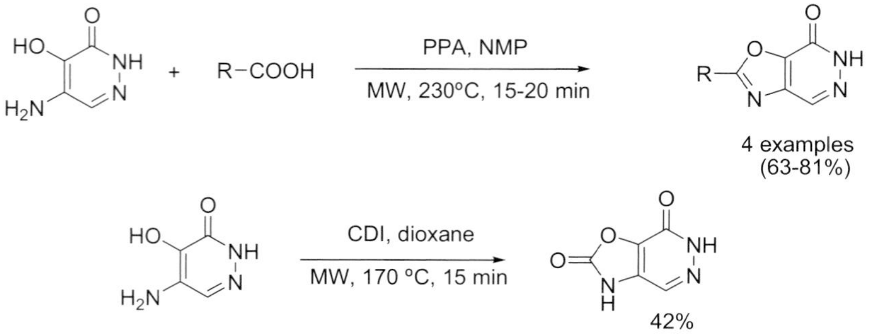

Scheme 6.214 Preparation of oxazolopyridazinones.

obtained when the reaction mixture was exposed to microwave irradiation at 210–250 °C for 10–15 min. The addition of a base or a Lewis acid was not necessary. By choosing a set of six 2-aminophenols and eight acid chlorides, a library of 48 benzoxazoles was prepared in good to excellent yields.

A somewhat similar method for the preparation of 1,3-oxazolo[4,5-d]pyridazinones has been described by Ivachtchenko and coworkers (Scheme 6.214) [385]. Here, the key intermediate 5-amino-4-hydroxy-3(2H)-pyridazinone was treated with 1.5 equivalents of a carboxylic acid in 1-methyl-2-pyrrolidone in the presence of polyphosphoric acid (PPA). The desired oxazolopyridazinones were obtained in good yields by microwave heating at 230 °C for 15–20 min. Reaction of the same precursor with 1,1'-carbonyldiimidazole (CDI) in dioxane at 170 °C furnished the 1,3-oxazolo[4,5-d]pyridazine-2(3H)-7(6H)-dione in 42% yield. Here, the conventional reaction conditions (dioxane, reflux) failed and did not provide any isolable product.

6.24.4.5 Thiazoles

The group of Bolognese has disclosed a synthesis of thiazolidin-4-ones by condensation of benzylidene-anilines and mercaptoacetic acid (Scheme 6.215 a) [386]. The authors found that microwave heating of an equimolar mixture of the two components in benzene at 30 °C for just 10 min provides excellent yields of the thiazolidin-4-one heterocycles. Surprisingly, when the same transformation was carried out at reflux temperature (80 °C), much longer reaction times (2 h) were required and the products were obtained in significantly lower yields (25–69%).

More recently, Miller and coworkers have reported a one-pot protocol for the preparation of thiazolidin-4-ones by condensation of aromatic aldehydes, amines, and mercaptoacetic acid in ethanol (Scheme 6.215 b) [387]. The optimized procedure involved microwave irradiation of a mixture of the amine hydrochloride, aldehyde,

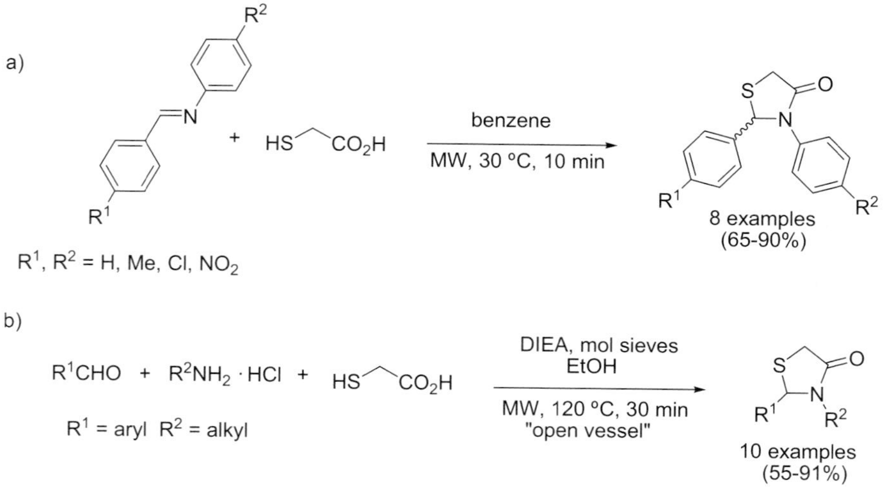

Scheme 6.215 Preparation of thiazolidin-4-ones.

and mercaptoacetic acid (molar ratio 1:2:3) in the presence of 1.25 equivalents of N,N-diisopropylethylamine (DIEA) base in ethanol at 120 °C for 30 min at atmospheric pressure.

Almqvist and coworkers have developed a two-step synthesis of optically active 2-pyridones via thiazolines (Scheme 6.216) [388]. Thus, heating a suspension of (R)-cysteine methyl ester hydrochloride with 2 equivalents of an imino ether and 2 equivalents of triethylamine base in 1,2-dichloroethane at 140 °C for 3 min furnished the desired thiazolines in near quantitative yield with limited racemization. Purification by filtration through a short silica gel column and concentration of the filtrate gave a crude product, which was used directly in the next step. Thus, after

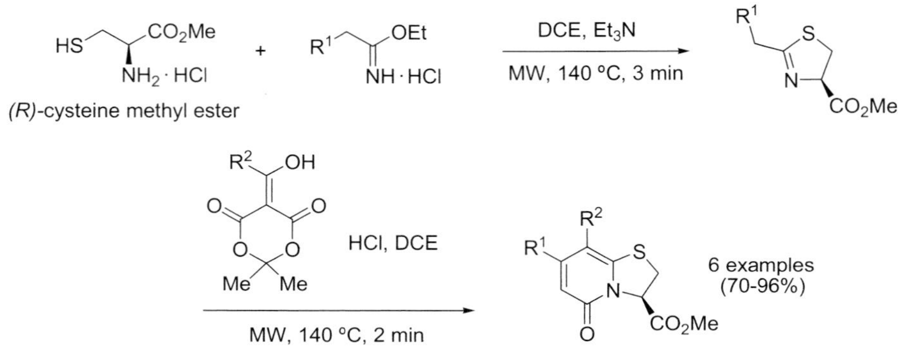

Scheme 6.216 Preparation of thiazolines and 2-pyridones.

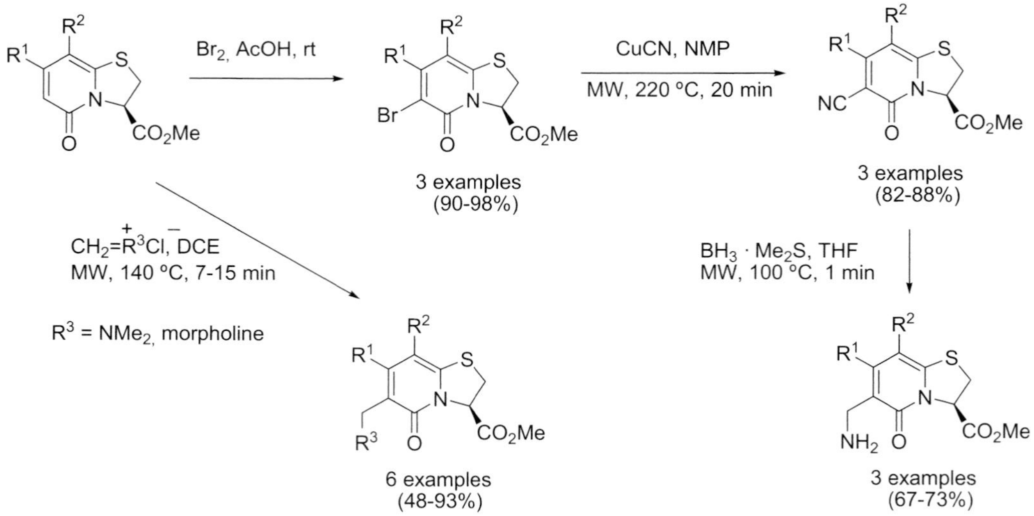

Scheme 6.217 Functionalization of 2-pyridones.

the addition of a pre-saturated solution of hydrochloric acid in 1,2-dichloroethane containing 1.5 equivalents of an acyl Meldrum acid derivative, the resulting solution was again heated under microwave conditions to 140 °C for 2 min. Applying this two-step protocol, a small array of six bicyclic 2-pyridinones was prepared in 70–96% yield in only 5 min of microwave irradiation time. The optical purity of the heterocyclic products proved to be slightly lower (78–88% ee) than that obtained by the conventional protocol (75–97% ee), but the total reaction time was reduced from 2 days to 5 min.

The same group subsequently described the rapid functionalization of the bicyclic 2-pyridone ring, in particular various aminomethylation pathways (Scheme 6.217)

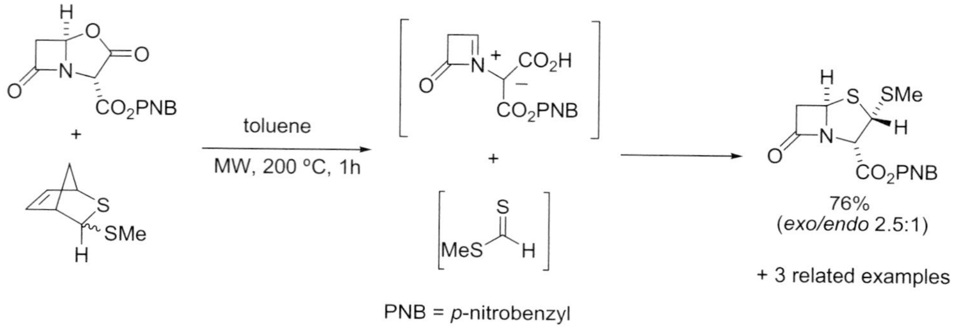

PNB = p-nitrobenzyl

Scheme 6.218 Synthesis of penem derivatives by azomethine ylide–thione cycloadditions.

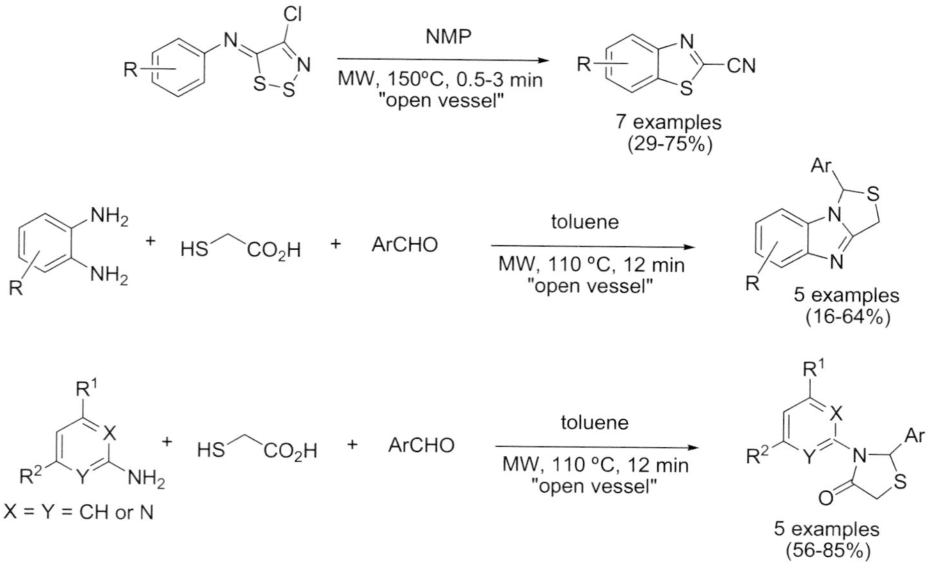

X = Y = CH or N

Scheme 6.219 Synthesis of benzothiazoles [391, 392], thiazolobenzimidazoles [370], and heteroaryl-substituted thiazolidinones [370].

[389]. Primary amino methylene substituents were introduced by a sequence of cyanodehalogenation and subsequent reduction of the resulting nitrile with borane dimethyl sulfide. To incorporate tertiary aminomethylene substituents into the 2-pyridone framework, a microwave-assisted Mannich reaction using preformed iminium salts proved to be effective.

Gallagher and coworkers have elaborated a rather complex strategy for the synthesis of fused thiazolines of the penem type (bicyclic β-lactams) based on the generation of β-lactam-based azomethine ylides [390]. In the example shown in Scheme 6.218, an oxazolidinone precursor reacts through sequential ring cleavage to give an azomethine ylide dipole, which then undergoes spontaneous cycloaddition with an *in situ* generated S-alkyl dithioformate dipolarophile to produce the target penem. The thermal process required 2 days to reach completion and this only produced a very modest 19% yield of the cycloadduct. Under microwave irradiation conditions utilizing toluene as solvent at 200 °C, a 76% isolated yield of the cycloadduct was obtained within 1 h. Somewhat lower yields were obtained when the transformation was carried out in ionic liquid-doped toluene or under open-vessel conditions.

Alternative methods for the microwave-assisted preparation of thiazole derivatives are summarized in Scheme 6.219.

6.24.5
Five-Membered Heterocycles with Three Heteroatoms

6.24.5.1 1,2,3-Triazoles
The 1,3-dipolar cycloaddition of azides to alkynes is a versatile route to 1,2,3-triazoles. Different combinations of substituents on the azide and on the alkyne allow the preparation of diverse *N*-substituted 1,2,3-triazoles. Katritzky and Singh have described the synthesis of *C*-carbamoyl-1,2,3-triazoles by microwave-induced cycloaddition of benzyl azides to acetylenic amides (Scheme 6.220) [393]. Employing equimolar mixtures of the azide and alkyne under solvent-free conditions, the authors were able to achieve good to excellent isolated product yields by microwave heating at 55–85 °C for 30 min. In general, the triazole products were obtained as mixtures of regioisomers. Control experiments carried out under thermal (oil bath)

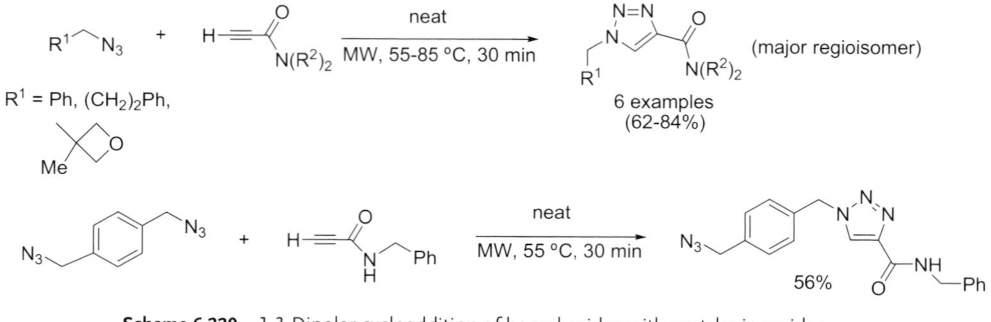

Scheme 6.220 1,3-Dipolar cycloaddition of benzyl azides with acetylenic amides.

conditions at 55–60 °C did not show any conversion even after 24 h. With bis-azides, besides the anticipated mono-cycloadducts, in some cases the bis-triazoles could also be obtained (not shown) [394]. Other microwave-assisted 1,3-dipolar cycloaddition reactions of azides with symmetrical alkynes have been studied by Savin and coworkers (see Scheme 4.28) [395].

A significant advance in this area has been the development of a copper(I)-catalyzed variation of this classical 1,3-dipolar cycloaddition process. For terminal alkynes, this method leads to a regiospecific coupling of the two reaction partners. The Kappe group has exploited a microwave-assisted version of this copper(I)-catalyzed azide–alkyne ligation process ("click chemistry") for the preparation of 6-(1,2,3-triazol-1-yl)-dihydropyrimidones (Scheme 6.221) [396]. Here, a suitable heterocyclic azide intermediate (obtained by microwave-assisted azidation) was treated with phenyl acetylene in *N,N*-dimethylformamide employing 2 mol% of copper(II) sulfate/sodium ascorbate as catalyst precursor. After completion of the cycloaddition process, the triazole product could be precipitated in high yield (73%) and purity by addition to ice/water. For the model reaction depicted in Scheme 6.221, full conversion at room temperature required 1 h. By carrying out the same reaction utilizing controlled microwave heating at 80 °C, complete conversion was achieved within 1 min. A library of 27 6-(1,2,3-triazol-1-yl)-dihydropyrimidones was prepared with four points of diversity.

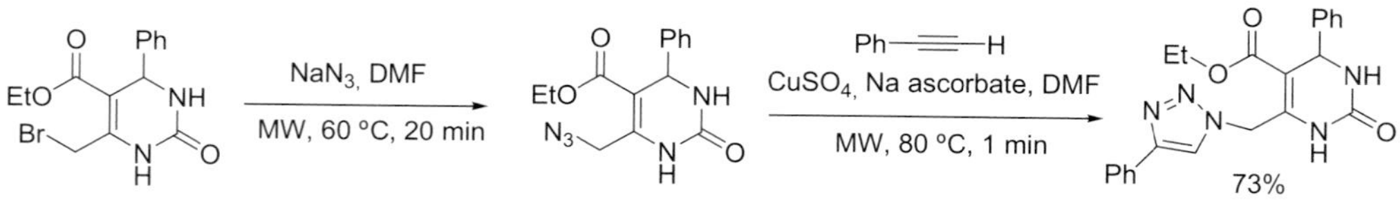

Scheme 6.221 Copper(I)-catalyzed azide–alkyne ligation.

For certain substrates, Fokin, Van der Eycken, and coworkers subsequently discovered that the azidation and ligation steps can be carried out in a one-pot fashion, thereby simplifying the overall protocol (Scheme 6.222) [397]. This procedure eliminates the need to handle organic azides, as they are generated *in situ*. Other applications of microwave-assisted copper(I)-catalyzed azide–alkyne ligations ("click chemistry") have been reported [398].

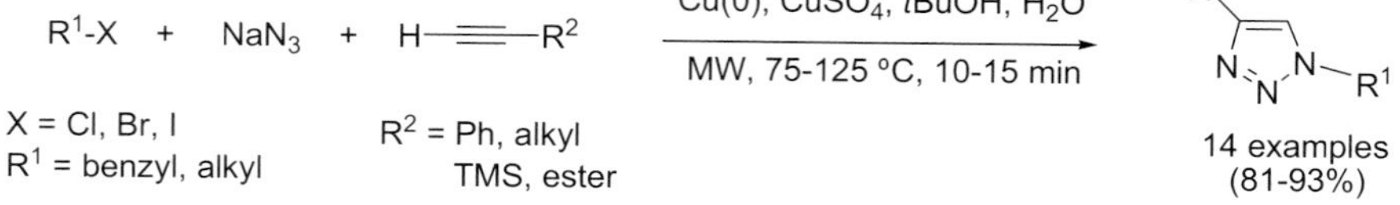

Scheme 6.222 Copper(I)-catalyzed azide–alkyne ligation in a one-pot fashion.

6.24.5.2 **1,2,4-Oxadiazoles**

The 1,2,4-oxadiazole class of heterocycles has been shown to possess a variety of CNS (central nervous system) related activities. A group from ArQule and Pfizer has reported the rapid synthesis of 1,2,4-oxadiazoles based on the coupling of amidoximes with carboxylic acids in the presence of O-benzotriazol-1-yl-N,N,N′,N′-tetramethyluronium hexafluorophosphate (HBTU) (Scheme 6.223) [399]. The authors made extensive use of statistical design of experiments (DoE) methods (see Section 5.3.4) in optimizing the protocol. The optimum conditions involved the utilization of equimolar amounts of carboxylic acid, amidoxime, and HBTU, and 2.35 equivalents of N,N-diisopropylethylamine (DIEA) base in N,N-dimethylformamide as the solvent. Microwave irradiation of this mixture for 2 min at 191 °C led to >90% conversion for the majority of the 30 library compounds studied. Similar results were achieved by Santagada and coworkers in an independent investigation of the same general transformation [400].

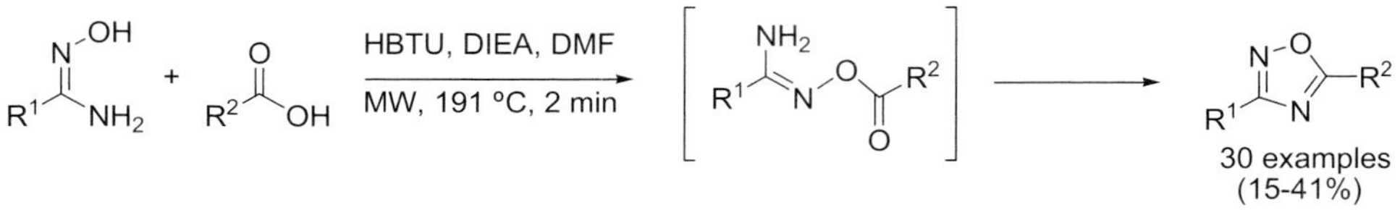

Scheme 6.223 Synthesis of 1,2,4-oxadiazoles.

6.24.5.3 **1,3,4-Oxadiazoles**

In the context of preparing potential inhibitors of histone deacetylase, Vasudevan and a team from Abbott have described the cyclization of 1,2-diacylhydrazides to 1,3,4-oxadiazoles with Burgess' reagent under microwave conditions (150 °C, 15 min) (Scheme 6.224 a) [232]. A different approach was chosen by Natero and coworkers, who prepared 2-chloromethyl-1,3,4-oxadiazoles by treatment of acyl hydrazides with 1-chloro-2,2,2-trimethoxyethane (Scheme 6.224 b) [401]. Here, the reagent was used as solvent and the mixture was heated by microwave irradiation at 160 °C for 5 min.

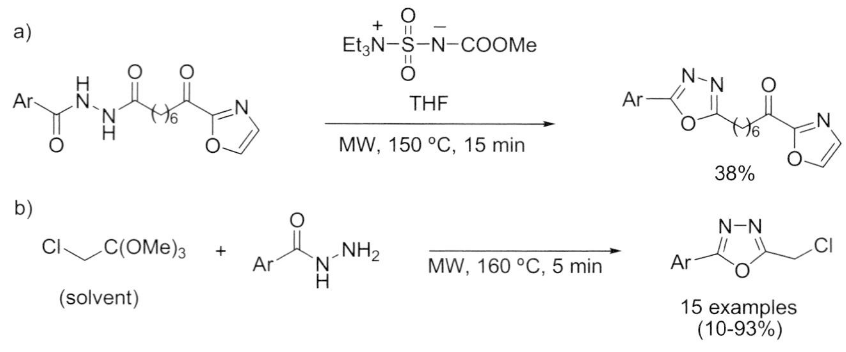

Scheme 6.224 Synthesis of 1,3,4-oxadiazoles.

6.24.5.4 1,3,2-Diazaphospholidines

A method for the preparation of 1,3,2-diazaphospholidine heterocycles has been described by Deng and Chen (Scheme 6.225) [402]. The authors found that treating hindered 1,2-diamino substrates such as α-amino acid amides with tris(diethylamino)phosphine as reagent/solvent under open-vessel microwave conditions at 250 °C for 1 min furnished a trivalent phosphorus intermediate. Subsequent thiation of this intermediate with elemental sulfur in refluxing benzene provided the target 1,3,2-diazaphospholidin-4-ones in good overall yields. The yields were much improved compared to those achieved by standard thermal methods.

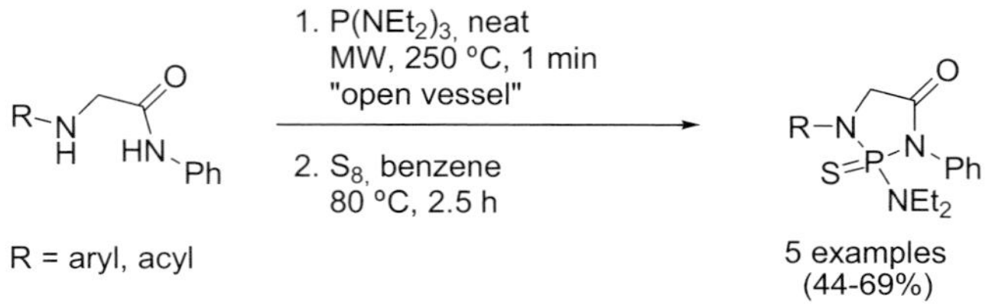

Scheme 6.225 Synthesis of 1,3,2-diazaphospholidin-4-ones.

6.24.6
Five-Membered Heterocycles with Four Heteroatoms

In 2004, three research groups independently described the synthesis of (hetero)aryl tetrazoles by cycloadditions of appropriate aromatic nitriles with the reagent system trimethylsilyl azide/dibutyltin oxide (TMS-N$_3$/DBTO) (Scheme 6.226). Lukayanov and colleagues used various nicotinonitriles as starting materials (Scheme 6.226 a) [403] and obtained moderate yields of the tetrazole products by employing 4 equivalents of trimethylsilyl azide and 0.3 equivalents of dibutyltin oxide in anhydrous dioxane at 140 °C for 8 h. Under these conditions, full conversion was not achieved and 15–63% of the starting nitrile was recovered. Schulz and colleagues have reported the preparation of aryltetrazole boronate esters under similar reaction conditions (Scheme 6.226 b) [404]. While the cycloaddition proved to be problematic with the free boronic acids, high conversions were achieved with boronic acids protected as pinacol esters (prepared *in situ*). Typically, the nitrile was reacted with 2 equivalents of trimethylsilyl azide and 0.1 equivalents of dibutyltin oxide in 1,2-dimethoxyethane (DME) at 150 °C for 10 min. After the addition of more of the TMS-N$_3$/DBTO reagent mixture, the vessel was heated for a further 10 min at 150 °C to complete the reaction. This provided a high isolated yield of the aryltetrazole boronates, which were subsequently used in microwave-assisted Suzuki couplings with aryl bromides (see Section 6.1.2). In a more complex variation of the same reaction, the group of Frejd has described the synthesis of fused tetrazole derivatives via a tandem cycloaddition and *N*-allylation reaction sequence (Sche-

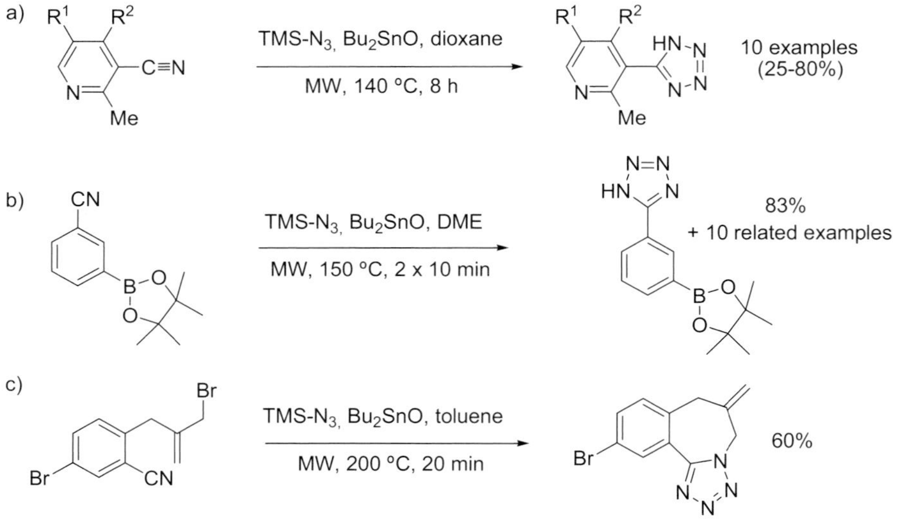

a)

b)

c)

Scheme 6.226 Synthesis of tetrazoles.

me 6.226 c) [405]. Here, 5 equivalents of trimethylsilyl azide and 1 equivalent of dibutyltin oxide were employed (toluene, 200 °C, 20 min).

6.24.7
Six-Membered Heterocycles with One Heteroatom

6.24.7.1 Pyridines

The Bohlmann–Rahtz synthesis of trisubstituted pyridines from β-aminocrotonates and an ethynyl ketone has found application in the preparation of a variety of heterocycles based on the substituted pyridine motif. Bagley and coworkers have developed a microwave-assisted modification of this one-pot heteroannulation method that is best conducted in dimethyl sulfoxide at 170 °C for 20 min, providing the desired pyridines in 24–94% yield (Scheme 6.227) [406, 407]. Typically, 2 equivalents of the β-aminocrotonates were employed.

Another well-known method for the preparation of heterocycles is the Hantzsch dihydropyridine synthesis. In 2001, Öhberg and Westman presented a microwave-

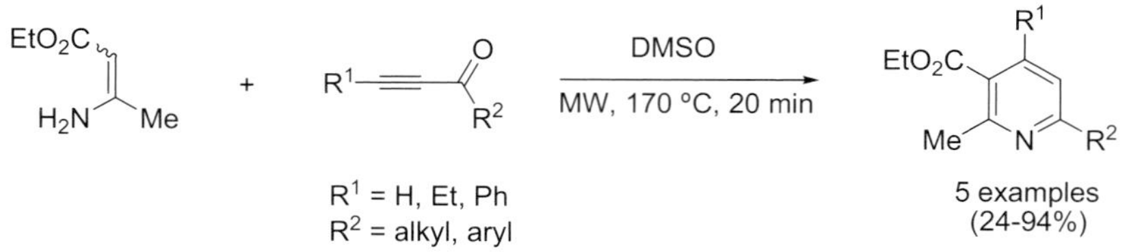

R^1 = H, Et, Ph
R^2 = alkyl, aryl

Scheme 6.227 The Bohlmann–Rahtz synthesis of pyridines.

assisted Hantzsch dihydropyridine synthesis that allowed the rapid preparation of heterocycles of this type in a multicomponent, one-pot fashion (Scheme 6.228 a) [408]. Thus, suitable aliphatic or (hetero)aromatic aldehydes were reacted with 5 equivalents of a β-ketoester and 4 equivalents of concentrated aqueous ammonia, which was used both as a reagent and as a solvent. The best yields were obtained by exposing the reaction mixture to microwave heating at 140–150 °C for 10–15 min. In order to prepare a diverse set of products, six different aldehydes and four different β-ketoesters or 1,3-dicarbonyl substrates were used. All 24 of the resulting compounds were formed in moderate to good yields. A somewhat related multicomponent transformation for the synthesis of 5-deaza-5,8-dihydropterins (pyrido[2,3-d]pyrimidines) has been described by Bagley and Singh (Scheme 6.228 b) [409]. Heating 2,6-diaminopyrimidin-4-one with an aldehyde (2 equivalents) and a suitable CH-acidic carbonyl compound (2 equivalents) in dimethyl sulfoxide in the presence of 20 mol% of zinc bromide as Lewis acid at 160 °C for 20 min provided good yields of the desired fused dihydropyridines.

The Kappe group has described a multicomponent, one-pot, two-step pathway to 3,5,6-substituted 2-pyridones (Scheme 6.229) [410]. In the first step, equimolar mix-

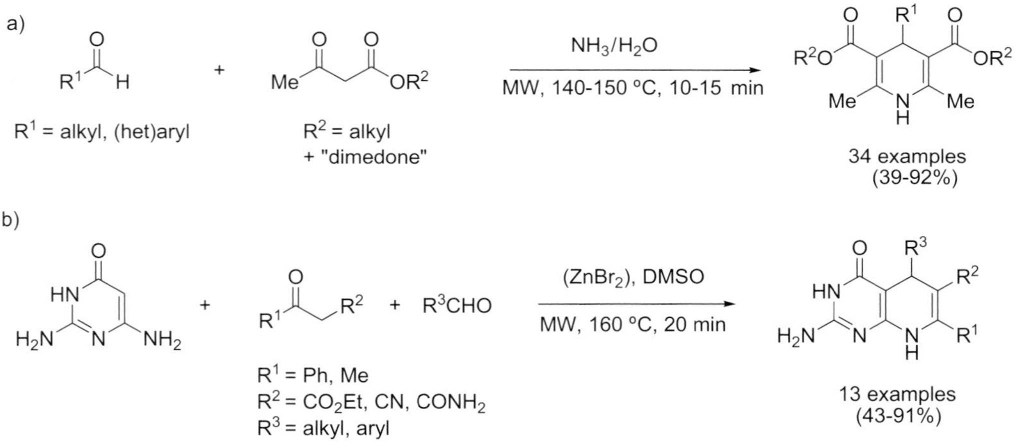

Scheme 6.228 Hantzsch dihydropyridine synthesis.

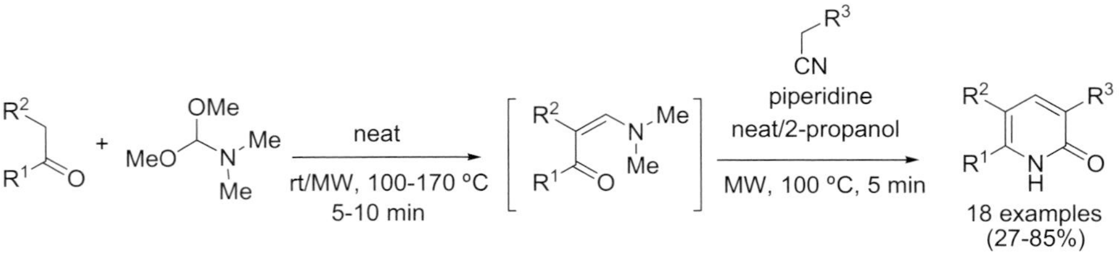

Scheme 6.229 Three-component synthesis of pyridones.

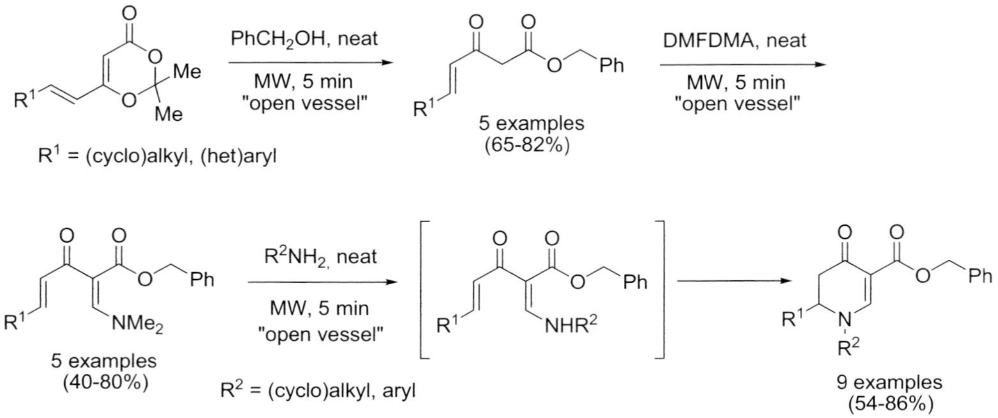

Scheme 6.230 Multistep preparation of 2,3-dihydropyridin-4-ones.

tures of a CH-acidic carbonyl compound and *N,N*-dimethylformamide dimethyl acetal (DMFDMA) were reacted to form the corresponding enamines, either at room temperature or under microwave conditions. After the addition of a methylene-active nitrile (1 equivalent), 2-propanol as solvent, and a catalytic amount of piperidine base, the reaction mixture was heated at 100 °C for 5 min under microwave conditions. In most cases, the desired heterocyclic product precipitated directly after cooling of the reaction mixture and could be collected by filtration.

In Scheme 6.230, the multistep synthesis of 2,3-dihydro-4-pyridones is highlighted [411]. The pathway described by Panunzio and coworkers starts from a dioxin-4-one precursor, which is reacted with 2 equivalents of benzyl alcohol under solvent-free microwave conditions to furnish the corresponding *β*-diketo benzyl esters. Subsequent treatment with 1 equivalent of *N,N*-dimethylformamide dimethyl acetal (DMFDMA), again under solvent-free conditions, produces an enamine, which is then cyclized with an amine building block (1.1 equivalents) to produce the desired 4-pyridinone products. All microwave protocols were conducted under open-vessel conditions using power control.

Many of the traditional condensation reactions leading to heterocycles require high temperatures, and conventional reaction conditions very often involve heating of the reactants in an oil, metal, or sand bath for many hours or even days. One example published by Kappe and coworkers is illustrated in Scheme 6.231 a, namely the formation of 4-hydroxyquinolin-2(1*H*)-ones from anilines and malonic esters [412]. The corresponding conventional thermal protocol involves heating of the two components in equimolar amounts in an oil bath at 220–300 °C for several hours (without solvent), whereas similarly high yields can be obtained by microwave heating at 250 °C for 10 min. Here, it was essential to use open-vessel technology, since two equivalents of a volatile by-product (ethanol) are formed that under normal (atmospheric pressure) conditions are simply distilled off and therefore removed from the equilibrium (see Section 4.3.1). Similar results were previously obtained by Lange and colleagues (Scheme 6.231 b) [413].

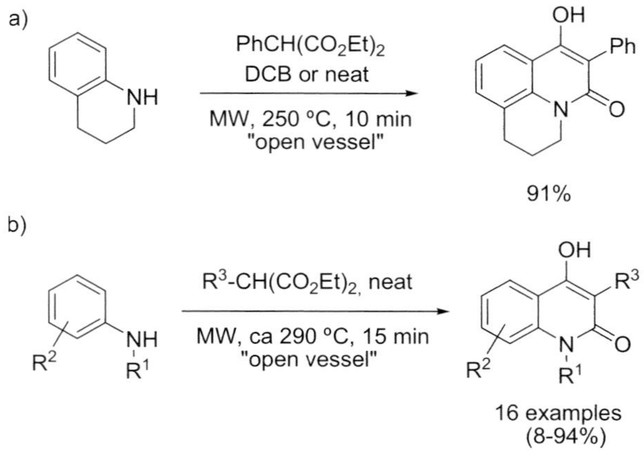

a)

PhCH(CO₂Et)₂
DCB or neat
→
MW, 250 °C, 10 min
"open vessel"

91%

b)

R³-CH(CO₂Et)₂, neat
→
MW, ca 290 °C, 15 min
"open vessel"

16 examples
(8-94%)

Scheme 6.231 Formation of 4-hydroxyquinolin-2(1*H*)-ones from anilines and malonic esters.

The Pictet–Spengler reaction is another important reaction principle for the generation of (fused) pyridine systems. Several groups have independently reported on microwave-assisted Lewis acid or ionic liquid-mediated Pictet–Spengler cyclizations involving iminium ion intermediates. Srinivasan and Ganesan have described a Pictet–Spengler approach to the tetrahydro-β-carboline ring system, utilizing the one-pot condensation of tryptophan methyl ester or tryptamine with aliphatic and aromatic aldehydes (Scheme 6.232) [414]. The most active catalytic system, in particular for transformations involving tryptophan, was found to be a combination of 10 mol% of ytterbium(III) triflate and 50 mol% of the ionic liquid 1-butyl-3-methyl-imidazolium chloroaluminate salt ([bmim]Cl/AlCl₃). Employing 1.2 equivalents of the aldehyde building block, very high yields of products were obtained under microwave irradiation in dichloromethane at 100–120 °C for 30–60 min.

In a closely related approach, Yen and Chu have reported the preparation of tetra-hydro-β-carbolinediketopiperazines employing a three-step Pictet–Spengler, Schotten–Baumann, and intramolecular ester amidation sequence (Scheme 6.233) [415]. Throughout the synthesis, the ionic liquid 1-butyl-2,3-dimethylimidazolium hexa-fluorophosphate (bdmimPF₆) was employed. In a typical experiment, (S)-tryptophan methyl ester was dissolved in a 1:1 mixture of the ionic liquid and tetrahydrofuran

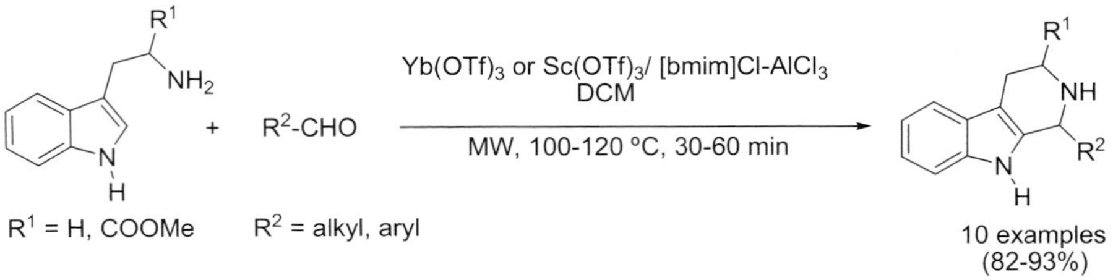

Yb(OTf)₃ or Sc(OTf)₃/ [bmim]Cl-AlCl₃
DCM
→
MW, 100-120 °C, 30-60 min

R¹ = H, COOMe   R² = alkyl, aryl

10 examples
(82-93%)

Scheme 6.232 Lewis acid-catalyzed Pictet–Spengler reactions.

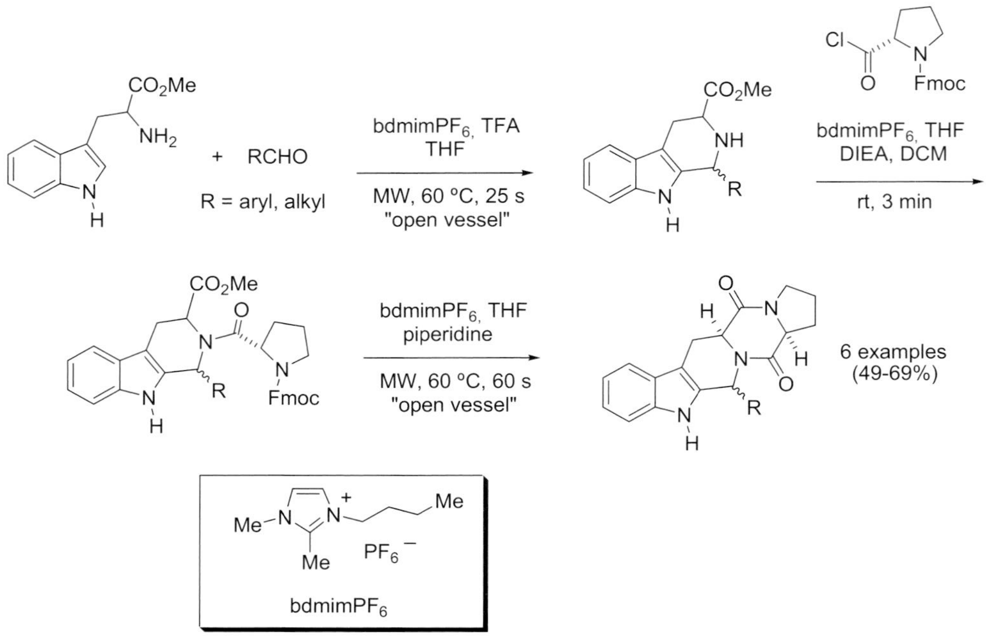

Scheme 6.233 Ionic liquid-mediated preparation of tetrahydro-β-carbolinediketopiperazines.

(THF) containing 10% (v/v) of trifluoroacetic acid (TFA). After microwave irradiation for 25 s at 60 °C, the mixture was concentrated to dryness, 5 equivalents each of N,N-diisopropylethylamine (DIEA) and (S)-proline acid chloride in bdmimPF$_6$/ THF were added, and the resulting mixture was stirred at room temperature for 3 min. Finally, after *in situ* deprotection with 20% piperidine (v/v) in bdmimPF$_6$/ THF at 60 °C for 1 min in the microwave reactor, the target compounds were obtained as mixtures of diastereomers.

Grieco and coworkers have independently described the same type of Pictet–Spengler cyclization reactions involving tryptophan methyl ester and aldehydes, but using methanol as solvent and hydrochloric acid as a catalyst (microwave irradiation, 50 °C, 20–50 min) [416]. Moderate to good product yields were obtained.

Yen and Chu subsequently also disclosed a related Pictet–Spengler reaction involving tryptophan and ketones for the preparation of 1,1-disubstituted indole alkaloids [417]. In the approach shown in Scheme 6.234, tryptophan was reacted with numerous ketones (12 equivalents) in toluene in the presence of 10 mol% of trifluoroacetic acid catalyst. Using microwave irradiation at 60 °C under open-vessel conditions, the desired products were obtained in high yields. Compared to transformations carried out at room temperature, reaction times were typically reduced from days to minutes. Subsequent treatment with isocyanates or isothiocyanates led to tetrahydro-β-carbolinehydantoins.

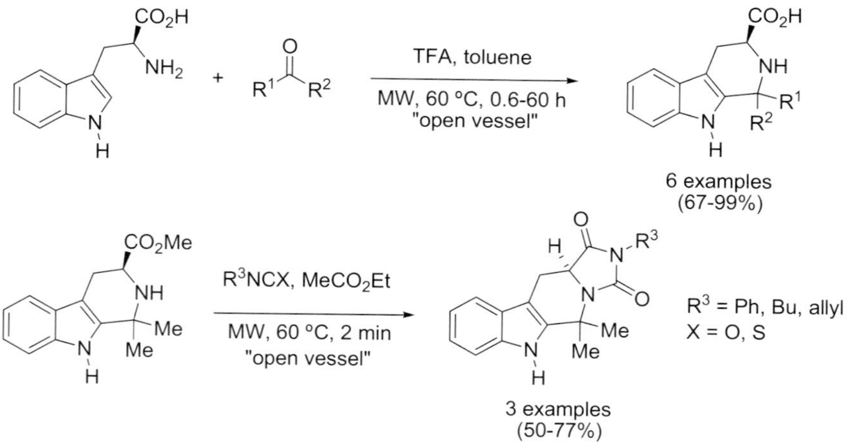

Scheme 6.234 Trifluoroacetic acid-mediated Pictet–Spengler reactions.

Other types of *N*-acyliminium ion-based cyclizations that are assisted by microwave irradiation are highlighted in Scheme 6.235 [418].

As reported by Padwa and coworkers, exposing a suitable nitrofuran precursor to microwave irradiation in 1-methyl-2-pyrrolidone (NMP) in the presence of 2,6-lutidine catalyst provided 1,4-dihydro-2*H*-benzo[4,5]furo[2,3-*c*]pyridin-3-one as the major product in 36% yield (Scheme 6.236) [419]. In contrast, under thermal conditions, removal of the *tert*-butyl group was observed as the major reaction pathway.

The Skraup cyclization is another reaction principle that provides rapid access to the quinoline moiety. Theoclitou and Robinson have published the preparation of a 44-member library based on the 2,2,4-trisubstituted 1,2-dihydroquinoline scaffold by the Lewis acid-catalyzed cyclization of substituted anilines or aminoheterocycles with appropriate ketones (Scheme 6.237) [420]. The best results were obtained using 10 mol% of scandium(III) triflate as a catalyst in acetonitrile as solvent at

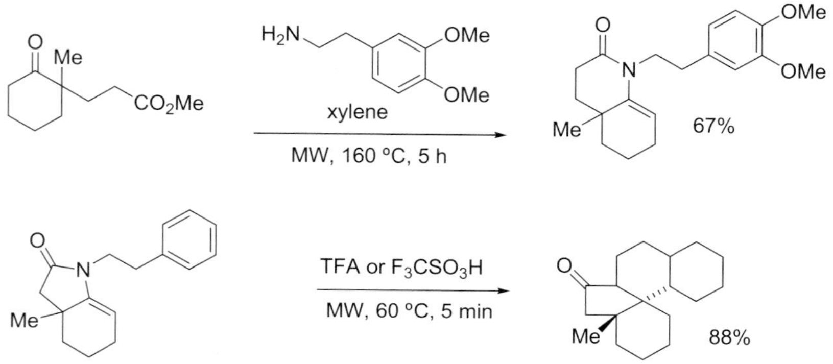

Scheme 6.235 *N*-Acyliminium ion-based cyclizations.

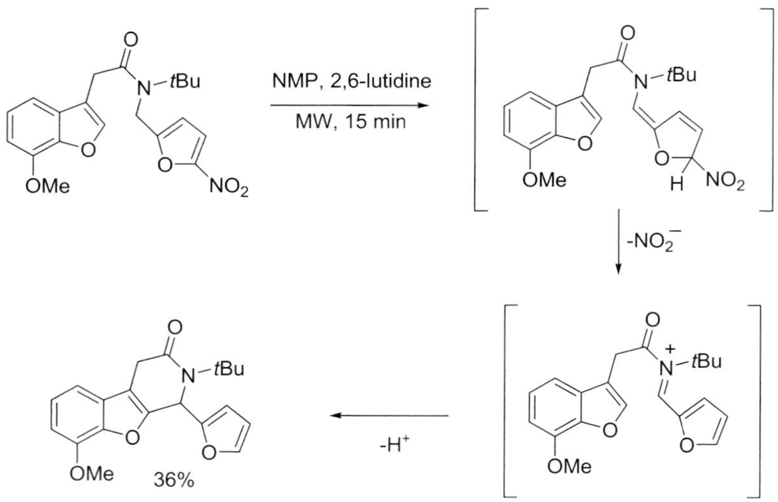
NMP, 2,6-lutidine
MW, 15 min
N-tBu
OMe
NO2
-NO2−
-H+
36%

Scheme 6.236 *N*-Acyliminium ion-based cyclizations leading to fused pyridones.

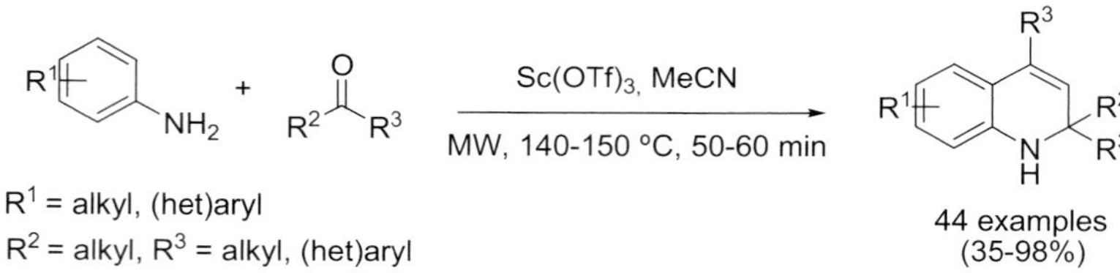
Sc(OTf)3, MeCN
MW, 140-150 °C, 50-60 min
R1 = alkyl, (het)aryl
R2 = alkyl, R3 = alkyl, (het)aryl
44 examples
(35-98%)

Scheme 6.237 Skraup dihydroquinoline synthesis.

140–150 °C. A variety of anilines and unsymmetrical ketones (5 equivalents) can be used to broaden the scope of the substituents, particularly five-membered ring heterocycles, to yield a variety of fused dihydropyridines.

The Friedländer reaction is the acid- or base-catalyzed condensation of an *ortho*-acylaniline with an enolizable aldehyde or ketone. Hénichart and coworkers have described microwave-assisted Friedländer reactions for the synthesis of indolizino[1,2-*b*]quinolines, which constitute the heterocyclic core of camptothecin-type antitumor agents (Scheme 6.238) [421]. The process involved the condensation of *ortho*-aminobenzaldehydes (or imines) with tetrahydroindolizinediones to form the quinoline structures. Employing 1.25 equivalents of the aldehyde or imine component in acetic acid as solvent provided the desired target compounds in 57–91% yield within 15 min. These transformations were carried out under open-vessel conditions at the reflux temperature of the acetic acid solvent.

The Pfitzinger reaction of isatins with α-methylene carbonyl compounds is widely used for the synthesis of substituted quinoline-4-carboxylic acids. Ivachtchenko and colleagues have recently reported on the Pfitzinger reaction in a series of 5-sulfamoylisatins (Scheme 6.239) [422]. Treatment of 5-sulfamoylisatins with diethyl mal-

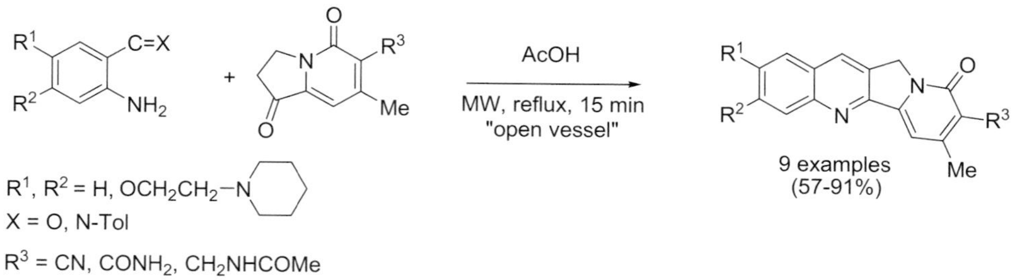

Scheme 6.238 Friedländer synthesis of indolizino[1,2-b]quinolines.

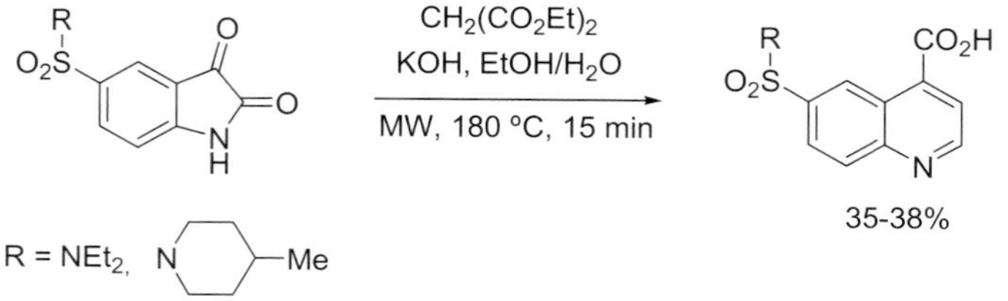

Scheme 6.239 Synthesis of 6-sulfamoylquinoline-4-carboxylic acids under Pfitzinger conditions.

onate under basic conditions (ethanol/water) surprisingly led to 6-sulfamoylquinoline-4-carboxylic acids, instead of the anticipated Pfitzinger products, namely the 2-oxo-1,2-dihydroquinoline-4-carboxylic acids (not shown). A careful mechanistic investigation involving ^{13}C-labeled ethanol demonstrated that the key step in this process is the reaction of *in situ* generated acetaldehyde (from the solvent) with the hydrolytically cleaved isatin ring. Moderate yields of the quinoline products were isolated after 15 min of microwave irradiation of the isatin precursors in 2.5 N potassium hydroxide solution at 180 °C.

The generation of a library of 2-aminoquinoline derivatives has been described by Wilson and colleagues (Scheme 6.240) [423]. The process involved microwave irradiation of the secondary amine and aldehyde components to form an enamine (1,2-dichloroethane, 180 °C, 3 min) and subsequent addition of the resulting crude enamine to a 2-azidobenzophenone derivative (0.8 equivalents) and further microwave heating for 7 min at the same temperature.

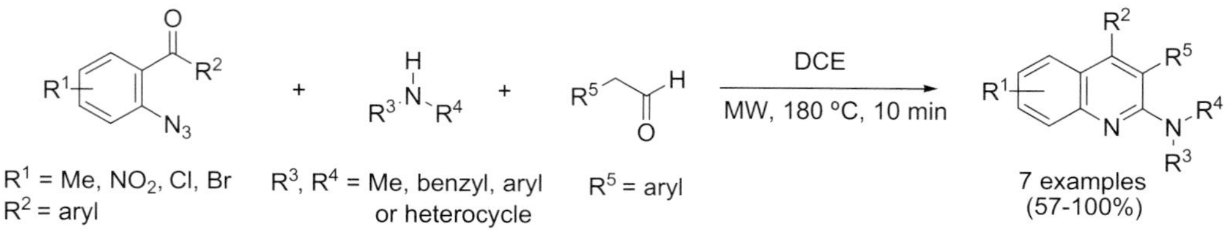

Scheme 6.240 Synthesis of a 2-aminoquinoline library by a three-component reaction.

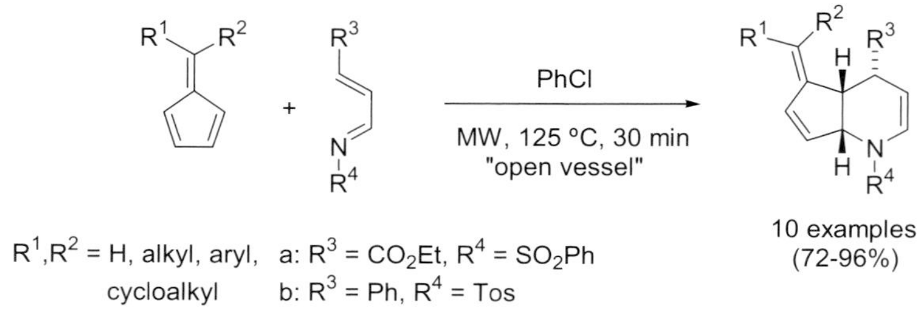
PhCl
MW, 125 °C, 30 min
"open vessel"
10 examples
(72-96%)
R1,R2 = H, alkyl, aryl, cycloalkyl
a: R3 = CO2Et, R4 = SO2Ph
b: R3 = Ph, R4 = Tos

Scheme 6.241 Azadiene Diels–Alder cycloaddition of fulvenes.

A general hetero-Diels–Alder cycloaddition of fulvenes with azadienes to furnish tetrahydro-[1]pyrindines has been described by Hong and coworkers (Scheme 6.241; see also Scheme 6.92) [424]. A solution of the azadiene and fulvene (1.2 equivalents) precursors in chlorobenzene was heated under open-vessel microwave irradiation for 30 min at 125 °C to provide the target compounds in excellent yields and with exclusive regio- and diastereoselectivity. Performing the reactions under conventional conditions or under microwave irradiation in different solvents provided significantly reduced yields.

A multicomponent assembly of pyrido-fused tetrahydroquinolines has been accomplished by Lavilla and coworkers in a one-pot process by the interaction of dihydroazines, aldehydes, and anilines (Scheme 6.242) [425]. The reactions were conducted with 20 mol% of scandium(III) triflate as a catalyst in dry acetonitrile in the presence of 4 Å molecular sieves, employing equimolar amounts of the building blocks. This protocol provided the cycloadducts shown in Scheme 6.242 in 80% yield as a 2:1 mixture of diastereoisomers following microwave irradiation at 80 °C for 5 min. The same reaction at room temperature required 12 h to reach completion.

Moody and coworkers have employed a "biomimetic" hetero-Diels–Alder–aromatization sequence for the construction of the 2,3-dithiazolepyridine core unit in amythiamicin D and related thiopeptide antibiotics (Scheme 6.243 a) [426]. The key cycloaddition reaction between the azadiene and enamine components was carried out by microwave irradiation at 120 °C for 12 h and gave the required 2,3,6-tris(thiazolyl)pyridine intermediate in a moderate 33% yield. Coupling of the remaining building blocks then completed the first total synthesis of the thiopeptide antibiotic

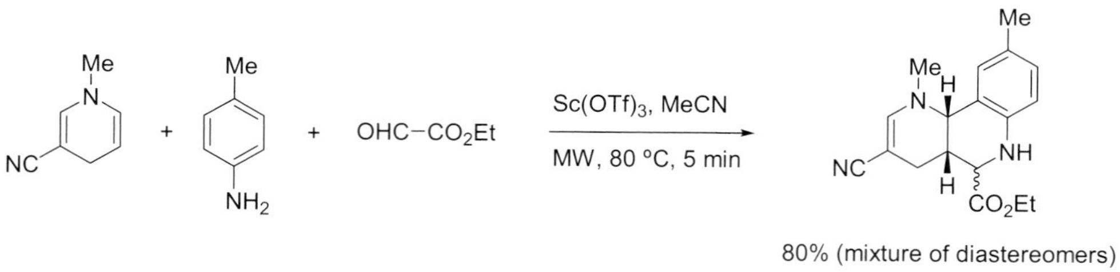
OHC−CO2Et
Sc(OTf)3, MeCN
MW, 80 °C, 5 min
80% (mixture of diastereomers)

Scheme 6.242 Multicomponent synthesis of pyrido-fused tetrahydroquinolines.

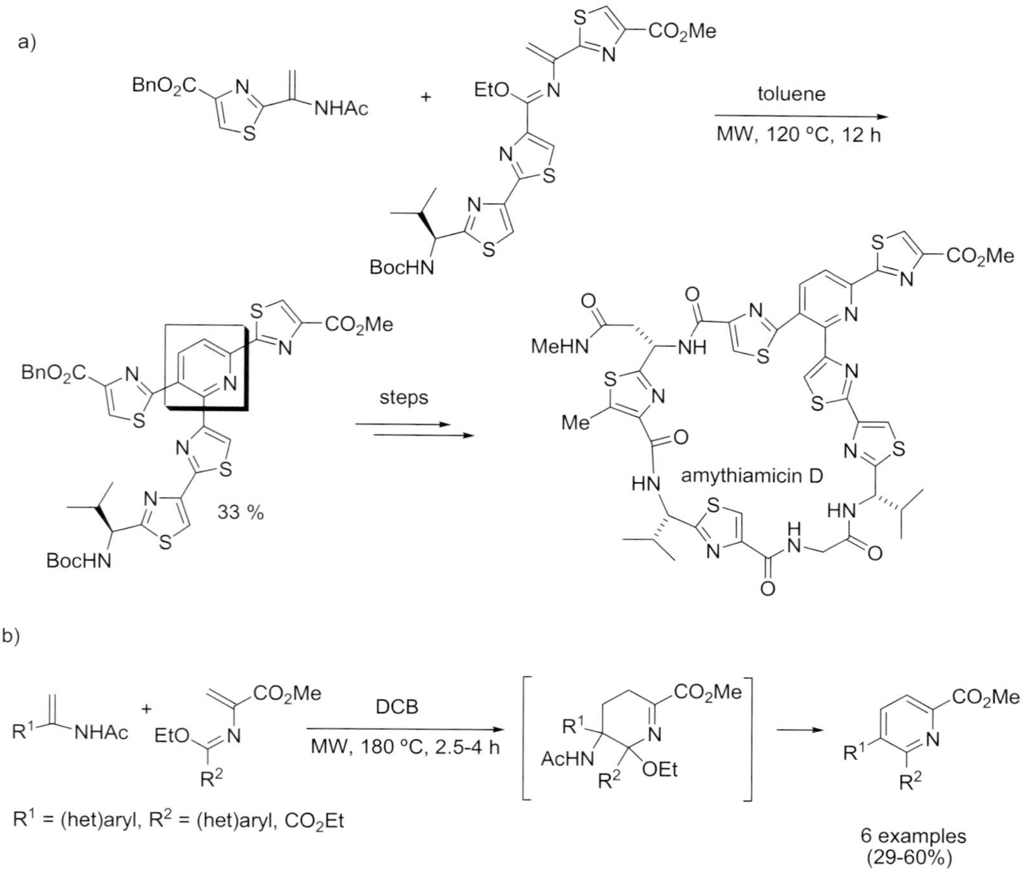

a)

b)

R¹ = (het)aryl, R² = (het)aryl, CO₂Et

6 examples
(29-60%)

Scheme 6.243 Biomimetic hetero-Diels–Alder–aromatization sequences.

amythiamicin D. In previous work, the authors established the concept of this bio-mimetic approach with a series of model structures (Scheme 6.243 b) [427]. Here, performing the reaction in 1,2-dichlorobenzene typically gave the best results.

6.24.7.2 Pyrans

Both natural and non-natural compounds with a $2H,5H$-pyrano[4,3-b]pyran-5-one skeleton are of interest in medicinal chemistry. Several natural products, such as the pyripyropenes, incorporate this bicyclic ring system. The group of Beifuss has described an efficient microwave-promoted domino synthesis of the $2H,5H$-pyrano[4,3-b]pyran-5-one skeleton by condensation of α,β-unsaturated aldehydes with 4-hydroxy-6-methyl-$2H$-pyran-2-one (Scheme 6.244) [428]. It is assumed that in the presence of an amino acid catalyst a Knoevenagel condensation occurs first, which is then followed by a 6π-electron electrocyclization to the pyran ring. While the conventional thermal protocol required a reaction time of up to 25 h (refluxing ethyl

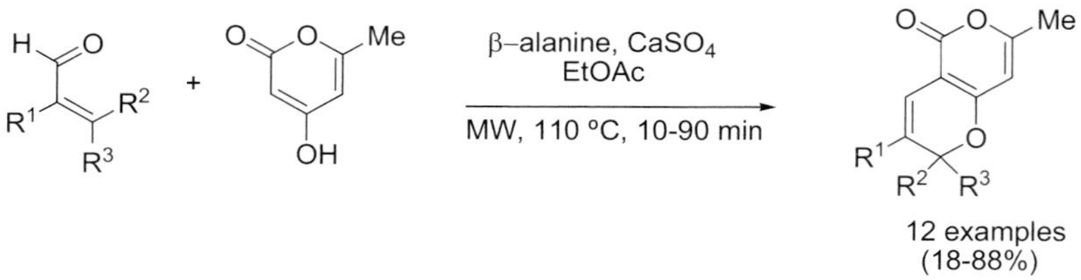

Scheme 6.244 Domino Knoevenagel condensation/6π-electron electrocyclization in the synthesis of 2H,5H-pyrano[4,3-b]pyran-5-ones.

acetate), the microwave method employing 1.1 equivalents of the pyran-2-one and 0.5 equivalents each of β-alanine as catalyst and calcium sulfate as dehydrating agent could be run much more efficiently. The best yields for a variety of diverse aldehyde building blocks were obtained at 110 °C within 10–90 min, although in some cases somewhat higher yields were achieved by the conventional thermal method (80 °C).

The sodium bromide catalyzed three-component cyclocondensation of aryl aldehydes, CH-acidic nitriles, and dimedone under solvent-free conditions has been studied by Devi and Bhuyan (Scheme 6.245) [429]. Utilizing equimolar amounts of the building blocks and 20 mol% of sodium bromide as catalyst, microwave irradia-

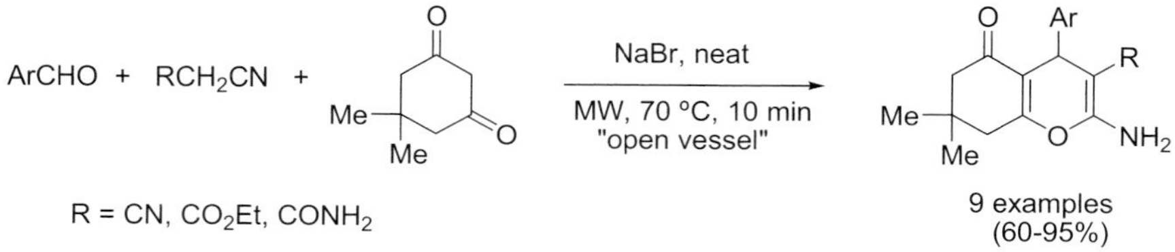

Scheme 6.245 Multicomponent condensation in the synthesis of tetrahydrobenzo[b]pyrans.

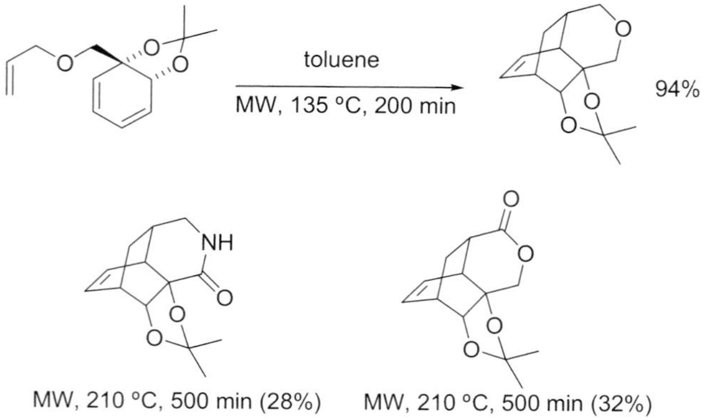

Scheme 6.246 Intramolecular Diels–Alder cyclization of biodihydroxylated benzoic acid derivatives.

tion for 10 min at 70 °C produced the anticipated tetrahydrobenzo[*b*]pyrans in good to excellent yields.

A pyran ring is formed in the intramolecular Diels–Alder cycloaddition of alkene-tethered enantiopure (1*S*,2*R*)-1,2-dihydroxycyclohexa-3,5-diene-1-carboxylic acid derivatives (derived from the biodihydroxylation of benzoic acid). For the three cases illustrated in Scheme 6.246, Mihovilovic and colleagues found that moderate to high yields of the desired cycloadducts could be obtained by exposing a solution of the precursor to microwave irradiation at 135–210 °C for extended periods of time [430]. In all cases, the yields achieved by the microwave protocol were higher than those achieved in conventional runs (refluxing toluene, 3–7 days).

6.24.8
Six-Membered Heterocycles with Two Heteroatoms

6.24.8.1 Pyrimidines

2-Substituted pyrimidines can be obtained from propargylic alcohols and amidines through a tandem oxidation–heteroannulation. As demonstrated by Bagley and co-workers, both benzamidine and acetamidine react with 1-phenyl-2-propyn-1-ol in the presence of either manganese dioxide or *ortho*-iodoxybenzoic acid (IBX) as *in situ* oxidizing agent to provide the corresponding pyrimidines (Scheme 6.247 a) [431]. A yield of up to 84% was obtained by exposing the alcohol, the amidine hydrochloride salt (1.2 equivalents), and sodium carbonate as base (2.4 equivalents) to an excess of the oxidant in acetonitrile at 120 °C for 40 min. In subsequent work, the same group expanded on this protocol and showed that a more diverse set of substituted pyrimidines could be obtained under similar reaction conditions by starting directly from the corresponding alkynones (Scheme 6.247 b) [432, 433]. This method also allows the preparation of 2-amino-substituted pyrimidines using guanidines as starting materials.

An important multicomponent transformation for the synthesis of dihydropyrimidines is the Biginelli reaction, which involves the acid-catalyzed condensation

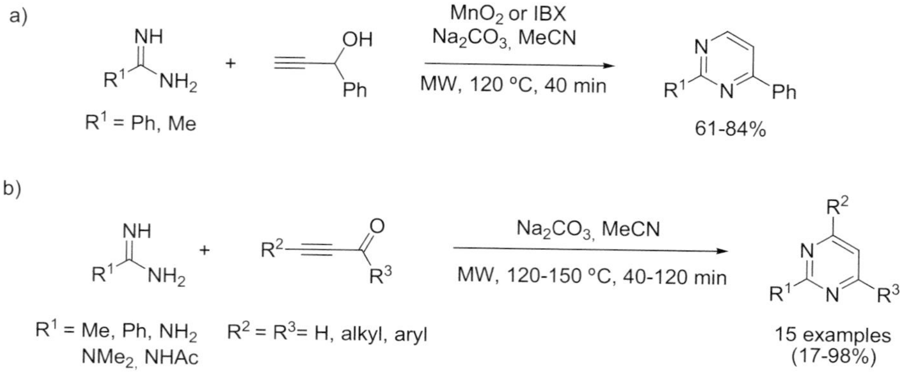

Scheme 6.247 Synthesis of 2,4,6-trisubstituted pyrimidines from amidine and alkyne building blocks.

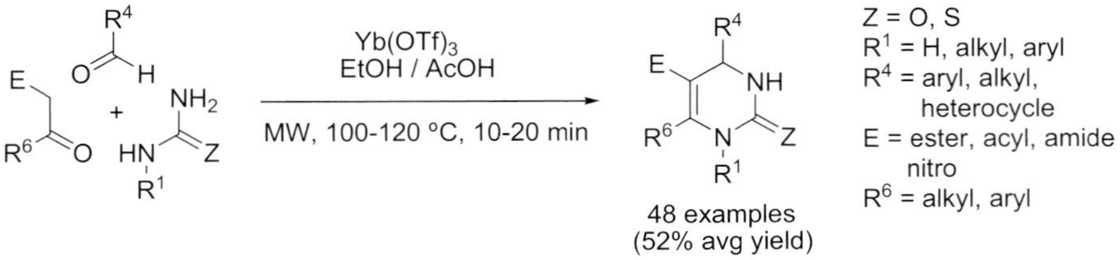

Scheme 6.248 Biginelli three-component synthesis of dihydropyrimidines.

of aldehydes, CH-acidic carbonyl components, and urea-type building blocks (Scheme 6.248) [434, 435]. Under conventional conditions, this MCR typically requires heating for several hours under reflux conditions (ca. 80 °C) in a solvent such as ethanol, whereas microwave-assisted protocols can be completed within 10–20 min and provide improved product yields. A detailed description of microwave-assisted Biginelli reactions elaborated by Stadler and Kappe is given in Section 5.3.

A multistep microwave-assisted route to 2-amino-3,4,5,6-tetrahydropyrimidines has been disclosed by Wellner and coworkers (Scheme 6.249) [436]. The central basic scaffold was constructed solely through the application of microwave-assisted chemistry, without any need for activating agents or protecting group manipulations. The initially required diaminoaryl ether building block was synthesized by nucleophilic aromatic substitution from fluorobenzenes (1.1 equivalents) and 1,3-diaminopropan-2-ol. With sodium hydride (1.2 equivalents) as a non-nucleophilic base and *N,N*-dimethylacetamide (DMA) as solvent, moderate to good yields of the diaminoaryl ethers were obtained under microwave conditions (170 °C, 4 min). Addition of carbon disulfide at room temperature led to spontaneous formation of a dithiocarbonate intermediate, which was transformed by thermolytic cleavage and hydrogen sulfide extrusion (140 °C, 4 min) to the cyclic thioureas. The latter crystal-

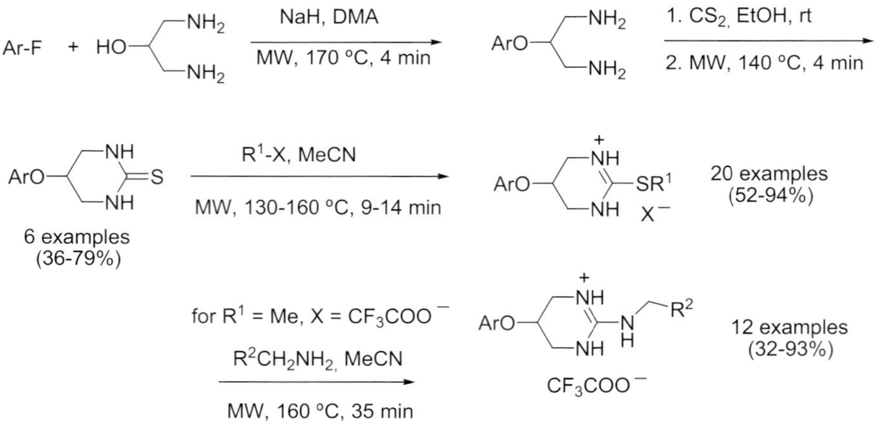

Scheme 6.249 Preparation of cyclic isothioureas and guanidines.

lized spontaneously and could be collected by filtration without any need for further purification. Further *S*-alkylation with alkyl halides (2 equivalents) in acetonitrile (130–160 °C, 9–14 min) provided the isothiouronium salts, which, after transforming the halide salts to trifluoroacetate salts, could be further reacted with primary amines (1.1 equivalents), again using acetonitrile as solvent and applying microwave conditions (160 °C, 35 min).

Another multistep protocol that initially involves the formation of fused pyrimidines (quinazolines) has been described by Besson and coworkers in the context of synthesizing 8*H*-quinazolino[4,3-*b*]quinazolin-8-ones via double Niementowski condensation reactions (Scheme 6.250) [437]. In the first step of the sequence, an anthranilic acid was condensed with formamide (5.0 equivalents) under open-vessel microwave conditions (Niementowski condensation). Subsequent chlorination with excess POCl$_3$, again under open-vessel conditions, produced the anticipated 4-chloroquinazoline derivatives, which were subsequently condensed with anthranilic acids in acetic acid to produce the tetracyclic 8*H*-quinazolino[4,3-*b*]quinazolin-8-one target structures. The final condensation reactions were completed within 20 min under open-vessel reflux conditions (ca. 105 °C), but not surprisingly could also be performed within 10 min by sealed-vessel heating at 130 °C.

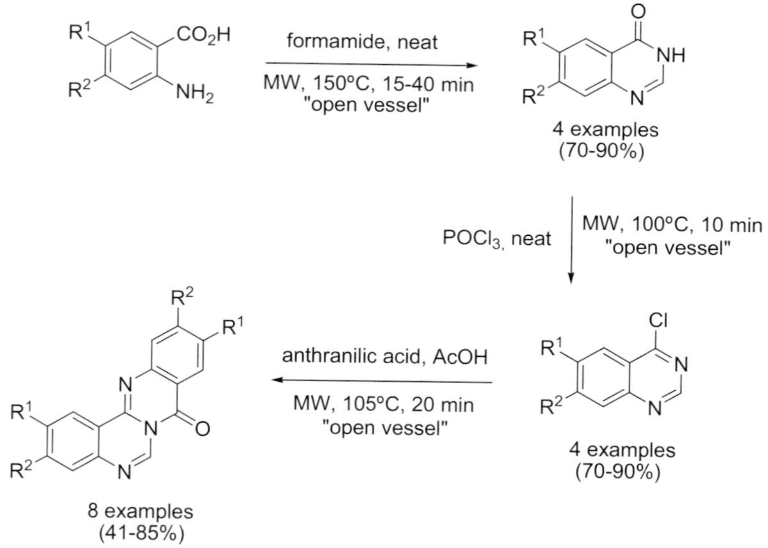

Scheme 6.250 Formation of 8*H*-quinazolino[4,3-*b*]quinazolin-8-ones through double Niementowski condensation.

The same authors have described a related Niementowski condensation for the preparation of 3*H*-nitroquinazolin-4-ones. Subsequent manipulation of this structure led to 8*H*-thiazolo[5,4-*f*]quinazolin-9-ones through a series of open-vessel microwave-assisted transformations, as indicated in Scheme 6.251 [205, 438].

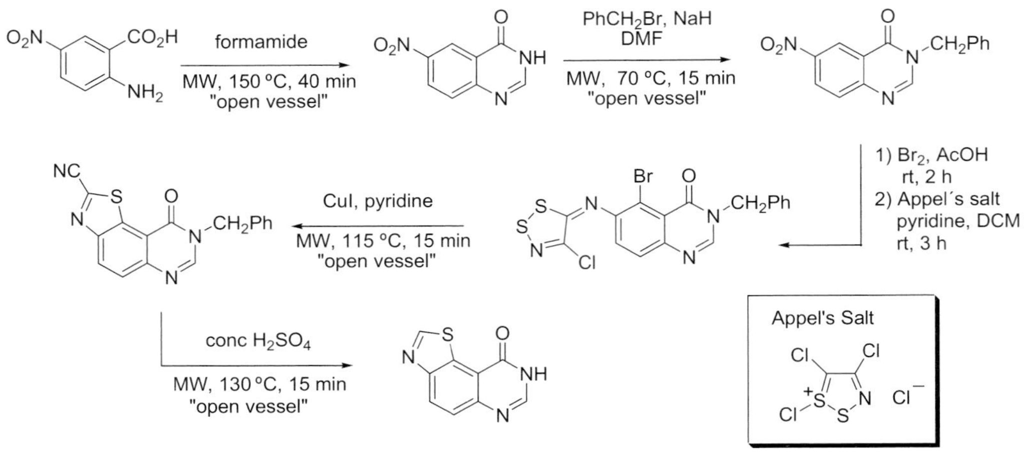

Scheme 6.251 Formation of 8*H*-thiazolo[5,4-*f*]quinazolin-9-ones.

A one-pot preparation of pyrrolo[1,2-*a*]quinazoline libraries with three points of diversification by condensation of α-cyano-ketones and 2-hydrazino-benzoic acids has been developed by Hulme and coworkers (Scheme 6.252) [439]. The protocol simply involved heating a solution of equimolar amounts of the two building blocks in acetic acid at 150 °C for 5 min. In many cases, the final products precipitated directly from the reaction mixture. In such cases, simple washing with diethyl ether yielded the products in >95% purity. A 63-member library was prepared by employing seven α-cyano-ketones and nine 2-hydrazino-benzoic acids.

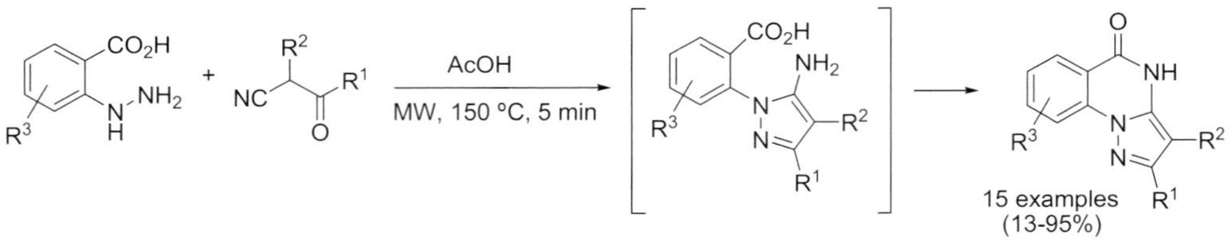

Scheme 6.252 Formation of pyrrolo[1,2-*a*]quinazolines.

A simple, efficient, and high-yielding synthesis of quinazolin-4-ylamines and thieno[3,2-*d*]pyridin-4-ylamines based on the condensation of appropriately functionalized *N'*-(2-cyanophenyl)-*N*,*N*-dimethylformamidines and primary amines has been reported by Han and coworkers (Scheme 6.253) [440]. Optimization of the reaction parameters resulted in the use of acetonitrile/acetic acid as a solvent mixture and of 1.2 equivalents of the requisite amine. In general, microwave heating at 160 °C for 10 min provided excellent product yields.

The group of Borrell has described a variety of different microwave-assisted approaches to biologically active pyrido[2,3-*d*]pyrimidin-7(8*H*)-ones [441–443]. In the multicomponent method outlined in Scheme 6.254 a, the target structures were obtained in a one-pot fashion by cyclocondensation of α,β-unsaturated esters, amidines or guanidines, and CH-acidic nitriles (malononitrile or ethyl cyanoacetate) [441, 442]. Employing sodium methoxide as base in methanol as solvent, low to excellent yields were obtained by microwave heating at 100–140 °C for 10 min. Typically, the resulting pyrido[2,3-d]pyrimidin-7(8H)-ones, possessing four centers for diversification, crystallized directly from the reaction mixture in high purity. In a modification of this strategy (Scheme 6.254 b) [443], the same authors subsequently

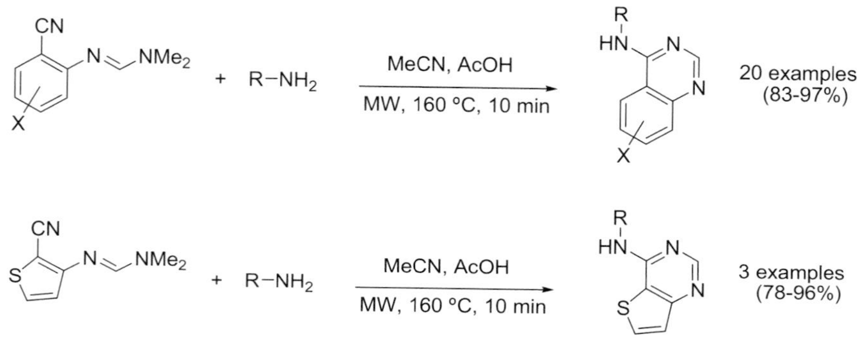

Scheme 6.253   Formation of aminoquinazoline and thieno[3,2-d]pyrimidine derivatives.

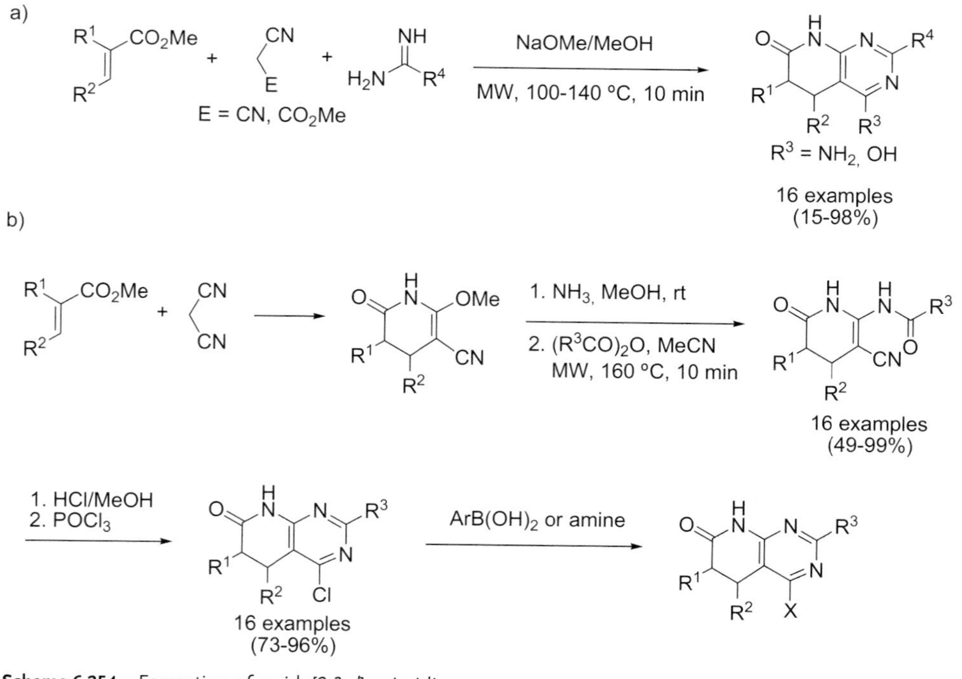

Scheme 6.254   Formation of pyrido[2,3-d]pyrimidines.

disclosed a stepwise protocol that allowed the isolation of 2-methoxy-6-oxo-1,4,5,6-tetrahydropyridine-3-carbonitrile intermediates under conventional reaction conditions, and these were subsequently transformed to 2-aminopyridones by treatment with ammonia at room temperature. Microwave-assisted acylation of the amino group with carboxylic acid anhydrides (R^3) in acetonitrile at 160 °C for 10 min furnished 2-acylaminopyridones, which were subsequently cyclized with hydrochloric acid in methanol to the corresponding 4-oxopyrido[2,3-d]pyrimidines in almost quantitative yield. Further incorporation of a chloro substituent by treatment with phosphorus oxychloride provided an ideal substrate for further decoration of the pyrido[2,3-d]pyrimidine scaffold by means of Suzuki chemistry or nucleophilic displacement reactions with amines. For both types of transformations, rapid microwave-assisted protocols were utilized, allowing access to a diverse set of pyrido[2,3-d]pyrimidines with four centers allowing for diversity.

Several other microwave-assisted approaches to fused pyrimidines are summarized in Scheme 6.255 [444–446].

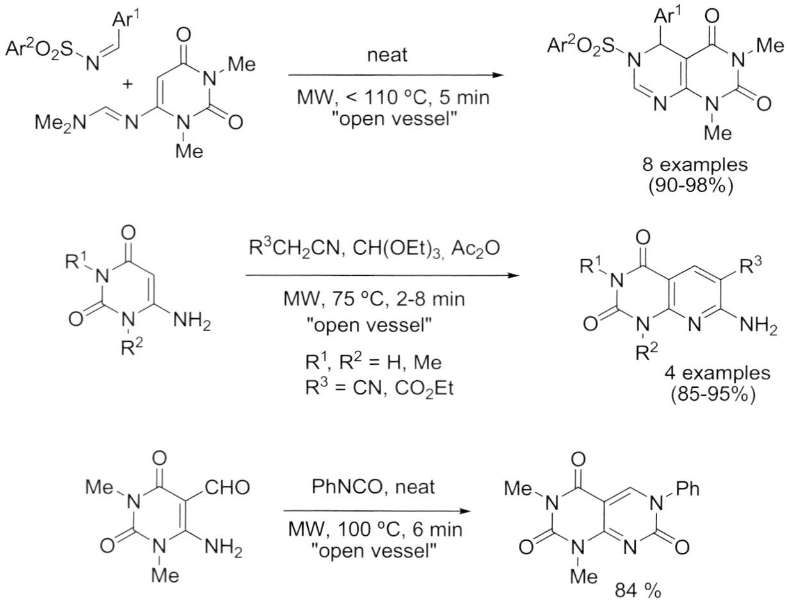

Scheme 6.255 Formation of bicyclic pyrimidines.

Various other examples in this chapter have already highlighted how N,N-dimethylformamide dimethyl acetal can be efficiently utilized as a synthon for the construction of heterocyclic rings (see Schemes 6.189, 6.194, 6.195, 6.229, and 6.230). Westman and coworkers have described a two-step method for the generation of a wide variety of heterocyclic scaffolds, based on the initial formation of alkylaminopropenones and alkylaminopropenoates from N,N-dimethylformamide diethyl acetal (DMFDEA) and the corresponding CH-acidic carbonyl compounds (Scheme 6.256)

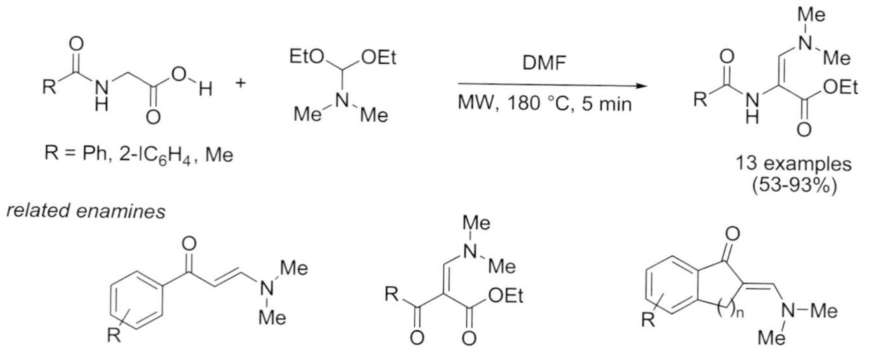

Scheme 6.256 Formation of alkylaminopropenones and alkylaminopropenoates.

[447]. Treatment of the CH-acidic carbonyl compound with 1.5–2.5 equivalents of DMFDEA in *N,N*-dimethylformamide at 180 °C resulted in full conversion to the enamine synthons within 5 min. The enamines were obtained in 53–93% yield (based on LC-MS analysis) and were used without further purification in the next step.

For the preparation of the anticipated heterocyclic library compounds (Scheme 6.257), solutions of the prepared enamine synthons were split and diluted with an appropriate solvent, and 1.2 equivalents of a dinucleophile (hydrazine, hydroxylamine, amidines; see Scheme 6.257 for more complex building blocks) was added. Subsequent exposure to microwave conditions in acetic acid/DMF mixtures

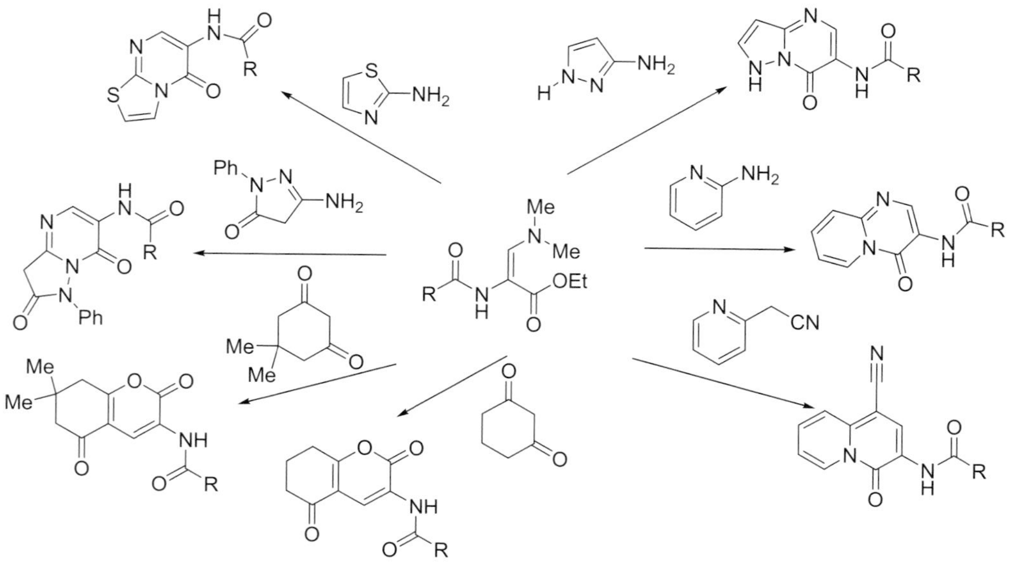

Scheme 6.257 Reaction of enamine synthons with dinucleophiles.

at 180 °C for a further 5 min furnished the desired target compounds in moderate to good overall yields [447]. More than 100 compounds were prepared using this strategy, and examples are highlighted in Scheme 6.257.

In a subsequent article, the same group described a cascade sequence involving (triphenylphosphoranylidene)ethenone ($Ph_3PC=C=O$) as a versatile building block for the construction of heterocycles and other unsaturated amides (Scheme 6.258) [448]. Typically, the phosphoranylidene reagent was reacted with a carbonyl compound possessing a hydroxyl, amine or thiol group at an α- or β-position to form five- or six-membered heterocycles (Scheme 6.258 b). The reactions were performed at 180 °C for 5–8 min in 1,2-dichloroethene (DCE) and resulted in 54–97% product yields (LC-MS). Alternatively, the phosphoranylidene reagent (1.5 equivalents) could also be used in a multicomponent reaction with an aldehyde and an amine to form α,β-unsaturated amides under similar reaction conditions (Scheme 6.258 a) [448].

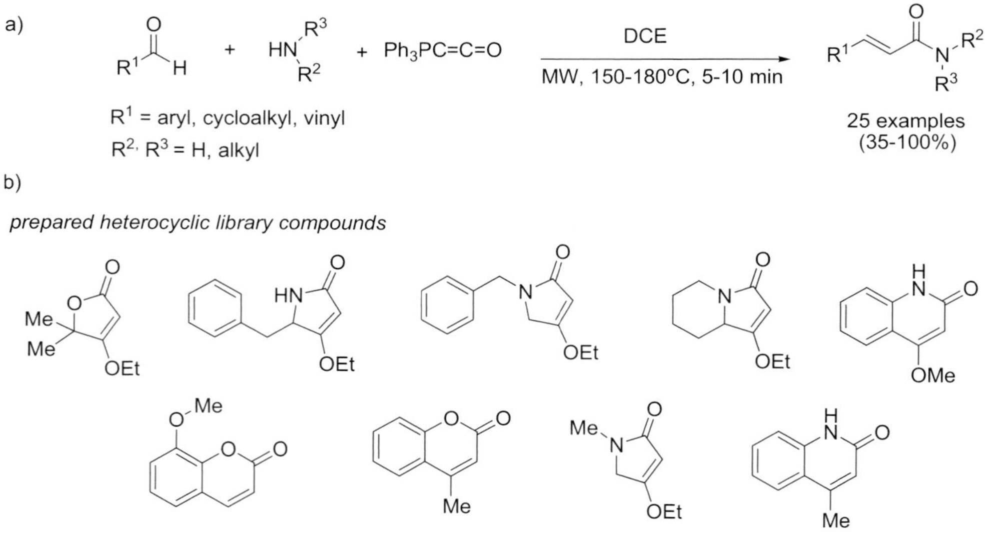

a)

R^1 = aryl, cycloalkyl, vinyl

$R^{2,} R^3$ = H, alkyl

25 examples
(35-100%)

b)

prepared heterocyclic library compounds

Scheme 6.258 Reactions of the (triphenylphosphoranylidene)ethenone reagent.

6.24.8.2 Pyrazines

ortho-Quinodimethane derivatives are reactive dienes and can be generated *in situ* by a number of methods. The inter- and intramolecular Diels–Alder reactions of these compounds form the basis of syntheses of a wide range of target molecules. Díaz-Ortiz and colleagues have described the generation of a pyrazine *ortho*-quinodimethane derivative that undergoes cycloaddition to non-activated dienophiles under microwave conditions (Scheme 6.259) [449]. Heating of 2,3-bis(dibromomethyl)pyrazine with 5 equivalents of sodium iodide in the presence of a small amount of *N,N*-dimethylformamide as solvent and 3.6 equivalents of an alkyne or enamine (not shown) dienophile resulted in the formation of quinoxaline cycloadducts in

38–69% yield within 10–15 min at 90 °C. A higher reaction temperature led to decomposition of the *ortho*-quinodimethane intermediate before reaction with the dienophile and accordingly to a decrease in the yield.

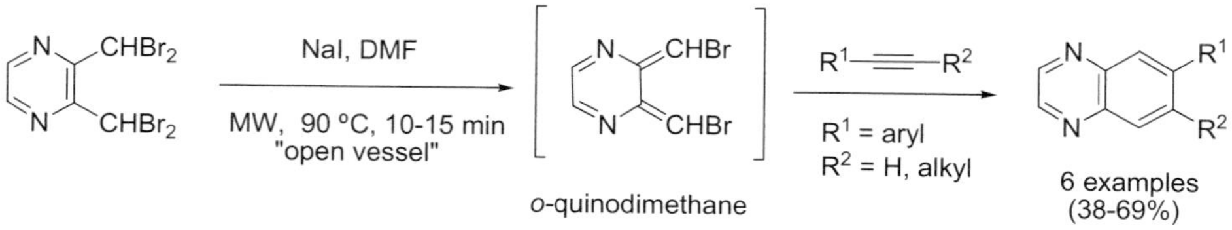

Scheme 6.259 Diels–Alder cycloadditions of pyrazine *ortho*-quinodimethane with dienophiles.

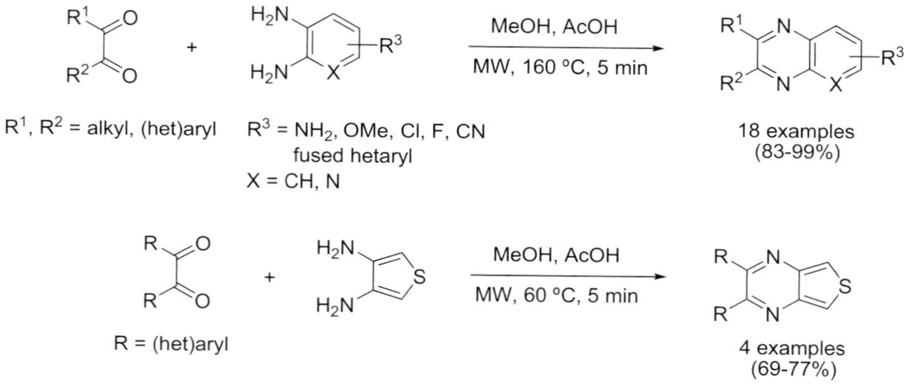

Scheme 6.260 Synthesis of quinoxalines and heterocycle-fused pyrazines.

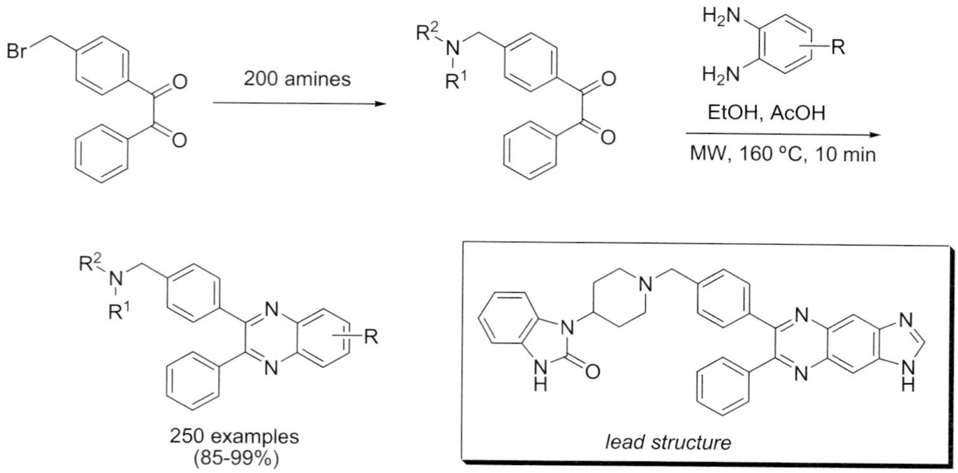

Scheme 6.261 Synthesis of quinoxalines as allosteric Akt kinase inhibitors.

A different approach to quinoxalines and heterocycle-fused pyrazines has been described by the Lindsley group, based on the cyclocondensation of 1,2-diketones and aryl/heteroaryl 1,2-diamines (Scheme 6.260) [450]. Optimized reaction conditions involved heating an equimolar mixture of the diketone and diamine components for 5 min at 160 °C in a 9:1 methanol/acetic acid solvent mixture, which furnished the substituted quinoxalines in excellent yields. This approach could also be applied equally successfully to the synthesis of heteroaryl pyrazines, such as pyrido[2,3-*b*]pyrazines and thieno[3,4-*b*]pyrazines. The same group has employed 1,2-diketone building blocks for the preparation of other heterocyclic structures (see Schemes 6.198, 6.268, and 6.269).

An application of this synthetic strategy by the same group led to the development of a series of potent and selective allosteric Akt (protein kinase B/PKB) kinase inhibitors that induced apoptosis in tumor cells and inhibited Akt phosphorylation *in vivo* (Scheme 6.261) [451].

6.24.8.3 Oxazines

In the context of synthesizing libraries of thiophene derivatives of potential therapeutic interest, Rault and coworkers have described the preparation of thieno[3,2-*d*]-[1,3]oxazine-2,4-diones (thiaisatoic anhydride) using a microwave approach (Scheme 6.262) [452]. The synthesis of the thiaisatoic anhydrides started from the corresponding aminoesters. Alkaline hydrolysis was performed in 7 N potassium hydroxide (in water/ethanol, 1:3) under microwave irradiation for 1 h. After cooling to 0 °C, gaseous phosgene was bubbled through the solution for 30 min with stirring, leading to precipitation of the desired thiaisatoic anhydride analogues in yields of 66–86%. The transformations were carried out on a 15 gram scale.

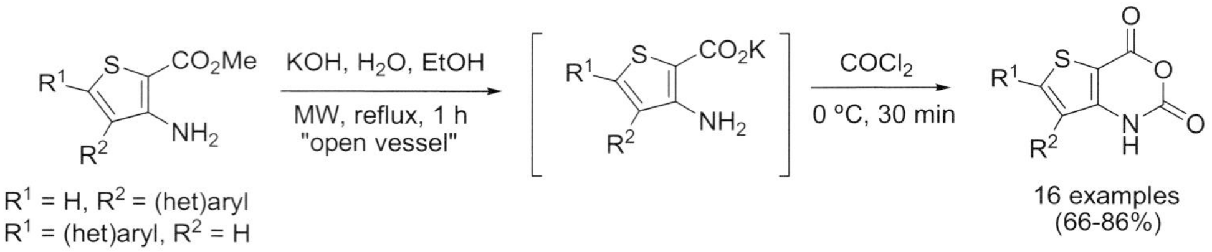

R^1 = H, R^2 = (het)aryl
R^1 = (het)aryl, R^2 = H

Scheme 6.262 Synthesis of thiaisatoic anhydrides.

A solvent-free synthesis of substituted spiroindolinonaphth[2,1-*b*][1,4]oxazines through condensation of 2-methylene-1,3,3-trimethylindoline derivatives with 1-nitroso-2-naphthol under microwave irradiation has been described by Fedorova and colleagues (Scheme 6.263) [453]. In a typical reaction, an equimolar mixture of the two starting materials was irradiated at 65–110 °C for 15 min to produce the desired spiroindolinonaphth[2,1-*b*][1,4]oxazines, which are useful as photochromic compounds. In a related procedure, addition of morpholine to the reaction mixture led to the formation of the corresponding 6′-amino-functionalized spiroindolinonaphth[2,1-*b*][1,4]oxazines, which exhibit a strong hypsochromic color shift (not shown) [453].

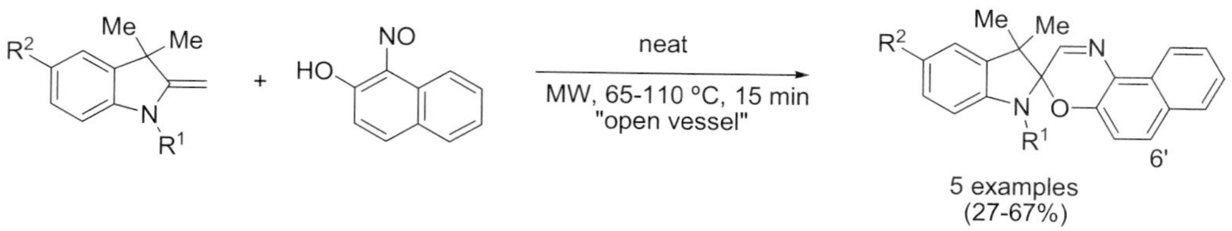

Scheme 6.263 Synthesis of spiroindolinonaphth[2,1-*b*][1,4]oxazines.

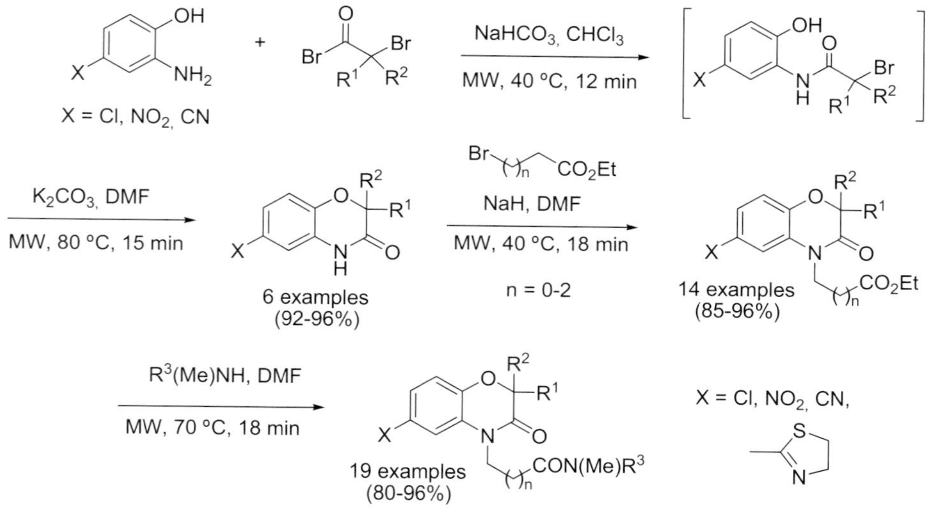

Scheme 6.264 Multistep synthesis of substituted benzoxazine derivatives.

The group of Caliendo and coworkers has described a multistep synthesis of benzoxazine libraries, in the context of a search for compounds with vasorelaxant activity related to cromakalim [454]. As highlighted in Scheme 6.264, all of the required synthetic manipulations were carried out under microwave irradiation conditions. In all cases, a reduction in reaction time as compared to the corresponding thermal protocols was reported.

6.24.8.4 Thiazines

Phenothiazines are well-known as intermediates for pharmaceuticals, and are also active as insecticides and antioxidants. These compounds are usually prepared by the thiation of diphenylamines with elemental sulfur. In this context, the group of Toma has elaborated a synthesis of 3-phthalimidophenothiazine, as shown in Scheme 6.265 [455]. Using a variety of high-boiling solvents under conventional thermal reflux conditions, low isolated yields of the desired product were obtained. The highest conversion and isolated product yield (55%) was achieved by microwave irradiation of a mixture of the starting *N*-(4-phenylaminophenyl)phthalimide with

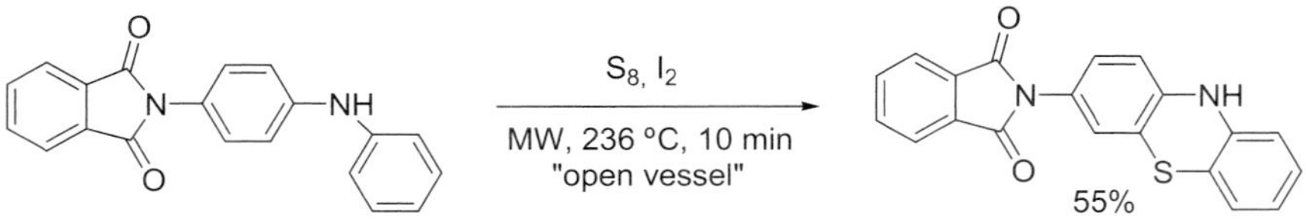

Scheme 6.265 Synthesis of phenothiazines.

5 equivalents of elemental sulfur and 6.25 mol% of iodine as catalyst at 236 °C for 10–20 min. This reaction was run on a 15 gram scale, and subsequent deprotection with hydrazine hydrate in ethanol provided 3-aminophenothiazine (not shown).

6.24.9
Six-Membered Heterocycles with Three Heteroatoms

Lee and Rana have reported on the preparation of a 20-member library of 4,6-diamino-2,2-dimethyl-1,2-dihydro-1-phenyl-1,3,5-triazines, which are established inhibitors of dihydrofolate reductase (DHFR). The authors utilized a one-pot, three-component method, involving the microwave-assisted condensation of N-cyanoguanidine, acetone, and an aromatic amine (Scheme 6.266) [456]. The reaction was carefully optimized with respect to temperature and reaction time. The most suitable conditions involved heating a solution of the aniline with 1.1 equivalents of N-cyanoguanidine (dicyandiamide) in acetone as solvent, containing 1 equivalent of hydrochloric acid, at 90 °C for 35 min. After leaving the solution to stand at 4 °C overnight, the desired triazine products precipitated as their hydrochloride salts. In general, the microwave-assisted method provided the triazine compounds in improved yields and with higher purities compared to the conventional thermal method (acetone, reflux, 22 h).

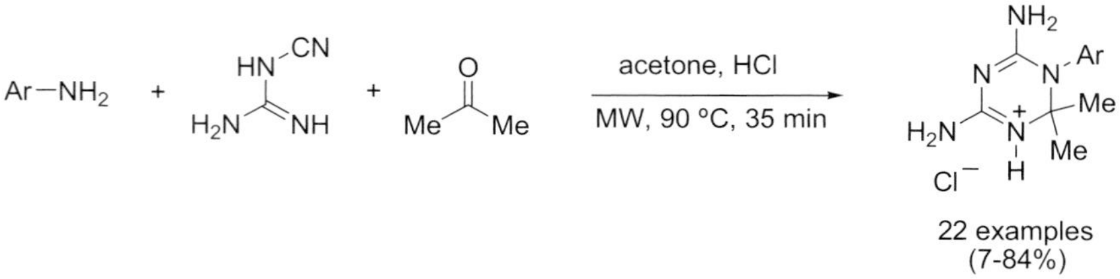

Scheme 6.266 Synthesis of 4,6-diamino-2,2-dimethyl-1,2-dihydro-1-phenyl-1,3,5-triazines.

A somewhat related condensation involving cyanoguanidine and aryl nitriles under basic conditions leading to 6-aryl-2,4-diamino-1,3,5-triazines has been described by Peng and Song (Scheme 6.267) [457]. Here, a benzonitrile derivative was reacted with 1.1 equivalents of N-cyanoguanidine (dicyandiamide) in the ionic liquid 1-butyl-3-methylimidazolium hexafluorophosphate (bmimPF$_6$) as reaction medium in the presence of 20 mol% of powdered potassium hydroxide as catalyst. The highest yields were obtained after 10–15 min of irradiation at 130 °C. In a pre-

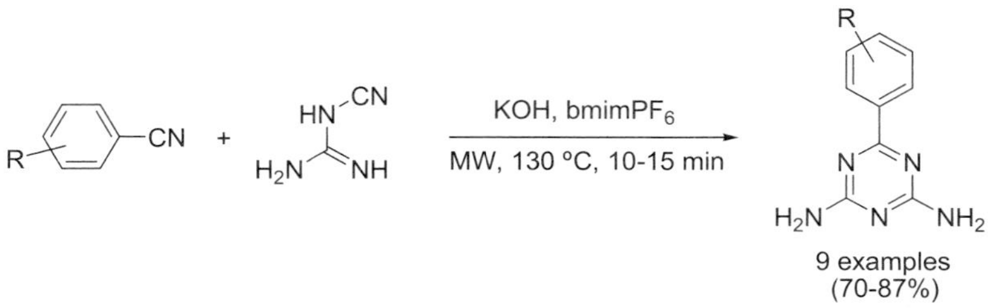

Scheme 6.267 Synthesis of 6-aryl-2,4-diamino-1,3,5-triazines.

heated oil bath, the same transformation required 8 h to reach completion at the same apparent reaction temperature. The ionic liquid could be recycled at least five times.

In addition to the aforementioned syntheses involving 1,2-diketones as building blocks, Zhao and his colleagues have described the preparation of 1,2,4-triazine libraries by the condensation of 1,2-diketones with acyl hydrazides and ammonium acetate (Scheme 6.268) [458]. Microwave heating of equimolar mixtures of the two starting materials with 10 equivalents of ammonium acetate in acetic acid to 180 °C for 5 min resulted in the formation of the anticipated 1,2,4-triazine products in excellent yield for all of the 48 examples studied. For most of the library compounds, the desired products precipitated directly from the reaction mixture upon cooling. The conventional protocols involving refluxing in acetic acid provided much lower product yields and required reaction times of 6–24 h.

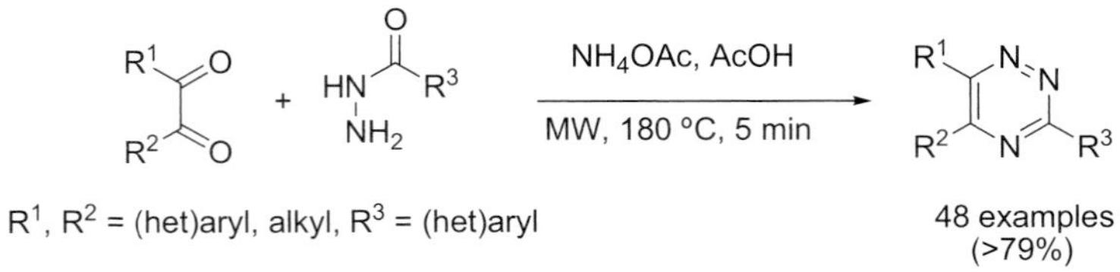

Scheme 6.268 Synthesis of 3,5,6-trisubstituted-1,2,4-triazines.

An application of this triazine synthesis by the same authors toward the preparation of the basic canthine tetracyclic alkaloid skeleton is highlighted in Scheme 6.269 [459]. Here, a suitable "indole-tethered" acyl hydrazide was prepared "on demand" in quantitative yield in a rapid microwave protocol as the hydrazide was shown to slowly decompose at room temperature. Treatment of the acyl hydrazide with 1,2-diketones under the standard conditions indicated in Scheme 6.268 (180 °C, 5 min) provided the anticipated triazine-tethered indole products, along with a tetracyclic canthine analogue formed by intramolecular inverse-electron-demand hetero-Diels–Alder cycloaddition. By increasing the reaction temperature to 220 °C and prolonging the reaction time to 40 min, the desired canthine derivatives could be obtained directly in moderate to good yields.

An even more complex pathway involving inverse-electron-demand Diels–Alder reactions between imidazoles and 1,2,4-triazines linked by a tri- or tetramethylene

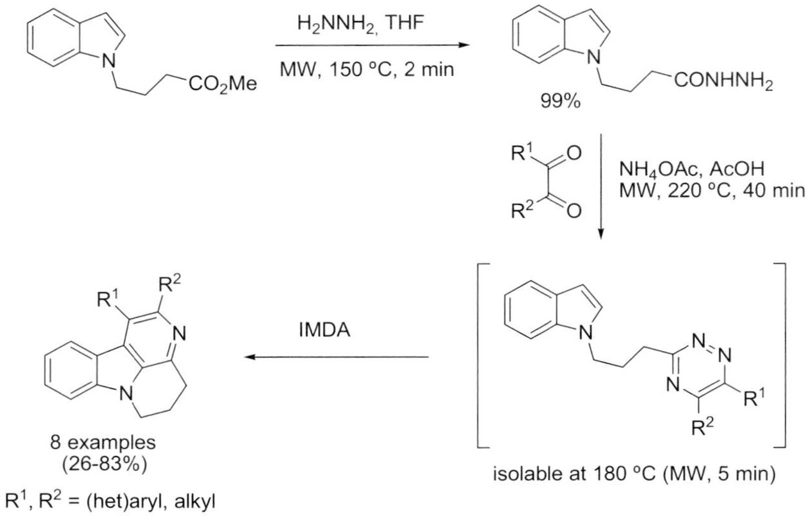

Scheme 6.269 One-pot preparation of the tetracyclic canthine alkaloid skeleton.

tether between the imidazole N1 position and the triazine C3, produced 1,2,3,4-tetrahydro-1,5-naphthyridines or 2,3,4,5-tetrahydro-1*H*-pyrido[3,2-*b*]azepines, respectively (Scheme 6.270). The sequence carried out by Snyder and coworkers is believed to proceed through cycloaddition with subsequent loss of nitrogen, followed by stepwise loss of a nitrile [460]. The best conditions for this cycloaddition were found to involve microwave irradiation at 210 °C for 20 min in 1,2-dichlorobenzene in the presence of 15 equivalents of ammonium acetate (trimethylene tether). The yields under these conditions were comparable to those obtained after refluxing in diphenyl ether under thermal heating (259 °C, 80 min, 89% yield). Ammonium acetate is believed to act as an energy-transfer reagent in these reactions.

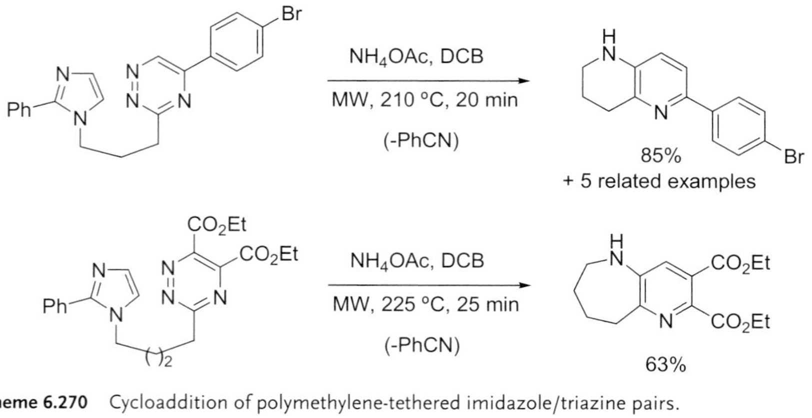

Scheme 6.270 Cycloaddition of polymethylene-tethered imidazole/triazine pairs.

6.24.10
Larger Heterocyclic and Polycyclic Ring Systems

The preparation of 5,11-dihydrobenzo[e]pyrido[3,2-b][1,4]diazepin-6-one by cyclocondensation of 3-amino-2-chloropyridine and ethyl 2-aminobenzoate in the presence of the strong base potassium tert-butoxide has been described by Holzgrabe and Heller (Scheme 6.271) [231]. Microwave heating of an equimolar mixture of the two starting materials with 3 equivalents of base in dry dioxane under an argon atmosphere at 100 °C for 2.5 h provided a 42% yield of the tricyclic product, which was subsequently used as a starting material for the synthesis of diazepinone analogues of the muscarinic receptor antagonist AFDX-384 (see Scheme 6.156).

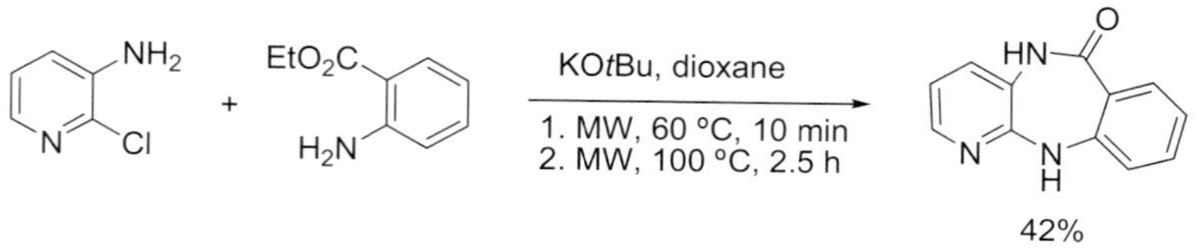

Scheme 6.271   Preparation of 5,11-dihydrobenzo[e]pyrido[3,2-b][1,4]diazepin-6-one.

Vasudevan and his group have used an intramolecular transamidation to generate a collection of bicyclic fused azepinones, wherein the substitution patterns of up to six positions could be varied, four of these in a stereospecific manner [461]. In the example highlighted in Scheme 6.272, the β-lactam precursor (generated by imine formation and subsequent [2+2] cycloaddition of a bifunctional amine component) is heated in N,N-dimethylformamide solution at 200 °C for 40 min to induce ring-opening/ring-expansion, thereby furnishing the desired 1,4-diazepin-5-one in 70% yield. This methodology has been used to synthesize a library of 120 1,4-diazepin-5-ones and other related ring systems.

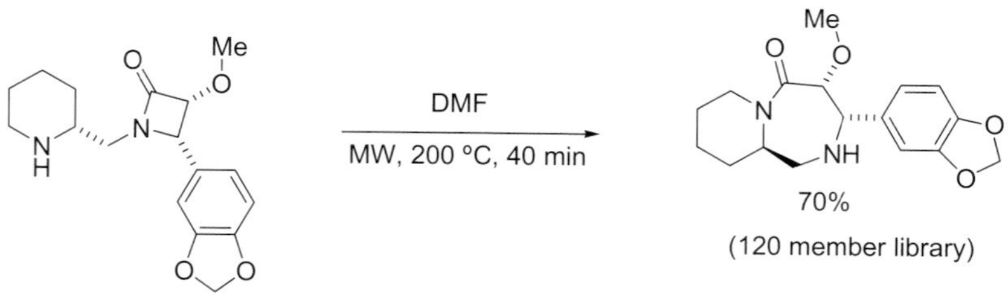

Scheme 6.272   Preparation of bicyclic fused azepinones by intramolecular transamidation.

The construction of a diazaadamantane skeleton under microwave conditions has been explored by Ivachtchenko and colleagues (Scheme 2.273) [462]. Cleavage of semi-natural tetrahydro-(−)-cytisine in acidic methanol provided the corresponding free diamine ester, which was used directly in the next step without purification. Thus, 1,1′-carbonyldiimidazole (CDI) in a water/methanol mixture was added at

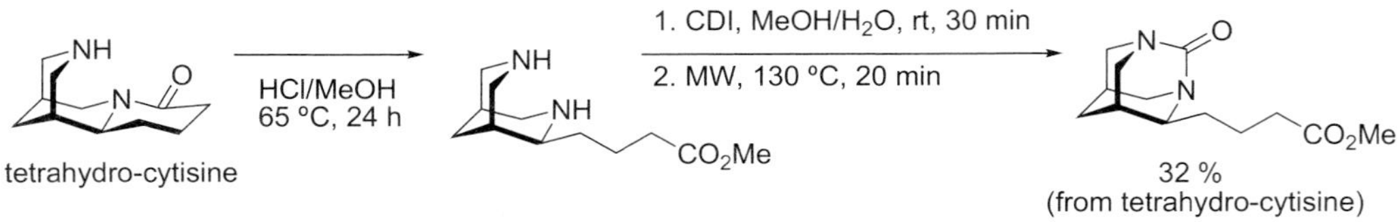

Scheme 6.273 Preparation of diazaadamantane skeletons.

room temperature to yield an *N*-acylimidazole intermediate (not shown), which was cyclized to the target diazaadamantane system upon microwave irradiation at 130 °C for 20 min. The reaction under microwave conditions was higher yielding and led to fewer side-products as compared to the thermal run.

Corroles are porphyrin analogues that lack one *meso* carbon bridge. Collman and Decréau have shown that the classical Gross synthesis of corroles from aldehydes and pyrroles adsorbed on an inorganic solid support such as alumina can be efficiently carried out under microwave irradiation (Scheme 6.274) [463]. Compared to conventional heating, the microwave technique afforded an increase in corrole yields of 20–50% and led to noticeably cleaner reactions. The general conditions involved mixing the aldehyde component with 3 equivalents of pyrrole and oven-dried basic alumina, and then heating the mixture by microwave irradiation at 120–200 °C for 2–20 min.

The preparation of metallophthalocyanines under solvent-free conditions from 1,2-phthalonitrile or phthalic anhydride and urea in the presence of metal templates

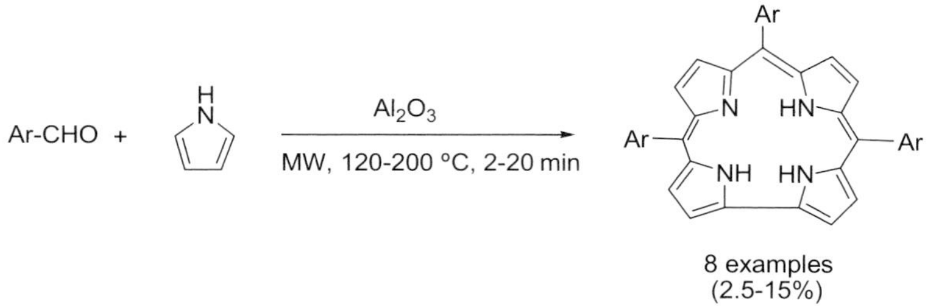

Scheme 6.274 Preparation of corroles from aromatic aldehydes and pyrrole.

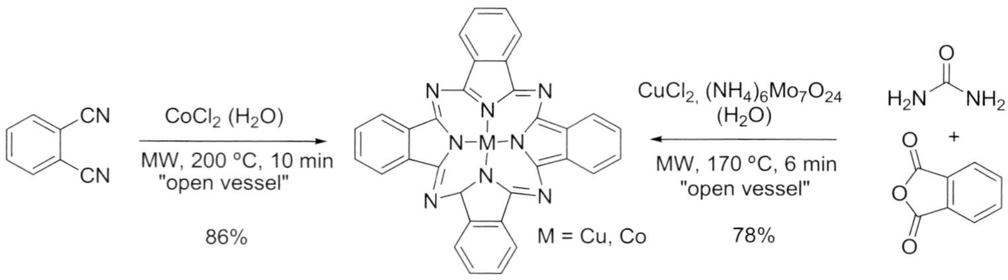

Scheme 6.275 Preparation of metallophthalocyanines.

has been reported by Bogdal and coworkers (Scheme 6.275) [464]. Among the many conditions screened by the authors, the use of either copper(II) or cobalt(II) chloride in the presence of small amounts of water provided the best results.

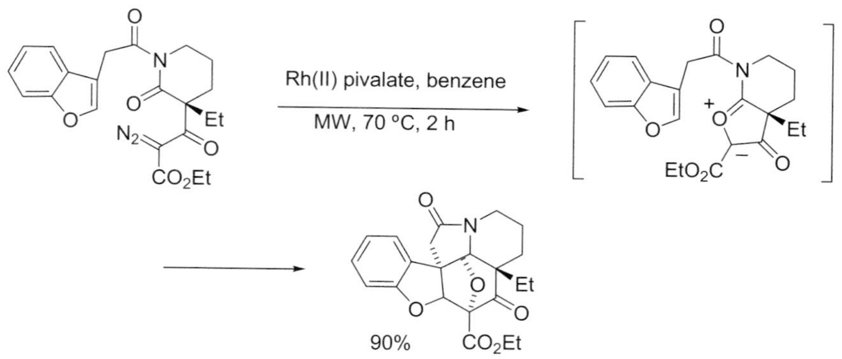

Scheme 6.276 An intramolecular [3+2] cycloaddition reaction of a push-pull dipole across a heteroaromatic π-system.

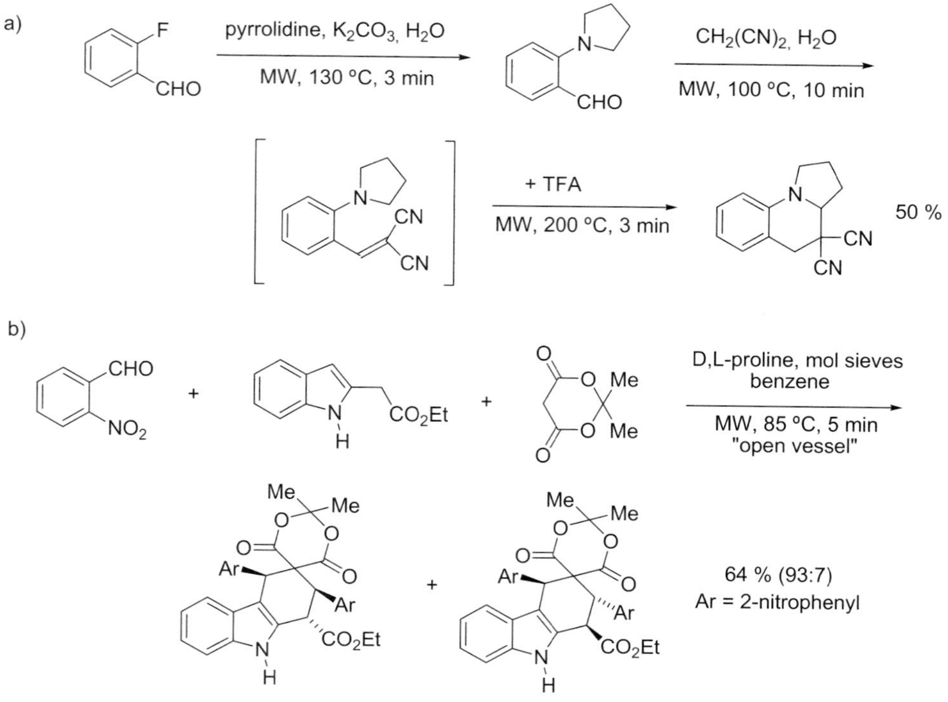

Scheme 6.277 Preparation of polycyclic and spiroheterocyclic structures.

Mejía-Oneto and Padwa have explored intramolecular [3+2] cycloaddition reactions of push-pull dipoles across heteroaromatic π-systems induced by microwave irradiation [465]. The push-pull dipoles were generated from the rhodium(II)-catalyzed reaction of a diazo imide precursor containing a tethered heteroaromatic ring. In the example shown in Scheme 6.276, microwave heating of a solution of the diazo imide precursor in dry benzene in the presence of a catalytic amount of rhodium(II) pivalate and 4 Å molecular sieves for 2 h at 70 °C produced a transient cyclic carbonyl ylide dipole, which spontaneously underwent cycloaddition across the tethered benzofuran π-system to form a pentacyclic structure related to alkaloids of the vindoline type.

Reaction pathways leading to other polycyclic structures and spiroheterocycles are summarized in Scheme 6.277 [466, 467].

References

[1] General organic synthesis: R. A. Abramovitch, *Org. Prep. Proced. Int.* **1991**, *23*, 685–711; S. Caddick, *Tetrahedron* **1995**, *51*, 10403–10432; P. Lidström, J. Tierney, B. Wathey, J. Westman, *Tetrahedron* **2001**, *57*, 9225–9283; B. L. Hayes, *Aldrichimica Acta* **2004**, *37*, 66–77; C. O. Kappe, *Angew. Chem. Int. Ed.* **2004**, *43*, 6250–6284.

[2] C. R. Strauss, R. W. Trainor, *Aust. J. Chem.* **1995**, *48*, 1665–1692; C. R. Strauss, *Aust. J. Chem.* **1999**, *52*, 83–96.

[3] Open-vessel technology (MORE): A. K. Bose, B. K. Banik, N. Lavlinskaia, M. Jayaraman, M. S. Manhas, *Chemtech* **1997**, *27*, 18–24; A. K. Bose, M. S. Manhas, S. N. Ganguly, A. H. Sharma, B. K. Banik, *Synthesis* **2002**, 1578–1591.

[4] Cycloaddition reactions: A. de la Hoz; A. Díaz-Ortis, A. Moreno, F. Langa, *Eur. J. Org. Chem.* **2000**, 3659–3673.

[5] Heterocycle synthesis: J. Hamelin, J.-P. Bazureau, F. Texier-Boullet, in *Microwaves in Organic Synthesis* (Ed.: A. Loupy), Wiley-VCH, Weinheim, **2002**, pp 253–294 (Chapter 8); T. Besson, C. T. Brain, in *Microwave-Assisted Organic Synthesis* (Eds.: P. Lidström, J. P. Tierney), Blackwell Publishing, Oxford, **2005** (Chapter 3); Y. Xu, Q.-X. Guo, *Heterocycles* **2004**, *63*, 903–974.

[6] Radiochemistry: N. Elander, J. R. Jones, S.-Y. Lu, S. Stone-Elander, *Chem. Soc. Rev.* **2000**, 239–250; S. Stone-Elander, N. Elander, *J. Label. Compd. Radiopharm.* **2002**, *45*, 715–746.

[7] Homogeneous transition metal catalysis: M. Larhed, C. Moberg, A. Hallberg, *Acc. Chem. Res.* **2002**, *35*, 717–727; K. Olofsson, M. Larhed, in *Microwave-Assisted Organic Synthesis* (Eds.: P. Lidström, J. P. Tierney), Blackwell Publishing, Oxford, **2005**.

[8] Medicinal chemistry: J. L. Krstenansky, I. Cotterill, *Curr. Opin. Drug Discovery Dev.* **2000**, *4*, 454–461; M. Larhed, A. Hallberg, *Drug Discovery Today* **2001**, *6*, 406–416; B. Wathey, J. Tierney, P. Lidström, J. Westman, *Drug Discovery Today* **2002**, *7*, 373–380; N. S. Wilson, G. P. Roth, *Curr. Opin. Drug Discov. Dev.* **2002**, *5*, 620–629; C. D. Dzierba, A. P. Combs, in *Ann. Rep. Med. Chem.* (Ed.: A. M. Doherty), Academic Press, **2002**, Vol. 37, pp 247–256.

[9] Combinatorial chemistry: A. Lew, P. O. Krutznik, M. E. Hart, A. R. Chamberlin, *J. Comb. Chem.* **2002**, *4*, 95–105; C. O. Kappe, *Curr. Opin. Chem. Biol.* **2002**, *6*, 314–320; P. Lidström, J. Westman, A. Lewis, *Comb. Chem. High-Throughput Screen.* **2002**, *5*, 441–458; H. E. Blackwell, *Org. Biomol. Chem.* **2003**, *1*, 1251–1255; F. Al-Obeidi, R. E. Austin, J. F. Okonya, D. R. S. Bond, *Mini-Rev. Med. Chem.* **2003**, *3*, 449–460; K. M. K. Swamy, W.-B. Yeh, M.-J. Lin, C.-M. Sun, *Curr. Med. Chem.* **2003**, *10*, 2403–2423; C. O. Kappe (Ed.), *Microwaves in Combinatorial and High-Throughput Synthesis*, Kluwer, Dordrecht, **2003** (a special issue of *Mol. Diversity* **2003**, *7*, pp 95–307).

[10] For online resources on microwave-assisted organic synthesis (MAOS), see: www.maos. net.

[11] A. Loupy (Ed.), *Microwaves in Organic Synthesis*, Wiley-VCH, Weinheim, **2002**.

[12] B. L. Hayes, *Microwave Synthesis: Chemistry at the Speed of Light*, CEM Publishing, Matthews, NC, **2002**.

[13] P. Lidström, J. P. Tierney (Eds.), *Microwave-Assisted Organic Synthesis*, Blackwell Publishing, Oxford, **2005**.

[14] I. P. Beletskaya, A. V. Cheprakov, *Chem. Rev.* **2000**, *100*, 3009–3066.

[15] M. Larhed, A. Hallberg, *J. Org. Chem.* **1996**, *61*, 9582–9584.

[16] A. Stadler, B. H. Yousefi, D. Dallinger, P. Walla, E. Van der Eycken, N. Kaval, C. O. Kappe, *Org. Process Res. Dev.* **2003**, *7*, 707–716.

[17] K. S. A. Vallin, P. Emilsson, M. Larhed, A. Hallberg, *J. Org. Chem.* **2002**, *67*, 6243–6246.

[18] G. K. Datta, K. S. A. Vallin, M. Larhed, *Mol. Diversity* **2003**, *7*, 107–114.

[19] For a recent review, see: A. F. Littke, G. C. Fu, *Angew. Chem. Int. Ed.* **2002**, *41*, 4176–4211.

[20] M. R. Netherton, G. C. Fu, *Org. Lett.* **2001**, *3*, 4295–4298.

[21] W. A. Herrmann, V. P. H. Bohm, C. P. Reisinger, *J. Organomet. Chem.* **1999**, *576*, 23–41.

[22] L. Botella, C. Nájera, *Tetrahedron Lett.* **2004**, *45*, 1833–1836.

[23] L. Botella, C. Nájera, *Tetrahedron* **2004**, *60*, 5563–5570.

[24] A. Svennebring, P. Nilsson, M. Larhed, *J. Org. Chem.* **2004**, *69*, 3345–3349.

[25] M. M. S. Andappan, P. Nilsson, M. Larhed, *Mol. Diversity* **2003**, *7*, 97–106.

[26] M. M. S. Andappan, P. Nilsson, H. von Schenck, M. Larhed, *J. Org. Chem.* **2004**, *69*, 5212–5218.

[27] T. Flessner, V. Ludwig, H. Siebeneicher, E. Winterfeld, *Synthesis* **2002**, 1373–1378.

[28] V. Gracias, J. D. Moore, S. W. Djuric, *Tetrahedron Lett.* **2004**, *45*, 417–420.

[29] L. F. Tietze, J. M. Wiegand, C. Vock, *J. Organomet. Chem.* **2003**, *687*, 346–352.

[30] L. F. Tietze, J. M. Wiegand, C. Vock, *Eur. J. Org. Chem.* **2004**, 4107–4112.

[31] U. S. Sørensen, E. Pombo-Villar, *Helv. Chim. Acta* **2004**, *87*, 82–89.

[32] N. Miyaura, A. Suzuki, *Chem. Rev.* **1995**, *95*, 2457–2483.

[33] N. E. Leadbeater, M. Marco, *Org. Lett.* **2002**, *4*, 2973–2976.

[34] N. E. Leadbeater, M. Marco, *J. Org. Chem.* **2003**, *68*, 888–892.

[35] L. Bai, J.-X. Wang, Y. Zhang, *Green Chem.* **2003**, *5*, 615–617.

[36] N. E. Leadbeater, M. Marco, *Angew. Chem. Int. Ed.* **2003**, *42*, 1407–1409.

[37] N. E. Leadbeater, M. Marco, *J. Org. Chem.* **2003**, *68*, 5660–5667.

[38] C.-J. Li, *Angew. Chem. Int. Ed.* **2003**, *42*, 4856–4858.

[39] O. Navarro, H. Kaur, P. Mahjoor, S. P. Nolan, *J. Org. Chem.* **2004**, *69*, 3173–3180.

[40] Y. Gong, W. He, *Org. Lett.* **2002**, *4*, 3803–3805.

[41] P. Appukkuttan, A. B. Orts, R. P. Chandran, J. L. Goeman, J. Van der Eycken, W. Dehan, E. Van der Eycken, *Eur. J. Org. Chem.* **2004**, 3277–3285.

[42] M. Melucci, G. Barbarella, M. Zambianchi, A. Bongini, *J. Org. Chem.* **2004**, *69*, 4821–4828.

[43] S. P. Miller, J. B. Morgan, F. J. Nepveux, J. P. Morken, *Org. Lett.* **2004**, *6*, 131–133.

[44] R. R. Poondra, P. M. Fischer, N. J. Turner, *J. Org. Chem.* **2004**, *69*, 6920–6922.

[45] G. Luo, L. Chen, G. S. Pointdexter, *Tetrahedron Lett.* **2002**, *43*, 5739–5742.

[46] Y. Gong, W. He, *Heterocycles* **2004**, *62*, 851–856.

[47] N. Kaval, K. Bisztray, W. Dehaen, C. O. Kappe, E. Van der Eycken, *Mol. Diversity* **2003**, *7*, 125–134.

[48] P. Holmberg, D. Sohn, R. Leideborg, P. Caldirola, P. Zlatoidsky, S. Hanson, N. Mohell, S. Rosqvist, G. Nordvall, A. M. Johansson, R. Johansson, *J. Med. Chem.* **2004**, *47*, 3927–3930.

[49] M. G. Organ, S. Mayer, F. Lepifre, B. N'Zemba, J. Khatri, *Mol. Diversity* **2003**, *7*, 211–227.

[50] T. Y. H. Wu, P. G. Schultz, S. Ding, *Org. Lett.* **2003**, *5*, 3587–3590.

[51] D. Nöteberg, W. Schaal, E. Hamelink, L. Vrang, M. Larhed, *J. Comb. Chem.* **2003**, *5*, 456–464.

[52] K. Esmark, I. Feierberg, S. Bjelic, E. Hamelink, F. Hackett, M. J. Blackman, J. Hultén, B. Samuelsson, J. Åqvist, A. Hallberg, *J. Med. Chem.* **2004**, *47*, 110–112.

[53] R. Kurukulasuriya, B. K. Sorensen, J. T. Link, J. R. Patel, H.-S. Jae, M. X. Winn, J. R. Rohde, N. D. Grihalde, C. W. Lin, C. A. Ogiela, A. L. Adler, C. A. Collins, *Bioorg. Med. Chem. Lett.* **2004**, *14*, 2047–2050.

[54] R. A. Tromp, S. van Ameijde, C. Pütz, C. Sundermann, B. Sundermann, J. K. von Frijtag Drabe Künzel, A. P. IJzerman, *J. Med. Chem.* **2004**, *47*, 5441–5450.

[55] Y. Wan, C. Wallinder, B. Plouffe, H. Beaudry, A. K. Mahalingam, X. Wu, B. Johansson, M. Holm, M. Botoros, A. Karlén, A. Pettersson, F. Nyberg, L. Fändriks, N. Gallo-Payet, A. Hallberg, M. Alterman, *J. Med. Chem.* **2004**, *47*, 5995–6008.

[56] Q. J. Zhou, K. Worm, R. E. Dolle, *J. Org. Chem.* **2004**, *69*, 5147–5148.

[57] J. W. Han, J. C. Castro, K. Burgess, *Tetrahedron Lett.* **2003**, *44*, 9359–9362.

[58] B. S. Nehls, U. Asawapirom, S. Füldner, E. Preis, T. Farrell, U. Scherf, *Adv. Funct. Mater.* **2004**, *14*, 352–356.

[59] W. Solodenko, U. Schön, J. Messinger, A. Glinschert, A. Kirschning, *Synlett* **2004**, 1699–1702.

[60] Y. Wang, D. R. Sauer, *Org. Lett.* **2004**, *6*, 2793–2796.

[61] W. M. Seganish, P. DeShong, *Org. Lett.* **2004**, *6*, 4379–4381.

[62] K. Sonogashira, *J. Organomet. Chem.* **2002**, *653*, 46–49.

[63] M. Erdélyi, A. Gogoll, *J. Org. Chem.* **2001**, *66*, 4165–4169.

[64] M. Erdélyi, V. Langer, A. Karlén, A. Gogoll, *New J. Chem.* **2002**, *26*, 834–843.

[65] O. Š. Miljanic, K. P. C. Vollhardt, G. D. Whitener, *Synlett* **2003**, 29–34.

[66] E. Petricci, M. Radi, F. Corelli, M. Botta, *Tetrahedron Lett.* **2003**, *44*, 9181–9184.

[67] J. Zhu, A. Germain, J. A. Porco, *Angew. Chem. Int. Ed.* **2004**, *43*, 1239–1243.

[68] C. R. Hopkins, N. Collar, *Tetrahedron Lett.* **2004**, *45*, 8631–8633.

[69] P. H. Kwan, M. J. MacLachlan, T. M. Swager, *J. Am. Chem. Soc.* **2004**, *126*, 8638–8639.

[70] A. Khan, S. Hecht, *Chem. Commun.* **2004**, 300–301.

[71] N. E. Leadbeater, M. Marco, B. J. Tominack, *Org. Lett.* **2003**, *5*, 3919–3922.

[72] P. Appukkuttan, W. Dehaen, E. Van der Eycken, *Eur. J. Org. Chem.* **2003**, 4713–4716.

[73] H. He, Y.-J. Wu, *Tetrahedron Lett.* **2004**, *45*, 3237–3239.

[74] J. K. Stille, *Angew. Chem. Int. Ed. Engl.* **1986**, *25*, 508–523.

[75] Y. Zhang, O. A. Pavlova, S. I. Chefer, A. W. Hall, V. Kurian, L. L. Brown, A. S. Kimes, A. G. Mukhin, A. G. Horti, *J. Med. Chem.* **2004**, *47*, 2453–2465.

[76] D. J. O'Neill, L. Shen, C. Prouty, B. R. Conway, L. Westover, J. Z. Xu, H.-C. Zhang, B. E. Maryanoff, W. V. Murray, K. T. Demarest, G.-H. Kuo, *Bioorg. Med. Chem.* **2004**, *12*, 3167–3185.

[77] P. Rashatasakhon, A. D. Ozdemir, J. Willis, A. Padwa, *Org. Lett.* **2004**, *6*, 917–920.

[78] E.-i. Negishi, in *Metal-Catalyzed Cross-Coupling Reactions* (Eds.: F. Diederich, P. J. Stang), Wiley-VCH, New York, **1998**.

[79] J. Hassan, M. Sévignon, C. Gozzi, E. Schulz, *Chem. Rev.* **2002**, *102*, 1359–1470.

[80] L. Öhberg, J. Westman, *Synlett* **2001**, 1893–1896.

[81] P. Stanetty, M. Schnürch, M. D. Mihovilovic, *Synlett* **2003**, 1862–1864.

[82] P. Walla, C. O. Kappe, *Chem. Commun.* **2004**, 564–565.

[83] I. Mutule, E. Suna, *Tetrahedron Lett.* **2004**, *45*, 3909–3912.

[84] K. Krascsenicsová, P. Walla, P. Kasáka, G. Uray, C. O. Kappe, M. Putala, *Chem. Commun.* **2004**, 2606–2607.

[85] E. Bentz, M. G. Moloney, S. M. Westaway, *Tetrahedron Lett.* **2004**, *45*, 7395–7397.

[86] B. H. Lipshutz, B. Frieman, *Tetrahedron* **2004**, *60*, 1309–1316.

[87] B. H. Lipshutz, B. Frieman, H. Birkedal, *Org. Lett.* **2004**, *6*, 2305–2308.

[88] S. L. Wiskur, A. Korte, G. C. Fu, *J. Am. Chem. Soc.* **2004**, *126*, 82–83.

[89] B. H. Lipshutz, S. S. Pfeiffer, T. Tomioka, K. Noson, in *Titanium and Zirconium in Organic Synthesis* (Ed.: I. Marek), Wiley-VCH, Weinheim, **2002**, p. 110–148.

[90] P. Wipf, J. Janjic, C. R. J. Stephenson, *Org. Biomol. Chem.* **2004**, *2*, 443–445.

[91] P. Wipf, C. R. J. Stephenson, K. Okumura, *J. Am. Chem. Soc.* **2003**, *125*, 14694–14695.

[92] N.-F. K. Kaiser, A. Hallberg, M. Larhed, *J. Comb. Chem.* **2002**, *4*, 109–111.

[93] J. Wannberg, M. Larhed, *J. Org. Chem.* **2003**, *68*, 5750–5753.

[94] M. A. Herrero, J. Wannberg, M. Larhed, *Synlett* **2004**, 2335–2338.

[95] J. Georgsson, A. Hallberg, M. Larhed, *J. Comb. Chem.* **2003**, *5*, 456–458.

[96] X. Wu, A. K. Mahalingam, Y. Wan, M. Alterman, *Tetrahedron Lett.* **2004**, *45*, 4635–4638.

[97] X. Wu, P. Nilsson, M. Larhed, *J. Org. Chem.* **2005**, *70*, 346–349.

[98] Y. Wan, M. Alterman, M. Larhed, A. Hallberg, *J. Org. Chem.* **2002**, *67*, 6232–6235.

[99] Y. Wan, M. Alterman, M. Larhed, A. Hallberg, *J. Comb. Chem.* **2003**, *5*, 82–84.

[100] P.-A. Enquist, P. Nilsson, M. Larhed, *Org. Lett.* **2003**, *5*, 4875–4878.

[101] M. Larhed, C. Moberg, A. Hallberg, *Acc. Chem. Res.* **2002**, *35*, 717–727

[102] N.-F. K. Kaiser, U. Bremberg, M. Larhed, C. Moberg, A. Hallberg, *J. Organomet. Chem.* **2000**, *603*, 2–5.

[103] U. Bremberg, S. Lutsenko, N.-F. K. Kaiser, M. Larhed, A. Hallberg, C. Moberg, *Synthesis* **2000**, 1004–1008.

[104] N.-F. K. Kaiser, U. Bremberg, M. Larhed, C. Moberg, A. Hallberg, *Angew. Chem. Int. Ed.* **2000**, *39*, 3596–3598.

[105] O. Belda, N.-F. Kaiser, U. Bremberg, M. Larhed, A. Hallberg, C. Moberg, *J. Org. Chem.* **2000**, *65*, 5868–5870.

[106] O. Belda, C. Moberg, *Synthesis* **2002**, 1601–1606.

[107] B. M. Trost, N. G. Andersen, *J. Am. Chem. Soc.* **2002**, *124*, 14320–14321.

[108] O. Belda, S. Lundgren, C. Moberg, *Org. Lett.* **2003**, *5*, 2275–2278.

[109] P. Nilsson, H. Gold, M. Larhed, A. Hallberg, *Synthesis* **2002**, 1611–1614.

[110] S. Lutsenko, C. Moberg, *Tetrahedron: Asymmetry* **2001**, *12*, 2529–2532.

[111] D. A. Alonso, C. Nájera, M. C. Pacheco, *J. Org. Chem.* **2004**, *69*, 1615–1619.

[112] L. S. Liebeskind, J. Srogl, *Org. Lett.* **2002**, *4*, 979–982.

[113] A. Lengar, C. O. Kappe, *Org. Lett.* **2004**, *6*, 771–774.

[114] A. Zhang, J. L. Neumeyer, *Org. Lett.* **2003**, *5*, 201–203.

[115] A. Zhang, W. Xiong, J. M. Bidlack, J. E. Hilbert, B. I. Knapp, M. P. Wentland, J. L. Neumeyer, *J. Med. Chem.* **2004**, *47*, 165–174.

[116] M. Alterman, A. Hallberg, *J. Org. Chem.* **2000**, *65*, 7984–7989.

[117] R. K. Arvela, N. E. Leadbeater, *J. Org. Chem.* **2003**, *68*, 9122–9125.

[118] A. Gopalsamy, K. Lim, G. Ciszewski, K. Park, J. W. Ellingboe, J. Bloom, S. Insaf, J. Upeslacis, T. S. Mansour, G. Krishnamurthy, M. Damarla, Y. Pyatski, D. Ho, A. Y. M. Howe, M. Orlowski, B. Feld, J. O'Connell, *J. Med. Chem.* **2004**, *47*, 6603–6608.

[119] G. Vo-Thanh, H. Lahrache, A. Loupy, I.-J. Kim, D.-H. Chang, C.-H. Jun, *Tetrahedron* **2004**, *60*, 5539–5543.

[120] A. R. Muci, S. L. Buchwald, *Top. Curr. Chem.* **2002**, *219*, 131–209; J. P. Wolfe, S. Wagaw, J. F. Marcoux, S. L. Buchwald, *Acc. Chem. Res.* **1998**, *31*, 805–818.

[121] J. F. Hartwig, *Angew. Chem. Int. Ed. Engl.* **1998**, *37*, 2046–2067.

[122] Y. Wan, M. Alterman, A. Hallberg, *Synthesis* **2002**, 1597–1600.

[123] S. Antane, *Synth. Commun.* **2003**, *33*, 2147–2149.

[124] T. Wang, D. R. Magnin, L. G. Hamann, *Org. Lett.* **2003**, *5*, 897–900.

[125] T. Ulrich, F. Giraud, *Tetrahedron Lett.* **2003**, *44*, 4207–4211.

[126] C. T. Brain, J. T. Steer, *J. Org. Chem.* **2003**, *68*, 6814–6816.

[127] T. A. Jensen, X. Liang, D. Tanner, N. Skjaerbaek, *J. Org. Chem.* **2004**, *69*, 4936–4947.

[128] A. J. McCarroll, D. A. Sandham, L. R. Titcomb, A. K. de K. Lewis, F. G. N. Cloke, B. P. Davies, A. P. de Santana, W. Hiller, S. Caddick, *Mol. Diversity* **2003**, *7*, 115–123.

[129] B. U. W. Maes, K. T. J. Loones, G. L. F. Lemiére, R. A. Dommisse, *Synlett* **2003**, 1822–1825.

[130] B. U. W. Maes, K. T. J. Loones, S. Hostyn, G. Diels, *Tetrahedron* **2004**, *60*, 11559–11564.

[131] G. Burton, P. Cao, G. Li, R. Rivero, *Org. Lett.* **2003**, *5*, 4373–4376.

[132] M. Harmata, X. Hong, S. K. Ghosh, *Tetrahedron Lett.* **2004**, *45*, 5233–5236.

[133] A. W. Thomas, S. V. Ley, *Angew. Chem. Int. Ed.* **2003**, *42*, 5400–5449; K. Kunz, U. Scholz, D. Ganzer, *Synlett* **2003**, 2428–2439.

[134] Y.-J. Wu, H. He, A. L'Heureux, *Tetrahedron Lett.* **2003**, *44*, 4217–4218.

[135] Y.-J. Wu, H. He, *Synlett* **2003**, 1789–1790.

[136] H. He, Y.-J. Wu, *Tetrahedron Lett.* **2003**, *44*, 3445–3446.

[137] J. H. M. Lange, L. J. F. Hofmeyer, F. A. S. Hout, S. J. M. Osnabrug, P. C. Verveer, C. G. Kruse, R. Feenstra, *Tetrahedron Lett.* **2002**, *43*, 1101–1104.

[138] D. Macmillan, D. W. Anderson, *Org. Lett.* **2004**, *6*, 4659–4662.

[139] A. Stadler, C. O. Kappe, *Org. Lett.* **2002**, *4*, 3541–3544.

[140] A. Fürstner, G. Seidel, *Org. Lett.* **2002**, *4*, 541–543.

[141] P. Appukkuttan, E. Van der Eycken, W. Dehaen, *Synlett* **2003**, 1204–1206.

[142] R. H. Grubbs, S. Chang, *Tetrahedron* **1998**, *54*, 4413–4450; A. Fürstner, *Angew. Chem. Int. Ed.* **2000**, *39*, 3012–3043.

[143] K. G. Mayo, E. H. Nearhoof, J. J. Kiddle, *Org. Lett.* **2002**, *4*, 1567–1570.

[144] S. Garbacia, B. Desai, O. Lavastre, C. O. Kappe, *J. Org. Chem.* **2003**, *68*, 9136–9139.

[145] G. Vo Thanh, A. Loupy, *Tetrahedron Lett.* **2003**, *44*, 9091–9094.

[146] S. Varry, C. Gauzy, F. Lamaty, R. Lazaro, J. Martinez, *J. Org. Chem.* **2000**, *65*, 6787–6790.

[147] C. Yang, W. V. Murray, L. J. Wilson, *Tetrahedron Lett.* **2003**, *44*, 1783–1786.

[148] R. Grigg, W. Martin, J. Morris, V. Sridharan, *Tetrahedron Lett.* **2003**, *44*, 4899–4901.

[149] D. Balan, H. Adolfsson, *Tetrahedron Lett.* **2004**, *45*, 3089–3092.

[150] V. Declerck, P. Ribière, J. Martinez, F. Lamaty, *J. Org. Chem.* **2004**, *69*, 8372–8381.

[151] J. Efskind, K. Undheim, *Tetrahedron Lett.* **2003**, *44*, 2837–2839.

[152] A. Fürstner, F. Stelzer, A. Rumbo, H. Krause, *Chem. Eur. J.* **2002**, *8*, 1856–1871.

[153] S. S. Salim, R. K. Bellingham, R. C. D. Brown, *Eur. J. Org. Chem.* **2004**, 800–806.

[154] C. Schultz-Fademrecht, P. H. Deshmukh, K. Malagu, P. A. Procopiou, A. G. M. Barrett, *Tetrahedron* **2004**, *60*, 7515–7524.

[155] P. L. Pauson, *Tetrahedron* **1985**, *41*, 5855–5860.

[156] S. Fischer, U. Groth, M. Jung, A. Schneider, *Synlett* **2002**, 2023–2026.

[157] M. Iqbal, N. Vyse, J. Dauvergne, P. Evans, *Tetrahedron Lett.* **2002**, *43*, 7859–7862.

[158] M. Iqbal, Y. Li, P. Evans, *Tetrahedron* **2004**, *60*, 2531–2538.

[159] For reviews, see: G. Dyker, *Angew. Chem. Int. Ed.* **1999**, *38*, 1699–1712; F. Kakiuchi, S. Murai, *Acc. Chem. Res.* **2002**, *35*, 826–834; V. Ritleng, C. Sirlin, M. Pfeffer, *Chem. Rev.* **2002**, *102*, 1731–1769.

[160] K. L. Tan, A. Vasudevan, R. G. Bergman, J. A. Ellman, A. J. Souers, *Org. Lett.* **2003**, *5*, 2131–2134.

[161] M. J. Gaunt, A. S. Jessiman, P. Orsini, H. R. Tanner, D. F. Hook, S. V. Ley, *Org. Lett.* **2003**, *5*, 4819–4822.

[162] M. E. Jung, A. Maderna, *J. Org. Chem.* **2004**, *69*, 7755–7757.

[163] J. M. Mejía-Oneto, A. Padwa, *Tetrahedron Lett.* **2004**, *45*, 9115–9118.

[164] B. H. Lipshutz, C. C. Caires, P. Kuipers, W. Chrisman, *Org. Lett.* **2003**, *5*, 3085–3088.

[165] E. J. Hutchinson, W. J. Kerr, E. J. Magennis, *Chem. Commun.* **2002**, 2262–2263.

[166] P. I. Dosa, G. D. Whitener, K. P. C. Vollhardt, *Org. Lett.* **2002**, *4*, 2075–2078.

[167] K. R. Carter, *Macromolecules* **2002**, *35*, 6757–6759.

[168] I. R. Baxendale, A.-I. Lee, S. V. Ley, *J. Chem. Soc., Perkin Trans.* **2002**, 1850–1857.

[169] T. Durand-Reville, L. B. Gobbi, B. L. Gray, S. V. Ley, J. S. Scott, *Org. Lett.* **2002**, *4*, 3847–3850.

[170] G. Nordmann, S. L. Buchwald, *J. Am. Chem. Soc.* **2003**, *125*, 4978–4979.

[171] T. Yamamoto, Y. Wada, H. Enokida, M. Fujimoto. K. Nakamura, S. Yanagida, *Green Chem.* **2003**, *5*, 690–692.

[172] C. J. Bennett, S. T. Caldwell, D. B. McPhail, P. C. Morrice, G. G. Duthie, R. C. Hartley, *Bioorg. Med. Chem.* **2004**, *12*, 2079–2098.

[173] J. M. Cid, J. L. Romera, A. A. Trabanco, *Tetrahedron Lett.* **2004**, *45*, 1133–1136.

[174] R. Schobert, G. J. Gordon, G. Mullen, R. Stehle, *Tetrahedron Lett.* **2004**, *45*, 1121–1124.

[175] J. A. Farand, I. Denissova, L. Barriault, *Heterocycles* **2004**, *62*, 735–748.

[176] E. L. O. Sauer, L. Barriault, *J. Am. Chem. Soc.* **2004**, *126*, 8569–8575.

[177] E. L. O. Sauer, L. Barriault, *Org. Lett.* **2004**, *6*, 3329–3332.

[178] L. Morency, L. Barriault, *Tetrahedron Lett.* **2004**, *45*, 6105–6107.

[179] C. E. McIntosh, I. Martínez, T. V. Ovaska, *Synlett* **2004**, 2579–2581.

[180] B. M. Trost, O. R. Thiel, H.-C. Tsui, *J. Am. Chem. Soc.* **2003**, *125*, 13155–13164.

[181] P. S. Baran, D. P. O'Malley, A. L. Zografos, *Angew. Chem. Int. Ed.* **2004**, *43*, 2674–2677.

[182] D. C. Braddock, S. M. Ahmad, G. T. Douglas, *Tetrahedron Lett.* **2004**, *45*, 6583–6587.

[183] S. G. Sudrik, S. P. Chavan, K. R. S. Chandrakumar, S. Pal, S. K. Date, S. P. Chavan, H. R. Sonawane, *J. Org. Chem.* **2002**, *67*, 1574–1579.

[184] B. S. Patil, G.-R. Vasanthakumar, V. V. S. Babu, *J. Org. Chem.* **2003**, *68*, 7274–7280.

[185] S. K. Das, K. A. Reddy, J. Roy, *Synlett* **2003**, 1607–1610.

[186] B. M. Trost, M. L. Crawley, *J. Am. Chem. Soc.* **2002**, *124*, 9328–9329.

[187] D. C. G. A. Pinto, A. M. S. Silva, L. M. P. M. Almeida, J. R. Carrillo, A. Díaz-Ortiz, A. de la Hoz, J. A. S. Cavaleiro, *Synlett* **2003**, 1415–1418.

[188] M. S. Leonard, P. J. Carroll, M. M. Joullié, *J. Org. Chem.* **2004**, *69*, 2526–2531.

[189] A. Loupy, F. Maurel, A. Sabatié-Gogová, *Tetrahedron* **2004**, *60*, 1683–1691.

[190] N. E. Leadbeater, H. M. Torenius, *J. Org. Chem.* **2002**, *67*, 3145–3148.

[191] B.-C. Hong, Y.-J. Shr, J.-H. Liao, *Org. Lett.* **2002**, *4*, 663–666.

[192] I.-H. Chen, J.-N. Young, S. J. Yu, *Tetrahedron* **2004**, *60*, 11903–11909.

[193] J. P. Eddolls, M. Iqbal, S. M. Roberts, M. G. Santoro, *Tetrahedron* **2004**, *60*, 2539–2550.

[194] M. Iqbal, Y. Li, P. Evans, *Tetrahedron* **2004**, *60*, 2531–2538.

[195] E. Van der Eycken, P. Appukkuttan, W. De Borggraeve, W. Dehaen, D. Dallinger, C. O Kappe, *J. Org. Chem.* **2002**, *67*, 7904–7909.

[196] N. Kaval , W. Dehaen, C. O. Kappe, E. Van der Eycken, *Org. Biomol. Chem.* **2004**, *2*, 154–156.

[197] N. Kaval, J. Van der Eycken, J. Caroen, W. Dehaen, G. A. Strohmeier, C. O. Kappe, E. Van der Eycken, *J. Comb. Chem.* **2003**, *5*, 560–568.

[198] C. J. Tucker, M. E. Welker, C. S. Day, M. W. Wright, *Organometallics* **2004**, *23*, 2257–2262.

[199] P. Dupau, R. Epple, A. A. Thomas, V. V. Fokin, K. B. Sharpless, *Adv. Synth. Catal.* **2002**, *344*, 421–433.

[200] J. S. Clark, M.-R. Clarke, J. Clough, A. J. Blake, C. Wilson, *Tetrahedron Lett.* **2004**, *45*, 9447–9450.

[201] J. Freitag, M. Nüchter, B. Ondruschka, *Green Chem.* **2003**, *5*, 291–295.

[202] M. Takahashi, K. Oshima, S. Matsubara, *Tetrahedron Lett.* **2003**, *44*, 9201–9203.

[203] M. C. Bagley, J. W. Dale, X. Xiong, J. Bower, *Org. Lett.* **2003**, *5*, 4421–4424.

[204] N. Stiasni, C. O. Kappe, *ARKIVOC* **2002**, (viii), 71–79.

[205] F.-R. Alexandre, A. Berecibar, R. Wrigglesworth, T. Besson, *Tetrahedron Lett.* **2003**, *44*, 4455–4458.

[206] S. Mazumder, D. D. Laskar, D. Prajapati, M. K. Roy, *Chem. Biodiversity* **2004**, *1*, 925–929.

[207] A. Steinreiber, A. Stadler, S. F. Mayer, K. Faber, C. O. Kappe, *Tetrahedron Lett.* **2001**, *42*, 6283–6286.

[208] L. R. Lampariello, D. Piras, M. Rodriquez, M. Taddei, *J. Org. Chem.* **2003**, *68*, 7893–7895.

[209] I. T. Raheem, S. N. Goodman, E. N. Jacobsen, *J. Am. Chem. Soc.* **2004**, *126*, 706–707.

[210] E. Söderberg, J. Westman, S. Oscarson, *J. Carbohydr. Chem.* **2001**, *20*, 397–410.

[211] F. Mathew, K. N. Jayaprakash, B. Fraser-Reid, J. Mathew, J. Scicinski, *Tetrahedron Lett.* **2003**, *44*, 9051–9054.

[212] G. J. S. Lohman, P. Seeberger, *J. Org. Chem.* **2004**, *69*, 4081–4093.

[213] M. L. Paoli, S. Piccini, M. Rodriquez, A. Sega, *J. Org. Chem.* **2004**, *69*, 2881–2883.

[214] L. Ballell, J. A. F. Joosten, F. Ait el Maate, R. M. J. Liskamp, R. J. Pieters, *Tetrahedron Lett.* **2004**, *45*, 6685–6687.

[215] V. Roy, L. Colombeau, R. Zerrouki, P. Krausz, *Carbohydr. Res.* **2004**, *339*, 1829–1831.

[216] K.-S. Ko, C. J. Zea, N. L. Pohl, *J. Am. Chem. Soc.* **2004**, *126*, 13188–13189.

[217] M. Bejugam, S. L. Flitsch, *Org. Lett.* **2004**, *6*, 4001–4004.

[218] M. Abramov, A. Marchand, A. Calleja-Marchand, P. Herdewijn, *Nucl. Nucleot. Nucl. Acids* **2004**, *23*, 439–455.

[219] S. K. Das, *Synlett* **2004**, 915–932.

[220] A. Corsaro, U. Chiacchio, V. Pistara, G. Romeo, *Curr. Org. Chem.* **2004**, *8*, 511–538.

[221] F. Lehmann, Å. Pilotti, K. Luthman, *Mol. Diversity* **2003**, *7*, 145–152.

[222] N. E. Leadbeater, H, M. Torenius, H. Tye, H. *Mol. Diversity* **2003**, *7*, 135–144.

[223] L. Shi, Y.-Q. Tu, M. Wang, F.-M. Zhang, C.-A. Fan, *Org. Lett.* **2004**, *6*, 1001–1004.

[224] Y. Ju, C.-J. Li, R. S. Varma, *QSAR Comb. Sci.* **2004**, *23*, 891–894.

[225] N. J. McLean, H. Tye, M. Whittaker, *Tetrahedron Lett.* **2004**, *45*, 993–995.

[226] O. I. Zbruyev, N. Stiasni, C. O. Kappe, *J. Comb. Chem.* **2003**, *5*, 145–148.

[227] M. G. Saulnier, K. Zimmermann, C. P. Struzynski, X. Sang, U. Velaparthi, M. Wittman, D. B. Frennesson, *Tetrahedron Lett.* **2004**, *45*, 397–399.

[228] Y. Ju, R. S. Varma, *Green Chem.* **2004**, *6*, 219–221.

[229] J. L. Romera, J. M. Cid, A. A. Trabanco, *Tetrahedron Lett.* **2004**, *45*, 8797–8800.

[230] L. D. Jennings, K. W. Foreman, T. S. Rush, D. H. H. Tsao, L. Mosyak, Y. Li, M. N. Sukhdeo, W. Ding, E. G. Dushin, C. H. Kenny, S. L. Moghazeh, P. J. Petersen, A. V. Ruzin, M. Tuckman, A. G. Sutherland, *Bioorg. Med. Chem. Lett.* **2004**, *14*, 1427–1431.

[231] U. Holzgrabe, E. Heller, *Tetrahedron* **2003**, *59*, 781–787.

[232] A. Vasudevan, Z. Ji, R. R. Frey, C. K. Wada, D. Steinman, H. R. Heyman, Y. Guo, M. L. Curtin, J. Guo, J. Li, L. Pease, K. B. Glaser, P. A. Marcotte, J. J. Bouska, S. K. Davidsen, M. R. Michaelides, *Bioorg. Med. Chem. Lett.* **2003**, *13*, 3909–3913.

[233] L. Cai, F. T. Chin, V. W. Pike, H. Toyama, J.-S. Liow, S. S. Zoghbi, K. Modell, E. Briard, H. U. Shetty, K. Sinclair, S. Donohue, D. Tipre, M.-P. Kung, C. Dagostin, D. A. Widdowson, M. Green, W. Gao, M. M. Herman, M. Ichise, R. B. Innis, *J. Med. Chem.* **2004**, *47*, 2208–2218.

[234] J. Alcázar, G. Diels, B. Schoentjes, *QSAR Comb. Sci.* **2004**, *23*, 906–910.

[235] J. Sarju, T. N. Danks, G. Wagner, *Tetrahedron Lett.* **2004**, *45*, 7675–7677.

[236] M. Radi, E. Petricci, G. Maga, F. Corelli, M. Botta, *J. Comb. Chem.* **2005**, *7*, 117–122.

[237] Y.-J. Cherng, *Tetrahedron* **2002**, *58*, 887–890.

[238] Y.-J. Cherng, *Tetrahedron* **2002**, *58*, 1125–1129.

[239] Y.-J. Cherng, *Tetrahedron* **2002**, *58*, 4931–4935.

[240] G. Luo, L. Chen, G. S. Pointdexter, *Tetrahedron Lett.* **2002**, *43*, 5739–5742.

[241] A. G. Takvorian, A. P. Combs, *J. Comb. Chem.* **2004**, *6*, 171–174.

[242] R. Arienzo, D. E. Clark, S. Cramp, S. Daly, H. J. Dyke, P. Lockey, D. Norman, A. G. Roach, K. Stuttle, M. Tomlinson, M. Wong, S. P. Wren, *Bioorg. Med. Chem. Lett.* **2004**, *14*, 4099–4102.

[243] C. R. Hopkins, K. Neuenschwander, A. Scotese, S. Jackson, T. Nieduzak, H. Pauls, G. Liang, K. Sides, D. Cramer, J. Cairns, S. Maignand, M. Mathieu, *Bioorg. Med. Chem. Lett.* **2004**, *14*, 4819–4823.

[244] A. J. Souers, D. Wodka, J. Gao, J. C. Lewis, A. Vasudevan, R. Gentles, S. Brodjian, B. Dayton, C. A. Ogiela, D. Fry, L. E. Hernandez, K. C. Marsh, C. A. Collins, P. R. Kym, *Bioorg. Med. Chem. Lett.* **2004**, *14*, 4873–4877.

[245] A. J. Souers, D. Wodka, J. Gao, J. C. Lewis, A. Vasudevan, S. Brodjian, B. Dayton, C. A. Ogiela, D. Fry, L. E. Hernandez, K. C. Marsh, C. A. Collins, P. R. Kym, *Bioorg. Med. Chem. Lett.* **2004**, *14*, 4883–4886.

[246] G. Priem, M. S. Anson, S. J. F. Macdonald, B. Pelotier, I. B. Campbell, *Tetrahedron Lett.* **2002**, *43*, 6001–6003.

[247] S. Narayan, T. Seelhammer, R. E. Gawley, *Tetrahedron Lett.* **2004**, *45*, 757–759.

[248] Y. Boursereau, I. Coldham, *Bioorg. Med. Chem. Lett.* **2004**, *14*, 5841–5844.

[249] M. Loung, A. Loupy, S. Marque, A. Petit, *Heterocycles* **2004**, *63*, 297–308.

[250] P. G. Steel, C. W. T. Teasdale, *Tetrahedron Lett.* **2004**, *45*, 8977–8980.

[251] A. A. Nikitenko, Y. E. Raifeld, T. Z. Wang, *Bioorg. Med. Chem. Lett.* **2001**, *11*, 1041–1044.

[252] Y.-J. Cherng, *Tetrahedron* **2000**, *56*, 8287–8289.

[253] G. R. Brown, A. J. Foubister, C. A. Roberts, S. L. Wells, R. Wood, *Tetrahedron Lett.* **2001**, *42*, 3917–3919.

[254] M. Mečiarova, J. Podlesná, S. Toma, *Monatsh. Chem.* **2004**, *135*, 419–423.

[255] I. Niculescu-Duvaz, I. Scanlon, D. Niculescu-Duvaz, F. Friedlos, J. Martin, R. Marais, C. J. Springer, *J. Med. Chem.* **2004**, *47*, 2651–2658.

[256] J. I. Levin, M. T. Du, *Synth. Commun.* **2002**, *32*, 1401–1406.

[257] L. Shi, M. Wang, C.-An. Fan, F.-M. Zhang, Y.-Q. Tu, *Org. Lett.* **2003**, *5*, 3515–3517.

[258] G. Xu, Y.-G. Wang, *Org. Lett.* **2004**, *6*, 985–987.

[259] N. A. Swain, R. C. D. Brown, G. Bruton, *Chem. Commun.* **2002**, 2042–2043.

[260] N. A. Swain, R. C. D. Brown, G. Bruton, *J. Org. Chem.* **2004**, *69*, 122–129.

[261] F. Lake, C. Moberg, *Eur. J. Org. Chem.* **2002**, 3179–3188.

[262] V. Alikhani, D. Beer, D. Bentley, I. Bruce, B. M. Cuenoud, R. A. Fairhurst, P. Gedeck,

S. Haberthuer, C. Hayden, D. Janus, L. Jordan, C. Lewis, K. Smithies, E. Wissler, *Bioorg. Med. Chem. Lett.* **2004**, *14*, 4705–4710.

[263] K. B. Lindsay, S. G. Pyne, *Synlett* **2004**, 779–782.

[264] K. B. Lindsay, S. G. Pyne, *Tetrahedron Lett.* **2004**, *60*, 4173–4176.

[265] J. Fawcett, G. A. Griffith, J. M. Percy, E. Uneyama, *Org. Lett.* **2004**, *6*, 1277–1280.

[266] G. F. Busscher, S. Groothuys, R. de Gelder, F. R. J. T. Rutjes, F. L. van Delft, *J. Org. Chem.* **2004**, *69*, 4477–4481.

[267] V. Bailliez, A. Olesker, J. Cleophax, *Tetrahedron* **2004**, *60*, 1079–1085.

[268] L. D. Jennings, K. W. Foreman, T. S. Rush, D. H. H. Tsao, L. Mosyak, S. L. Kincaid, M. N. Sukhdeo, A. G. Sutherland, W. Ding, C. H. Kenny, C. L. Sabus, H. Liu, E. G. Dushin, S. L. Moghazeh, P. Labthavikul, P. J. Petersen, M. Tuckmand, A. V. Ruzin, *Bioorg. Med. Chem.* **2004**, *12*, 5115–5131.

[269] V. A. Chebanov, C. Reidlinger, H. Kanaani, C. Wentrup, C. O. Kappe, G. Kollenz, *Supramol. Chem.* **2004**, *16*, 121–127.

[270] S. Blanchard, I. Rodriguez, C. Tardy, B. Baldeyrou, C. Bailly, P. Colson, C. Houssier, S. Léonce, L. Kraus-Berthier, B. Pfeiffer, P. Renard, A. Pierré, P. Caubére, G. Guillaumet, *J. Med. Chem.* **2004**, *47*, 978–987.

[271] Y. Zhang, C.-J. Li, *Tetrahedron Lett.* **2004**, *45*, 7581–7584.

[272] D. Mimeau, A.-C. Gaumont, *J. Org. Chem.* **2003**, *68*, 7016–7022.

[273] A. Vasudevan, M. K. Verzal, *Synlett* **2004**, 631–634.

[274] D. Mimeau, O. Delacroix, A.-C. Gaumont, *Chem. Commun.* **2003**, 2928–2929.

[275] L. Ackermann, L. T. Kaspar, C. J. Gschrei, *Org. Lett.* **2004**, *6*, 2515–2518.

[276] M. C. Bagley, K. Chapaneri, C. Glover, E. A. Merritt, *Synlett* **2004**, 2615–2617.

[277] L. B. Schenkel, J. A. Ellman, *Org. Lett.* **2004**, *6*, 3621–3624.

[278] S. Collina, G. Loddo, A. Barbieri, L. Linati, S. Alcaro, P. Chimenti, O. Azzolina, *Tetrahedron Asymm.* **2004**, *15*, 3601–3608.

[279] A. K. Bose, S. N. Ganguly, M. S. Manhas, V. Srirajan, A. Bhattacharjee, S. Rumthao, A. H. Sharma, *Tetrahedron Lett.* **2004**, *45*, 1179–1181.

[280] S. A. Shackelford, M. B. Anderson, L. C. Christie, T. Goetzen, M. C. Guzman, M. A. Hananel, W. D. Kornreich, H. Li, V. P. Pathak, A. K. Rabinovich, R. J. Rajapakse, L. K. Truesdale, S. M. Tsank, H. N. Vazir, *J. Org. Chem.* **2003**, *68*, 267–275.

[281] J. Xie, A. B. Comeau, C. T. Seto, *Org. Lett.* **2004**, *6*, 83–86.

[282] P. A. Plé, T. P. Green, L. F. Hennequin, J. Curwen, M. Fennell, J. Allen, C. Lambert- van der Brempt, G. Costello, *J. Med. Chem.* **2004**, *47*, 871–887.

[283] L. Paolini, E. Petricci, F. Corelli, M. Botta, *Synthesis* **2003**, 1039–1042.

[284] X. Peng, A. Zhang, N. S. Kula, R. J. Baldessarini, J. L. Neumeyer, *Bioorg. Med. Chem. Lett.* **2004**, *14*, 5635–5639.

[285] T. Inagaki, T. Fukuhara, S. Hara, *Synthesis* **2003**, 1157–1159.

[286] S. Kobayashi, A. Yoneda, T. Fukuhara, S. Hara, *Tetrahedron Lett.* **2004**, *45*, 1287–1289.

[287] C. H. M. Amijs, G. P. M. van Klink, G. van Koten, *Green Chem.* **2003**, 470–474.

[288] R. K. Arvela, N. E. Leadbeater, *Synlett* **2003**, 1145–1148.

[289] A. R. Katritzky, S. Majumder, R. Jain, *ARKIVOC* **2003**, (xiii), 74–79.

[290] A. Bengtson, A. Hallberg, M. Larhed, *Org. Lett.* **2002**, *4*, 1231–1233.

[291] H. Kiyota, D. J. Dixon, C. K. Luscombe, S. Hettstedt, S. V. Ley, *Org. Lett.* **2002**, *4*, 3223–3226.

[292] M. C. Bagley, J. W. Dale, R. L. Jenkins, J. Bower, *Chem. Commun.* **2004**, 102–103.

[293] A. Pouilhés, Y. Langlois, A. Chiaroni, *Synlett* **2003**, 1488–1490.

[294] P. S. Baran, J. M. Richter, *J. Am. Chem. Soc.* **2004**, *126*, 7450–7451.

[295] S. J. Coats, M. J. Schulz, J. R. Carson, E. E. Codd, D. J. Hlasta, P. M. Pitis, D. J. Stone, S.-P. Zhang, R. W. Colburn, S. L. Dax, *Biorg. Med. Chem. Lett.* **2004**, *14*, 5493–5498.

[296] V. Santagada, F. Frecentese, F. Fiorino, D. Cirillo, E. Perissutti, B. Severino, S. Terracciano, G. Caliendo, *QSAR Comb. Sci.* **2004**, *23*, 903–905.

[297] T. Gustafsson, M. Schou, F. Almqvist, J. Kihlberg, *J. Org. Chem.* **2004**, *69*, 8694–8701.

[298] J.-M. Cloarec, A. B. Charette, *Org. Lett.* **2004**, *6*, 4731–4734.

[299] X. Liao, G. S. V. Raghavan, V. A. Yaylayan, *Tetrahedron Lett.* **2002**, *43*, 45–48.

[300] C. Villa, E. Mariani, A. Loupy, C. Grippo, G. C. Grossi, A. Bargagna, *Green Chem.* **2003**, *5*, 623–626.

[301] Q. Su, A. B. Beeler, E. Lobkovsky, J. A. Porco, J. S. Panek, *Org. Lett.* **2003**, *5*, 2149–2152.

[302] E. Veverková, M. Mečiarová, Š. Toma, J. Balko, *Monatsh. Chem.* **2003**, *134*, 1215–1219.

[303] R. Nicewonger, A. Fowke, K. Nguyen, L. Ditto, J. Caserta, M. Harris, C. M. Baldino, *Mol. Diversity* **2003**, *7*, 247–252.

[304] B. Miriyala, J. S. Williamson, *Tetrahedron Lett.* **2003**, *44*, 7957–7959.

[305] J. J. Chen, S. V. Deshpande, *Tetrahedron Lett.* **2003**, *44*, 8873–8876.

[306] S. Caddick, J. D. Wilden, H. D. Bush, D. B. Judd, *QSAR Comb. Sci.* **2004**, *23*, 895–898.

[307] D. P. Curran, Q. Zhang, *Adv. Synth. Catal.* **2003**, *345*, 329–332.

[308] G. Bélanger, F. Lévesque, J. Pâquet, G. Barbe, *J. Org. Chem.* **2005**, *70*, 291–296.

[309] K. Olofsson, S.-Y. Kim, M. Larhed, D. P. Curran, A. Hallberg, *J. Org. Chem.* **1999**, *64*, 4539–4541.

[310] C. Wetter, A. Studer, *Chem. Commun.* **2004**, 174–175.

[311] A. Teichert, K. Jantos, K. Harms, A. Studer, *Org. Lett.* **2004**, *6*, 3477–3480.

[312] C. Ericsson, L. Engman, *J. Org. Chem.* **2004**, *69*, 5143–5146.

[313] C. Goretzki, A. Krlej, C. Steffens, H. Ritter, *Macromol. Rapid Commun.* **2004**, *25*, 513–516.

[314] H. Zhang, U. S. Schubert, *Macromol. Rapid Commun.* **2004**, *25*, 1225–1230.

[315] F. Wiesbrock, R. Hoogenboom, U. S. Schubert, *Macromol. Rapid Commun.* **2004**, *25*, 1739–1764.

[316] F. A. Jaipuri, M. F. Jofre, K. A. Schwarz, N. L. Pohl, *Tetrahedron Lett.* **2004**, *45*, 4149–4152.

[317] E. Wellner, H. Sandin, L. Pääkkönen, *Synthesis* **2002**, 223–226.

[318] J. Wettergren, A. B. E. Minidis, *Tetrahedron Lett.* **2003**, *44*, 7611–7612.

[319] M. Belema, V. N. Nguyen, F. C. Zusi, *Tetrahedron Lett.* **2004**, *45*, 1693–1697.

[320] M. Karramkam, F. Hinnen, M. Berrehouma, C. Hlavacek, F. Vaufrey, C. Halladin, J. A. McCarron, V. W. Pike, F. Dollé, *Bioorg. Med. Chem.* **2003**, *11*, 2769–2782.

[321] M. Karramkam, F. Hinnen, F. Vaufrey, F. Dollé, *J. Label. Compd. Radiopharm.* **2003**, *46*, 979–992.

[322] F. Wüst, T. Kniess, *J. Label. Compd. Radiopharm.* **2003**, *46*, 699–713.

[323] S.-Y. Lu, F. T. Chin, J. A. McCarron, V. W. Pike, *J. Label. Compd. Radiopharm.* **2004**, *47*, 289–297.

[324] T. Persson, S. K. Johansen, L. Martiny, M. Giskov, J. Nielsen, *J. Label. Compd. Radiopharm.* **2004**, *47*, 627–634.

[325] S. H. Park, H. J. Gwon, K. B. Park, *Chem. Lett.* **2004**, *33*, 1278–1279.

[326] S. H. Park, H. J. Gwon, J. S. Park, K. B. Park, *QSAR Comb. Sci.* **2004**, *23*, 868–874.

[327] M. Yamamoto, K. Oshima, S. Matsubara, *Chem. Lett.* **2004**, *33*, 846–847.

[328] S. A. de Keczer, T. S. Lane, M. R. Masjedizadeh, *J. Label. Compd. Radiopharm.* **2004**, *47*, 733–740.

[329] V. Derdau, *Tetrahedron Lett.* **2004**, *45*, 8889–8893.

[330] S. Mayer, D. M. Daigle, E. D. Brown, J. Khatri, M. G. Organ, *J. Comb. Chem.* **2004**, *6*, 776–782.

[331] J. Wu, H. Wu, S. Wei, W.-M. Dai, *Tetrahedron Lett.* **2004**, *45*, 4401–4404.

[332] O. Correc, K. Guillou, J. Hamelin, L. Paquin, F. Texier-Boullet, L. Toupet, *Tetrahedron Lett.* **2004**, *45*, 391–395.

[333] J. Hamelin, A. Soudi, H. Benhaoua, *Synthesis* **2003**, 2185–2188.

[334] F. Borrelli, C. Campagnuolo, R. Capasso, E. Fattorusso, O. Taglialatela-Scafati, *Eur. J. Org. Chem.* **2004**, 3227–3232.

[335] Y.-S. Hon, T.-R. Hsu, C.-Y. Chen, Y.-H. Lin, F.-J. Chang, C.-H. Hsieh, P.-H. Szu, *Tetrahedron* **2003**, *59*, 1509–1520.

[336] I. Barrios, P. Camps, M. Comes-Franchini, D. Muñoz-Torrero, A. Ricci, L. Sánchez, *Tetrahedron Lett.* **2003**, *59*, 1971–1979.

[337] M. E. McDonnell, S.-P. Zhang, N. Nasser, A. E. Dubin, S. L. Dax, *Bioorg. Med. Chem. Lett.* **2004**, *14*, 531–534.

[338] E. Rábarova, P. Koiš, M. Lácová, A. Krutošíková, *ARKIVOC* **2004** (i), 110–112.

[339] G. Torres, W. Torres, J. A. Prieto, *Tetrahedron* **2004**, *60*, 10245–10251.

[340] M. R. Linder, J. Podlech, *Org. Lett.* **2001**, *3*, 1849–1851.

[341] A. Mortoni, M. Martinelli, U. Piarulli, N. Regalia, S. Gagliardi, *Tetrahedron Lett.* **2004**, *45*, 6623–6627.

[342] G. Minetto, L. F. Raveglia, M. Taddei, *Org. Lett.* **2004**, *6*, 389–392.

[343] A. R. Bharadwaj, K. A. Scheidt, *Org. Lett.* **2004**, *6*, 2465–2468.

[344] D. Tejedor, D. González-Cruz, F. García-Tellado, J. J. Marrero-Tellado, M. López Rodríguez, *J. Am. Chem. Soc.* **2004**, *126*, 8390–8391.

[345] H. Tye, M. Whittaker, *Org. Biomol. Chem.* **2004**, *2*, 813–815.

[346] A. Díaz-Ortiz, A. de la Hoz, M. Antonia Herrero, P. Prieto, A. Sánchez-Migallón, F. P. Cossio, A. Arrieta, S. Vivanco, C. Foces-Foces, *Mol. Diversity* **2003**, *7*, 175–180.

[347] G. Bashiardes, I. Safir, A. S. Mohamed, F. Barbot, J. Laduranty, *Org. Lett.* **2003**, *5*, 4915–4918.

[348] J. Pospíšil, M. Potáček, *Eur. J. Org. Chem.* **2004**, 710–716.

[349] N. S. Wilson, C. R. Sarko, G. P. Roth, *Tetrahedron Lett.* **2001**, *42*, 8939–8941.

[350] T. Lipińska, *Tetrahedron Lett.* **2004**, *45*, 8831–8834.

[351] J. Siu, I. R. Baxendale, S. V. Ley, *Org. Biomol. Chem.* **2004**, *2*, 160–167.

[352] S. Liu, M. A. Haller, H. Ma, L. R. Dalton, S.-H. Jang, A. K.-Y. Jen, *Adv. Mater.* **2003**, *15*, 603–607.

[353] C. J. H. Morton, R. L. Riggs, D. M. Smith, N. J. Westwood, P. Lightfoot, A. M. Z. Slawin, *Tetrahedron* **2005**, *61*, 727–738.

[354] J. R. Boot, G. Brace, C. L. Delatour, N. Dezutter, J. Fairhurst, J. Findlay, P. T. Gallagher, I. Hoes, S. Mahadevan, S. N. Mitchell, R. E. Rathmell, S. J. Richards, R. G. Simmonds, L. Wallace, M. A. Whatton, *Bioorg. Med. Chem. Lett.* **2004**, *14*, 5395–5399.

[355] D. Allen, O. Callaghan, F. L. Cordier, D. R. Dobson, J. R. Harris, T. M. Hotten, W. M. Owton, R. E. Rathmell, V. A. Wood, *Tetrahedron Lett.* **2004**, *45*, 9645–9647.

[356] G. Giacomelli, A. Porcheddu, M. Salaris, M. Taddei, *Eur. J. Org. Chem.* **2003**, 537–541.

[357] V. Molteni, M. M. Hamilton, L. Mao, C. M. Crane, A. P. Termin, D. M. Wilson, *Synthesis* **2002**, 1669–1674.

[358] C. D. Cox, M. J. Breslin, B. J. Mariano, *Tetrahedron Lett.* **2004**, *45*, 1489–1493.

[359] J. L. Delgado, P. de la Cruz, V. López-Arza, F. Langa, D. B. Kimball, M. M. Haley, Y. Araki, O. Ito, *J. Org. Chem.* **2004**, *69*, 2661–2668.

[360] J. L. Delgado, P. de la Cruz, F. Langa, A. Urbina, J. Casado, J. T. López Navarrete, *Chem. Commun.* **2004**, 1734–1735.

[361] S. E. Wolkenberg, D. D. Wisnoski, W. H. Leister, Y. Wang, Z. Zhao, C. W. Lindsley, *Org. Lett.* **2004**, *6*, 1453–1456.

[362] R. B. Sparks, A. P. Combs, *Org. Lett.* **2004**, *6*, 2473–2475.

[363] S. M. Ireland, H. Tye, M. Whittaker, *Tetrahedron Lett.* **2003**, *44*, 4369–4371.

[364] C. M. Coleman, J. M. D. MacElroy, J. F. Gallagher, D. F. O'Shea, *J. Comb. Chem.* **2002**, *4*, 87–93.

[365] J. Pospíšil, M. Potáček, *Heterocycles* **2004**, *63*, 1165–1173.

[366] Y. J. Kim, R. S. Varma, *Tetrahedron Lett.* **2004**, *45*, 7205–7208.

[367] G. H. Merriman, L. Ma, P. Shum, D. McGarry, F. Volz, J. S. Sabol, A. Gross, Z. Zhao, D. Rampe, L. Wang, F. Wirtz-Brugger, B. A. Harris, D. Macdonald, *Bioorg. Med. Chem. Lett.* **2005**, *15*, 435–438.

[368] N. Boufatah, A. Gellis, J. Maldonado, P. Vanelle, *Tetrahedron* **2004**, *60*, 9131–9137.

[369] S. Ferro, A. Rao, M. Zappalà, A. Chimirri, M. L. Barreca, M. Witvrouw, Z. Debyser, P. Monforte, *Heterocycles* **2004**, *63*, 2727–2734.

[370] A. Rao, A. Chimirri, S. Ferro, A. M. Monforte, P. Monforte, M. Zappalá, *ARKIVOC* **2004** (v), 147–155.

[371] B. Dymock, X. Barril, M. Beswick, A. Collier, N. Davies, M. Drysdale, A. Fink, C. Fromont, R. E. Hubbard, A. Massey, A. Surgenor, L. Wright, *Bioorg. Med. Chem. Lett.* **2004**, *14*, 325–328.

[372] J. Ilaš, S. Pečar, J. Hockemeyer, H. Euler, A. Kirfel, C. E. Müller, *J. Med. Chem.* **2005**, *48*, 2108–2114.

[373] G. Giacomelli, L. De Luca, A. Porcheddu, *Tetrahedron* **2003**, *59*, 5437–5440.

[374] L. Bernardi, B. F. Bonini, M. Comes-Franchini, M. Fochi, M. Folegatti, S. Grilli, A. Mazzanti, A. Ricci, *Tetrahedron: Asymm.* **2004**, *15*, 245–250.

[375] B. Desai, T. N. Danks, G. Wagner, *Dalton Trans.* **2003**, 2544–2549.

[376] B. Desai, T. N. Danks, G. Wagner, *Dalton Trans.* **2004**, 166–171.

[377] B. M. Trost, K. Dogra, M. Franzini, *J. Am. Chem. Soc.* **2004**, *126*, 1944–1945.

[378] M. C. Pirrung, L. N. Tumey, A. L. McClerren, C. R. H. Raetz, *J. Am. Chem. Soc.* **2003**, *125*, 1575–1586.

[379] S. Crosignani, A. C. Young, B. Linclau, *Tetrahedron Lett.* **2004**, *45*, 9611–9615.

[380] A. R. Katritzky, C. Cai, K. Suzuki, S. K. Singh, *J. Org. Chem.* **2004**, *69*, 811–814.

[381] F. García-Tellado, A. Loupy, A. Petit, A. L. Marrero-Terrero, *Eur. J. Org. Chem.* **2003**, 4387–4391.

[382] N. Kuhnert, T. N. Danks, *Green Chem.* **2001**, *3*, 68–70.

[383] F. Diwischek, E. Heller, U. Holzgrabe, *Monatsh. Chem.* **2003**, *134*, 1105–1111.

[384] R. S. Pottorf, N. K. Chadha, M. Katkevics, V. Ozola, E. Suna, H. Ghane, T. Regberg, M. R. Player, *Tetrahedron Lett.* **2003**, *44*, 175–178.

[385] E. B. Frolov, F. J. Lakner, A. V. Khvat, A. V. Ivachtchenko, *Tetrahedron Lett.* **2004**, *45*, 4693–4696.

[386] A. Bolognese, G. Correale, M. Manfra, A. Lavecchia, E. Novellino, V. Barone, *Org. Biomol. Chem.* **2004**, *2*, 2809–2813.

[387] V. Gududuru, V. Nguyen, J. T. Dalton, D. D. Miller, *Synlett* **2004**, 2357–2358.

[388] H. Emtenäs, C. Taflin, F. Almqvist, *Mol. Diversity* **2003**, *7*, 165–169.

[389] N. Pemberton, V. Åberg, H. Almstedt, A. Westermark, F. Almqvist, *J. Org. Chem.* **2004**, *69*, 7830–7835.

[390] S. Sanchez, J. H. Bateson, P. J. O'Hanlon, T. Gallagher, *Org. Lett.* **2004**, *6*, 2781–2783.

[391] S. Frére, V. Thiéry, C. Bailly, T. Besson, *Tetrahedron* **2003**, *59*, 773–779.

[392] S. Frére, V. Thiéry, T. Besson, *Synth. Commun.* **2003**, *33*, 3789–3798.

[393] A. R. Katritzky, S. K. Singh, *J. Org. Chem.* **2002**, *67*, 9077–9079.

[394] A. R. Katritzky, Y. Zhang, S. K. Singh, P. J. Steel, *ARKIVOC* **2003**, *(xv)*, 47–64.

[395] K. A. Savin, M. Robertson, D. Gernert, S. Green, E. J. Hembre, J. Bishop, *Mol. Diversity* **2003**, *7*, 171–174.

[396] B. Khanetskyy, D. Dallinger, C. O. Kappe, *J. Comb. Chem.* **2004**, *7*, 884–892.

[397] P. Appukkuttan, W. Dehan, V. V. Fokin, E. Van der Eycken, *Org. Lett.* **2004**, *7*, 4223–4225.

[398] D. Ermolat'ev, W. Dehaen, E. Van der Eycken, *QSAR Comb. Sci.* **2004**, *24*, 915–918.

[399] M. D. Evans, J. Ring, A. Schoen, A. Bell, P. Edwards, D. Berthelot, R. Nicewonger, C. M. Baldino, *Tetrahedron Lett.* **2003**, *44*, 9337–9341.

[400] V. Santagada, F. Frecentese, E. Perissutti, D. Cirillo, S. Terracciano, G. Caliendo, *Bioorg. Med. Chem. Lett.* **2004**, *14*, 4491–4493.

[401] R. Natero, D. O. Koltun, J. A. Zablocki, *Synth. Commun.* **2004**, *34*, 2523–2529.

[402] S.-L. Deng, R.-Y. Chen, *Monatsh. Chem.* **2004**, *135*, 1113–1119.

[403] I. V. Bliznets, A. A. Vasil'ev, S. V. Shorshnev, A. E. Stepanov, S. M. Lukyanov, *Tetrahedron Lett.* **2004**, *45*, 2571–2573.

[404] M. J. Schulz, S. J. Coats, D. J. Hlasta, *Org. Lett.* **2004**, *6*, 3265–3268.

[405] F. Ek, S. Manner, L.-G. Wistrand, T. Frejd, *J. Org. Chem.* **2004**, *69*, 1346–1352.

[406] M. C. Bagley, R. Lunn, X. Xiong, *Tetrahedron Lett.* **2002**, *43*, 8331–8334.

[407] M. C. Bagley, X. Xiong, *Org. Lett.* **2004**, *6*, 3401–3404.

[408] L. Öhberg, J. Westman, *Synlett* **2001**, 1296–1298.

[409] M. C. Bagley, N. Singh, *Synlett* **2002**, 1718–1720.

[410] N. Yu. Gorobets, B. Yousefi, C. O. Kappe, *Tetrahedron* **2004**, *60*, 8633–8644.

[411] M. Panunzio, M. A. Lentini, E. Campana, G. Martelli, E. Tamanini, P. Vicennati, *Synth. Commun.* **2004**, *34*, 345–359.

[412] A. Stadler, S. Pichler, G. Horeis, C. O. Kappe, *Tetrahedron* **2002**, *58*, 3177–3183.

[413] J. H. M. Lange, P. C. Verveer, S. J. M. Osnabrug, G. M. Visser, *Tetrahedron Lett.* **2001**, *42*, 1367–1369.

[414] N. Srinivasan, A. Ganesan, *Chem. Commun.* **2003**, 916–917.

[415] Y.-H. Yen, Y.-H. Chu, *Tetrahedron Lett.* **2004**, *45*, 8137–8140.

[416] P. Campiglia, I. Gomez-Monterrey, T. Lama, E. Novellino, P. Grieco, *Mol. Diversity* **2004**, *8*, 427–430.

[417] F.-M. Kuo, M.-C. Tseng, Y.-H Yen, Y.-H. Chu, *Tetrahedron* **2004**, *60*, 12079–12088.

[418] A. Padwa, H. I. Lee, P. Rashatasakhon, M. Rose, *J. Org. Chem.* **2004**, *69*, 8209–8218.

[419] K. R. Crawford, S. K. Bur, C. S. Straub, A. Padwa, *Org. Lett.* **2003**, *5*, 3337–3340.

[420] M.-E. Theoclitou, L. A. Robinson, *Tetrahedron Lett.* **2002**, *43*, 3907–3910.

[421] A. Perzyna, R. Houssin, D. Barby, J.-P. Hénichart, *Synlett* **2002**, 2077–2079.

[422] A. V. Ivachtchenko, A. V. Khvat, V. V. Kobak, V. M. Kysil, C. T. Williams, *Tetrahedron Lett.* **2004**, *45*, 5473–5476.

[423] N. S. Wilson, C. R. Sarko, G. P. Roth, *Tetrahedron Lett.* **2002**, *43*, 581–583.

[424] B.-C. Hong, J.-L. Wu, A. K. Gupta, M. S. Hallur, J.-H. Liao, *Org. Lett.* **2004**, *6*, 3453–3456.

[425] I. Carranco, J. L. Díaz, O. Jiménez, M. Vendrell, F. Albericio, M. Royo, R. Lavilla, *J. Comb. Chem.* **2005**, *7*, 33–41.

[426] R. A. Hughes, S. P. Thomson, L. Alcaraz, C. J. Moody, *Chem. Commun.* **2004**, 946–948.

[427] C. J. Moody, R. A. Hughes, S. P. Thompson, L. Alcaraz, *Chem. Commun.* **2002**, 1760–1761.

[428] H. Leutbecher, J. Conrad, I. Klaiber, U. Beifuss, *QSAR Comb. Sci.* **2004**, *23*, 899–902.

[429] I. Devi, P. J. Bhuyan, *Tetrahedron Lett.* **2004**, *45*, 8625–8627.

[430] M. D. Mihovilovic, H. G. Leisch, K. Mereiter, *Tetrahedron Lett.* **2004**, *45*, 7087–7090.

[431] M. C. Bagley, D. D. Hughes, H. M. Sabo, P. H. Taylor, X. Xiong, *Synlett* **2003**, 1443–1446.

[432] M. C. Bagley, D. D. Hughes, P. H. Taylor, *Synlett* **2003**, 259–261.

[433] M. C. Bagley, D. D. Hughes, M. C. Lubinu, E. A. Merritt, P. H. Taylor, N. C. O. Tomkinson, *QSAR Comb. Sci.* **2004**, *23*, 859–867.

[434] A. Stadler, C. O. Kappe, *J. Comb. Chem.* **2001**, *3*, 624–630.

[435] C. O. Kappe, A. Stadler, in *Combinatorial Chemistry, Part B* (Eds.: B. B. Bunin, G. Morales), Elsevier Sciences, **2003**, pp 197–223.

[436] H. Sandin, M.-L. Swanstein, E. Wellner, *J. Org. Chem.* **2004**, *69*, 1571–1580.

[437] F.-R. Alexandre, A. Berecibar, R. Wrigglesworth, T. Besson, *Tetrahedron* **2003**, *59*, 1413–1419.

[438] F.-R. Alexandre, A. Berecibar, T. Besson, *Tetrahedron Lett.* **2002**, *43*, 3911–3913.

[439] T. E. Vasquez, T. Nixey, B. Chenera, V. Gore, M. Bartberger, Y. Sun, C. Hulme, *Mol. Diversity* **2003**, *7*, 161–164.

[440] D. S. Yoon, Y. Han, T. M. Stark, J. C. Haber, B. T. Gregg, S. B. Stankovich, *Org. Lett.* **2004**, *6*, 4775–4778.

[441] N. Mont, J. Teixidó, J. I. Borrell, C. O. Kappe, *Tetrahedron Lett.* **2003**, *44*, 5385–5388.

[442] N. Mont, J. Teixidó, J. I. Borrell, C. O. Kappe, *Mol. Diversity* **2003**, *7*, 153–159.

[443] N. Mont, L. Fernández-Megido, J. Teixidó, C. O. Kappe, J. I. Borrell, *QSAR Comb. Sci.* **2004**, *23*, 836–849.

[444] M. Gohain, D. Prajapati, B. J. Gogoi, J. S. Sandhu, *Synlett* **2004**, 1179–1182.

[445] I. Devi, P. J. Bhuyan, *Synlett* **2004**, 283–286.

[446] I. Devi, H. N. Borah, P. J. Bhuyan, *Tetrahedron Lett.* **2004**, *45*, 2405–2408.

[447] J. Westman, R. Lundin, J. Stalberg, M. Ostbye, A. Franzen, A. Hurynowicz, *Comb. Chem. High-Throughput Screen.* **2002**, *5*, 565–570.

[448] J. Westman, K. Orrling, *Comb. Chem. High-Throughput Screen.* **2002**, *5*, 571–574.

[449] A. Díaz-Ortiz, A. de la Hoz, A. Moreno, P. Prieto, R. León, M. A. Herrero, *Synlett* **2002**, 2037–2038.

[450] Z. Zhao, D. D. Wisnoski, S. E. Wolkenberg, W. H. Leister, Y. Wang, C. W. Lindsley, *Tetrahedron Lett.* **2004**, *45*, 4873–4876.

[451] C. W. Lindsley, Z. Zhao, W. H. Leister, R. G. Robinson, S. F. Barnett, D. Defeo-Jones, R. E. Jones, G. D. Hartman, J. R. Huff, H. E. Huber, M. E. Duggan, *Bioorg. Med. Chem. Lett.* **2005**, *15*, 761–764.

[452] F.-X. Le Foulon, E. Braud, F. Fabis, J.-C. Lancelot, S. Rault, *Tetrahedron* **2003**, *59*, 10051–10057.

[453] A. V. Koshkin, O. A. Fedorova, V. Lokshin, R. Guglielmetti, J. Hamelin, F. Texier-Boullet, S. P. Gromov, *Synth. Commun.* **2004**, *34*, 315–322.

[454] G. Caliendo, E. Perissutti, V. Santagada, F. Fiorino, B. Severino, D. Cirillo, R. d'Emmanuele di Villa Bianca, L. Lippolis, A. Pinto, R. Sorrentino, *Eur. J. Med. Chem.* **2004**, *39*, 815–826.

[455] M. Mečiarova, S. Toma, P. Magdolen, *Synth. Commun.* **2003**, *33*, 3049–3054.

[456] H.-K. Lee, T. M. Rana, *J. Comb. Chem.* **2004**, *6*, 504–508.

[457] Y. Peng, G. Song, *Tetrahedron Lett.* **2004**, *45*, 5313–5316.

[458] Z. Zhao, W. H. Leister, K. A. Strauss, D. D. Wisnoski, C. W. Lindsley, *Tetrahedron Lett.* **2003**, *44*, 1123–1128.

[459] C. W. Lindsley, D. D. Wisnoski, Y. Wang, W. H. Leister, Z. Zhao, *Tetrahedron Lett.* **2003**, *44*, 4495–4498.

[460] B. R. Lahue, S.-M. Lo, Z.-K. Wan. G. H. C. Woo, J. K. Snyder, *J. Org. Chem.* **2004**, *69*, 7171–7182.

[461] A. Vasudevan, C. I. Villamil, S. W. Djuric, *Org. Lett.* **2004**, *6*, 3361–3364.

[462] A. Ivachtchenko, A. Khvat, S. E. Tkachenko, Y. B. Sandulenko, V. Y. Vvedensky, *Tetrahedron Lett.* **2004**, *45*, 6733–6736.

[463] J. P. Collman, R. A. Decréau, *Tetrahedron Lett.* **2003**, *44*, 1207–1210.

[464] A. Burczyk, A. Loupy, D. Bogdal, A. Petit, *Tetrahedron* **2005**, *61*, 179–188.

[465] J. M. Mejía-Oneto, A. Padwa, *Org. Lett.* **2004**, *6*, 3241–3244.

[466] N. Kaval, W. Dehaen, P. Mátyus, E. Van der Eycken, *Green Chem.* **2004**, *6*, 125–127.

[467] F. Cochard, M. Laronze, P. Sigaut, J. Sapi, J.-Y. Laronze, *Tetrahedron Lett.* **2004**, *45*, 1703–1707.

[468] R. K. Arvela, N. E. Leadbeater, M. S. Sangi, V. A. Williams, P. Granados, R. D. Singer, *J. Org. Chem.* **2005**, *70*, 161–168.

[469] H. Mohan, E. Gemma, K. Ruda, S. Oscarson, *Synlett* **2003**, 1255–1256.

[470] A. Stadler, H. von Schenck, K. S. A. Vallin, M. Larhed, A. Hallberg, *Adv. Synth. Catal.* **2004**, *346*, 1773–1781.

[471] R. B. Bedford, C. P. Butts, T. E. Hurst, P. Lindström, *Adv. Synth. Catal.* **2004**, *346*, 1627–1630.

[472] C. Nájera, J. Gil-Moltó, S. Karlström, *Adv. Synth. Catal.* **2004**, *346*, 1798–1811.

[473] K. Ovotrup, A. S. Andersson, J.-P. Mayer, A. S. Jepsen, M. B. Nielsen, *Synlett* **2004**, 2818–2820.

[474] H. M. L. Davies, R. E. J. Beckwith, *J. Org. Chem.* **2004**, *69*, 9241–9247.

[475] G. W. Bluck, N. B. Carter, S. C. Smith, M. D. Turnbull, *J. Fluorine Chem.* **2004**, *125*, 1873–1877.

[476] S. Marque, H. Snoussi, A. Loupy, N. Plé, A. Turck, *J. Fluorine Chem.* **2004**, *125*, 1847–1851.

[477] P. Conti, M. De Amici, G. Grazioso, G. Roda, F. F. Barberis Negra, B. Nielsen, T. B. Stensbol, U. Madsen, H. Bräuner-Osborne, K. Frydenvang, G. De Sarro, L. Toma, C. De Micheli, *J. Med. Chem.* **2004**, *47*, 6740–6748.

[478] V. L. M. Silva, A. M. S. Silva, D. C. G. A. Pinto, J. A. S. Cavaleiro, T. Patonay, *Synlett* **2004**, 2717–2720.

[479] A. N. French, J. Cole, T. Wirth, *Synlett* **2004**, 2291–2294.

7
Literature Survey Part B: Combinatorial Chemistry and High-Throughput Organic Synthesis

7.1
Solid-Phase Organic Synthesis

7.1.1
Combinatorial Chemistry and Solid-Phase Organic Synthesis

Combinatorial chemistry, the art and science of rapidly synthesizing and testing potential lead compounds for any desired property, has turned out to be one of the most promising approaches in drug discovery [1]. The fundamental meaning of combinatorial synthesis is the ability to generate large numbers of chemical compounds in a short time. Therefore, it is utilized to generate libraries of potential lead compounds, which can immediately be screened for biological efficiency. Whereas chemistry in the past has been characterized by slow, steady, and painstaking work, combinatorial chemistry has changed the characteristics of chemical research and permitted a level of productivity thought impossible a few years ago.

One of the key technologies used in combinatorial chemistry is solid-phase organic synthesis (SPOS) [2], originally developed by Merrifield in 1963 for the synthesis of peptides [3]. In SPOS, a molecule (scaffold) is attached to a solid support, for example a "polymer resin" (Fig. 7.1). In general, resins are insoluble base polymers with a "linker" molecule attached. Often, spacers are included to reduce steric hindrance by the bulk of the resin. Linkers, on the other hand, are functional moieties, which allow the attachment and cleavage of scaffolds under controlled conditions. Subsequent chemistry is then carried out on the molecule attached to the support until, at the end of the often multistep synthesis, the desired molecule is released from the support.

Solid-phase organic synthesis shows several advantages compared with classical protocols in solution. In order to accelerate reactions and to drive them to completion, a large excess of reagents can be used, as these can easily be removed by filtration and subsequent washing of the solid support. Different loading capacities of the polymeric supports allow the application of the high-dilution principle (hyperentropic effect) to achieve full conversion. If necessary, several reaction cycles can be performed in order to convert all of the starting material. Thus, even the final purification of the desired products is simplified, as by-products formed in solution do

Microwaves in Organic and Medicinal Chemistry. C. Oliver Kappe, Alexander Stadler
Copyright © 2005 WILEY-VCH Verlag GmbH & Co. KGaA, Weinheim
ISBN: 3-527-31210-2

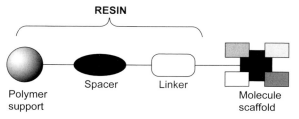

Fig. 7.1 The concept of solid-phase organic synthesis.

not influence the generation of the target molecule. Nowadays, a large number of linkers and various classical or microwave-assisted cleavage conditions are available, allowing efficient processing of various chemical transformations.

In addition, SPOS can easily be automated using appropriate robotics for both filtration and evaporation of the volatiles from the reaction mixture to obtain the cleaved product. Furthermore, SPOS can be applied to the powerful "split-and-mix" strategy, which has proved to be an important tool in combinatorial chemistry [7].

The above mentioned strategy for generating peptide molecules on polystyrene beads was developed by Merrifield, the pioneer of solid-phase synthesis, in 1963 [3]. Since the early 1970s, this technique has been successfully applied to non-peptide syntheses as well. During the 1980s, various protocols for performing solid-phase chemistry were introduced, including reactions on cellulose paper [4], on functionalized polypropylene pins [5], and the so-called "tea-bag" strategy, in which polypropylene mesh containers encapsulating polystyrene resin were utilized [6]. In the late 1980s, Furka and coworkers introduced the "split synthesis" strategy [7], which relies on the "one bead – one compound" concept, promising the delivery of millions of compounds produced simultaneously on beads in a short period of time. From the 1990s onwards, an ever growing number of publications covering numerous aspects of combinatorial synthesis have appeared, including reactions on soluble polymers or dendrimers carried out in homogeneous solution or on well-defined fluorous supports. Nowadays, a lot of common solution-phase reactions can be performed equally well by solid-phase methods, and a great variety of reagents, catalysts or scavengers have been attached to polymer supports.

With this background, a large number of organic reactions can be performed in a combinatorial manner, enabling the synthesis of various desired target molecules with increased efficiency and productivity.

7.1.2
Microwave Chemistry and Solid-Phase Organic Synthesis

Solid-phase organic synthesis (SPOS) exhibits several shortcomings, due to the nature of the heterogeneous reaction conditions. Nonlinear kinetic behavior, slow reactions, solvation problems, and degradation of the polymer support due to the long reaction times are some of the problems typically experienced in SPOS [2]. Any technique which is able to address these issues and to speed up the process of solid-phase synthesis is of considerable interest, particularly for research laboratories involved in high-throughput synthesis.

Not surprisingly, the benefits of microwave-assisted organic synthesis (MAOS) have also attracted interest from the combinatorial/medicinal chemistry community, where reaction speed is of great importance [8–12], and there is an ever growing number of publications reporting on rate enhancements in solid-phase organic synthesis utilizing microwaves. This attractive linkage between combinatorial processing and microwave heating is a logical consequence of the increased speed and effectiveness offered by the microwave approach.

7.1.2.1 Solvents for Microwave-Assisted Solid-Phase Organic Synthesis

The choice of solvent in microwave-assisted SPOS is absolutely critical. Ideally, the solvent should have: (1) good swelling properties for the resin involved, (2) a high boiling point if reactions are to be carried out at atmospheric pressure, (3) a high loss tangent (tan δ) for good interactions with microwaves, as the resins themselves are usually poor microwave absorbers, (4) high chemical stability and inertness to minimize side reactions. Clearly, solvents such as dichloromethane (tan $\delta = 0.047$), which are commonly used in SPOS under conventional conditions, may not be very useful in a microwave-assisted protocol. In these cases, a co-solvent with a high loss tangent (see Section 2.3) or an entirely different solvent system needs to be used. In general, solvents with a loss tangent > 0.1 are considered suitable for microwave dielectric heating. A summary of solvents useful for microwave-assisted solid-phase synthesis is given in Table 7.1.

For example, many microwave-assisted solid-phase coupling reactions utilize 1-methyl-2-pyrrolidone (NMP) or 1,2-dichlorobenzene (DCB) as the reaction solvent. The main reason for this is the high thermal stabilities of these solvents and their

Table 7.1 Swelling behavior, loss tangents (tan δ), and boiling points for solvents used in microwave-assisted solid-phase synthesis.

Solvent	PS swelling [a]	TentaGel swelling [a]	tan δ [b]	bp (°C)
N,*N*-dimethylformamide	5.2	4.4	0.161	153
N,*N*-dimethylacetamide	5.8	4.0	n.a.	166
1-methyl-2-pyrrolidone	6.4	4.4	0.275	202
Dimethyl sulfoxide	4.2	3.8	0.825	189
Tetrahydrofuran	6.0	4.0	0.047	65
1,4-dioxane	5.6	4.2	n.a.	106
1,2-dichloroethane	4.4	5.4	0.127	83
Chlorobenzene	n.a.	n.a.	0.101	132
1,2-dichlorobenzene	4.8	5.2	0.280	180
Nitrobenzene	4.3	4.8	0.589	211
Methanol	1.6	3.6	0.659	65
Water	1.6	3.6	0.123	100

a) Data from [2], PS = polystyrene.
b) Data from [13].

relatively high boiling points, which make it unnecessary to carry out reactions in specialized sealed vessels at elevated pressures. Furthermore, polystyrene resins typically show excellent swelling characteristics in both NMP and DCB (Table 7.1). Due to the polar nature of these solvents, a sufficiently strong absorption of microwave energy occurs. Indeed, extremely rapid heating profiles can be obtained with NMP or DCB inside a microwave cavity. In contrast to a solution-phase reaction, a high boiling point of a solvent is not considered a problem in the work-up/purification stage, since the desired target compound remains on the resin until the cleavage step, for which a lower boiling solvent can be used. Since most dedicated microwave reactors nowadays offer the convenience of performing reactions under sealed-vessel "autoclave-like" conditions (see Chapter 3), suitable lower boiling solvents such as tetrahydrofuran or 1,2-dichloroethane can also be used at high reaction temperatures.

7.1.2.2 Thermal and Mechanical Stability of Polymer Supports

As far as polymer supports for microwave-assisted SPOS are concerned, the use of cross-linked macroporous or microporous polystyrene (PS) resins has been most prevalent. In contrast to common belief, which states that the use of polystyrene resins limits reaction conditions to temperatures below 130 °C [14], it has been shown that these resins can withstand microwave irradiation for short periods of time, such as 20–30 min, even at 200 °C in solvents such as 1-methyl-2-pyrrolidone or 1,2-dichlorobenzene [15]. Standard polystyrene Merrifield resin shows thermal stability up to 220 °C without any degradation of the macromolecular structure of the polymer backbone, which allows reactions to be performed even at significantly elevated temperatures.

Macroporous resin beads, due to their mode of preparation, consist of a macroporous internal structure and highly cross-linked areas (>5%). The latter impart the resin with rigidity, whereas the porous areas provide a large internal surface for functionalization, even in the dry state. These macroporous polystyrene-based resins are subsequently modified in various manners, which render them compatible with numerous organic solvents. Furthermore, they show high resistance toward osmotic shock, but can be brittle when not manipulated carefully.

Microporous resins, like the well-known Merrifield resin, are only slightly cross-linked (1–2%) and show a more homogeneous network, as manifested in a glassy and transparent surface. In general, they are mechanically weak and are prone to damage. Furthermore, osmotic shock can occur when pre-swollen resin beads are introduced into a poor swelling solvent. The beads shrink rapidly under expulsion of the good swelling solvent and are subjected to stress, which leads to mechanical damage or considerable structural modifications. In addition, vigorous stirring over an extended period can cause breakdown of the polymeric surface. However, these microporous resins tolerate a broad range of reaction conditions and show chemical stability in the presence of a wide range of reagents, such as strong bases and acids and even weak oxidants, although strong oxidants and electrophilic reagents should in general be avoided.

TentaGel resins and cellulose membranes have also been used in microwave-assisted synthesis (see below), although some degradation has been observed during irradiation. It should be mentioned that most of the typically used polymer supports in SPOS do not possess a large number of polar functionalities and are therefore largely transparent to microwave energy. In general, due to the rather short reaction times of microwave solid-phase reactions, even magnetic stirring can be performed with these resins without mechanical degradation effects. In contrast, solid-phase reactions utilizing conventional thermal heating typically require longer reaction times, and therefore the mechanical degradation of the solid support caused by prolonged stirring is significant.

However, other polymer composite materials also popular in solid-phase synthesis, such as polyethylene or polypropylene "tea bags", lanterns, crowns, or plugs, are generally less suitable for high-temperature reactions (>160 °C). Therefore, microwave irradiation is typically not a very suitable tool to speed up reactions that utilize these materials as either a solid support or as containment for the solid support.

7.1.2.3 Special Equipment for Solid-Phase Synthesis

Although the examples of microwave-assisted solid-phase reactions presented in this chapter demonstrate that rapid synthetic transformations can in many cases be achieved using microwave irradiation, the possibility of high-speed synthesis does not necessarily mean that these processes can also be adapted to a truly high-throughput format. In the past few years, all commercial suppliers of microwave instrumentation for organic synthesis have moved towards combinatorial/high-throughput platforms for conventional solution-phase synthesis. Detailed descriptions of available systems for both single-mode and multimode reactors can be found in Chapter 3.

A recent development in this context is the Liberty™ system introduced by CEM in 2004 (see Fig. 3.25). This instrument is an automated microwave peptide synthesizer, equipped with special vessels, applicable for the unattended synthesis of up to 12 peptides employing 25 different amino acids. This tool offers the first commercially available dedicated reaction vessels for carrying out microwave-assisted solid-phase peptide synthesis. At the time of writing, no published work accomplished with this instrument was available.

However, several articles in the area of microwave-assisted parallel synthesis have described irradiation of 96-well filter-bottom polypropylene plates in conventional household microwave ovens for high-throughput synthesis [16–19]. While some authors have not reported any difficulties associated with the use of such equipment [19], others have experienced problems in connection with the thermal instability of the polypropylene material itself [17] and with respect to the creation of temperature gradients between individual wells upon microwave heating [17, 18]. While Teflon (or similar materials such as PFA) can eliminate the problem of thermal stability, the issue of bottom-filtration reaction vessels has not yet been adequately addressed.

An interesting article by O'Shea and coworkers described the construction and use of a parallel polypropylene reactor comprising cylindrical, expandable reaction vessels with porous frits at the bottom [20]. During the reaction, the piston of each individual vessel moves upwards, whereas during cooling the opposite downward motion occurs. On opening the bottom outlet, the solution is drained from the reaction mixture and the solid-support can be washed and prepared for cleavage. This work presented the very first description of reaction vessels for microwave-assisted synthesis that may be useful for carrying out solid-phase synthesis using bottom-filtration techniques in conjunction with microwave heating.

A prototype of a microwave reaction vessel that takes advantage of bottom filtration techniques was presented by Erdélyi and Gogoll in a more recent publication. Therein, the authors described the use of a modified reaction vessel (Fig. 7.2) for the Emrys instruments (see Section 3.5.1) with a polypropylene frit, suitable for the filtration/cleavage steps in their microwave-mediated solid-phase Sonogashira coupling (see Scheme 7.19) [21].

For general solid-phase reactions in a dedicated multimode instrument, an adaptable filtration unit is available from Anton Paar (see Fig. 3.18). This tool is connected to the appropriate reaction vessel by a simple screw cap and after turning over the vessel, the resin is filtered by applying a slight pressure of up to 5 bar. The resin can then be used for further reaction sequences or cleavage steps in the same reaction vessel without any material loss. However, at the time of writing, no applications of this system for solid-phase synthesis had been reported.

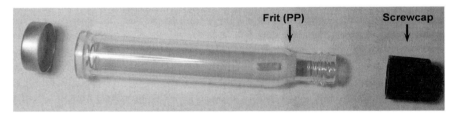

Fig. 7.2   Modified microwave vial for use in the Emrys series (reproduced with permission from [21]).

7.1.3
Peptide Synthesis and Related Examples

One of the first dedicated applications of microwaves in solid-phase chemistry was in the synthesis of small peptide molecules, as described by Wang and coworkers [22]. As a preliminary test, the authors coupled Fmoc-Ile and Fmoc-Val, respectively, with Gly-preloaded Wang resin using the corresponding symmetric anhydrides (Scheme 7.1).

The reactions were carried out rapidly in a modified domestic microwave oven with mountings in the side wall for sufficient inert gas supply. Furthermore, a dedicated custom-made solid-phase reaction vessel was employed under atmospheric pressure conditions.

The microwave protocol increased the reaction rate at least 2–3 fold, as conversion was only 60–80% after 6 min of conventional heating. This improved coupling efficiency was duplicated with numerous amino acid derivatives and a further two pep-

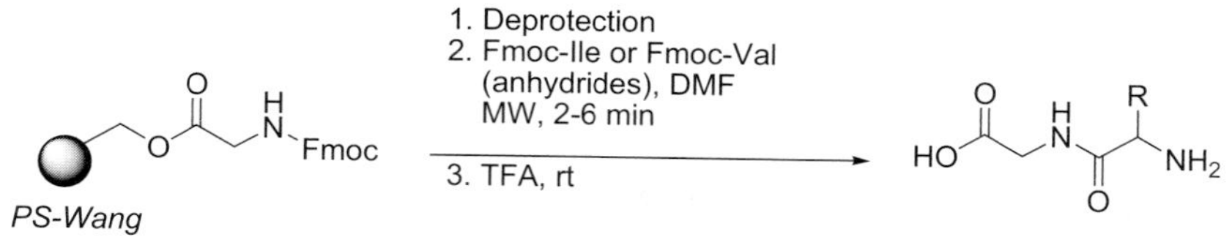

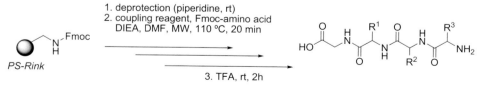

Scheme 7.1 Rapid solid-phase peptide coupling.

tide fragments were coupled with the Gly-Wang resin. These couplings were completed within 2 min, as verified by a quantitative ninhydrin assay.

For a more representative investigation, the authors synthesized a fragment of the acyl carrier protein ([65–74]ACP) using preformed active esters in N,N-dimethylformamide [22]. Each coupling step required only 4 min of irradiation in a modified domestic oven with an average coupling yield of >99% (Scheme 7.2).

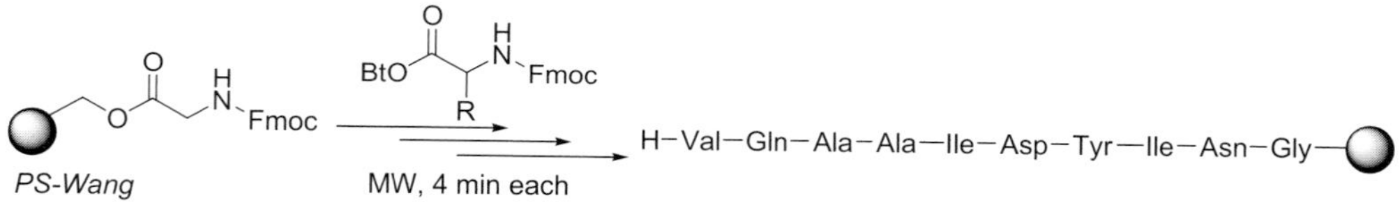

Scheme 7.2 Synthesis of [65–74]ACP employing stepwise coupling of amino acid esters.

Importantly, it was demonstrated that no significant racemization occurred in the course of peptide formation. Furthermore, the complete coupling of difficult peptide sequences could be accomplished within a few minutes, and it was determined that peptide fragments have higher reactivity than single amino acid derivatives under microwave irradiation conditions. However, the exact reaction temperature during the irradiation period was not determined.

In a related study, microwave irradiation was applied to the coupling of sterically hindered amino acids, leading to di- and tripeptides (Scheme 7.3) [23]. Erdélyi and Gogoll investigated a variety of common coupling reagents, for example benzotriazol-1-yl-oxytripyrrolidinophosphonium hexafluorophosphate (PyBOP), 2-(7-aza-1H-benzotriazol-1-yl)-1,1,3,3-tetramethyluronium hexafluoride (HATU), O-(benzotriazol-1-yl)-N,N,N′,N′-tetramethyluronium tetrafluoroborate (TBTU), and Mukaiyama's reagent for peptide synthesis. Under controlled microwave conditions with IR temperature measurement, coupling of the amino acids via the corresponding an-

Scheme 7.3 Efficient tripeptide synthesis employing various coupling reagents.

hydrides or *N*-hydroxybenzotriazole (HOBt)-activated esters was completed within 20 min without the need for double or triple coupling steps as in conventional protocols. The azobenzotriazole derivatives showed increased coupling efficiency at up to 110 °C. Above this temperature, decomposition of the reagents was indicated by a color change of the reaction mixtures. However, no degradation of the solid support was observed. Furthermore, both LC-MS and ^{1}H NMR confirmed the absence of racemization during the high-temperature treatment in the presence of *N,N*-diisopropylethylamine (DIEA) as base.

More recently, Somfai and coworkers have reported on the efficient coupling of a set of carboxylic acids suitable as potential scaffolds for peptide synthesis to a polymer-bound hydrazide linker [24]. Indole-like scaffolds were selected for this small library synthesis as these structures are found in numerous natural products showing interesting activities. The best results were obtained using 2-(7-aza-1*H*-benzotriazol-1-yl)-1,1,3,3-tetramethyluronium hexafluoride (HATU) and *N,N*-diisopropylethylamine (DIEA) in *N,N*-dimethylformamide as a solvent. Heating the reaction mixtures at 180 °C for 10 min furnished the desired products in high yields (Scheme 7.4). In this application, no Fmoc protection of the indole nitrogen is required.

Scheme 7.4   Coupling of an indolyl acid to polystyrene-bound hydrazide linker.

A notable increase in the overall speed of peptoid synthesis was reported by Kodadek and coworkers [25]. These authors presented a multistep protocol for the generation of various peptoids employing a domestic microwave oven (Scheme 7.5). Reaction times were drastically reduced, with less than 1 min being required for the coupling of each residue. With this protocol, nine different primary amines were used to generate various 9-residue homo-oligomers, one 20-residue homo-oligomer, and one 9-residue hetero-oligomer. These transformations were performed in *N,N*-di-

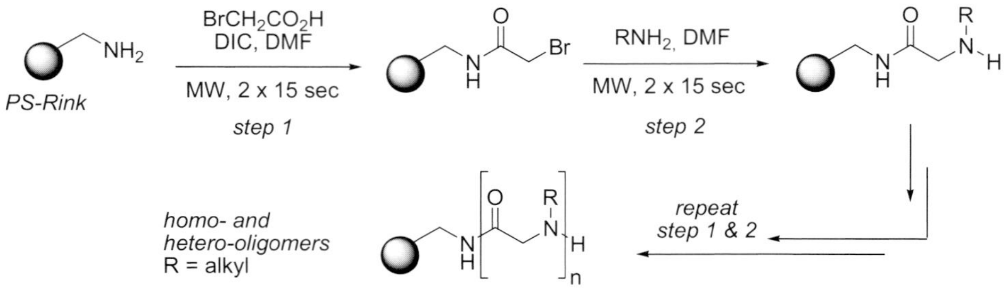

Scheme 7.5   Construction of peptoid sequences. Cleavage was carried out using trifluoroacetic acid (TFA) at room temperature.

methylformamide as a solvent, employing *N,N'*-diisopropylcarbodiimide (DIC) as activating agent in two runs of duration 15 s with manual stirring between the irradiation steps. In this case, the temperature of the solutions did not exceed 35 °C, as determined by inserting a thermometer after the second 15 s run. For comparative purposes, these couplings were also carried out by conventional thermal heating at 37 °C, which led to similar results. Both methods, however, provided better yields and purities than conventional room temperature couplings. In general, this protocol allows a convenient method for high-throughput peptoid synthesis, as microwave acceleration reduces the overall production time. For example, a 9-residue peptoid can be synthesized in 3 h, as compared to 20–32 h employing the standard protocol.

In another application, the group of Berteina-Raboin demonstrated the solid-supported synthesis of the indole core of melatonin analogues under microwave irradiation (Scheme 7.6) [26]. A benzoic acid derivative was coupled to Rink amide resin by

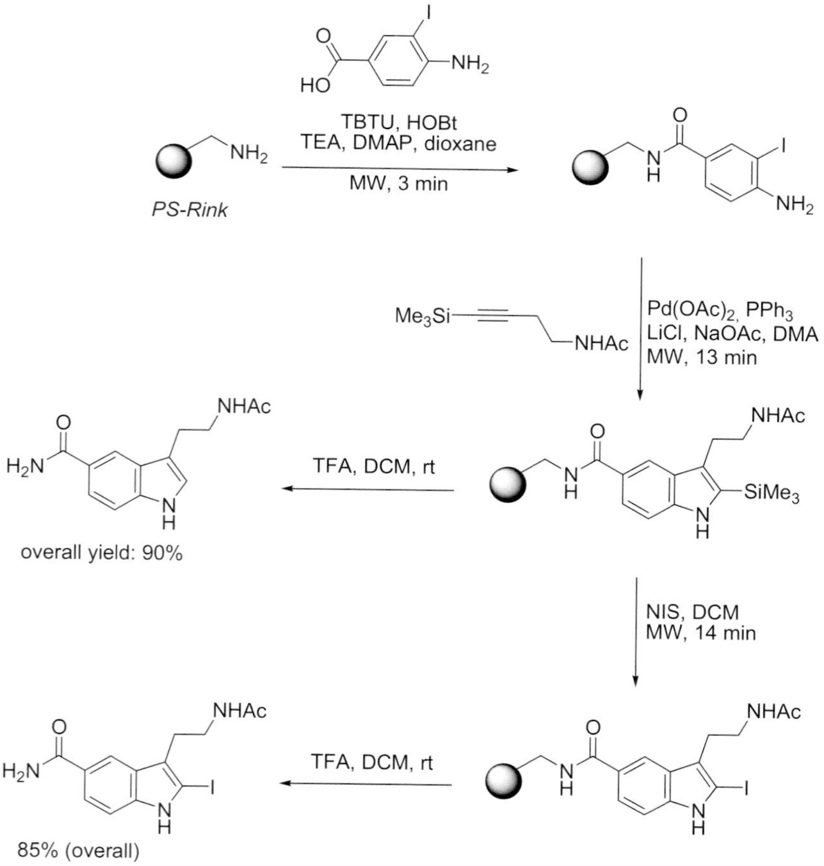

Scheme 7.6 Solid-phase synthesis of 5-carboxamido-*N*-acetyltryptamines.

using the peptide coupling reagents N-hydroxybenzotriazole (HOBt) and O-(benzo-triazol-1-yl)-N,N,N',N'-tetramethyluronium tetrafluoroborate (TBTU). Subsequent indole ring formation was carried out by palladium-catalyzed coupling directly on the solid phase. The corresponding iodo compound was synthesized by treatment of the polymer-bound indole derivative with N-iodosuccinimide (NIS). Finally, the desired compounds were released from the resin by using conventional trifluoro-acetic acid/dichloromethane cleavage cocktails at room temperature. Carrying out these reactions under microwave conditions led to a substantial increase in yields and to a significant reduction of the reaction time compared to conventional thermal heating, whereby each individual step requires 24–48 h. Due to the limited thermal stability of the solid support, the temperature during the reactions was kept below 140 °C through strict control of the power.

In a recent report, the group of Kusumoto presented a solid-phase synthesis of indol-2-ones based on radical cyclization [27]. A set of commercially available 2-bro-moanilines and acryloyl chloride derivatives was employed to generate a 40-member library of the corresponding 2-oxindoles, which are known to be important core structures for several drug candidate compounds (Scheme 7.7). N,N-Dimethylform-amide was found to be the best solvent for the radical cyclization, inducing a reagent concentration effect of the solid support. Interestingly, the radical-induced cycliza-tion with tributyltin hydride (Bu₃SnH) and 2,2'-azobisisobutyronitrile (AIBN) did not proceed at all under thermal conditions, and the microwave-mediated solution-phase cyclization also remained ineffective. However, after preparing the corre-sponding polymer-bound acrylamides through a multistep sequence, the radical

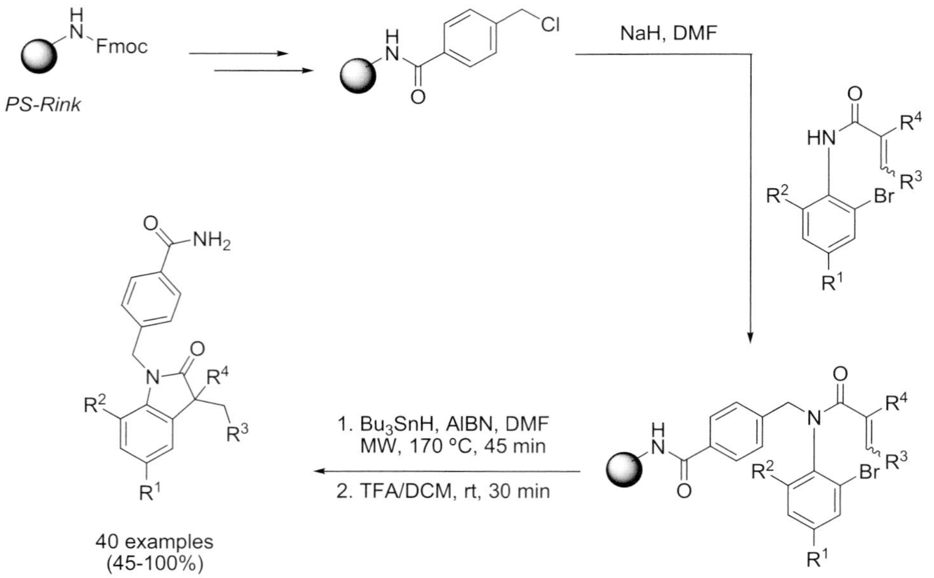

Scheme 7.7 Solid-phase synthesis of indol-2-ones through radical cyclization.

cyclization of these intermediates could be performed utilizing a dedicated single-mode microwave reactor. After conventional cleavage from the Rink amide resin employed by treatment with a trifluoroacetic acid/dichloromethane mixture, the desired compounds were obtained in moderate yields and purities.

7.1.4
Resin Functionalization

The functionalization of commercially available standard solid supports is of particular interest for combinatorial purposes to enable a broad range of reactions to be studied. Since these transformations usually require long reaction times under conventional thermal conditions, it was obvious to combine microwave chemistry with the art of resin functionalization.

As a suitable model reaction, the coupling of various substituted carboxylic acids to polymer resins has been investigated by Stadler and Kappe (Scheme 7.8) [28]. The resulting polymer-bound esters served as useful building blocks in a variety of further solid-phase transformations. In a preliminary experiment, benzoic acid was attached to Merrifield resin under microwave conditions within 5 min (Scheme 7.8 a). This functionalization was additionally used to determine the effect of microwave irradiation on the cleavage of substrates from polymer supports (see Section 7.1.10). The benzoic acid was quantitatively coupled within 5 min via its cesium salt utilizing standard glassware under atmospheric reflux conditions at 200 °C.

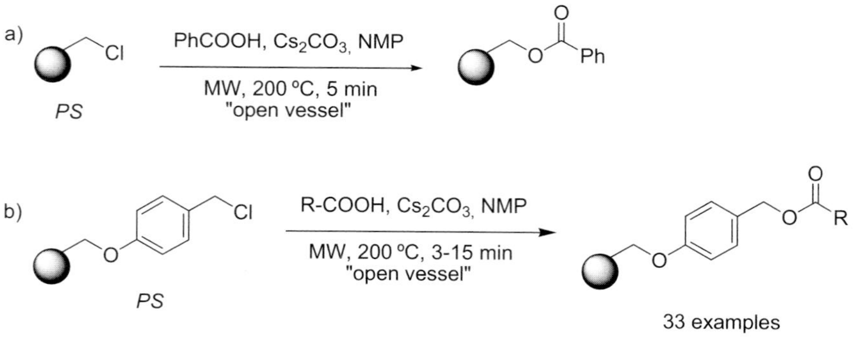

Scheme 7.8   Resin functionalization with carboxylic acids using (a) Merrifield resin and (b) chlorinated Wang resin as the polymer support.

In a more extensive study, 33 substituted carboxylic acids were coupled to chlorinated Wang resin employing an identical reaction protocol (Scheme 7.8 b). In the majority of cases, the microwave-mediated conversion reached at least 85% after 3–15 min at 200 °C. These microwave conditions represented a significant rate enhancement with respect to the conventional protocol, which required 24–48 h. The microwave protocol has additional benefits in comparison to the conventional method, as the amounts of acid and base equivalents can be reduced and potassium iodide is no longer needed as an additive [28]. While no attempt was made to optimize all 33 examples, a number of substituted benzoic acids were selected to compare their coupling behavior under microwave conditions with that seen when applying the thermally heated protocol. High loadings of the resin-bound esters could be obtained very rapidly; even sterically demanding acids were coupled successfully. Importantly, in all the examples given, the loadings accomplished after 15 min of microwave irradiation were actually higher than those achieved using the thermally heated protocol.

In a related study by the same authors, the effect of microwave irradiation on carbodiimide-mediated esterifications on a solid support was investigated, employing benzoic acid [29]. The carboxylic acid was activated using N,N'-diisopropylcarbodiimide (DIC) through the *O*-acyl isourea or the symmetrical anhydride protocol (Scheme 7.9). The isourea protocol was carried out in a dichloromethane/N,N-dimethylformamide mixture in sealed vessels, whereas the anhydride reactions were carried out in 1-methyl-2-pyrrolidinone (NMP) at atmospheric pressure.

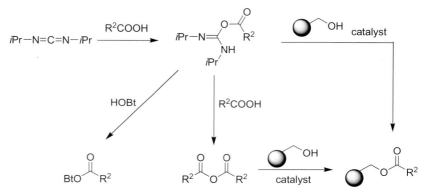

Scheme 7.9   Carbodiimide-mediated pathways for esterification reactions.

The isourea protocol showed some deficiencies, such as the fact that complete conversion could not be obtained due to unexpected side reactions at higher temperatures. The anhydride protocol proved superior to this method as it could be carried out quantitatively at 200 °C within 10 min under open-vessel conditions without the need for high-pressure vessels.

Song and coworkers have reported the synthesis of a series of functionalized Merrifield resins [30]. Utilizing a modified domestic microwave oven with a reflux condenser, the reaction rates were dramatically enhanced as compared to conventional methods, such that high conversions were achieved within 25 min (Scheme 7.10). Since the choice of solvent was crucial in this procedure, the authors used a solvent mixture to balance between the aspects of good resin-swelling properties and high microwave absorption efficiency (see Table 7.1). These microwave-mediated pathways provide convenient methods for rapid and efficient solid-phase synthesis, using PS-Merrifield resin as either a support or a scavenger (see Section 7.7).

An interesting approach to resin functionalization has been presented by the group of Yaylayan, who described microwave-assisted PEGylation of Merrifield

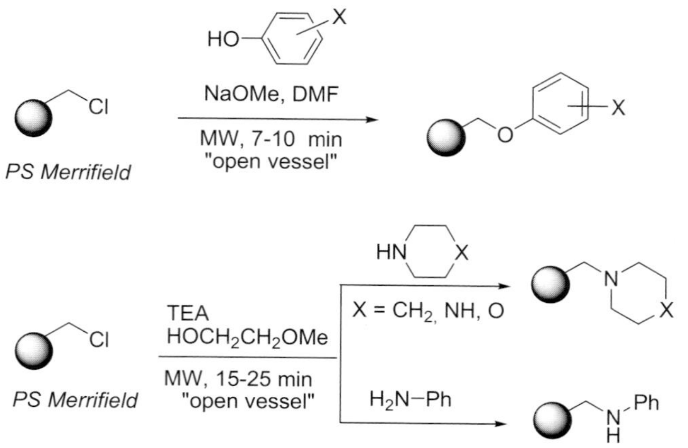

Scheme 7.10 Efficient preparation of functionalized resins for several solid-phase applications.

resin [31]. Treating commercially available polystyrene Merrifield resin with polyethylene glycol (PEG-200) at 170 °C for only 2 min afforded the corresponding hybrid polymer, which combined the advantages of both insoluble and soluble polymers (Scheme 7.11). The Merrifield resin was suspended in excess PEG, which served as the solvent and at the same time prevented cross-linking of the resin. The mixture was irradiated in a pulsed sequence for a total irradiation time of 2 min. The product was purified by subsequent washings with water, 10% hydrochloric acid, and methanol. However, it became apparent that the yield decreased with increasing molecular weight of the PEG and also with increasing chlorine-loading of the Merrifield resins.

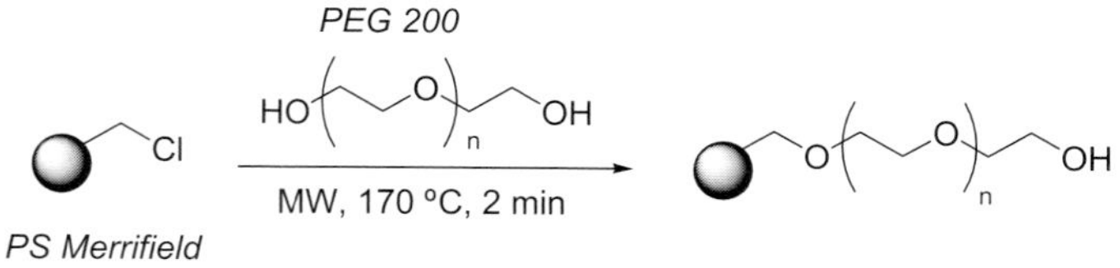

Scheme 7.11 PEGylation of Merrifield resin.

In a more recent study, Westman and Lundin have described solid-phase syntheses of aminopropenones and aminopropenoates en route to heterocycles [32]. Two different three-step methods for the preparation of these heterocycles were developed. The first method involved the formation of the respective ester from *N*-protected glycine derivatives and Merrifield resin (Scheme 7.12 a), while the second method involved the use of aqueous methylamine solution for functionalization of the solid support (Scheme 7.12 b). The desired heterocycles were obtained by treatment of the generated polymer-bound benzylamine with the requisite acetophenones under similar conditions to those shown in Scheme 7.12 a, utilizing 5 equivalents of *N,N*-dimethylformamide diethyl acetal (DMFDEA) as reagent. The final

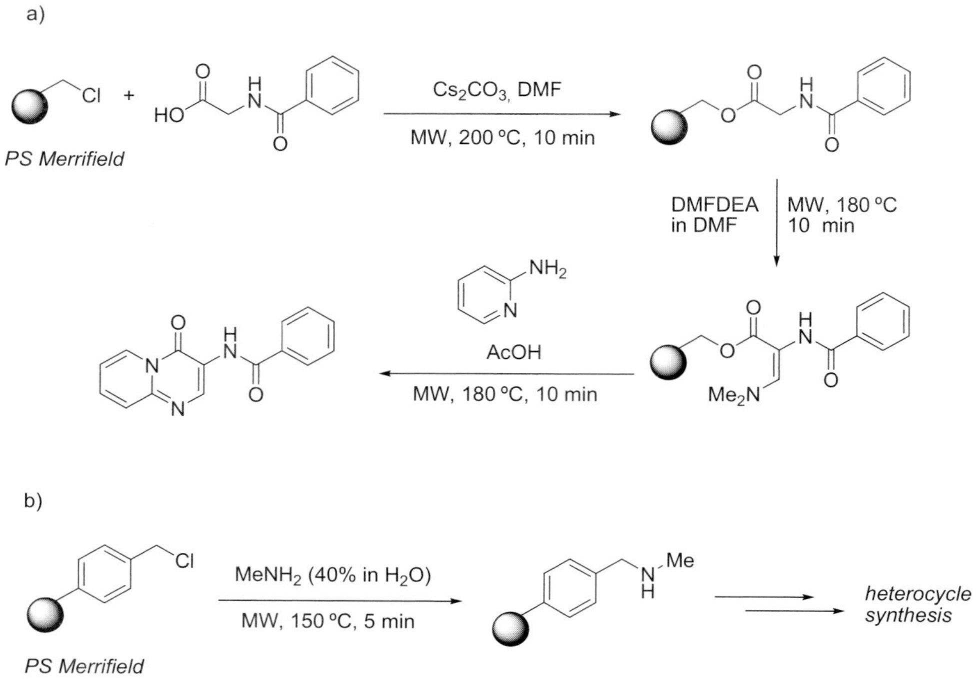

a)

b)

Scheme 7.12 Reaction strategies for the polymer-supported synthesis of dialkylaminopropenones.

step in the synthesis of the pyridopyrimidinones (Scheme 7.12 a) involved the release of the products from the solid support by intramolecular cyclization, whereupon the pure products were obtained.

In a dedicated combinatorial approach, Strohmeier and Kappe have reported the rapid parallel synthesis of polymer-bound enones [33]. This approach involved a two-step protocol utilizing initial high-speed acetoacetylation of Wang resin with a selection of common β-ketoesters (Scheme 7.13) and subsequent microwave-mediated Knoevenagel condensations with a set of 13 different aldehydes (see Section 7.3.6).

These transesterifications are believed to proceed through the initial formation of a highly reactive α-oxo ketene intermediate, with the elimination of the alcohol component of the acetoacetic ester being the limiting factor. Subsequent trapping of the

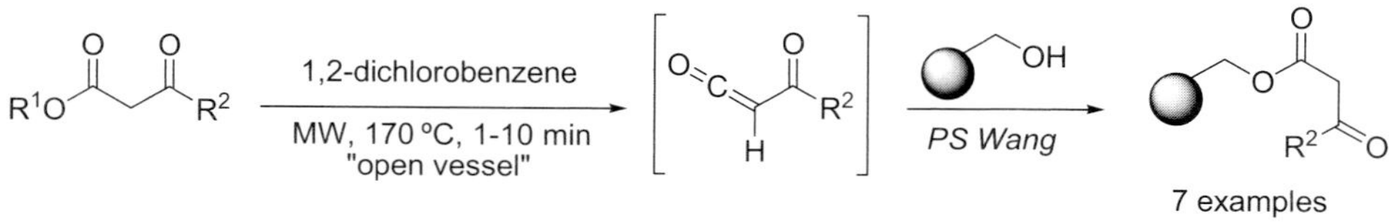

Scheme 7.13 Acetoacetylation reactions.

ketene intermediate affords the transacetoacetylated products. For better handling of the polymer support, the reactions can be carried out at atmospheric pressure in open PFA vessels, using 1,2-dichlorobenzene as solvent. Acetoacetylations were successfully performed within 1–10 min under these microwave conditions at 170 °C. Furthermore, the acetoacetylated products could be obtained in a parallel fashion in a single run of 10 min duration employing a multi-vessel rotor system. It is worth mentioning that these transesterifications need to be carried out under open-vessel conditions so that the alcohol by-product can be removed from the reaction mixture (see Section 4.3).

7.1.5
Transition Metal Catalysis

Palladium-catalyzed cross-coupling reactions constitute one of the cornerstones of modern organic synthesis. The combination of both microwave irradiation and solid-phase synthesis with such chemical transformations is therefore of great interest (see also Section 6.1). One of the first publications dealing with such reactions was presented by the group of Hallberg in 1996 [34]. Therein, the authors investigated the effect of microwave irradiation on Suzuki- and Stille-type cross-coupling reactions on a solid phase.

The reactions were carried out in sealed Pyrex tubes employing a prototype single-mode microwave cavity. The reagents were added to the resin-bound aryl halide under a nitrogen atmosphere and irradiated for the time periods indicated (Scheme 7.14). Rather short reaction times provided almost quantitative conversions, with minimal degradation of the solid support.

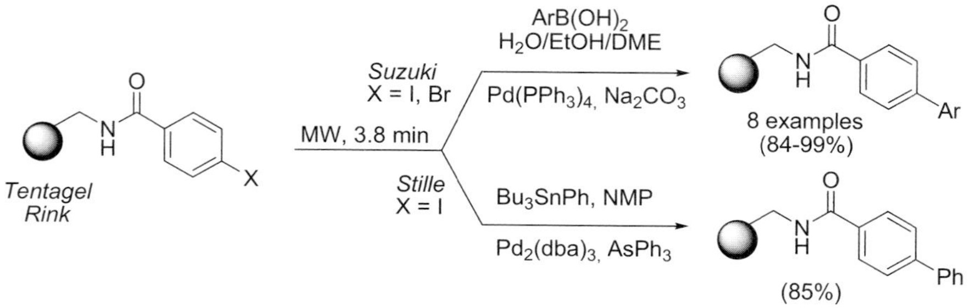

Scheme 7.14 Palladium-catalyzed cross-coupling reactions.

In a related study, the same group investigated molybdenum-catalyzed alkylations in solution and on a solid phase [35], demonstrating that microwave irradiation could also be applied to highly enantioselective reactions (Scheme 7.15). For these examples, commercially available and stable molybdenum hexacarbonyl [Mo(CO)$_6$] was used to generate the catalytic system *in situ*. The reactions in solution provided good yields (see Scheme 6.50). In contrast, the conversion rates for the solid-phase examples were rather poor. However, the enantioselectivity was excellent (>99% *ee*) for both the solution and solid-phase reactions.

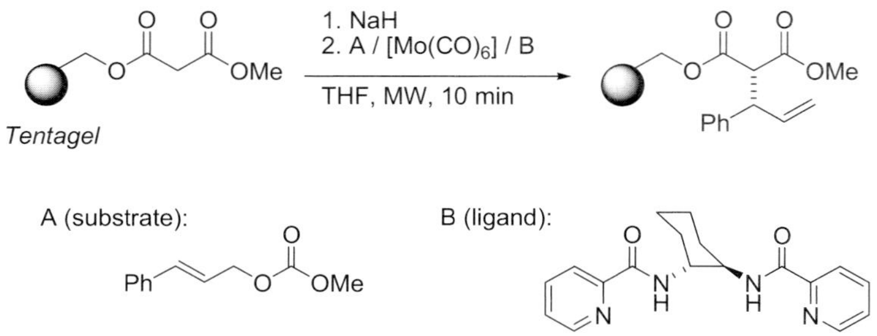

Scheme 7.15 Solid-phase molybdenum-catalyzed allylic alkylation.

In further studies, Hallberg and coworkers investigated the microwave-promoted preparation of tetrazoles employing organonitriles [36]. After establishing a solution-phase protocol, this was adapted for solid-phase examples (Scheme 7.16). Complete conversion to the corresponding nitriles was achieved within very short reaction times, at a maximum temperature of 175 °C. The nitriles were subsequently treated with sodium azide at 220 °C for 15 min to form the desired tetrazoles. Despite the rather high temperature, only negligible decomposition of the solid support was observed. It is worth noting that the formation of tetrazoles could easily be achieved in good yields as a one-pot reaction, eliminating the need for the tetrakis-(triphenylphosphine)palladium catalyst in the reaction mixture. Furthermore, reaction times were drastically reduced by using the microwave protocol; comparable yields were only achieved after 3–96 h using conventional thermal heating.

In a more recent study, Wannberg and Larhed reported solid-supported aminocarbonylations employing molybdenum hexacarbonyl as a solid source of carbon monoxide [37]. Carbon monoxide is smoothly liberated at the reaction temperature upon the addition of the strong base 1,8-diazabicyclo[2.2.2]octane (DBU). In this transfor-

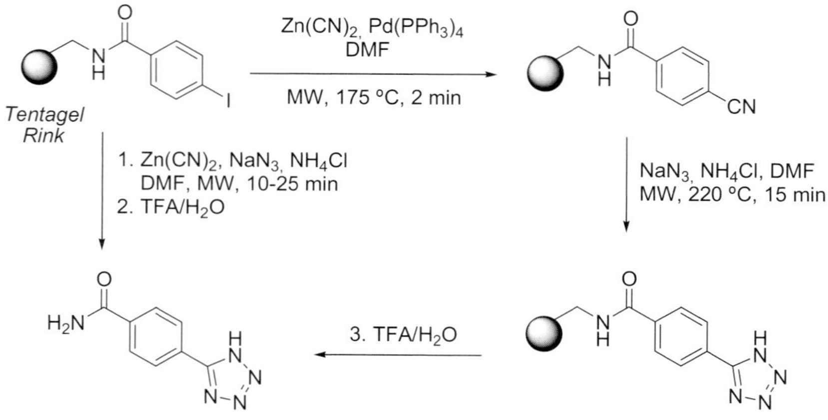

Scheme 7.16 Tetrazole synthesis on a solid phase.

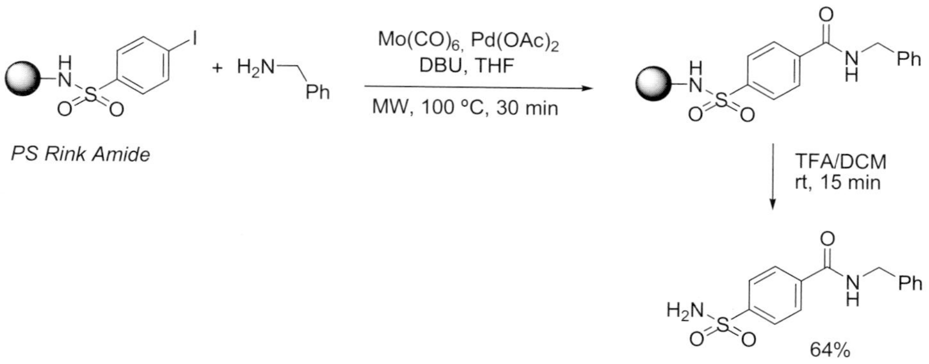

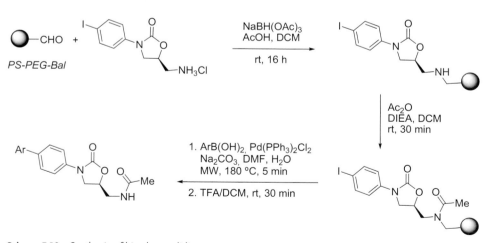

Scheme 7.17 Aminocarbonylations on a solid phase.

mation, 5 equivalents of Mo(CO)$_6$ was used and a 30-fold excess of DBU was added to the reaction mixture immediately prior to the irradiation. Cleavage with a conventional mixture of trifluoroacetic acid and dichloromethane furnished the desired sulfamoylbenzamide in good yield (Scheme 7.17).

Combs and coworkers have presented a study on the solid-phase synthesis of oxazolidinone antimicrobials by microwave-mediated Suzuki coupling [38]. A valuable oxazolidinone scaffold was coupled to Bal resin (PS-PEG resin with a 4-formyl-3,5-dimethoxyphenoxy linker) to afford the corresponding resin-bound secondary amine (Scheme 7.18). After subsequent acylation, the resulting intermediate was transformed to the corresponding biaryl compound by microwave-assisted Suzuki coupling. Cleavage with trifluoroacetic acid/dichloromethane yielded the desired target structures.

Scheme 7.18 Synthesis of biaryl oxazolidinones.

Another palladium-catalyzed carbon–carbon coupling that can be efficiently accelerated by microwave heating is the Sonogashira reaction, as demonstrated by Erdélyi and Gogoll [21]. Aryl bromides and iodides have been efficiently coupled with various terminal alkyne derivatives (Scheme 7.19). In a set of preliminary experiments, the authors determined the optimum conditions for full conversion of the substrate. It was found that 5 mol% of the palladium catalyst and 10 mol% of copper(I) iodide as co-catalyst in a mixture of diethylamine and *N*,*N*-dimethylformamide gave the best results. Depending on the substrates used, different irradiation times were required to achieve complete conversion. The reactions were performed in a dedicated single-mode instrument utilizing a modified reaction vessel (see Fig. 7.2) for simplified resin handling.

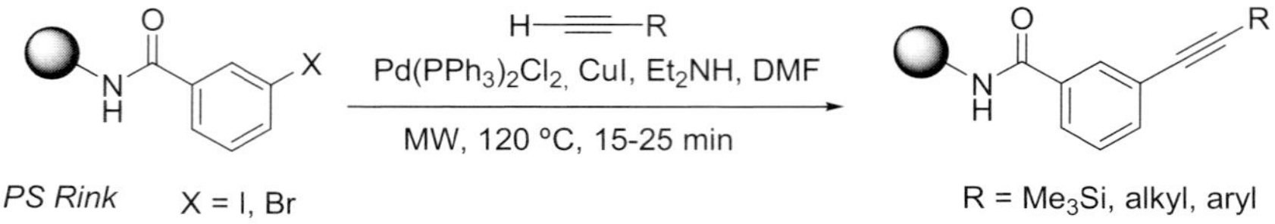

Scheme 7.19   Sonogashira couplings on a solid phase.

Recently, Walla and Kappe disclosed the first polymer-supported cross-coupling reaction involving aryl chlorides (see also Scheme 6.40) [39]. The authors performed rapid Negishi couplings utilizing organozinc reagents to prepare biaryl compounds from various aryl halides. As catalytic system, the highly reactive but air-stable tri-*tert*-butylphosphonium tetrafluoroborate and tris(dibenzylideneacetone)dipalladium(0) were employed (Scheme 7.20). Subsequent rapid cleavage under microwave conditions furnished the desired biaryl carboxylic acids in high yield and excellent purity.

The first examples of microwave-mediated solid-phase carbon–nitrogen cross-coupling reactions were reported in 1999 by the group of Combs [16], using a boronic

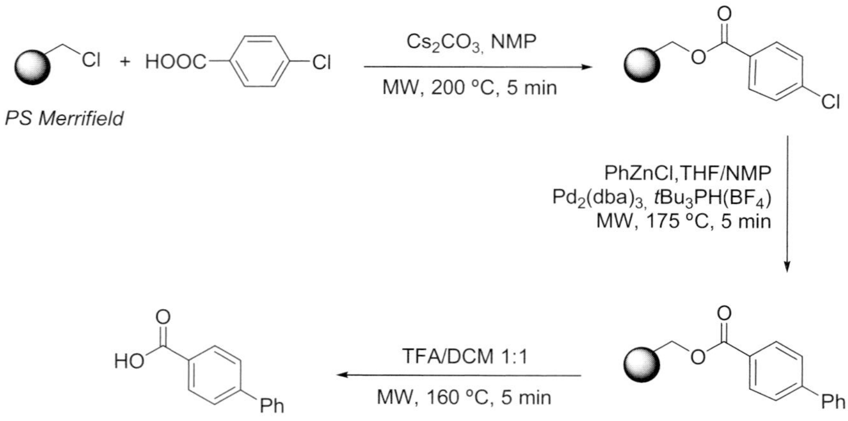

Scheme 7.20   Negishi coupling on a polymer support.

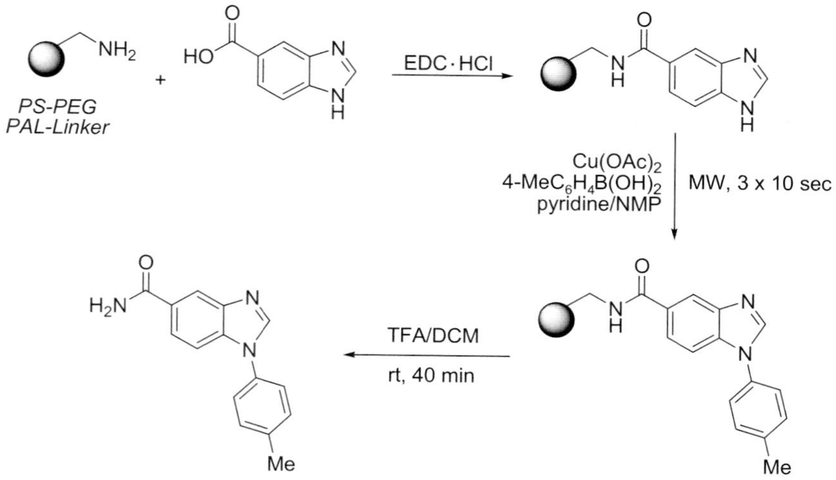

Scheme 7.21 Copper(II)-mediated *N*-arylation of polymer-bound benzimidazole.

acid and a copper(II) catalyst. The reactions were carried out in a domestic micro-wave oven at full power for 3 × 10 s. After 5 heating cycles with the addition of fresh reagents, no remaining starting benzimidazole amide could be detected after cleavage from the solid support (Scheme 7.21). This represented a reduction in the reaction time from 48 h under conventional heating at 80 °C to less than 5 min by microwave heating. However, it should be noted that both possible *N*-arylated regio-isomeric products were obtained by this microwave-heated procedure. To assess the versatility of this reaction on the solid support, several heterocyclic carboxylic acids were coupled to the PS-PEG resin (PAL linker). Applying the microwave conditions furnished the desired products in good yields and excellent purities (not shown).

A more recent publication by Weigand and Pelka has disclosed a polymer-bound Buchwald–Hartwig amination [40]. Activated, electron-deficient aryl halides were coupled with conventional PS Rink resin under microwave irradiation. Subsequent acidic cleavage afforded the desired aryl amines in moderate to good yields (Scheme 7.22). Commercially available Fmoc-protected Rink amide resin was sus-pended in 20% piperidine/*N,N*-dimethylformamide at room temperature for 30 min to achieve deprotection. After washing and drying, the resin was placed in a silylated microwave vessel and suspended in dimethoxyethane (DME)/*tert*-butanol

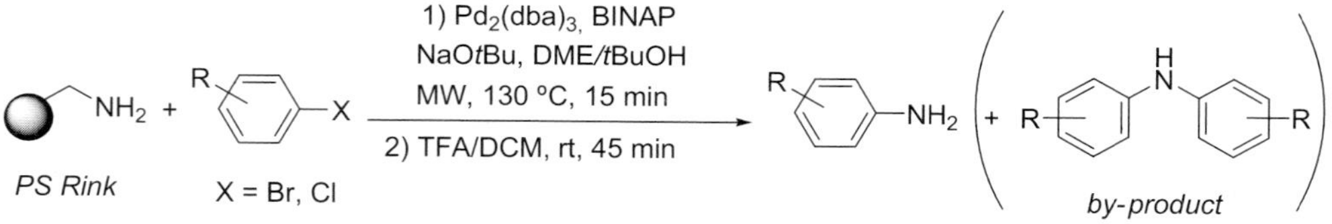

Scheme 7.22 Palladium-catalyzed Buchwald–Hartwig amination.

under argon atmosphere. After approximately 10 min, 10 equivalents of the aryl halide was added and the mixture was stirred under argon for an additional 10 min. Thereafter, 0.2 equivalents of the palladium catalyst, 0.3 equivalents of 2,2'-bis(diphenylphosphino)-1,1'-binaphthyl ligand (BINAP), and 10 equivalents of sodium *tert*-butoxide base (NaOtBu) were added, the vessel was sealed, and then submitted to microwave irradiation at 130 °C for 15 min. Finally, after several washings, the product was cleaved by employing conventional 5% trifluoroacetic acid in dichloromethane. After purification, the desired anilines were obtained in good yields and excellent purities.

Transition metal catalysis on solid supports can also be applied to indole formation, as shown by Dai and coworkers [41]. These authors reported a palladium- or copper-catalyzed procedure for the generation of a small indole library (Scheme 7.23), representing the first example of a solid-phase synthesis of 5-arylsulfamoyl-substituted indole derivatives. The most crucial step was the cyclization of the key polymer-bound sulfonamide intermediates. Whereas the best results for the copper-mediated cyclization were achieved using 1-methyl-2-pyrrolidinone (NMP) as solvent, the palladium-catalyzed variant required the use of tetrahydrofuran in order to achieve comparable results. Both procedures afforded the desired indoles in good yields and excellent purities [41].

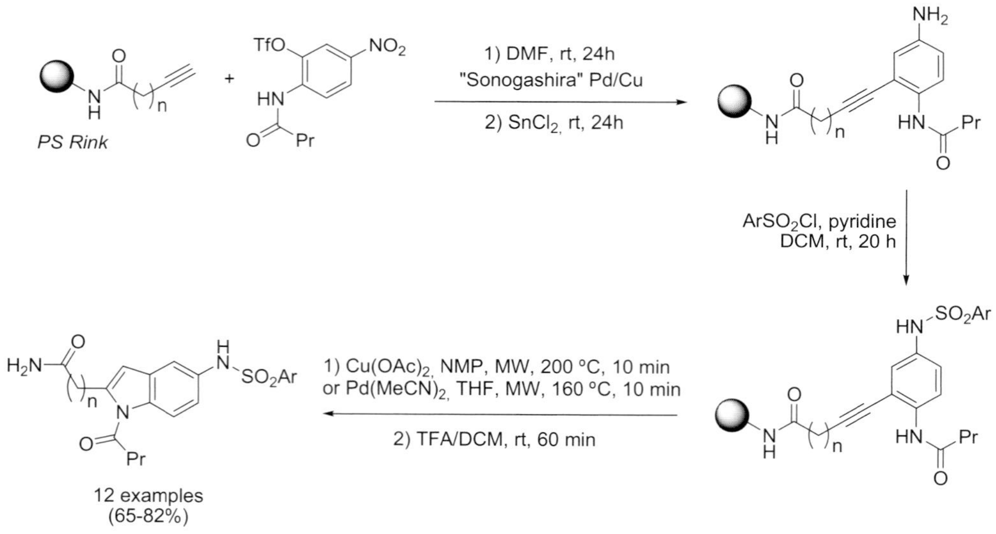

Scheme 7.23 Transition metal-catalyzed indole formation on a solid phase.

In a recent study, the group of Van der Eycken described the decoration of polymer-bound 2(1*H*)-pyrazinone scaffolds by performing various transition metal-catalyzed transformations [42]. The readily prepared pyrazinone was specifically decorated at the C3 position by employing microwave-mediated Suzuki, Stille, Sonogashira, and Ullmann protocols (Scheme 7.24), thereby introducing additional diver-

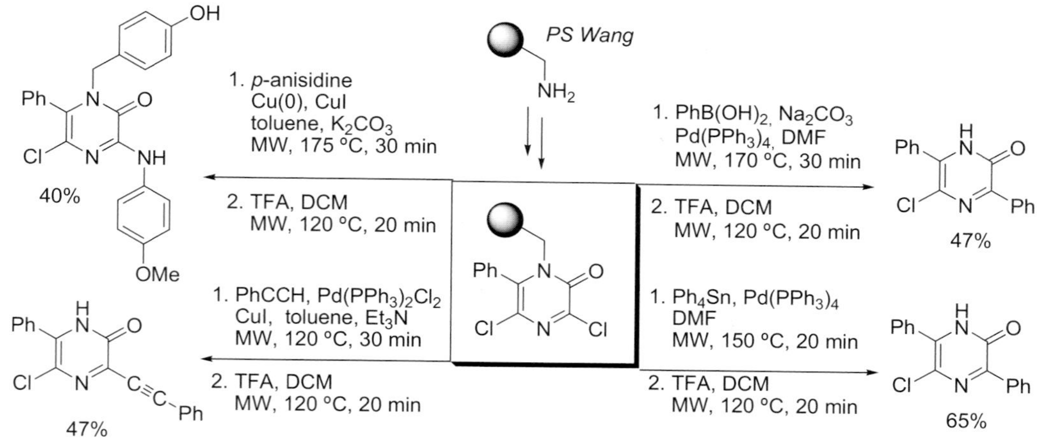

Scheme 7.24 Scaffold decoration of polymer-bound pyrazinone scaffolds.

sity. In all cases, smooth cleavage of the desired products was achieved by treatment of the resin with a 1:2 mixture of trifluoroacetic acid (TFA) and dichloromethane under microwave irradiation at 120 °C for 20 min.

7.1.6
Substitution Reactions

Another interesting field is that of microwave-mediated substitution reactions on solid phases. In this context, an innovative study has been presented by Wenschuh and coworkers [43]. Therein, the authors describe the synthesis of trisamino- and amino-oxy-1,3,5-triazines on cellulose and polypropylene membranes, applying SPOT-synthesis techniques (Scheme 7.25). This research demonstrates the usefulness of planar solid supports in addition to granulated polystyrene or PEG resins in microwave-assisted SPOS. The development of the SPOT-synthesis protocols allows the rapid generation of highly diverse spatially addressed single compounds under mild conditions. This SPOT-synthesis protocol required the investigation of suitable planar polymeric supports bearing an orthogonal ester-free linker system that were cleavable under dry conditions.

Several functionalized membranes could be synthesized by conventional methods at room temperature. In contrast, microwave heating was employed for both the synthesis of the triazine membrane and the practical generation of an 8000-member library of triazines bound to an amino-functionalized cellulose membrane (Scheme 7.26).

For the preparation of the triazine membranes, the entire solid support (cellulose or polypropylene membrane) was treated with a 5 M solution of the corresponding amine in 1-methyl-2-pyrrolidinone (NMP) and a 1 M solution of cesium phenolate in dimethyl sulfoxide (2 μL of each at one spot) and subsequently heated in a domestic microwave oven for 3 min. After washing the support successively with

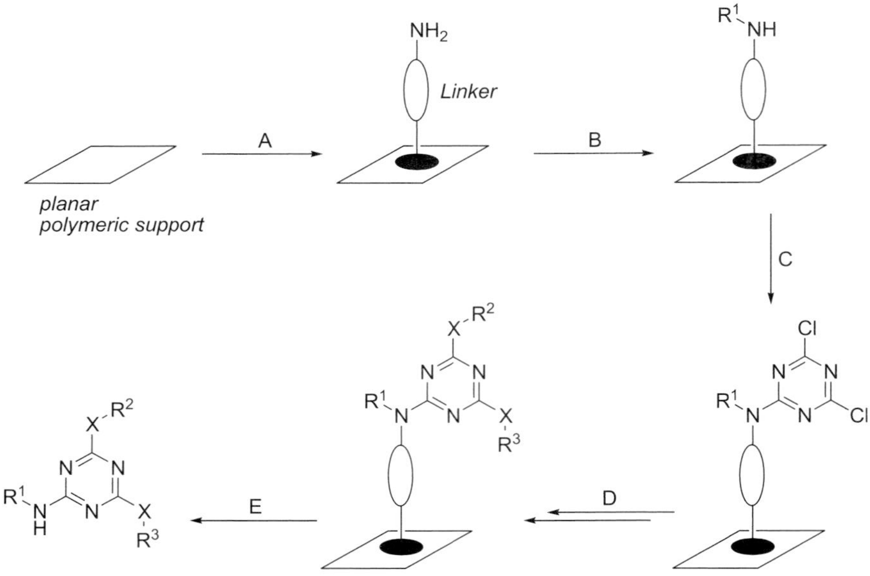

Scheme 7.25 General strategy for assembling triazines on planar surfaces:
(A) functionalization/linker attachment; (B) introduction of the first building block;
(C) attachment of cyanuric chloride; (D) stepwise chlorine substitution; (E) cleavage.

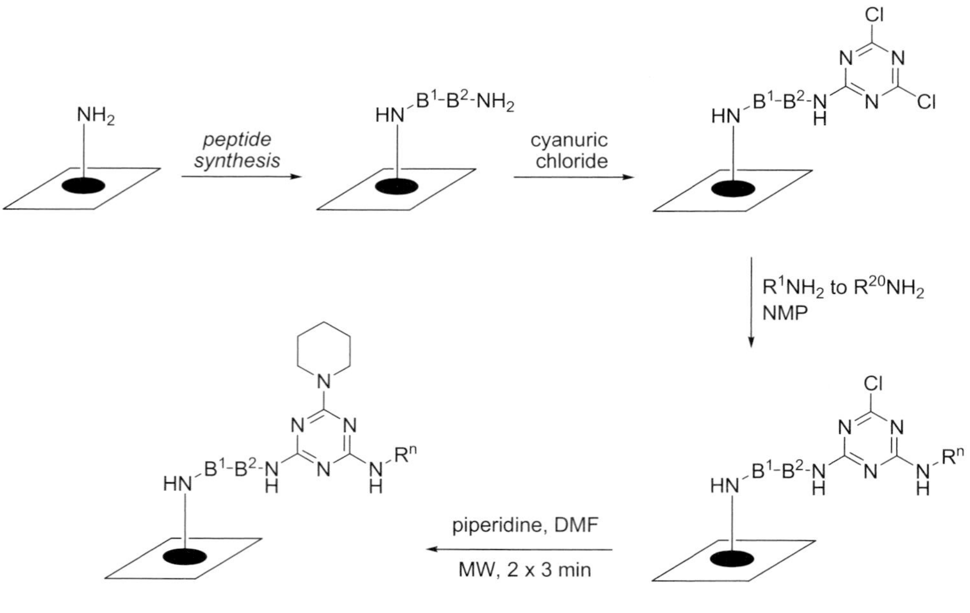

Scheme 7.26 Library generation on a cellulose membrane employing the SPOT technique.

N,N-dimethylformamide, methanol, and dichloromethane, the membrane was air-dried. All 400 possible dipeptides composed of the 20 proteinogenic L-amino acids (B^1 and B^2) were synthesized in 20 replica, as illustrated in Scheme 7.26. Subsequently, cyanuric chloride was attached to each dipeptide, and this was followed by selective substitution of one of the two chlorine atoms with 20 different amines. The second chlorine atom was ultimately replaced by piperidine under microwave-irradiation conditions. Thereafter, the resulting library of functionalized dipeptides was tested for binding in the murine IgG mab Tab2 assay, directly on the cellulose sheet. Cleavage of the generated compounds could be achieved under mild conditions by treatment with trifluoroacetic acid vapor, leaving the compounds adsorbed on the polymeric support. Since the described synthetic conditions could be applied to the parallel assembly of 8000 cellulose-bound triazines, the method could be of high potential for the parallel screening of small molecule compound libraries.

A similar approach has been described by the same authors for the synthesis of related cyclic peptidomimetics [44]. A set of ten nucleophiles was employed for the substitution of the chlorine atom of the cyclic triazinyl-peptide bound to the cellulose membrane. By virtue of the aforementioned rate enhancement effects for nucleophilic substitution of the solid-supported monochlorotriazines, these reactions could be rapidly carried out by microwave heating. All products were obtained in high purity, enabling systematic modification of the molecular properties of the cyclic peptidomimetics.

In general, this method was tested for the cyclization of peptides of various chain lengths. According to the SPOT-method described above [43], the N-terminal amino acids were attached to a photolinker-modified cellulose membrane (Scheme 7.27). 2,4,6-Trichloro-1,3,5-triazine was linked to the free N-termini of the peptides, and this was followed by deprotection of the Lys side chain. Cyclization was achieved by nucleophilic attack of the free amino group at the triazine moiety under basic conditions. Finally, the peptides were cleaved from the solid support by dry-state UV irradiation. For examples utilizing microwave heating, the remaining chlorine functionality on the triazine was substituted by a set of ten different nucleophiles immediately before the cleavage step [44].

In a recent study, this so-called SPOT synthesis was applied for the preparation of pyrimidines [45]. The group of Blackwell described primarily the appropriate support modification of commercially available cellulose sheets (Scheme 7.28). The initial introduction of the amine spacer was achieved within 15 min utilizing microwave irradiation, as compared to 6 h by conventional heating. The acid-cleavable Wang-type linker was attached by classical methods at ambient temperature.

The readily prepared support was then used for dihydropyrimidine and chalcone synthesis (Scheme 7.29). Thus, the modified support was activated prior to reaction by treatment with tosyl chloride. Solutions of the appropriate acetophenones were then spotted onto the membrane and the support was submitted to microwave irradiation for 10 min [45]. In the next step, several aryl aldehydes were attached under microwave irradiation to form a set of corresponding chalcones through a Claisen–Schmidt condensation.

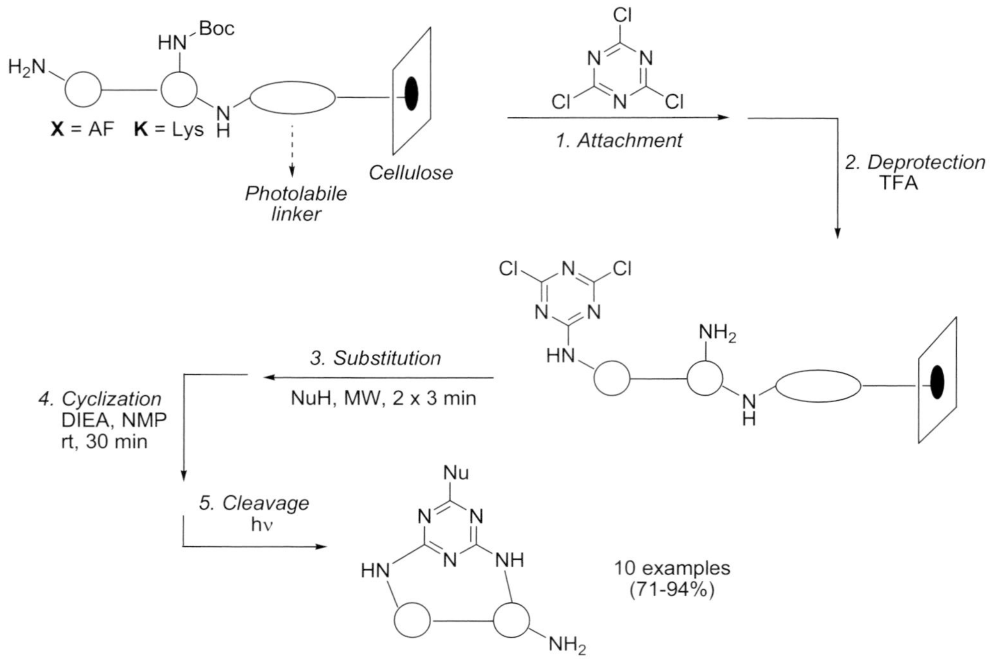

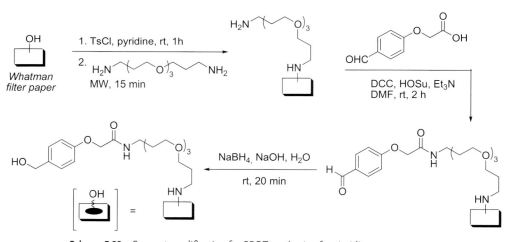

Scheme 7.27 Microwave-mediated synthesis and UV-induced cleavage of cyclic triazines on a cellulose membrane.

Scheme 7.28 Support modification for SPOT synthesis of pyrimidines.

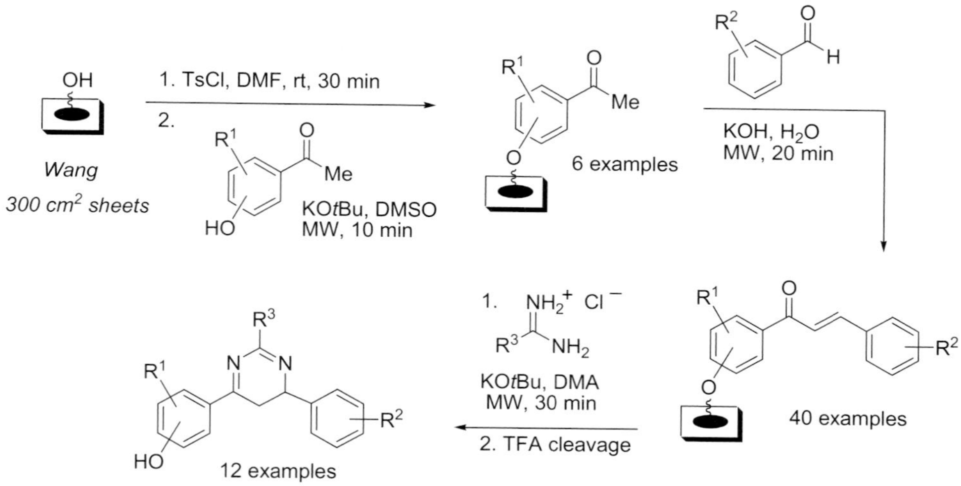

Scheme 7.29   Microwave-assisted SPOT synthesis of pyrimidines.

The successfully generated chalcones could be cleaved by treatment with trifluoro-acetic acid or used for the subsequent synthesis of pyrimidines [45]. Condensation of the polymer-bound chalcones with benzamidine hydrochloride under microwave irradiation for 30 min furnished the corresponding pyrimidines in good yields after TFA-induced cleavage. This new robust support/linker system for SPOT synthesis has been demonstrated to be compatible with a range of organic reactions and highly applicable for microwave conditions.

Very recently, an SPOS approach in DNA synthesis has been presented, involving a ten-step synthesis of a phosphoramidite building block of 1′-aminomethylthymi-dine starting from 2-deoxyribose [46]. Such oligonucleotides with chemical modifica-tions are routinely immobilized on surfaces to generate DNA microarrays. To facili-tate detection, the oligonucleotides are usually labeled with dyes or radioactive ele-ments. Grünefeld and Richert have described the microwave-mediated deprotection of an N-allyloxycarbonyl (alloc)-protected nucleoside and acylation with the residue of pyren-1-yl-butanoic acid (pyBA) (Scheme 7.30). To ensure complete conversion, the removal of the alloc protecting group was carried out on a support (controlled pore glass, cpg) under microwave conditions (80 °C, 10 min), as this step has proven to proceed sluggishly and is rather difficult under conventional conditions.

The readily prepared immobilized phosphoramidite could be used to efficiently synthesize oligodeoxyribonucleotides with modified thymidine residues. Whereas the effect of microwave irradiation on the deprotection by exposing the strand to tet-rakis-triphenylphosphine palladium(0) and diethylammonium bicarbonate was only small using dichloromethane as solvent, complete removal of the alloc group was achieved in N,N-dimethylformamide within 10 min at 80 °C (Scheme 7.30). After the reaction, the solid-supported product was washed with N,N-dimethylformamide and dichloromethane and dried, before being subjected to acylation. The coupling

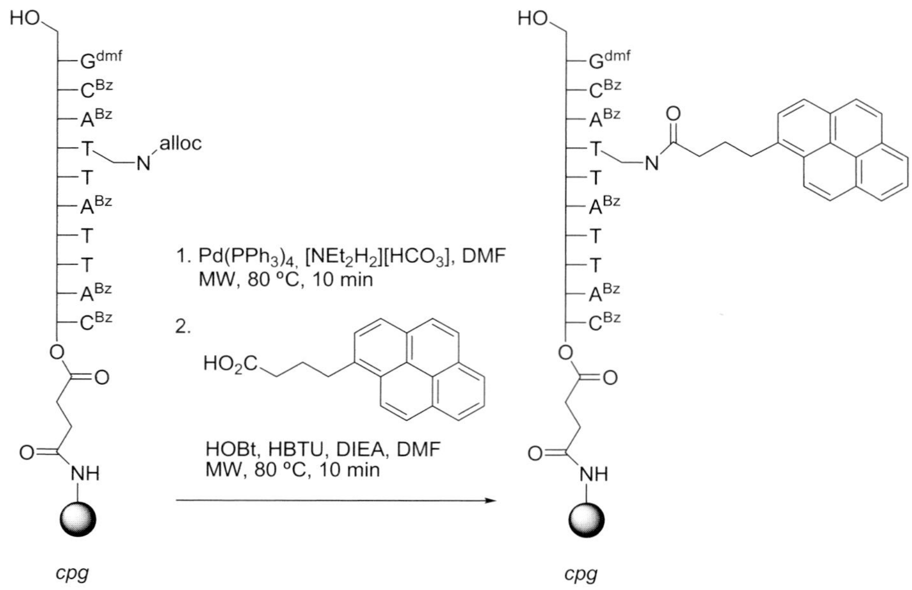

Scheme 7.30 Deprotection of oligonucleotides on controlled pore glass (cpg).

of the pyren-1-yl-butanoic acid (PyBA) was carried out under microwave irradiation at 80 °C for 10 min, employing a mixture of 1 equivalent of PyBA, 1 equivalent of 1-hydroxybenzotriazole (HOBt), and 0.9 equivalents of *O*-benzotriazol-1-yl-*N,N,N′,N′*-tetramethyluronium hexafluorophosphate (HBTU) together with 4.86 equivalents of *N,N*-diisopropylethylamine (DIEA) in *N,N*-dimethylformamide.

The progress of the deprotection and acylation was monitored by recording MALDI-TOF spectra of the crude products. In this particular case, microwave irradiation not only accelerated the palladium(0)-catalyzed deprotection and the ensuing amide formation, but also led to cleaner reactions. Apparently, the microwave heating breaks up any hindrance to the reaction sites such that the intermolecular reaction sites prevail and competing reactions are suppressed [46].

In a related study, Turnbull and coworkers described the attachment of carbohydrates to amino-derivatized glass slides. They found a significant rate enhancement when performing this step under microwave irradiation as compared to classical heating [47]. This method should be an efficient aid for the construction of functional carbohydrate array systems.

The group of Grieco has presented a method for efficiently performing macrocyclizations on a solid phase (Scheme 7.31) [48]. The preparation of the macrocyclic peptides required several standard transformations, which are not described in detail herein. The final intramolecular nucleophilic aromatic substitution step was carried out under microwave irradiation at 50 °C in a dedicated CombiCHEM system (see Fig. 3.9) utilizing microtiter plates in a multimode batch reactor. The cyclization product was obtained in good yield after a reaction time of 10 min and sub-

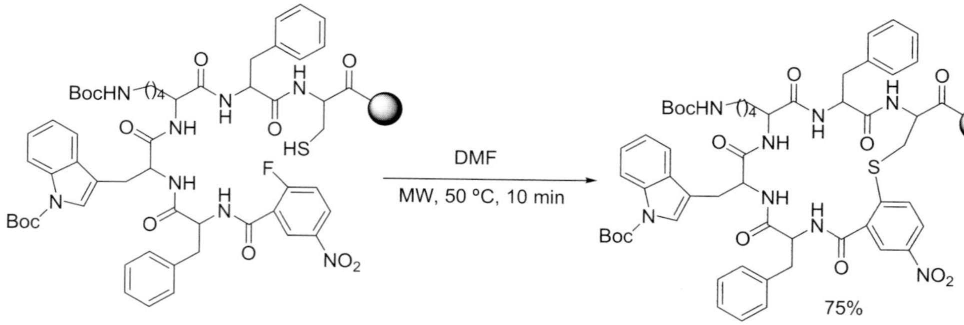

Scheme 7.31 Macrocyclization on a solid phase.

sequent cleavage, whereas with conventional heating the process required 16 h and furnished somewhat lower yields.

In a related study by the same group, the synthesis of cyclic peptides bearing non-natural thioether side chain bridges has been reported [49]. Instead of cyclization as described above, a microwave-mediated thioalkylation was performed (Scheme 7.32). The microwave reaction was again carried out in the above mentioned CombiCHEM system, employing the corresponding alkyl iodides and N,N-diisopropylethylamine (DIEA) as catalyst. The prepared precursor could be successfully cyclized following removal of the Fmoc and OFm protecting groups. Acidic cleavage utilizing trifluoroacetic acid furnished the desired products in good yield, whereas only moderate yields were achieved after conventional heating for several hours.

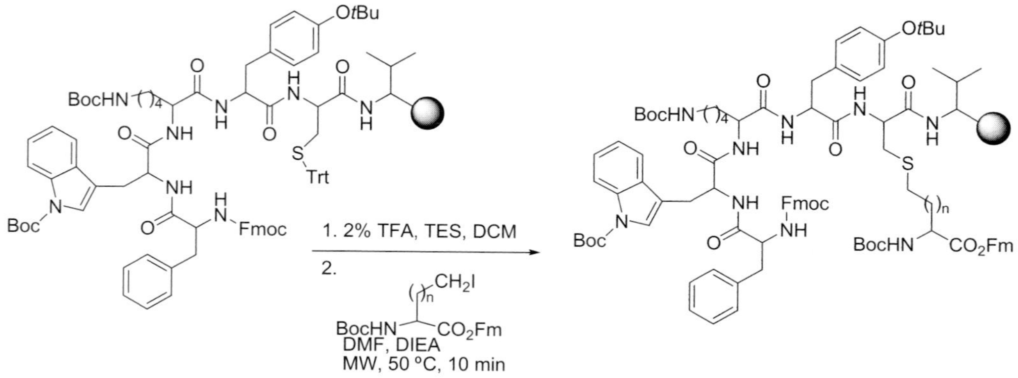

Scheme 7.32 Microwave-mediated thioalkylation to prepare precursors for macrocyclic peptide synthesis.

A different approach toward cyclic peptides has been presented by Leatherbarrow and coworkers, employing ring-closing metathesis (RCM) on a solid support [50]. The authors reported on the synthesis of conformationally strained cyclic peptides of the Bowman–Birk inhibitor type, which are naturally occurring serine protease

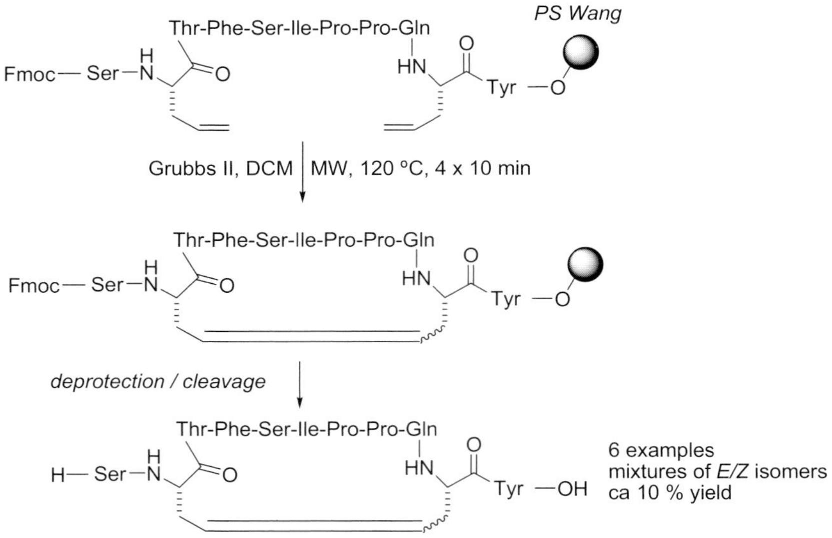

Scheme 7.33 Cyclic peptides by ring-closing metathesis.

inhibitors containing nine-residue disulfide-constrained loops. Leatherbarrow and his group demonstrated that the disulfide could be replaced by an all-carbon link employing RCM of a linear dienic peptide under microwave irradiation (Scheme 7.33). The reaction proceeded efficiently utilizing Grubbs' second-generation ruthenium benzylidene catalyst in dichloromethane and applying four cycles of microwave irradiation. Final cleavage was carried out by treatment with 95% trifluoroacetic acid, and the desired cyclic peptides were obtained in rather poor yields and as mixtures of *E/Z* isomers. Consequently, the true potencies of these compounds in enantiomerically pure form may be significantly higher than those quoted.

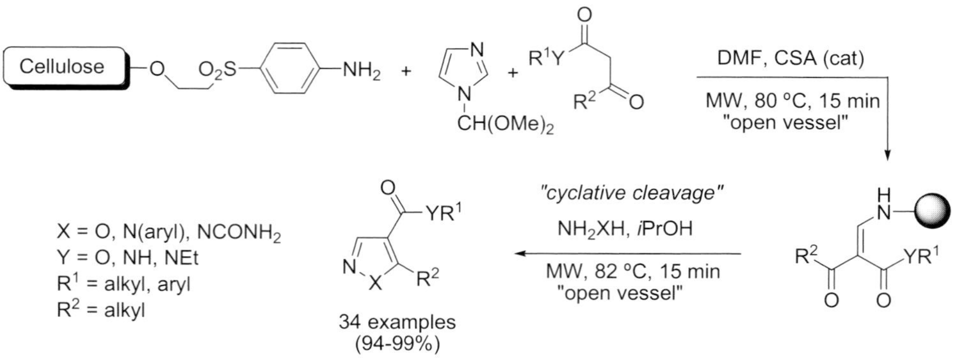

Scheme 7.34 Generation of pyrazole and oxazole libraries on cellulose beads.

The first report on the use of cellulose beads as a support for microwave-assisted SPOS was presented by the group of Taddei [51]. The authors generated a library of pyrazoles and isoxazoles, which are pharmacologically important heterocyclic scaffolds, through the *in situ* generation of polymer-bound enamines. The synthesis was performed using commercially available amino cellulose (Perloza VT-100) under mild conditions. Cellulose shows good swelling properties in the polar solvents used, is biodegradable, and furthermore, could be recycled for use in subsequent reactions. The cellulose beads utilized incorporated aminoaryl ethyl sulfone groups in flexible chains (Scheme 7.34). Initially, the solid support was treated with an excess of formyl imidazole and the corresponding *β*-keto compounds to generate cellulose-bound enaminones in a one-pot Bredereck-type condensation. The reaction was catalyzed by (+/–)-camphor-10-sulfonic acid (CSA) and carried out under microwave irradiation in an open vessel in order to allow the methanol formed to be removed from the reaction equilibrium [51].

Subsequent cyclization and cleavage from the support was achieved by microwave heating with several hydrazines or hydroxylamines in 2-propanol to afford the desired heterocyclic targets. The progress of the reaction was monitored by the disappearance of the carbonyl function on the polymeric support by a negative iron(III) chloride colorimetric test. Moreover, the cellulose-bound aniline could be recycled by simply washing with methanol and diethyl ether and then drying in vacuo, and could be re-used up to ten times without loss of efficiency or decreasing purity of the resulting compounds.

A microwave-assisted solid-phase synthesis of purines has been reported by the group of Al-Obeidi [52]. The heterocyclic scaffold was first attached to the acid-sensitive methoxybenzaldehyde (AMEBA)-linked polystyrene (Fig. 7.3) through an aromatic nucleophilic substitution reaction, performed by conventional heating in 1-methyl-2-pyrrolidinone (NMP) in the presence of *N,N*-diisopropylethylamine. The key nucleophilic aromatic substitution of the iodo substituent with primary and secondary amines was conducted by microwave heating for 30 min at 200 °C in 1-methyl-2-pyrrolidone (Scheme 7.35). This microwave heating step was carried out in a dedicated multimode oven employing Teflon reaction vessels [45]. The resin was subsequently washed with tetrahydrofuran and methanol and, after drying, the products were cleaved from the solid support using trifluoroacetic acid/water at 60 °C.

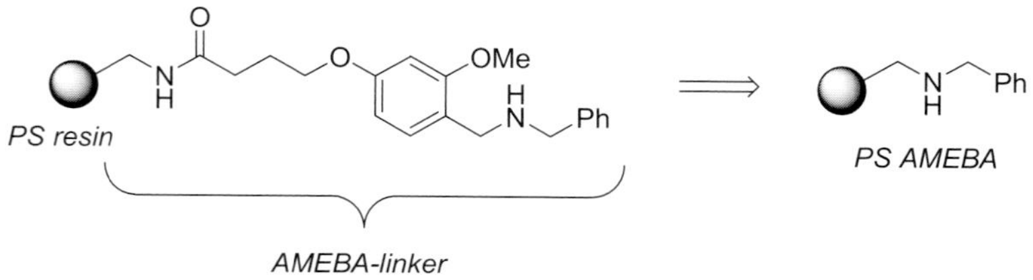

AMEBA-linker

Fig. 7.3 AMEBA-linked polystyrene resin.

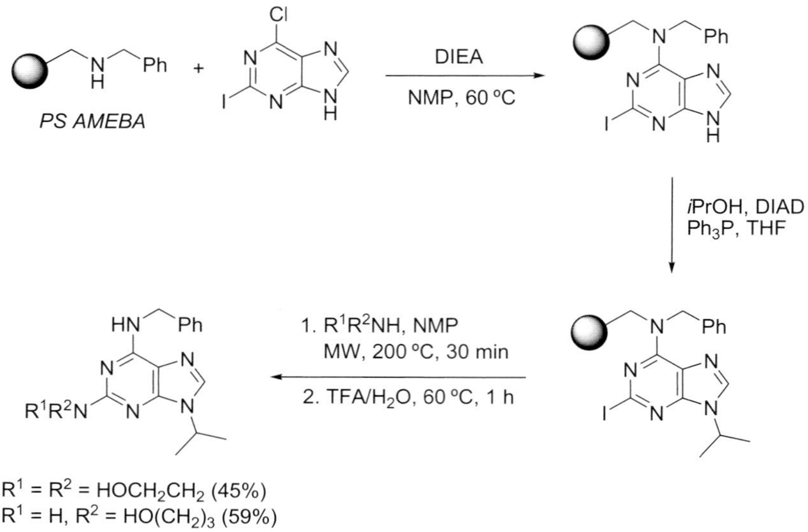

Scheme 7.35   Purine synthesis on a solid support.

7.1.7
Multicomponent Chemistry

Multicomponent reactions, in which three or more components come together to form a single product, have been the subject of considerable interest for several years. Since most of these reactions tolerate a wide range of building block combinations, they are frequently applied for combinatorial purposes.

The first solid-phase application of the Ugi four-component condensation, generating an 18-member acylamino amide library, was presented in 1999 by Nielsen and Hoel [53]. The authors described a library generation utilizing amino-functionalized PEG-polystyrene (Tentagel S RAM) as the solid support (Scheme 7.36). A set of three aldehydes, three carboxylic acids, and two isonitriles was used for the generation of the 18-member library.

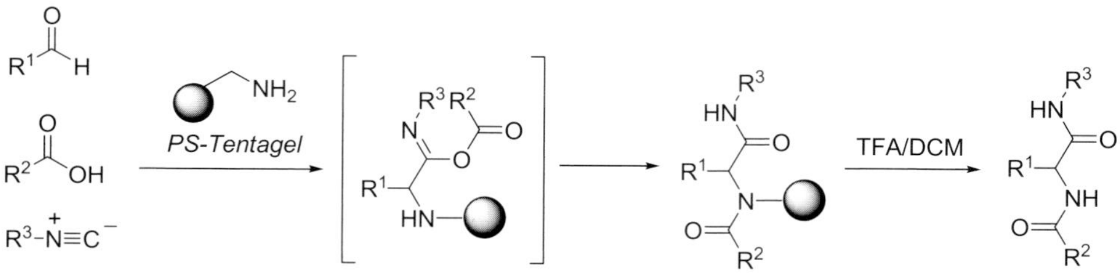

Scheme 7.36   Solid-supported Ugi condensation.

In a typical procedure, the PS-Tentagel Fmoc-protected amino resin was depro-tected using 20% piperidine in *N,N*-dimethylformamide. After placing the free amino resin in an appropriate microwave vial, it was swollen in a mixture of a 1 M solution of the appropriate aldehyde in dichloromethane and a 1 M solution of the corresponding carboxylic acid in methanol. After 30 min, a 1 M solution of the req-uisite isonitrile in dichloromethane was added to the pre-swollen resin mixture. The vial was flushed with nitrogen, sealed, and then irradiated for 5 min. Subsequent cleavage of the resins with trifluoroacetic acid/dichloromethane and evaporation of the solvents afforded the products in high purities, albeit in varying yields.

In a recent study, another method for microwave-assisted heterocycle synthesis leading to a small set of imidazole derivatives has been reported [54]. These pharma-ceutically important scaffolds were synthesized utilizing polymer-bound 3-*N,N*-(dimethylamino)isocyanoacrylate. This polymer support was easily prepared by treatment of [4-(bromomethyl)phenoxy]methyl polystyrene with a twofold excess of the appropriate isocyanoacrylate potassium salt in *N,N*-dimethylformamide (Scheme 7.37). The obtained intermediate was subsequently treated with *N,N*-di-methylformamide diethyl acetal (DMFDEA) in a mixture of tetrahydrofuran and eth-anol to generate the desired polymer-bound substrate.

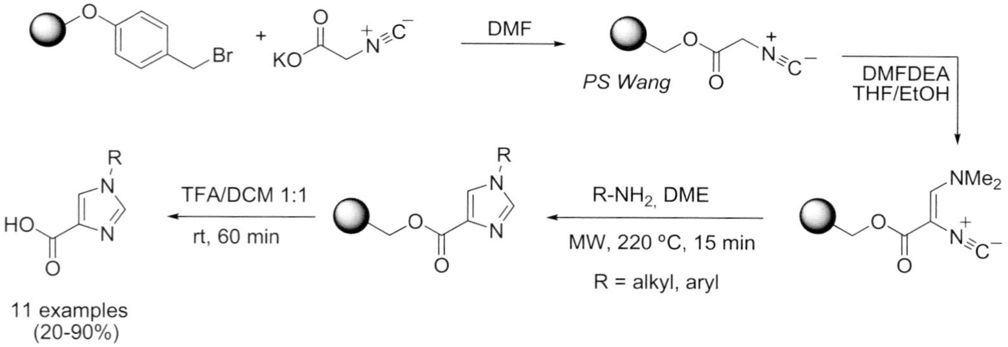

Scheme 7.37 Polymer-bound synthesis of 1-substituted 4-imidazolecarboxylates.

The best results in the imidazole synthesis were obtained by microwave-assisted reaction of an eightfold excess of the polymer-supported isonitrile suspended in 1,2-dimethoxyethane (DME) with the appropriate amines. Cleavage with 50% trifluoro-acetic acid in dichloromethane afforded the desired heterocyclic scaffolds in moder-ate yields.

Another interesting multicomponent reaction is the Gewald synthesis, leading to 2-acyl-3-aminothiophenes, which are of current interest since they are used com-mercially as dyes, form conducting polymers, and have shown extensive potential for pharmaceutical purposes. Earlier reports of the classical Gewald synthesis have described rather long reaction times by conventional heating and laborious purifica-tion of the resulting thiophenes.

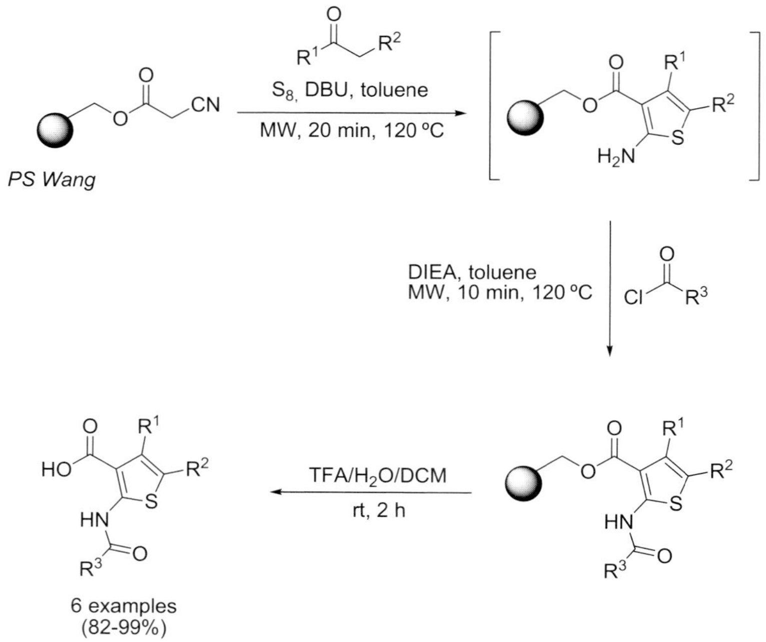

Scheme 7.38 One-pot Gewald synthesis on a polymeric support.

In view of these issues, Frutos Hoener and coworkers investigated a "one-pot" microwave-assisted Gewald synthesis on a solid support [55]. The reactions were carried out by employing commercially available cyanoacetic acid Wang resin as the solid support. Applying microwave irradiation, the overall two-step reaction procedure, including the acylation of the initially formed 2-aminothiophenes, could be performed in less than 1 h. This solid-phase "one-pot", two-step, microwave-promoted process constitutes an efficient route to 2-acylaminothiophenes (Scheme 7.38), requiring no filtration in between the two reaction steps. In total, 12 diverse aldehydes, ketones, and acylating agents were employed to generate the desired thiophene products in high yields and generally good purities.

7.1.8
Microwave-Assisted Condensation Reactions

As discussed in Section 7.1.4, polymer-bound acetoacetates can be used as precursors for the solid-phase synthesis of enones [33]. For these Knoevenagel condensations, the crucial step is to initiate enolization of the CH acidic component. In general, enolization can be initiated with a variety of catalysts (for example, piperidine, piperidinium acetate, ethylenediamine diacetate), but for the microwave-assisted procedure piperidinium acetate was found to be the catalyst of choice, provided that the temperature was kept below 130 °C. At higher reaction temperatures, there is significant cleavage of material from the resin.

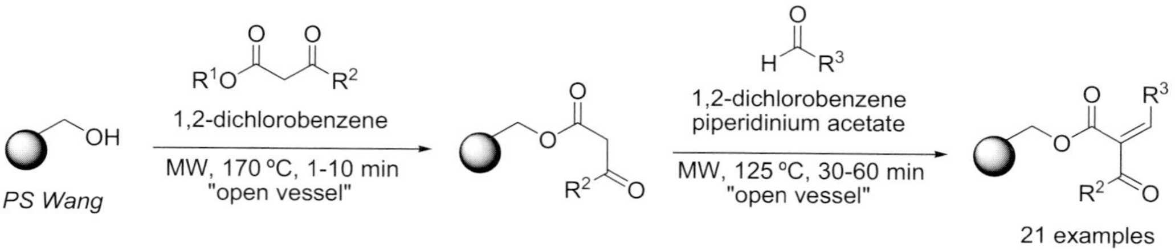

Scheme 7.39 Parallel synthesis of polymer-bound enones.

To ensure complete conversion for all examples of a 21-member library, irradiation times of 30–60 min were used (Scheme 7.39), employing a multi-vessel rotor system for parallel microwave-assisted synthesis (see Fig. 3.7). The results were confirmed by on-bead FTIR analysis, accurate weight-gain measurements of washed and dried resins, and post-cleavage analysis of the prepared enones.

These examples of Knoevenagel condensations illustrated that reaction times could be reduced from 1–2 days to 30–60 min by employing parallel microwave-promoted synthesis in open vessels, without affecting the purity of the resin-bound products.

Microwave-assisted Knoevenagel reactions have also been utilized in the preparation of resin-bound nitroalkenes [56]. The generation of various resin-bound nitroalkenes employing resin-bound nitroacetic acid has been described; the latter was condensed with a variety of aldehydes under microwave conditions (Scheme 7.40).

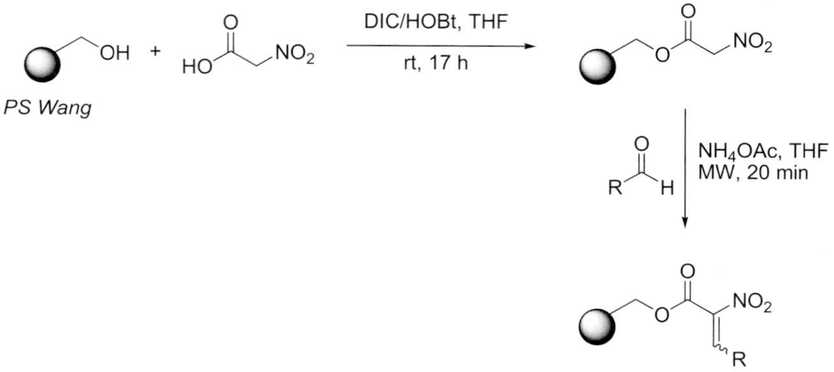

Scheme 7.40 Generation of polymer-bound nitroalkenes by Knoevenagel condensation.

In order to demonstrate the potential of these resin-bound products for combinatorial applications, the readily prepared nitroalkenes were subsequently employed in Diels–Alder reactions with 2,3-dimethylbutadiene [56]. In addition, the resin-bound nitroalkenes were also used in a "one-pot", three-component tandem [4+2]/[3+2] reaction with ethyl vinyl ether and styrene.

The group of Janda has presented a microwave-mediated oxazole synthesis utilizing β-ketoesters bound to a novel polymeric resin [57]. The desired polymer support was prepared by transesterification reactions of *tert*-butyl β-ketoesters and hydroxybutyl-functionalized *JandaJel* resin and subsequent standard diazo transfer. The resulting α-diazo β-ketoesters were employed for the synthesis of an array of oxazoles (Scheme 7.41).

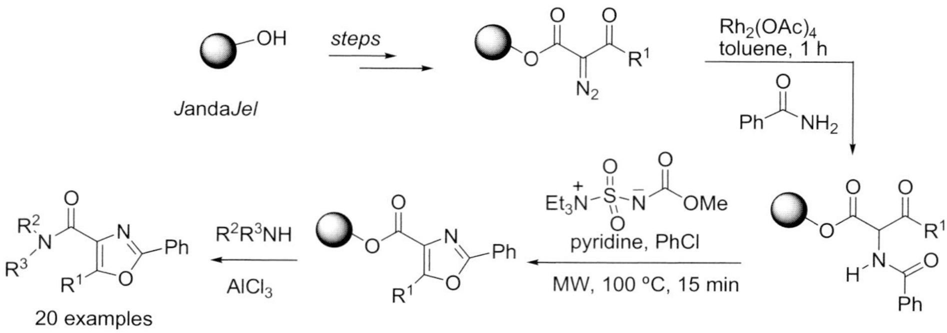

Scheme 7.41 Oxazole synthesis on functionalized *JandaJel*.

In a series of experiments, polymer-bound cyclization was investigated under both microwave irradiation and classical heating. In the microwave-assisted method, the use of 3 equivalents of Burgess' reagent and 20 equivalents of pyridine in chlorobenzene proved optimal. Due to the thermal sensitivity of the Burgess reagent, the temperature was kept rather low, but irradiation at 100 °C for 15 min furnished satisfying results for a range of oxazoles [57]. Cleavage from the solid support was achieved by a diversity-introducing amidation, leading to the corresponding oxazole amides in reasonable yields.

7.1.9
Rearrangements

Microwave-assisted rearrangement reactions on solid phases have not been discussed in the literature very often. To date, only two examples describing Claisen rearrangements have been reported [58, 59]. In the first study, Merrifield resin-bound *O*-allylsalicylic esters were rapidly rearranged to the corresponding *ortho*-allyl-salicylic esters under microwave heating [58] (Scheme 7.42). Acid-mediated cleavage of these resin-bound ester products afforded the corresponding *ortho*-allylsalicylic acids.

The resin-bound salicylic esters were suspended in *N,N*-dimethylformamide (DMF) and placed in an Erlenmeyer flask within a domestic microwave cavity. After microwave irradiation for 4–6 min (1 min cycles), the reaction mixture was allowed to cool to ambient temperature and the resin was collected by filtration and washed with methanol and dichloromethane. The desired compounds were subsequently cleaved with trifluoroacetic acid in dichloromethane. Removal of the solvent by evap-

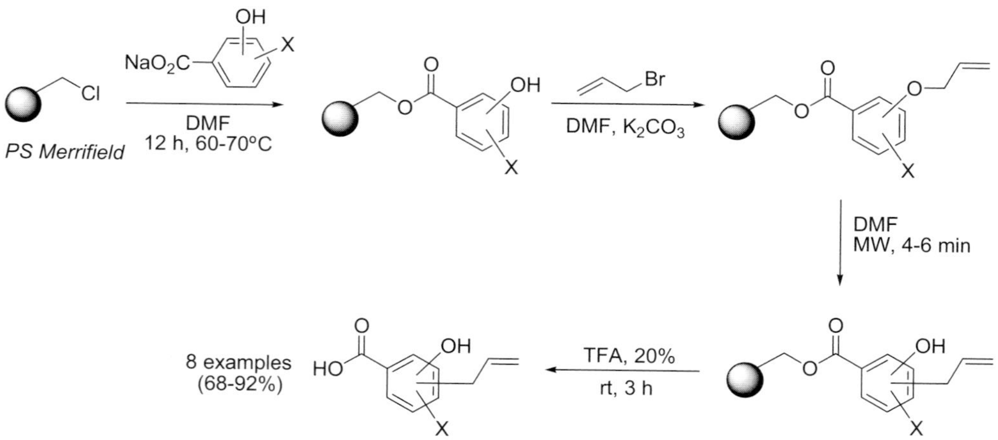

Scheme 7.42 Claisen rearrangement on a solid phase.

oration gave the corresponding acid products in high yields. Compared to conventional thermal heating (*N*,*N*-dimethylformamide, 140 °C), reaction times could be drastically reduced from 10–16 h to a few minutes using microwave flash heating and, moreover, higher yields of the products were obtained.

In a more recent report, a microwave-assisted tetronate synthesis was investigated. For this purpose, Schubert and Jagosch performed domino addition/Wittig olefinations of polymer-bound α-hydroxy esters with the cumulated phosphorus ylide $Ph_3P=C=C=O$ [59]. Employing α-hydroxy allyl esters gave either polymer-supported allyl tetronates or Claisen-rearranged tetronic acids, depending on the reaction conditions.

The desired immobilized α-hydroxy esters could be obtained by ring-opening of the corresponding glycidyl esters by OH-, NH- or SH-terminal polystyrenes of the Merrifield or Wang types (Scheme 7.43). Applying microwave irradiation (85 °C, 30 min), lithium perchlorate showed the highest efficiency in this ring-opening. The subsequent tandem Wittig reaction was carried out in tetrahydrofuran under

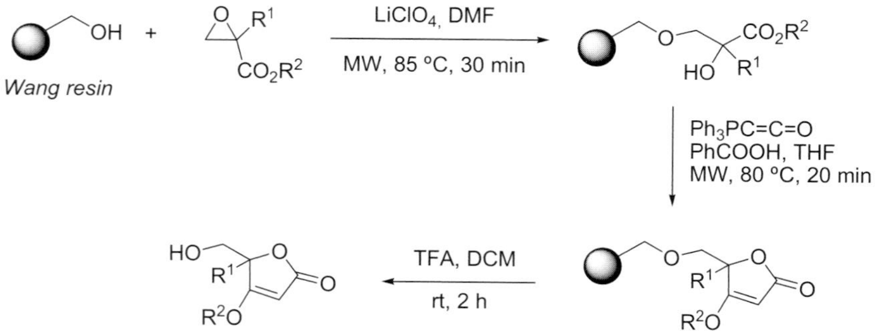

Scheme 7.43 Tetronate synthesis by domino addition/Wittig olefination.

microwave irradiation, employing catalytic amounts of benzoic acid. The formation of the polymer-bound tetronates was complete within 20 min of irradiation at 80 °C [59]. Quantitative cleavage of the tetronates was achieved by treatment with 50% trifluoroacetic acid in dichloromethane at room temperature for 2 h.

Allyl esters ($R^2 = CH_2C(R^3)=CH_2$) could be transformed to the corresponding 3-allyltetronic acids by maintaining microwave irradiation at 120 °C for 60 min [59]. However, cleavage of the tetronic acids under the above mentioned conditions remained somewhat troublesome. To obtain satisfactory amounts of product, the polymer-bound intermediates had to be protected at the O4 position so that cleavage could be accomplished with a mixture of trifluoroacetic acid/dichloromethane (1:9) (Scheme 7.44).

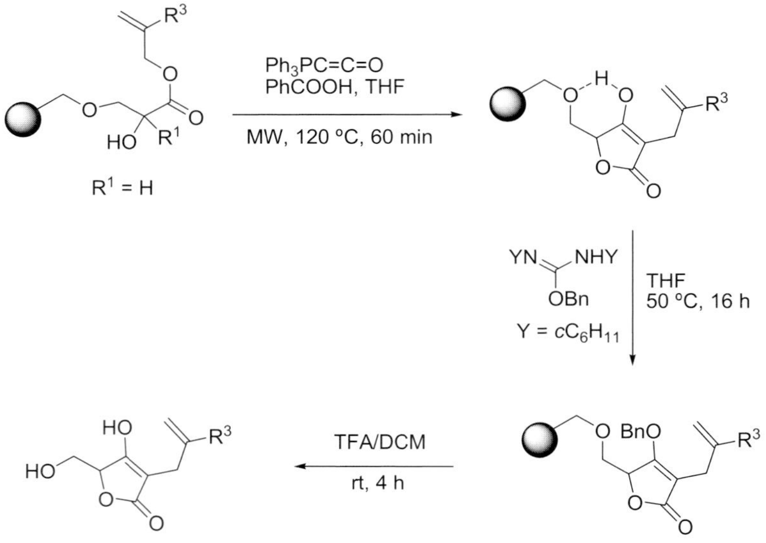

Scheme 7.44 Release of polymer-bound tetronic acids.

7.1.10
Cleavage Reactions

One of the key steps in combinatorial solid-phase synthesis is clearly the cleavage of the desired product from the solid support. A variety of cleavage protocols have been investigated, depending on the nature of the linker employed. A complete microwave-assisted protocol, involving attachment of the starting material to the solid support, scaffold preparation, scaffold decoration, and cleavage of the resin-bound product, would seem desirable.

A protocol for microwave-assisted acid-mediated resin cleavage has been presented by Stadler and Kappe [28]. Several resin-bound carboxylic acids (see Scheme 7.8) were cleaved from traditionally non-acid-sensitive Merrifield resin by

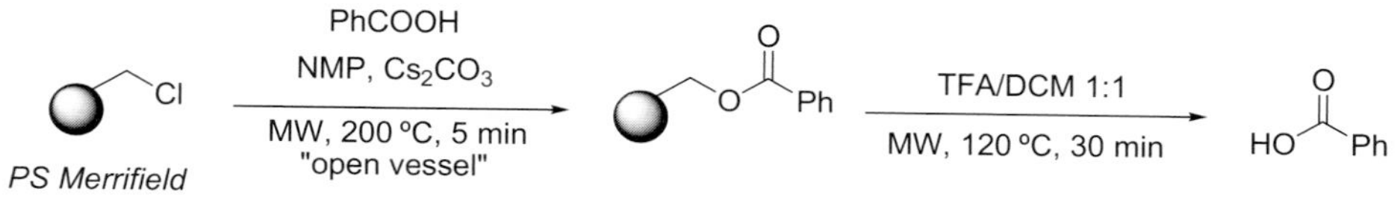

Scheme 7.45 Acidic cleavage under elevated pressure.

employing 50% trifluoroacetic acid in dichloromethane under microwave heating (Scheme 7.45).

In general, acidolysis of the Merrifield linker requires acids with high ionizing power, such as hydrogen fluoride, trifluoromethanesulfonic acid, or hydrogen bromide/acetic acid. Therefore, under conventional conditions, cleavage does not take place with trifluoroacetic acid. Employing microwave irradiation allows these cleavages to be performed at elevated pressure/temperature using sealed vessels. Thus, the resin-bound ester and the trifluoroacetic acid/dichloromethane mixture were placed in a 100-mL PFA reactor and heated with stirring for 30 min at 120 °C. Subsequent concentration of the filtrate to dryness furnished the benzoic acid in quantitative yield and with excellent purity. No degradation of the polymer support could be detected, even though the reaction conditions were rather harsh for solid-phase chemistry.

A different protocol was presented by Glass and Combs, elaborating the Kenner safety-catch principle for the generation of amide libraries [17, 18]. In another application of microwave-assisted resin cleavage, *N*-benzoylated alanine attached to 4-sulfamylbutyryl resin was cleaved (after activation of the linker with bromoacetonitrile employing Kenner's safety-catch principle) with a variety of amines (Scheme 7.46). Cleavage rates in dimethyl sulfoxide were investigated for *N,N*-diisopropylamine and aniline under various reaction conditions using both microwave (domestic oven) and conventional (oil bath) heating. The results showed that microwave heating did not accelerate the reaction rates over conventional heating when experiments were run at the same temperature (ca. 80 °C). However, using microwave

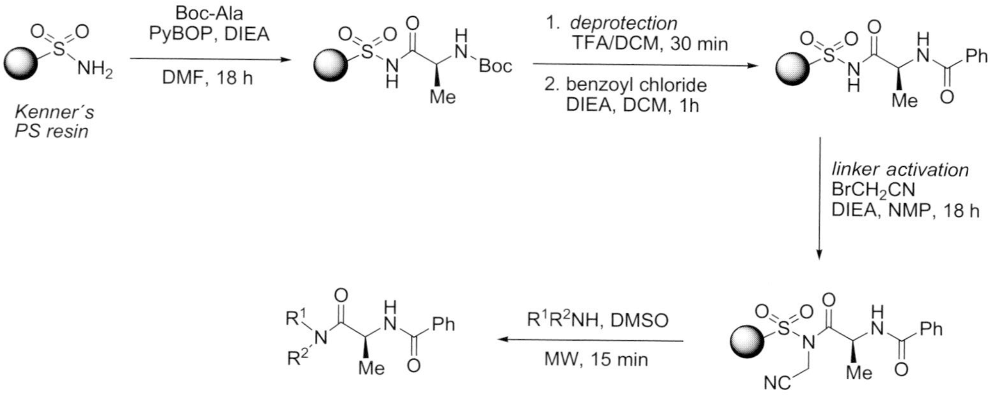

Scheme 7.46 Solid-supported amide synthesis employing the "safety-catch" principle.

heating, even cleavage with the normally unreactive aniline could be accomplished within 15 min at ca. 140 °C.

The microwave approach was used for the parallel synthesis of an 880-member library utilizing 96-well plates, employing ten different amino acids coupled to 4-sulf-amylbutyryl resin, each bearing a different acyl group, and using 88 diverse amines in the cleavage step. After linker activation, each resin was suspended in a solvent mixture of dichloromethane and *N,N*-dimethylformamide and split between 88 wells of an appropriate filter plate, using a 96-channel pipettor. Dimethyl sulfoxide was then added to each well as solvent, followed by a set of primary and secondary amines from a stock microtiter plate, again transferred using a 96-channel pipettor. Sets of four plates were placed in a domestic microwave oven and were heated at ca. 80 °C for 60 s. After the plates had cooled down, the solutions were drained from the wells into a collection microtiter plate and were combined with dimethyl sulfoxide resin washings from the respective wells to afford 10 mM solutions of the products in dimethyl sulfoxide for biological screening. Temperature gradients between the wells of the microtiter plates did not represent a significant issue for this type of chemistry.

In closely related work, similar solid-phase chemistry was employed by the same research group to prepare biaryl urea compound libraries by microwave-assisted Suzuki couplings followed by cleavage from the resin with amines (Scheme 7.47) [18]. The above described procedure enabled the generation of large biaryl urea compound libraries employing a simple domestic microwave oven.

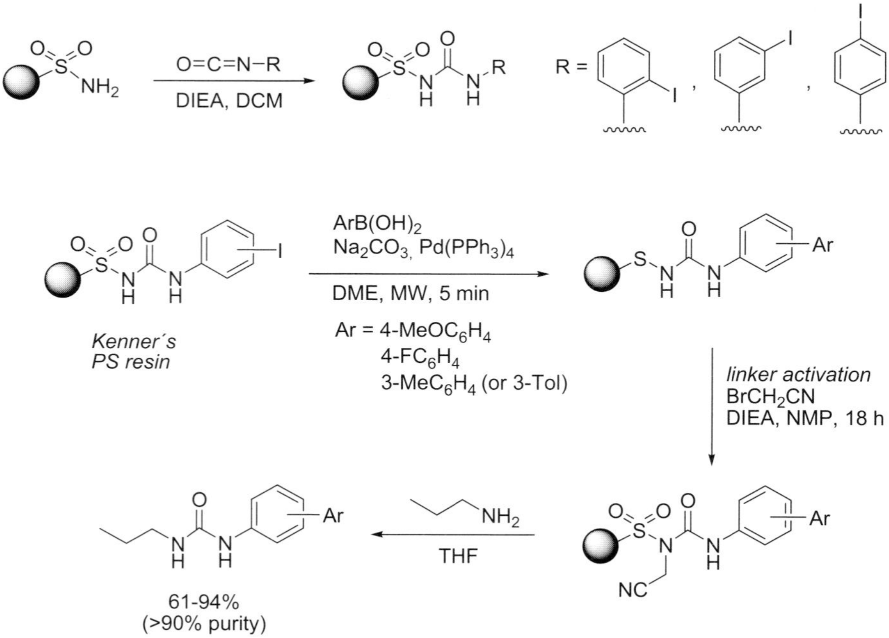

Scheme 7.47 Synthesis of urea derivatives on a solid support.

A recent study concerned the microwave-assisted parallel synthesis of di- and tri-substituted ureas utilizing dedicated 96-well plates in the CombiCHEM system [60]. In a typical procedure, modification of the Marshall resin utilized was achieved by treatment with *p*-nitrophenyl chloroformate and *N*-methylmorpholine (NMM) in dichloromethane at low temperatures. The resulting resin was further modified by attaching various amines to obtain a set of polymer-bound carbamates (Scheme 7.48).

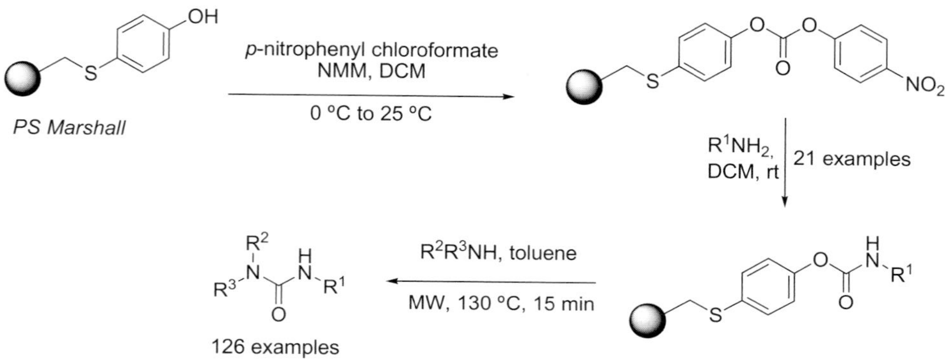

Scheme 7.48 Parallel synthesis of substituted ureas from thiophenoxy carbamate resins.

The immobilized carbamates (40 µmol) were transferred to a sealable 96-well Weflon plate, and admixed with 10 µmol each of various primary or secondary amines dissolved in 400 µL of anhydrous toluene. After sealing, the plate was irradiated in a multimode microwave instrument, first generating a ramp to reach 130 °C within 45 min and then holding this temperature for an additional 15 min. After cooling, the resins were filtered with the aid of a liquid handler and the filtrates were concentrated to obtain the desired substituted ureas in good purity and reasonable yields. Anilines reacted rather sluggishly and 2-substituted benzyl carbamates afforded somewhat inferior results.

In a study by the Kappe group, multidirectional cyclative cleavage transformations leading to bicyclic dihydropyrimidinones have been employed [15]. This approach required the synthesis of 4-chloroacetoacetate resin as the key starting material, which was prepared by microwave-assisted acetoacetylation of commercially available hydroxymethyl polystyrene resin under open-vessel conditions. This resin precursor was subsequently treated with urea and various aldehydes in a Biginelli-type multi-component reaction, leading to the corresponding resin-bound dihydropyrimidinones (Scheme 7.49). The desired furo[3,4-*d*]pyrimidine-2,5-dione scaffold was obtained by way of a novel protocol for cyclative release under microwave irradiation. The resin bearing the monocyclic pyrimidine intermediate was pre-swollen in *N*,*N*-dimethylformamide (DMF) in a sealed microwave vial and was then irradiated in a dedicated single-mode microwave cavity at 150 °C for 10 min. After cooling to ambient temperature, the resin was filtered off and washed with DMF to afford the corresponding furo[3,4-*d*]pyrimidine-2,5-dione in high purity.

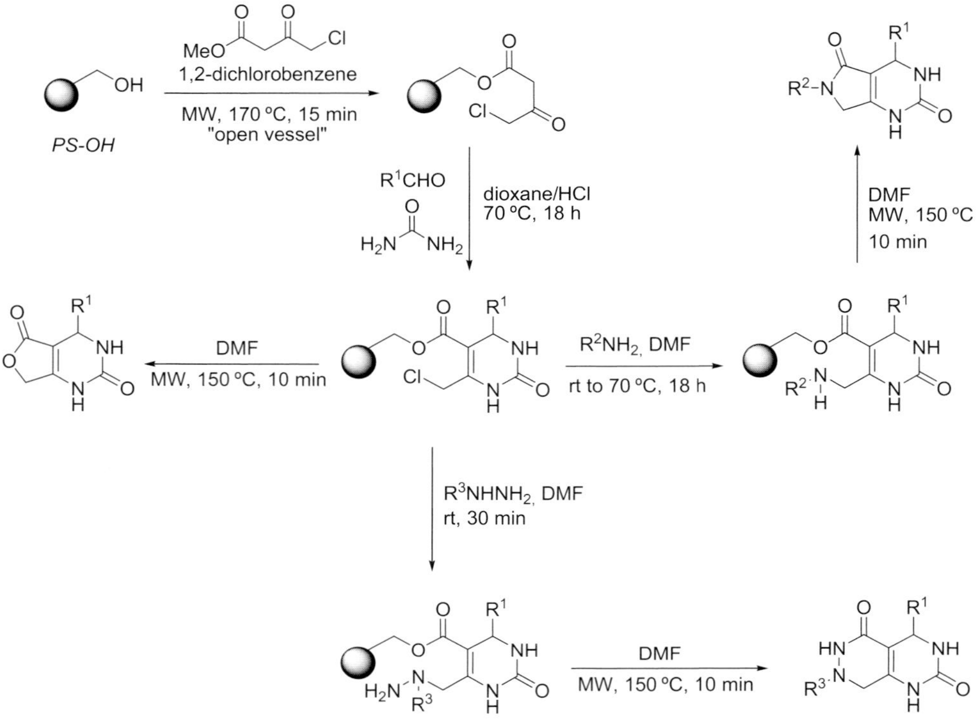

Scheme 7.49 Preparation of bicyclic dihydropyrimidinones employing cyclative cleavage.

Alternatively, pyrrolo[3,4-*d*]pyrimidine-2,5-diones were synthesized using the same pyrimidine resin precursor, which was first treated with a representative set of primary amines to substitute the chlorine. Subsequent cyclative cleavage was carried out as described previously, leading to the corresponding pyrrolopyrimidine-2,5-dione products in high purity but moderate yield.

Pyrimido[4,5-*d*]pyridazine-2,5-diones were synthesized in a similar manner, employing several hydrazines (R^3 = H, Me, Ph) for the nucleophilic substitution prior to cyclative cleavage. Due to the high nucleophilicity of the hydrazines, reaction times for the substitution step could be reduced to 30 min. In the case of phenylhydrazine, concomitant cyclization could not be avoided, which led to very low overall yields of the isolated products.

In an earlier report presented by Kurth and coworkers, a microwave-mediated intramolecular carbanilide cyclization to hydantoins was described (Scheme 7.50) [61]. Since the hydantoin moiety imparts a broad range of biological activities, several protocols involving both reactions in solution and on solid phases have been investigated. From preliminary studies, it was known that the carbanilide cyclization required rather long reaction times under conventional thermal heating (8–48 h). Therefore, it was obvious to investigate the effect of microwave irradiation upon this reaction. Reaction studies were carried out employing a single-mode microwave cav-

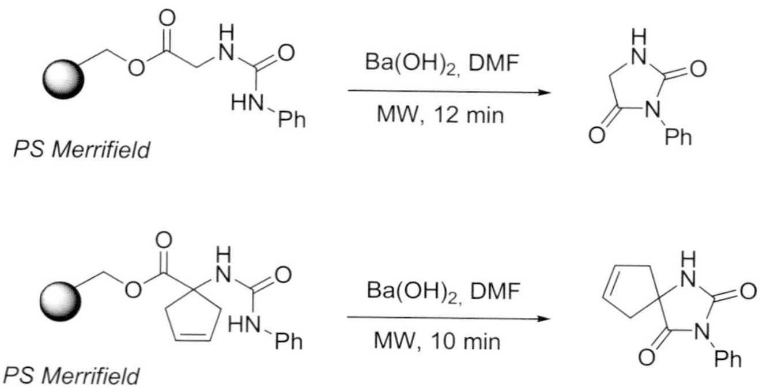

Scheme 7.50 Intramolecular carbanilide cyclization on a solid support.

ity with a variety of base and solvent combinations; barium hydroxide in N,N-dimethylformamide (DMF) proved to be the best combination for this cyclization reaction. Under these solution-phase conditions, carbanilides could be converted to the corresponding hydantoins in high yields within 2–7.5 min. With the appropriate solid-supported protocol, the carbanilide cyclization would act as a method for resin release of the hydantoins; reaction times could be drastically reduced to several minutes as compared to 48 h under thermal heating.

For this solid-phase approach, conventional iPrOCH$_2$-functionalized polystyrene resin (Merrifield linker) was employed. After attachment of the requisite substrate, the resin was pre-swollen in a solution of barium(II) hydroxide in N,N-dimethylformamide within an appropriate sealed microwave vial. The vial was heated in the microwave cavity for 5×2 min cycles (overall 10 min) with the reaction mixture being allowed to cool to room temperature in between irradiation cycles (Scheme 7.50), leading to comparatively modest isolated yields of hydantoins.

A more recent study by the group of Chassaing concerned the traceless solid-phase preparation of phthalimides by cyclative cleavage from conventional Wang resin (Scheme 7.51) [62]. In order to establish the optimum conditions for the cyclative cleavage step, ortho-phthalic acid was chosen as a model compound to be attached to the polystyrene resin. Applying a set of preliminary esterification methods, the Mitsunobu protocol gave the best results and highest purities of the desired intermediates.

For the cyclative cleavage step, it turned out that aprotic conditions were definitely superior to the use of protic media. Thus, employing N,N-dimethylformamide as solvent at somewhat elevated temperatures furnished the desired compounds in high yields and excellent purities. Having established the optimized conditions, various phthalic acids and amines were employed to prepare a set of phthalimides (Scheme 7.51). However, the nature of the amine was seen to have an effect on the outcome of the reaction. Benzyl derivatives furnished somewhat lower yields, probably due to the reduced activities of these amines. Aromatic amines could not be included in the study as auto-induced ring-closure occurred during the conversion of the polymer-bound phthalic acid.

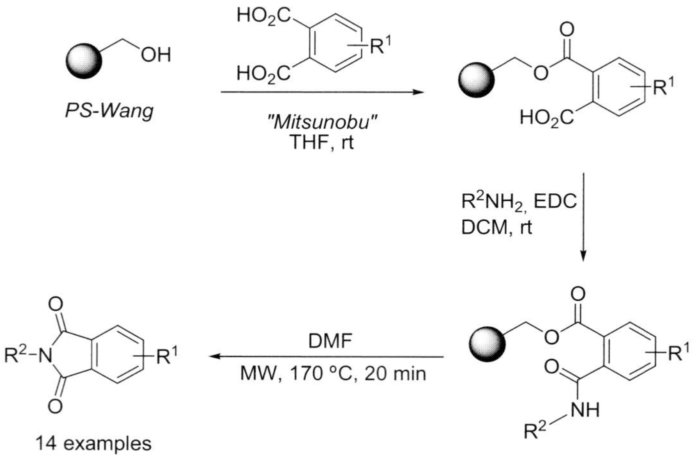

Scheme 7.51 Polymer-supported phthalimide synthesis.

7.1.11
Miscellaneous

Several other reaction types on solid supports have also been investigated utilizing microwave heating. For instance, in an early report, Yu and coworkers monitored the addition of resin-bound amines to isocyanates employing on-bead FTIR measurements in order to investigate the differences in reaction progress under microwave heating and thermal conditions [63].

The isocyanates were added to the respective resin-bound amines suspended in dichloromethane in open glass tubes. The resulting reaction mixtures were each irradiated in a single-mode microwave cavity for 2 min intervals (no temperature measurement given) (Scheme 7.52). After each step, samples were collected for on-bead FTIR analysis. Within 12 min (six irradiation cycles), each reaction had reached completion. Acid cleavage of the polymer-bound ureas furnished the corresponding hydrouracils.

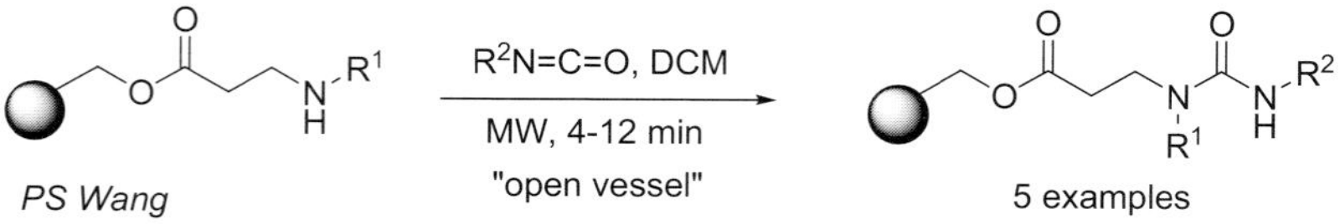

Scheme 7.52 Addition of isocyanates to resin-bound amines.

A very interesting approach toward solid-supported synthesis under microwave heating was introduced by Chandrasekhar and coworkers [64]. The authors developed a synthesis of *N*-alkyl imides on a solid phase under solvent-free conditions employing tantalum(V) chloride-doped silica gel as a Lewis acid catalyst (Scheme 7.53).

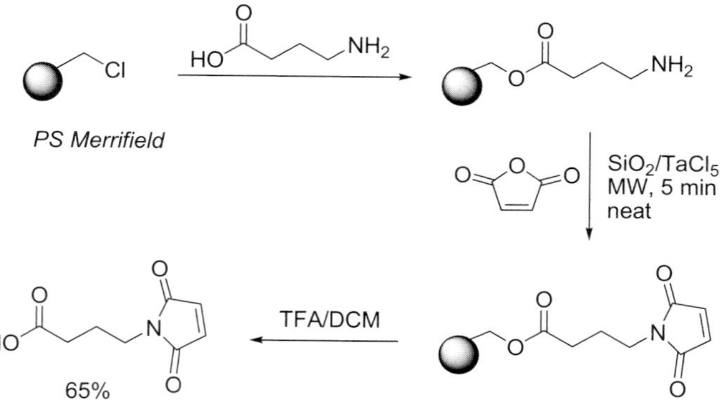

Scheme 7.53 Solvent-free preparation of N-alkyl imides.

Surprisingly, in this rather unusual method, dry and unswollen polystyrene resin is involved. The reaction was carried out employing a domestic microwave oven, performing 1 min irradiation cycles with thorough agitation after each step. Within 5–7 min, the reaction was complete, and the N-alkyl imide product was obtained in good yield after subsequent cleavage from the polystyrene resin/silica gel mixture employing trifluoroacetic acid/dichloromethane. In addition, employing two resin-bound amines and three different anhydrides in this solid-phase protocol, a set of six cyclic imides was synthesized in good yields.

Weik and Rademann have described the use of phosphoranes as polymer-bound acylation equivalents [65]. The authors chose a norstatine isostere as a synthetic target and employed classical polymer-bound triphenylphosphine in their studies (Scheme 7.54). Initial alkylation of the polymer-supported reagent was achieved with bromoacetonitrile under microwave irradiation. Simple treatment with triethyl-amine transformed the polymer-bound phosphonium salt into the corresponding stable phosphorane, which could be efficiently coupled with various protected amino acids. In this acylation step, the exclusion of water was crucial.

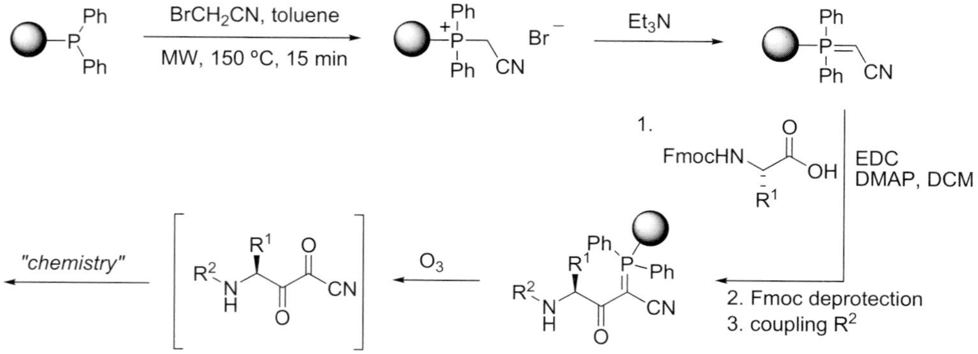

Scheme 7.54 Phosphoranes as polymer-bound acyl anion equivalents.

The best results were achieved by employing *N*-(3-dimethylaminopropyl)-*N*′-ethylcarbodiimide hydrochloride (EDC) as coupling agent. After Fmoc deprotection with piperidine in *N*,*N*-dimethylformamide, additional diversity could be introduced by acylation of the liberated amine position. Finally, the acyl cyano phosphoranes could be efficiently cleaved by ozonolysis at −78 °C or by utilizing freshly distilled 3,3-dimethyloxirane at room temperature [65]. The released compounds constituted highly activated electrophiles, which could be further converted *in situ* with appropriate nucleophiles.

In a more recent study of the Hallberg group, microwave irradiation was employed in the accelerated solid-phase synthesis of aryl triflates [66]. In addition to the solution-phase protocol (see Scheme 6.146), the authors also elaborated a rapid and convenient solid-phase procedure (Scheme 7.55). The use of *N*-phenyltriflimide as a triflating agent in microwave-mediated protocols is an appropriate choice, since it is a stable, crystalline agent, which often offers improved selectivity. A reduction of the reaction time from 3–8 h under conventional heating to only 6 min employing the microwave protocol has made this procedure more amenable to high-throughput synthesis. The use of the commercially available chlorotrityl linker as a solid support allows mild cleavage conditions to be employed to obtain the desired aryl triflates in good yields.

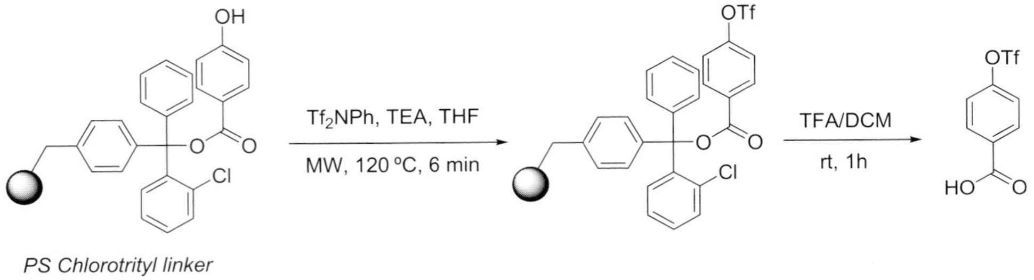

PS Chlorotrityl linker

Scheme 7.55 A solid-phase triflation procedure.

In addition to the ionic liquid-mediated procedure in solution (see Scheme 6.112), Leadbeater and coworkers also presented a solid-phase protocol for a one-pot Mannich reaction employing the above mentioned chlorotrityl linker [67]. In this approach, *p*-chlorobenzaldehyde and phenylacetylene were condensed with readily prepared immobilized piperazines (Scheme 7.56).

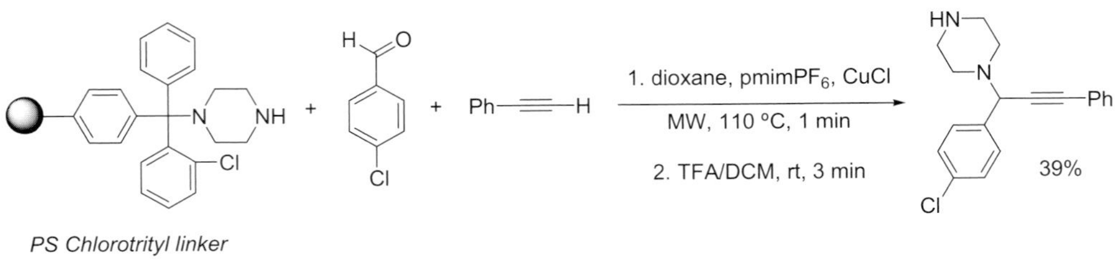

PS Chlorotrityl linker

Scheme 7.56 One-pot Mannich reaction utilizing polymer-supported piperazine.

The substrates were admixed with 50 mol% of copper(I) chloride and small amounts of 1-(2-propyl)-3-methylimidazolium hexafluorophosphate (pmimPF$_6$) in dioxane. The mixture was heated to 110 °C within 2 min and kept at this reaction temperature for an additional 1 min. After cooling to room temperature, the product was rapidly released from the polymer support employing 20% trifluoroacetic acid (TFA) in dichloromethane, furnishing the corresponding bis-TFA salt in moderate yield.

The group of Botta demonstrated the feasibility of their microwave-assisted iodination protocol (see Scheme 6.143 d) toward a polymer-supported substrate [68]. An appropriate pyrimidinone attached to conventional Merrifield polystyrene resin was suspended in N,N-dimethylformamide, treated with 2 equivalents of N-iodosuccinimide (NIS), and subjected to microwave irradiation for 3 min. Treatment of the polymer-bound intermediate with OXONE® released the desired 5-iodouracil in almost quantitative yield (Scheme 7.57).

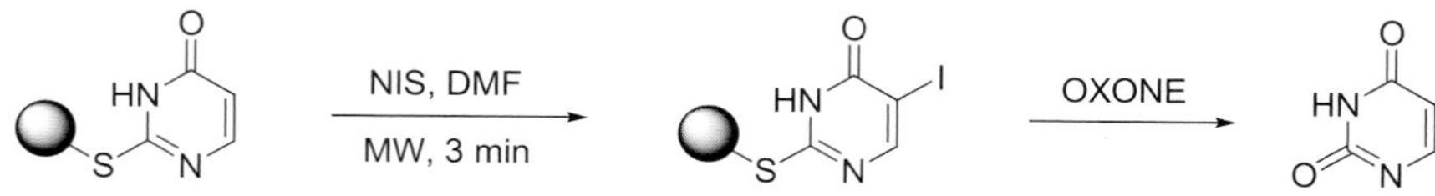

Scheme 7.57 Iodination of polymer-bound pyrimidinone.

Van der Eycken and coworkers have presented a study describing the microwave-assisted solid-phase Diels–Alder cycloaddition reaction of 2(1H)-pyrazinones with dienophiles [69]. After fragmentation of the resin-bound primary cycloadduct formed by Diels–Alder reaction of the 2(1H)-pyrazinone with an acetylenic dienophile, separation of the resulting pyridines from the pyridinone by-products was achieved by applying a traceless linking concept, whereby the pyridinones remained on the solid support with concomitant release of the pyridine products into solution (Scheme 7.58).

For this approach a novel, tailor-made and readily available linker, derived from inexpensive syringaldehyde, was developed [69]. This novel linker was produced by

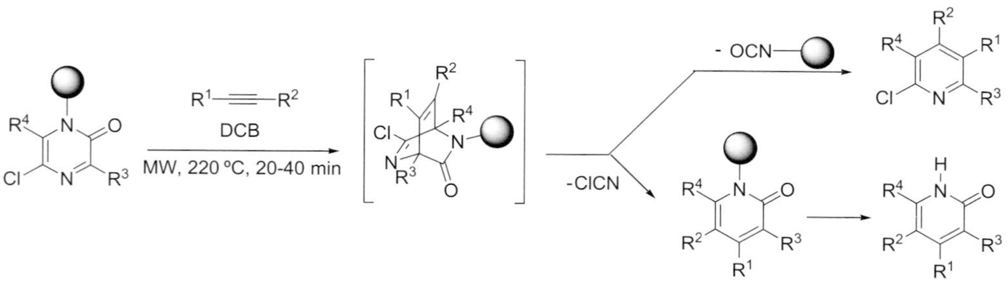

Scheme 7.58 General reaction sequence for 2(1H)-pyrazinone Diels–Alder cycloadditions with acetylenic dienophiles.

cesium carbonate-activated coupling of commercially available syringaldehyde to the Merrifield resin under microwave heating conditions. Subsequently, the aldehyde moiety was reduced at room temperature within 12 h, and the benzylic position was finally brominated by treatment with a large excess of thionyl bromide (10 equivalents) leading to the desired polymeric support (Scheme 7.59).

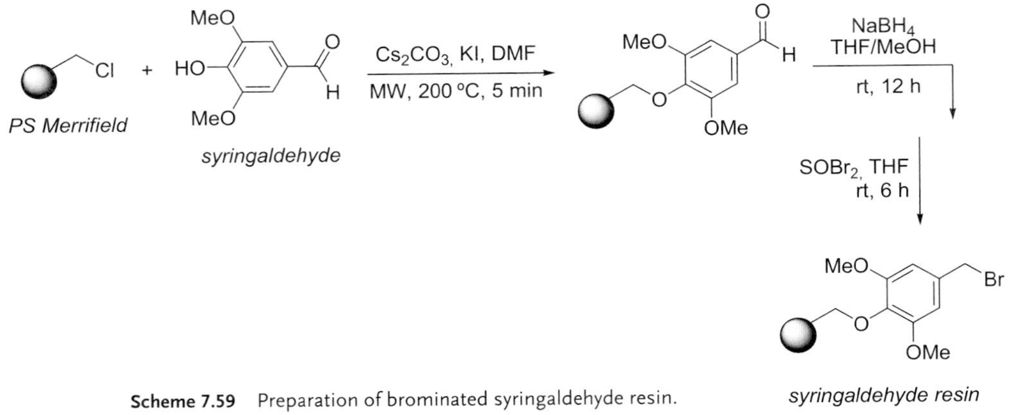

Scheme 7.59 Preparation of brominated syringaldehyde resin.

For the development of an appropriate strategy for cleavage from the novel syringaldehyde resin, the authors adapted a previously elaborated solution-phase model study on intramolecular Diels–Alder reactions for the solid-phase procedure (Scheme 7.60). The resulting pyridines could be easily separated from the polymer-bound by-products by employing a simple filtration step and subsequent evaporation of the solvent. The remaining resins were each washed and dried. After drying,

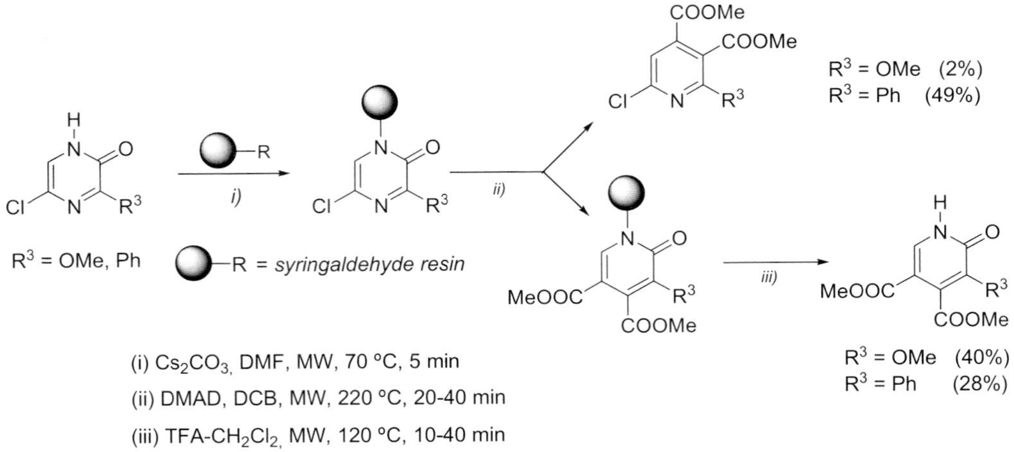

(i) Cs$_2$CO$_3$, DMF, MW, 70 °C, 5 min
(ii) DMAD, DCB, MW, 220 °C, 20-40 min
(iii) TFA-CH$_2$Cl$_2$, MW, 120 °C, 10-40 min

Scheme 7.60 Intermolecular 2(1H)-pyrazinone Diels–Alder reactions on a solid support.

the resins were each treated with trifluoroacetic acid/dichloromethane cleavage solutions under microwave irradiation to obtain the corresponding pyridinones.

Utilizing the novel syringaldehyde resin, smooth release from the support could be achieved upon microwave heating of a suspension of the resin-bound pyridinones in trifluoroacetic acid/dichloromethane (5:95) at 120 °C for just 10 min. The very mild cleavage conditions for this new linker, as well as its stability towards various reaction conditions and its easy accessibility, make it highly suitable for ongoing pyrazinone chemistry.

7.2
Soluble Polymer-Supported Synthesis

Besides solid-phase organic synthesis (SPOS) involving insoluble cross-linked polymer supports, chemistry on soluble polymer matrices, sometimes called liquid-phase organic synthesis, has recently emerged as a viable alternative. Problems associated with the heterogeneous nature of the ensuing chemistry and on-bead spectroscopic characterization in SPOS have led to the development of soluble polymers as alternative matrices for combinatorial library production. Synthetic approaches that utilize soluble polymers couple the advantages of homogeneous solution chemistry (high reactivity, lack of diffusion phenomena, and ease of analysis) with those of solid-phase methods (use of excess reagents and easy isolation and purification of products). Separation of the functionalized matrix is achieved by either solvent or heat precipitation, membrane filtration, or size-exclusion chromatography (Scheme 7.61).

Suitable soluble polymers for liquid-phase synthesis can be described as follows: their molecular weight is high enough for them to be crystalline at room temperature, they bear functional groups on their end termini or side chains, but in contrast

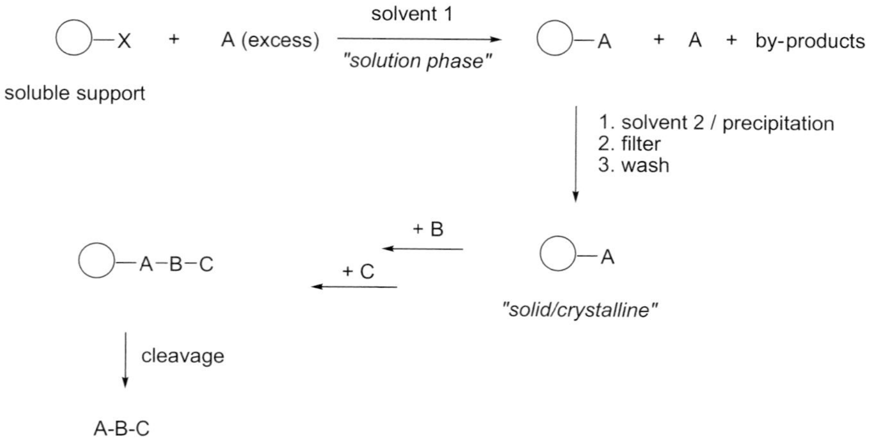

Scheme 7.61 Principles of syntheses on soluble polymer supports.

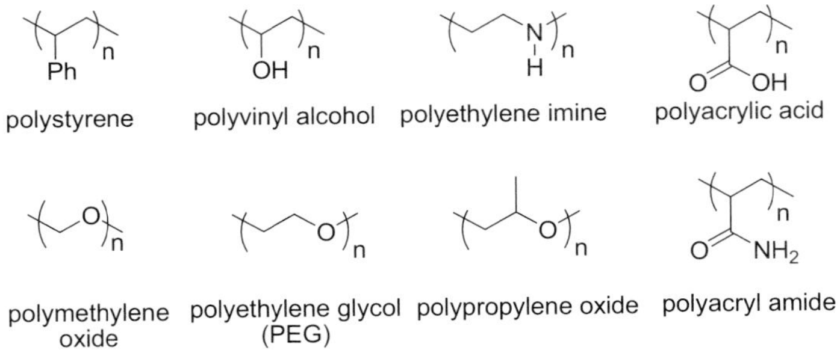

polystyrene   polyvinyl alcohol   polyethylene imine   polyacrylic acid

polymethylene   polyethylene glycol   polypropylene oxide   polyacryl amide
oxide   (PEG)

Fig. 7.4 Soluble polymers utilized in liquid-phase synthesis.

to the supports used in SPOS they are not cross-linked and therefore are soluble in several organic solvents (Fig. 7.4).

Several microwave-assisted protocols for soluble polymer-supported syntheses have been described. Among the first examples of so-called liquid-phase synthesis were aqueous Suzuki couplings. Schotten and coworkers presented the use of polyethylene glycol (PEG)-bound aryl halides and sulfonates in these palladium-catalyzed cross-couplings [70]. The authors demonstrated that no additional phase-transfer catalyst (PTC) is needed when the PEG-bound electrophiles are coupled with appropriate aryl boronic acids. The polymer-bound substrates were coupled with 1.2 equivalents of the boronic acids in water under short-term microwave irradiation in sealed vessels in a domestic microwave oven (Scheme 7.62). Work-up involved precipitation of the polymer-bound biaryl from a suitable organic solvent with diethyl ether. Water and insoluble impurities need to be removed prior to precipitation in order to achieve high recoveries of the products.

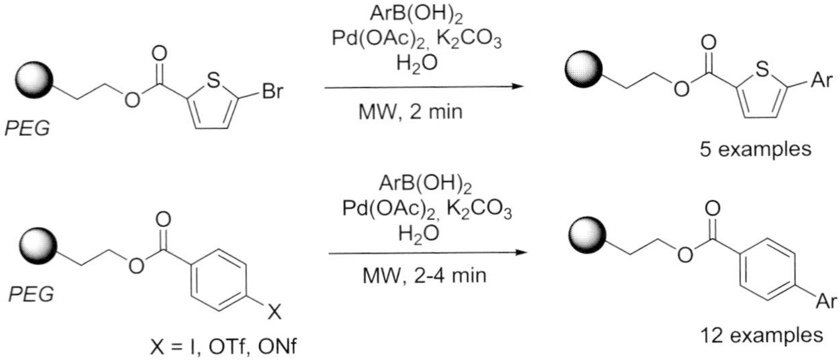

Scheme 7.62 PEG-supported Suzuki couplings.

Another palladium-catalyzed coupling reaction that has been successfully performed on soluble polymers is the Sonogashira coupling. Xia and Wang have presented an approach in which the PEG 4000 utilized simultaneously serves as polymeric support, solvent, and phase-transfer catalyst (PTC) in both the coupling and hydrolysis steps [71]. Polyethylene glycol (PEG)-bound 4-iodobenzoic acid could be readily reacted with several terminal alkynes under rapid microwave conditions (Scheme 7.63). Attachment of the 4-iodobenzoic acid was achieved by transesterification using *N,N'*-dicyclohexylcarbodiimide (DCC) and *N,N*-dimethylaminopyridine (DMAP) in dichloromethane. The coupling products were efficiently cleaved from the PEG support by simple saponification under microwave irradiation and subsequent acidification. The process time for this step was drastically reduced from 8 h at 50 °C (classical heating) to 2 min in an open beaker in a domestic microwave oven.

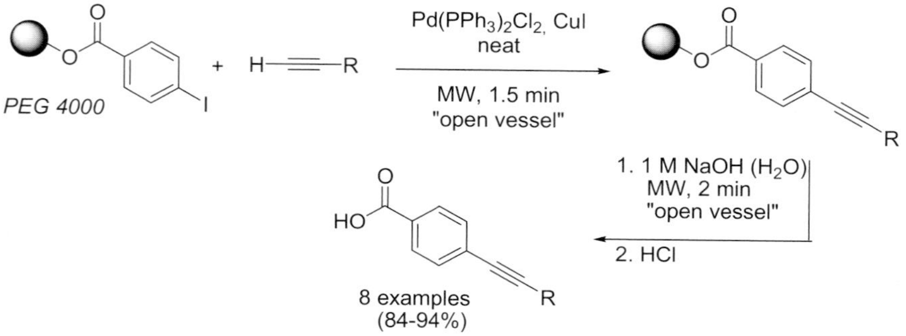

Scheme 7.63 Sonogashira couplings on a PEG support.

Another example where PEG played the role of polymeric support, solvent, and PTC was presented by the group of Lamaty [72]. In this study, a Schiff base-protected glycine was reacted with various electrophiles (RX) under microwave irradiation. No additional solvent was necessary to perform these reactions and the best results were obtained using cesium carbonate as an inorganic base (Scheme 7.64). After alkylation, the corresponding aminoesters were released from the polymer support by transesterification employing methanol in the presence of triethylamine.

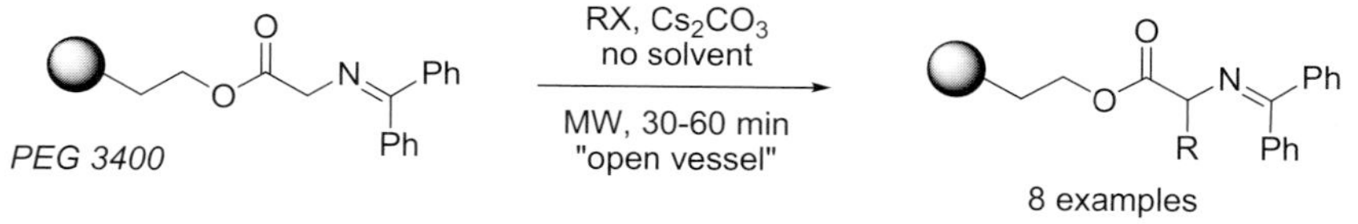

Scheme 7.64 Microwave-assisted alkylations on PEG.

A method for microwave-assisted transesterifications has been described by Vanden Eynde and Rutot [73]. The authors investigated the microwave-mediated derivatization of poly(styrene-co-allyl alcohol) as a key step in the polymer-assisted synthesis of heterocycles. Several *β*-ketoesters were employed in this procedure and multigram quantities of products were obtained when neat mixtures of the reagents in open vessels were subjected to microwave irradiation utilizing a domestic microwave oven (Scheme 7.65). The successful derivatization of the polymer was confirmed by IR, ^{1}H NMR, and ^{13}C NMR spectroscopic analyses. The soluble supports

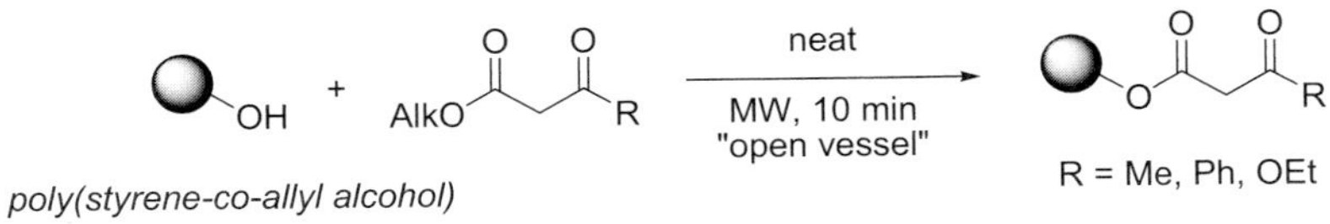

poly(styrene-co-allyl alcohol)

R = Me, Ph, OEt

Scheme 7.65 Transesterification of poly(styrene-co-allyl alcohol).

prepared were used for the preparation of various bicyclic heterocycles, such as pyr-azolo-pyridinediones and coumarins.

Another procedure for the preparation of valuable heterocyclic scaffolds has been presented by Xia and Wang, that is, the Biginelli condensation on a PEG support [74, 75]. Polymer-bound acetoacetate was prepared by reacting commercially available PEG 4000 with 2,2,6-trimethyl-4*H*-1,3-dioxin-4-one in refluxing toluene (Scheme 7.66). The quantitative conversion of the terminal hydroxyl groups of the PEG polymer was determined by ^{1}H NMR. The microwave-assisted cyclocondensa-tion was performed utilizing non-volatile polyphosphoric acid (PPA) as a catalyst. In a representative procedure, the polymer-bound acetoacetate was suspended with 2 equivalents each of urea and the requisite aldehyde, along with 2–3 drops of PPA, in a glass beaker. The open beaker was placed in a large container containing alu-mina as a heat sink in a domestic microwave oven [75]. During microwave heating, the PEG-bound substrate melted, ensuring a homogeneous reaction mixture. After the reaction, diethyl ether was added in order to precipitate the polymer-bound prod-ucts. The desired compounds were released by treatment with sodium methoxide (NaOMe) in methanol at room temperature. All of the dihydropyrimidines were obtained in high yields; purification was achieved by recrystallization from ethanol.

A related support frequently used for liquid-phase synthesis is methoxy-polyethyl-ene glycol (MeO-PEG). The group of Taddei has presented a general procedure for

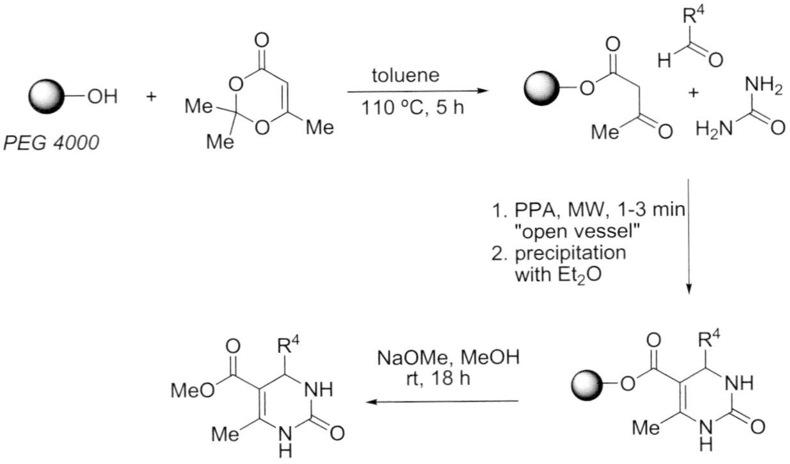

Scheme 7.66 One-pot Biginelli cyclocondensation on a PEG support.

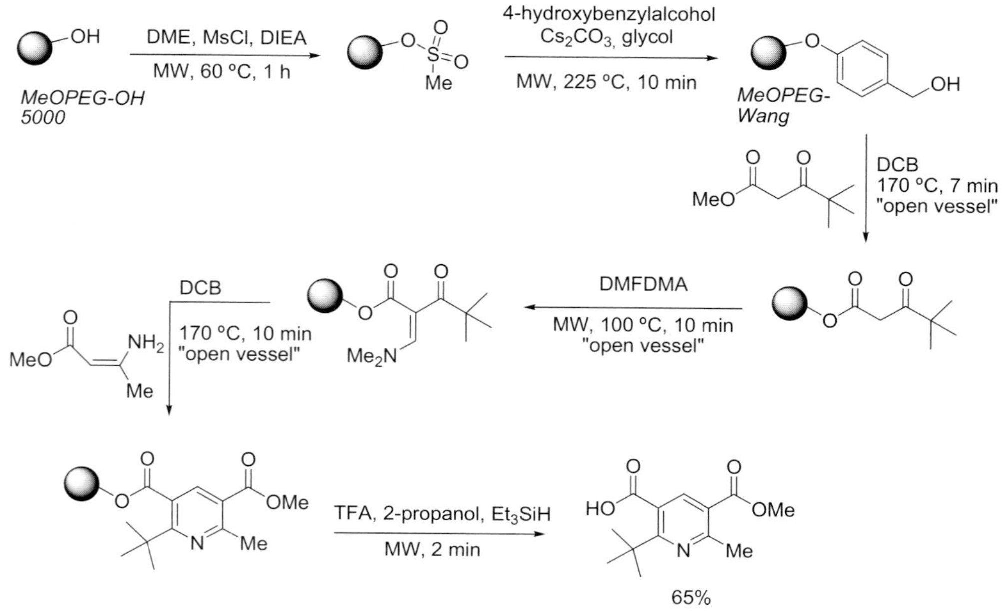

Scheme 7.67 Liquid-phase organic synthesis utilizing MeO-PEG.

the microwave-assisted synthesis of organic molecules on MeO-PEG [76]. The use of MeO-PEG under microwave conditions in open vessels simplifies the process of polymer-supported synthesis (Scheme 7.67) as the polymer-bound products precipitate on cooling after removal from the microwave oven. In addition, cleavage could be rapidly accomplished under microwave irradiation employing a 1:1 mixture of trifluoroacetic acid (TFA) and 2-propanol, furnishing the desired nicotinic acid in the form of the corresponding pyridinium trifluoroacetate.

Wu and Sun have presented a versatile procedure for the liquid-phase synthesis of 1,2,3,4-tetrahydro-β-carbolines [77]. After successful esterification of the MeO-PEG-OH utilized with Fmoc-protected tryptophan, one-pot cyclocondensations with various ketones and aldehydes were performed under microwave irradiation (Scheme 7.68). The desired products were released from the soluble support in good yields and high purity. The interest in this particular scaffold is due to the fact that the 1,2,3,4-tetrahydro-β-carboline pharmacophore is known to be an important structural element in several natural alkaloids, and that the template possesses multiple sites for combinatorial modifications. The microwave-assisted liquid-phase protocol furnished purer products than homogeneous protocols and product isolation/ purification was certainly simplified.

The desired polymer-bound tryptophan was rapidly generated under microwave irradiation, employing a classical esterification protocol using N,N'-dicyclohexylcarbodiimide (DCC) and a catalytic amount of N,N-dimethylaminopyridine (DMAP), followed by subsequent Fmoc deprotection (Scheme 7.68). Cyclocondensations with various carbonyl compounds were performed with catalytic amounts of p-toluenesulfonic acid (*p*TsOH) to obtain quantitative conversion within 15 min of microwave irradiation, as monitored by ^{1}H NMR. Finally, treatment of the polymer-bound heterocycles with 1 mol% of potassium cyanide (KCN) in methanol furnished the desired target structures in high yields. The crude products were found to be enriched with the respective *trans* isomers.

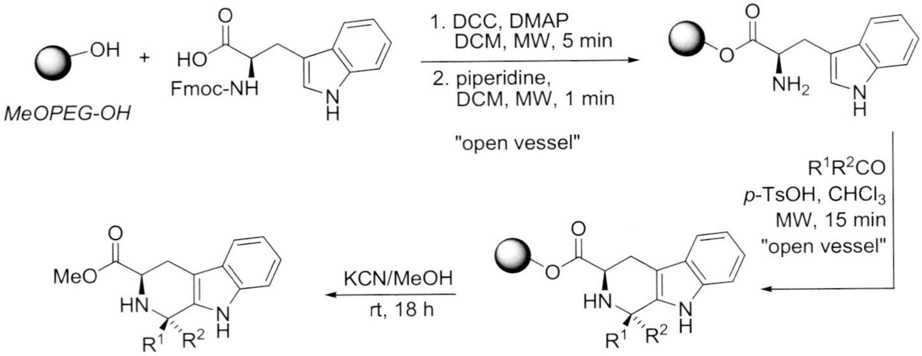

Scheme 7.68 Liquid-phase preparation of 1,2,3,4-tetrahydro-β-carbolines.

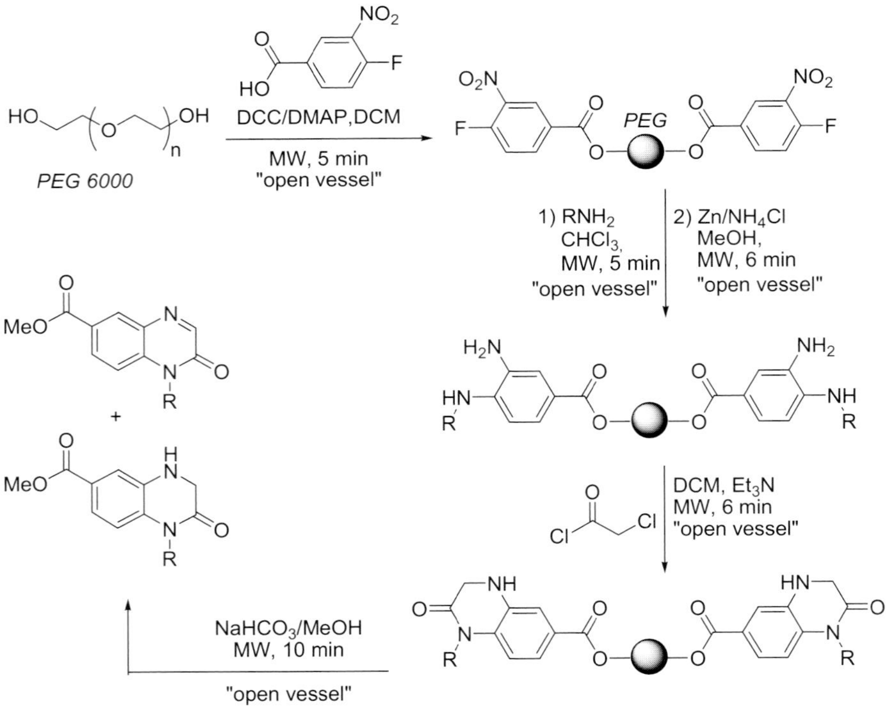

Scheme 7.69 Liquid-phase synthesis of quinoxalin-2-ones.

In a related study by the same group, the microwave-assisted liquid-phase synthesis of benzopiperazinones (quinoxalin-2-ones) was described [78]. The [6,6]-ring quinoxalinone system has also been reported to show promising pharmaceutical properties, and Sun and coworkers employed a hybrid strategy using both combinatorial and microwave synthesis to generate a quinoxalin-2-one library from readily available building blocks. Thus, 4-fluoro-3-nitrobenzoic acid was efficiently coupled with a commercially available polyethylene glycol (PEG-6000). Various primary amines were then condensed with the readily prepared immobilized *ortho*-fluoronitrobenzene through nucleophilic *ipso*-fluoro displacement. Subsequent reduction of the resulting *ortho*-nitroanilines furnished the corresponding *ortho*-phenylenediamines, which were then reacted with chloroacetyl chloride to form polymer-bound 1,2,3,4-tetrahydroquinoxalin-2-ones (Scheme 7.69). Each step was performed within a few minutes under microwave irradiation in a dedicated single-mode reactor; the PEG-bound intermediates were precipitated in ethanol and removed by filtration, leaving the by-products in the ethanolic phase. Finally, cleavage of the target molecules was achieved within 10 min under microwave irradiation employing a methanolic solution of sodium hydrogencarbonate ($NaHCO_3$), furnishing both the desired dihydroquinoxalinones and the corresponding oxidized compounds in varying ratios. On storage, the dihydro products were fully converted to the oxidized form within a few days [78].

Utilizing the same aryl fluoride linker on conventional MeOPEG polymer, these authors also presented a microwave-accelerated liquid-phase synthesis of benzimidazoles (Scheme 7.70) [79]. This bicyclic pharmacophore is an important and valuable structural element in medicinal chemistry, showing a broad spectrum of pharmacological activities, such as antihistaminic, antiparasitic, and antiviral effects.

Following the strategy described above involving *ipso*-fluoro displacement and subsequent reduction, the resulting *ortho*-phenylenediamines were treated with several aromatic isothiocyanates in the presence of *N,N'*-dicyclohexylcarbodiimide

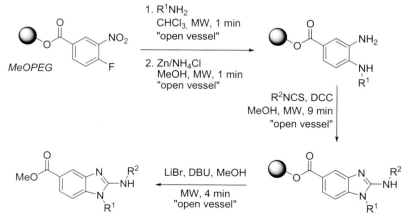

Scheme 7.70 Liquid-phase synthesis of benzimidazoles.

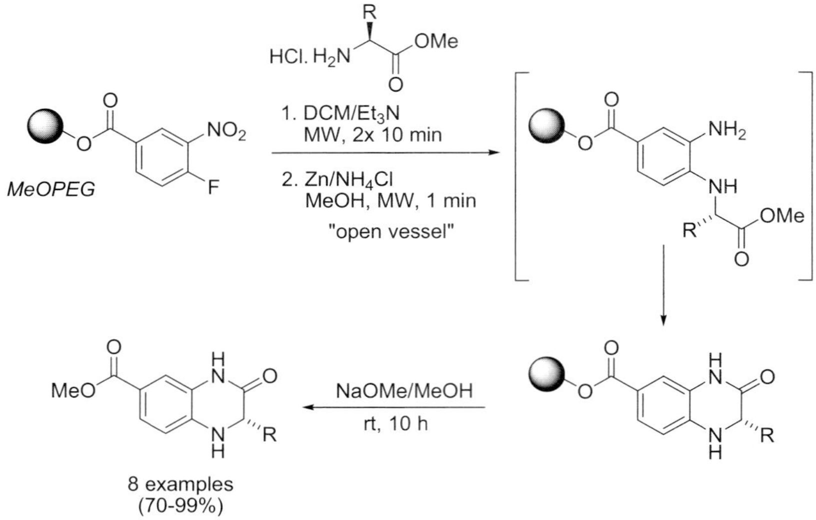

Scheme 7.71 Liquid-phase synthesis of chiral quinoxalinones.

(DCC) to form the corresponding PEG-bound benzimidazoles. Cleavage of the desired compounds was readily achieved by microwave-mediated transesterification (methanol) with lithium bromide and 1,8-diazabicyclo[5.4.0]undec-7-ene (DBU). MeOPEG-OH was separated by precipitation and filtration to furnish the crude products in high yields and good purities [79].

In a closely related study, Tung and Sun discussed the microwave-assisted liquid-phase synthesis of chiral quinoxalines [80]. Various L-α-amino acid methyl ester hydrochlorides were coupled to MeOPEG-bound *ortho*-fluoronitrobenzene by the aforementioned *ipso*-fluoro displacement method. Reduction under microwave irradiation resulted in spontaneous synchronous intramolecular cyclization to the corresponding 1,2,3,4-tetrahydroquinoxalin-2-ones (Scheme 7.71). Retention of the chiral moiety could not be monitored during the reaction, but after release of the desired products it was found that about 10% of the product had undergone racemization.

A variation of this method led to the generation of bis-benzimidazoles [81, 82]. The versatile immobilized *ortho*-phenylenediamine template was prepared as described above in several microwave-mediated steps. Additional *N*-acylation exclusively at the primary aromatic amine moiety was achieved utilizing the initially used 4-fluoro-3-nitrobenzoic acid at room temperature (Scheme 7.72). Various amines were used to introduce diversity through nucleophilic aromatic substitution. Cyclization to the polymer-bound benzimidazole was achieved by refluxing for several hours in a mixture of trifluoroacetic acid and chloroform. Individual steps at ambient temperature for selective reduction, cyclization with several aldehydes, and final detachment from the polymer support were necessary in order to obtain the desired bis-benzimidazoles. A set of 13 examples was prepared in high yields and good purities [81].

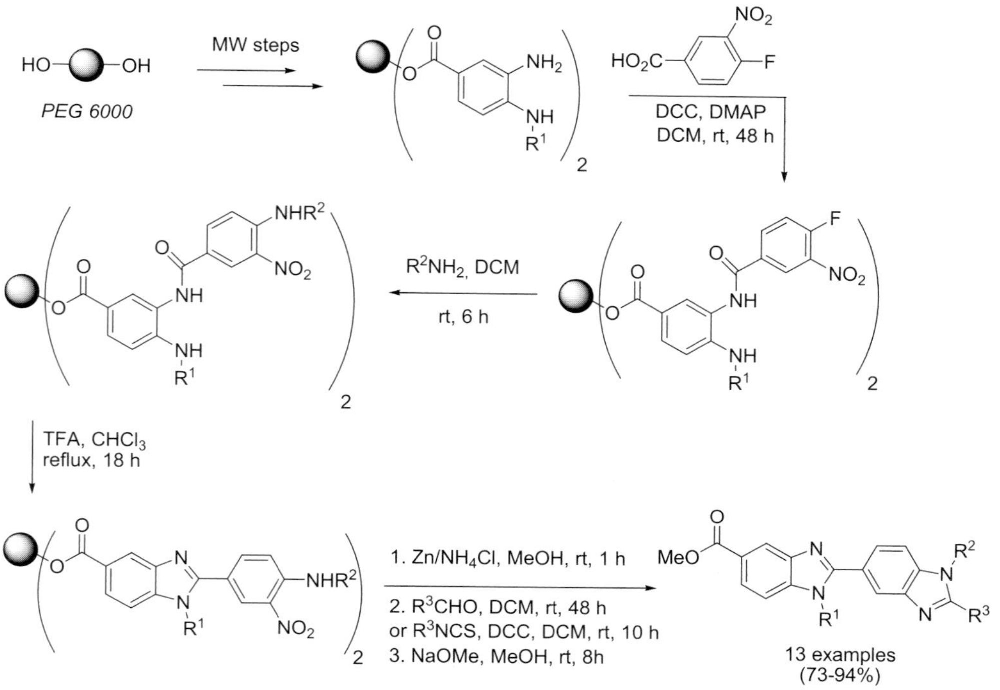

Scheme 7.72 Liquid-phase synthesis of bis-benzimidazoles.

The cyclization to the desired head-to-tail linked bis-benzimidazoles could also be performed utilizing aryl or alkyl isothiocyanates with N,N'-dicyclohexylcarbodiimide (DCC) [82]. Upon completion, the insoluble N,N'-dicyclohexylthiourea formed had to be removed by filtration and the desired PEG-bound products were precipitated by the addition of diethyl ether. The results were essentially the same as those of the cyclizations with the above mentioned aldehydes.

In a closely related, more recent study by the same group, mercury(II) chloride was utilized to catalyze the cyclization to the benzimidazoles [83].

In a further application of the bis-hydroxylated polymer support PEG-6000, the group of Sun presented several reports on the microwave-accelerated liquid-phase synthesis of thiohydantoins [84, 85]. These transformations were carried out in a dedicated single-mode instrument under open-vessel conditions (Scheme 7.73). Whereas efficient coupling of Fmoc-protected amino acids under DCC/DMAP activation was realized within 14 min in the microwave reactor, no conversion at all could be obtained after the same period of conventional heating (refluxing dichloromethane) [84]. After conventional deprotection employing 10% piperidine in dichloromethane at ambient temperature, various isothiocyanates were introduced under microwave irradiation. In order to drive the transformation to completion, 3 equivalents of the isothiocyanates had to be used. The resulting PEG-bound thiourea derivatives were successfully released from the support by cyclative cleavage under mild

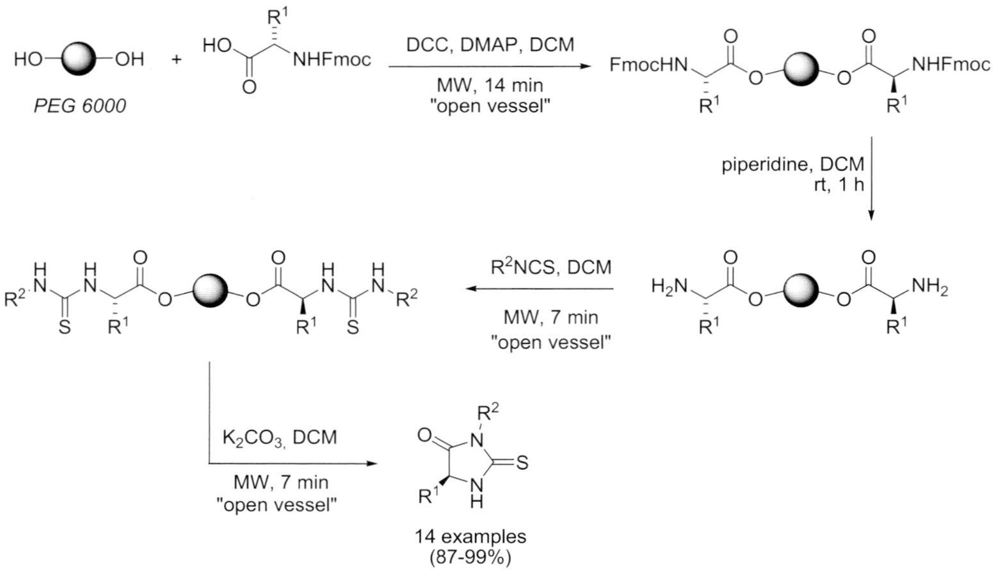

Scheme 7.73 Liquid-phase synthesis of thiohydantoins.

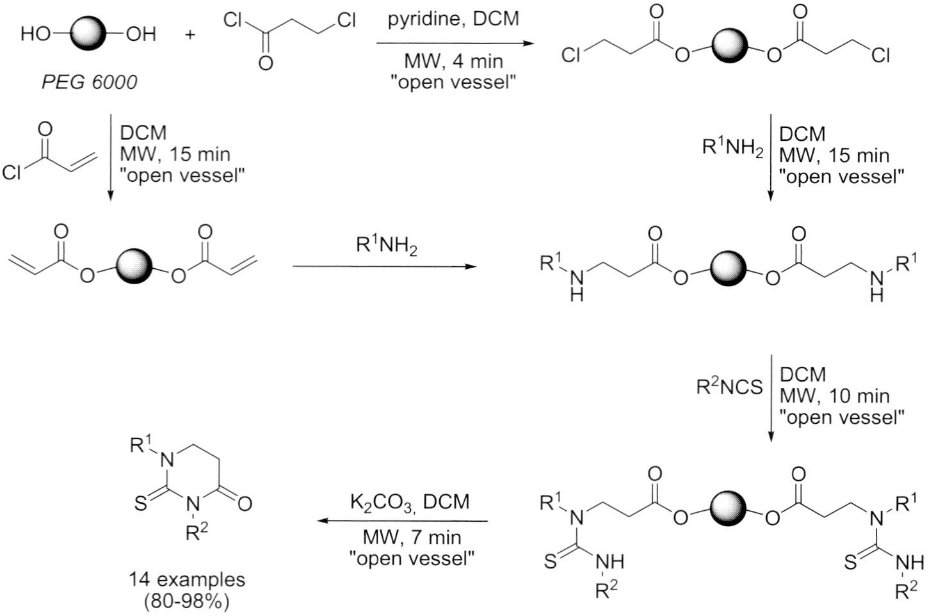

Scheme 7.74 Liquid-phase synthesis of 1,3-disubstituted thioxotetrahydropyrimidinones.

basic conditions employing potassium carbonate. This traceless cleavage protocol ensured that only the desired compounds were released from the soluble support. Removal of the polymer by precipitation and filtration afforded the heterocyclic compounds in high yields and excellent purities.

Additionally, the authors chose 3-chloropropionyl chloride as the immobilized building block in order to carry out a ring-expansion approach, which led to the generation of a 14-member library of thioxotetrahydropyrimidinones [85, 86]. The initially prepared polymer-bound chloropropionyl ester was efficiently transformed into the corresponding diamines by transamination utilizing several primary amines. These diamine intermediates could also be obtained by treatment of the pure polymeric support with acryloyl chloride and subsequent addition of the appropriate amines (Scheme 7.74).

In analogy to the thiohydantoin synthesis, the PEG-bound diamines have also been treated with various alkyl and aryl isothiocyanates and, after applying traceless cyclative cleavage, the desired thioxotetrahydropyrimidinones were obtained in excellent yields. In contrast to the thiohydantoin synthesis, purification of the products was more complicated if the excess of isothiocyanate amounted to more than 2.2 equivalents [85].

Finally, another related study from the Sun laboratory concerned the synthesis of hydantoins utilizing acryloyl chloride to prepare a suitable polymer support [87]. All steps were carried out under reflux conditions in a dedicated microwave instrument utilizing 50-mL round-bottomed flasks. Identical reactions under classical thermal heating did not proceed in the same time period.

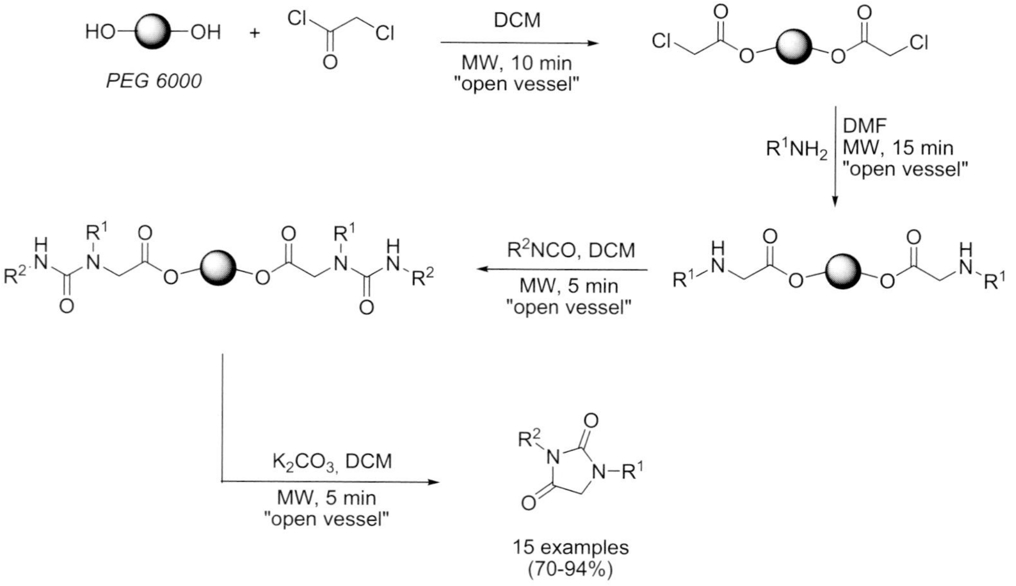

Scheme 7.75 Liquid-phase synthesis of hydantoins.

The soluble polymer support was dissolved in dichloromethane and treated with 3 equivalents of chloroacetyl chloride for 10 min under microwave irradiation. The subsequent nucleophilic substitution utilizing 4 equivalents of various primary amines was carried out in N,N-dimethylformamide as solvent. The resulting PEG-bound amines were reacted with 3 equivalents of aryl or alkyl isothiocyanates in dichloromethane to furnish the polymer-bound urea derivatives after 5 min of microwave irradiation (Scheme 7.75). After each step, the intermediates were purified by simple precipitation with diethyl ether and filtration, so as to remove by-products and unreacted substrates. Finally, traceless release of the desired compounds by cyclative cleavage was achieved under mild basic conditions within 5 min of microwave irradiation. The 1,3-disubstituted hydantoins were obtained in varying yields but high purity.

7.3
Fluorous Phase Organic Synthesis

Fluorous phase organic synthesis is a novel separation and purification technique for organic synthesis and process development. In this technique, organic molecules are rendered soluble in fluorocarbon solvents by the attachment of a suitable fluorocarbon group ("fluorous tag"). These hydrophobic perfluorinated hydrocarbon compounds are immiscible with water and many common organic solvents. Other physical properties, such as solubility in supercritical carbon dioxide (CO_2), make them ideal for clean organic syntheses. Fluorous phase organic synthesis has been applied in the immobilization of expensive metal catalysts, offering an ideal substitute for the more traditional solid phases used in combinatorial chemistry approaches. Most importantly, the immobilization of a wide variety of reagents through the attachment of perfluorinated tags enables easy and efficient isolation and recovery by liq-

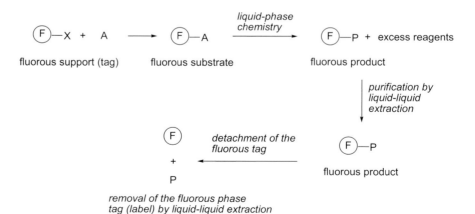

Fig. 7.5 The principle of fluorous synthesis.

uid-liquid and solid-liquid extractions (Fig. 7.5). At the desired stage of the synthesis, the fluorous label is cleaved and the product is rendered "organic" once more [88].

An early example of microwave-assisted fluorous synthesis was presented by the group of Hallberg, which involved palladium-catalyzed Stille couplings of fluorous tin reagents with aryl halides or triflates (Scheme 7.76) [89]. While the corresponding thermal process required one day to reach completion, the microwave-heated reactions (sealed vessels in monomode reactors) were complete within 2 min, with the additional benefit of reduced homocoupling of the tin reagent. The desired biaryl products were isolated in good yields and purities after a three-phase extraction. The same research group achieved similar results by utilizing so-called F-21 fluorous tags $(CH_2CH_2C_{10}F_{21})$ on the tin reagent [90].

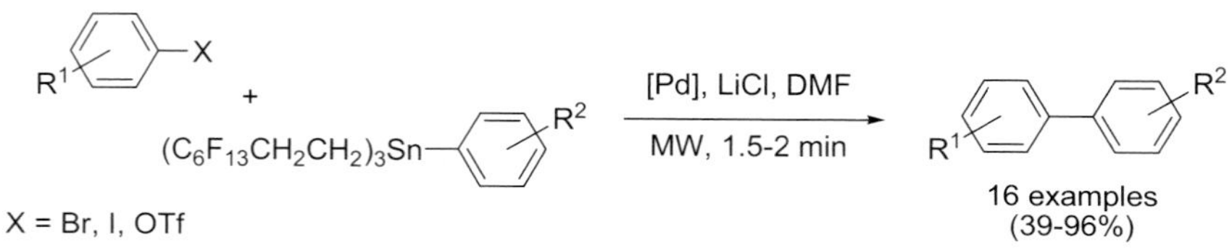

X = Br, I, OTf

16 examples
(39–96%)

Scheme 7.76 Fluorous Stille couplings.

In an additional application of fluorous chemistry, Hallberg's group has reported radical-mediated cyclizations utilizing benzotrifluoride (BTF) as a solvent under microwave irradiation [90]. In the presence of 2,2'-azobisisobutyronitrile (AIBN) as a radical initiator, the aryl iodide shown in Scheme 7.77 smoothly underwent microwave-mediated cyclization to the corresponding indole derivative in high isolated yield. The ability to promote highly fluorous reactions with microwave heating deserves special attention. With these highly fluorous tin reagents, microwave irradiation is more than an expedient means of reducing reaction times. Reactions conducted under traditional heating either did not work at all or did not work nearly as well. The advantage of microwave heating may be the rapid coalescence of the organic and fluorous phases to form a homogeneous solution [90].

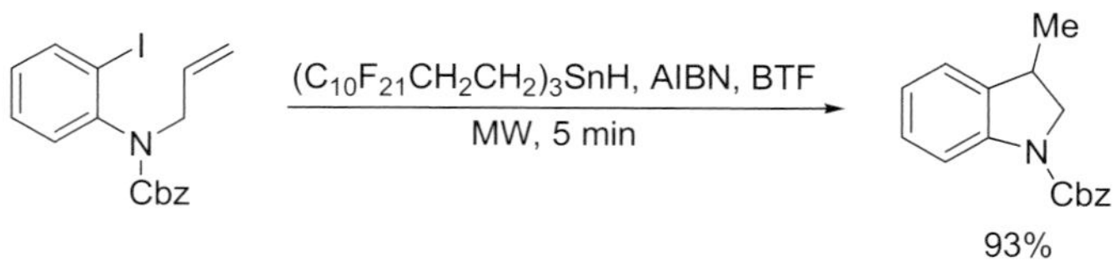

93%

Scheme 7.77 Radical cyclizations using fluorous tin reagents.

Zhang and coworkers have reported on a palladium-catalyzed carbon–sulfur cross-coupling of aryl perfluoroalkoxysulfonates with thiols (Scheme 7.78) [91]. The fluorous substrates were obtained from commercially available phenols by treatment with perfluorooctanesulfonyl fluoride $(C_8F_{17}SO_2F)$ under basic conditions. Various thiols were reacted with a slight excess of the perfluorinated sulfonates in a

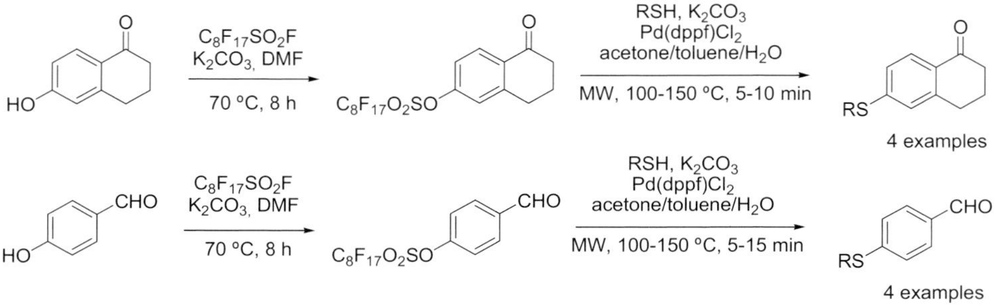

Scheme 7.78 Fluorous-phase palladium-catalyzed synthesis of aryl sulfides.

microwave-mediated Suzuki-type reaction employing 10 mol% of [1,1′-bis(diphenyl-phosphino)ferrocene]dichloropalladium(II) [Pd(dppf)Cl$_2$] as catalyst to furnish the corresponding aryl sulfides. Purification of the products after aqueous work-up was achieved by fluorous solid-phase extraction (F-SPE).

In a more recent study, Zhang and Nagashima have described a parallel synthesis approach towards an *N*-alkylated dihydropteridinone library utilizing fluorous amino acids [92]. The fluorous substrates were readily prepared from the corresponding *N*-Boc-amino acids by employing 3-(perfluorooctyl)propanol as the fluorinating agent. This was followed by deprotection under acidic conditions to afford the corresponding trifluoroacetic acid (TFA) salts of the amino acids. Four of the resulting trifluoroacetate salts were used without further purification in parallel displacement reactions utilizing 1.5 equivalents of 4,6-dichloro-5-nitropyrimidine (Scheme 7.79). After the addition of *N,N*-diisopropylethylamine (DIEA), these exothermic reactions were completed within 10 min. Subsequently, a second displacement was carried out by the addition of 3 equivalents of five different cyclic amines. The resulting 20 intermediates were purified by fluorous solid-phase extraction (F-SPE) and subjected to hydrogenation with platinum-on-charcoal for 12 h.

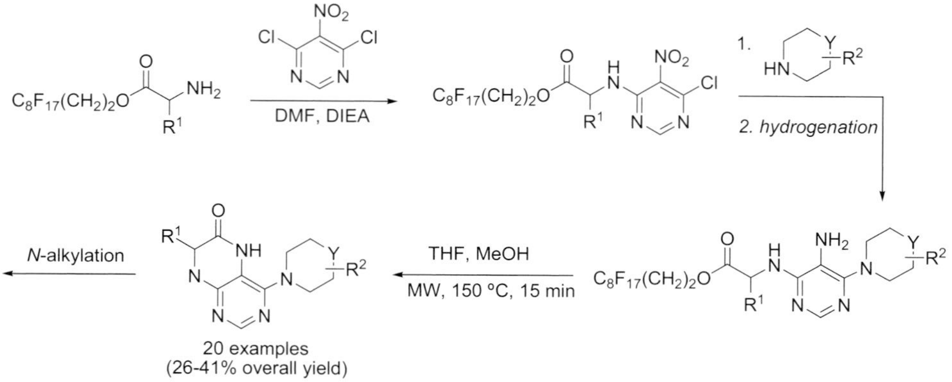

Scheme 7.79 Generation of dihydropteridinones by cyclative cleavage from a fluorous tag.

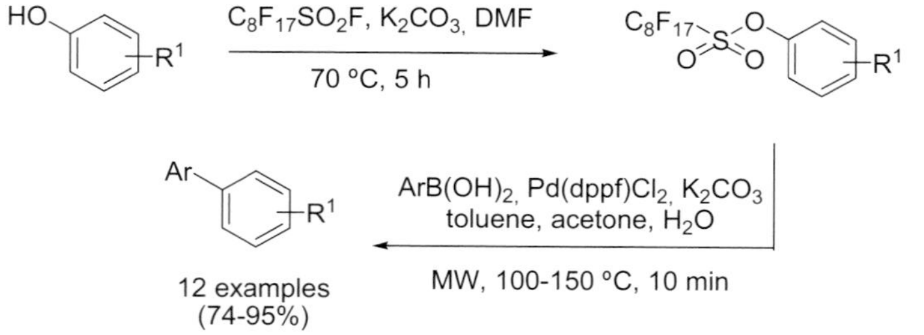

Scheme 7.80 Fluorous phase Suzuki-type couplings.

As only small amounts of the hydrogenation products underwent spontaneous cyclization, the compounds were heated in a sealed vessel under microwave irradiation. Due to the low solubility of the cyclized products at room temperature, they precipitated upon cooling and could be easily collected by filtration. The desired dihydropteridinones were obtained in high purity and could be readily used to generate a more diverse library by subsequent *N*-alkylation [92].

The same research group additionally presented further investigations on fluorous-phase palladium-catalyzed carbon–carbon couplings [93]. Fluorinated aryl octyl-sulfonates were reacted in slight excess with arylboronic acids under microwave con-

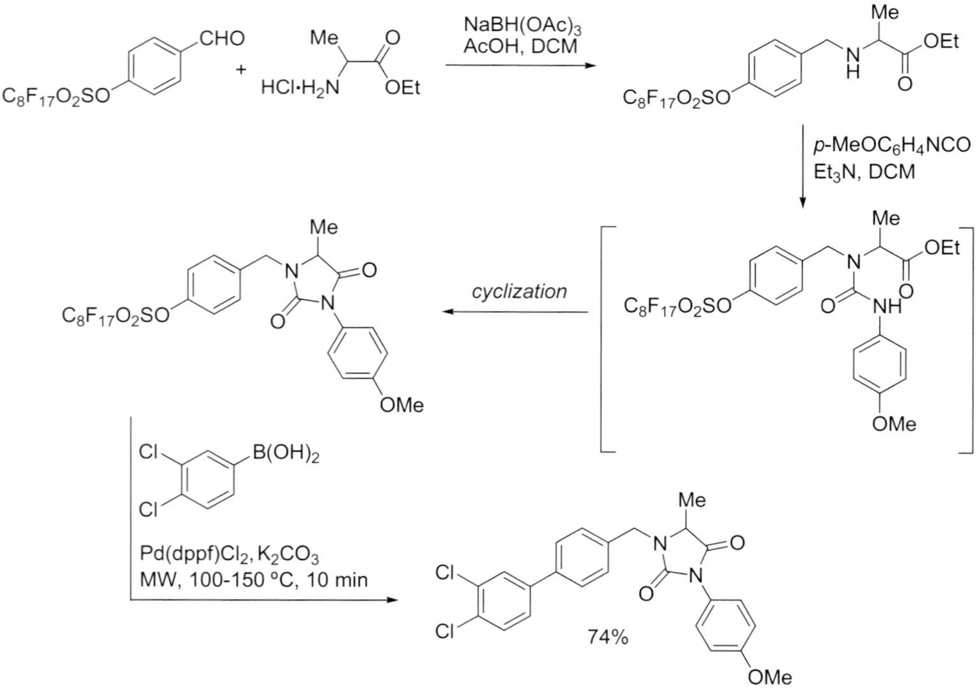

Scheme 7.81 Fluorous phase synthesis of *N,N'*-disubstituted hydantoins.

ditions to efficiently furnish the corresponding biaryls in a Suzuki-type cross-coupling (Scheme 7.80).

The described fluorous-tag strategy has also been applied to the synthesis of biaryl-substituted hydantoins (Scheme 7.81) [94]. 4-Hydroxybenzaldehyde was converted into the corresponding perfluorinated species, which was then subjected to a reductive amination. The resulting amine was treated with an isocyanate to produce the fluorous-tagged urea, which spontaneously cyclized to form the corresponding hydantoin. Finally, the fluorous tag was detached by a Suzuki-type carbon–carbon bond formation to furnish the desired target structure in good yield.

In a related approach from the same laboratory, the perfluorooctylsulfonyl tag was employed in a traceless strategy for the deoxygenation of phenols (Scheme 7.82) [94]. These reactions were carried out in a toluene/acetone/water (4:4:1) solvent mixture, utilizing 5 equivalents of formic acid and potassium carbonate/[1,1′-bis(diphenylphosphino)ferrocene]dichloropalladium(II) [Pd(dppf)Cl$_2$] as the catalytic system. After 20 min of irradiation, the reaction mixture was subjected to fluorous solid-phase extraction (F-SPE) to afford the desired products in high yields. This new traceless fluorous tag has also been employed in the synthesis of pyrimidines and hydantoins.

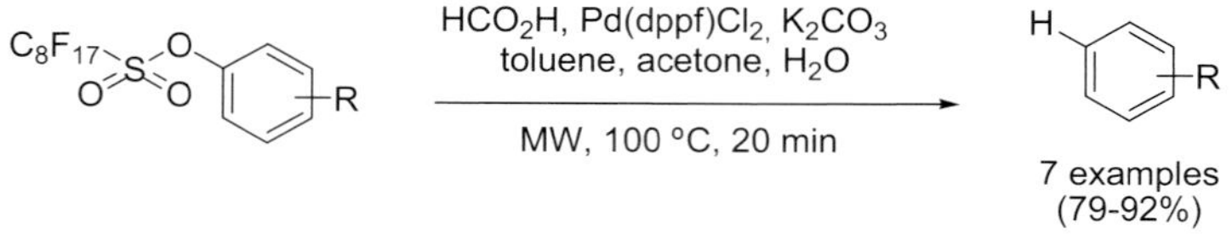

Scheme 7.82 Fluorous phase traceless deoxygenation of phenols.

Fluorous tags can also be introduced as acid-labile protecting groups in syntheses of both carboxamides and sulfonamides [95]. Ladlow and coworkers have presented the parallel generation of an 18-member library of biaryl carboxamides utilizing perfluorooctane-tagged iodopropane (C$_8$F$_{17}$(CH$_2$)$_3$I) to prepare the corresponding fluorous substrates (Scheme 7.83). The fluorous tag was ultimately removed by treatment with a

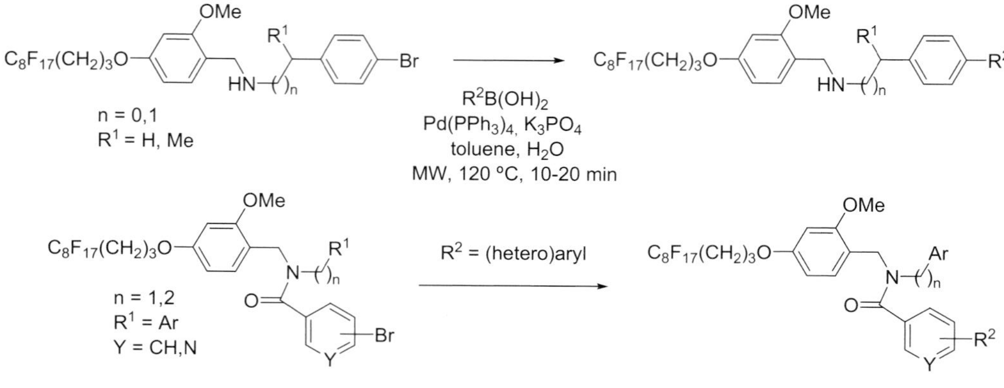

Scheme 7.83 Suzuki reactions utilizing fluorous-tagged acid-labile protecting groups.

mixture of trifluoroacetic acid, triethylsilane, and water (90:5:5). Subsequent purification by SPE afforded the desired products in good yields and excellent purities.

Furthermore, multicomponent reactions can also be performed under fluorous-phase conditions, as shown for the Ugi four-component reaction [96]. To improve the efficiency of a recently reported Ugi/de-Boc/cyclization strategy, Zhang and Tempest introduced a fluorous Boc group for amine protection and carried out the Ugi multicomponent condensation under microwave irradiation (Scheme 7.84). The desired fluorous condensation products were easily separated by fluorous solid-phase extraction (F-SPE) and deprotected by treatment with trifluoroacetic acid/tetrahydrofuran under microwave irradiation. The resulting quinoxalinones were purified by a second F-SPE to furnish the products in excellent purity. This methodology was also applied in a benzimidazole synthesis, employing benzoic acid as a substrate.

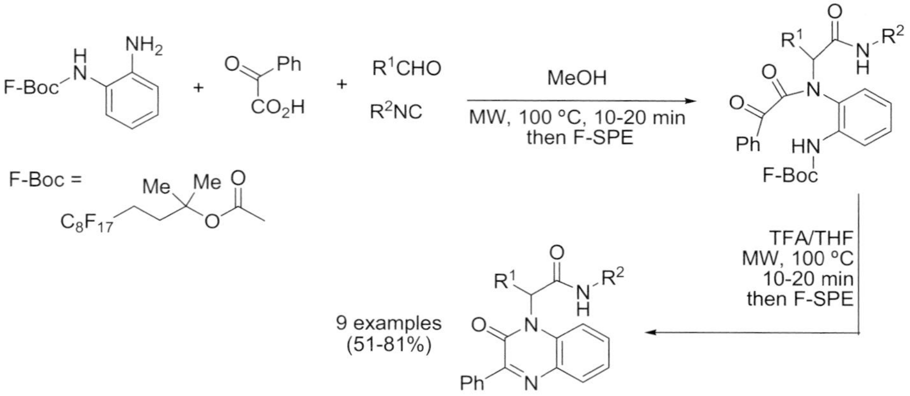

Scheme 7.84 Fluorous phase quinoxalinone synthesis.

In another related study, Lu and Zhang described the microwave-assisted fluorous-phase synthesis of a pyridine/pyrazine library employing a three-component condensation reaction [97]. Imidazo[1,2-a]pyridines/pyrazines are biologically interesting compounds, showing several activities such as antifungal, antibacterial, and benzodiazepine receptor antagonistic properties. As multicomponent reactions (MCRs) are powerful tools for the generation of sets of highly diverse molecules, this technique was applied in the fluorous-phase synthesis of the desired pyridines/pyrazines under microwave irradiation.

For this purpose, perfluorooctanesulfonyl-tagged benzaldehydes were reacted with 1.1 equivalents of a 2-aminopyridine (or 2-aminopyrazine), 1.2 equivalents of an isonitrile, and a catalytic amount of scandium(III) triflate [Sc(OTf)$_3$] under microwave irradiation in a mixture of dichloromethane and methanol (Scheme 7.85). A ramp time of 2 min was employed to achieve the pre-set temperature, and then the reaction mixture was maintained at the final temperature for a further 10 min. The fluorous tag constitutes a multifunctional tool in this reaction, protecting the phenol in the condensation step, being the phase tag for purification, and serving as an acti-

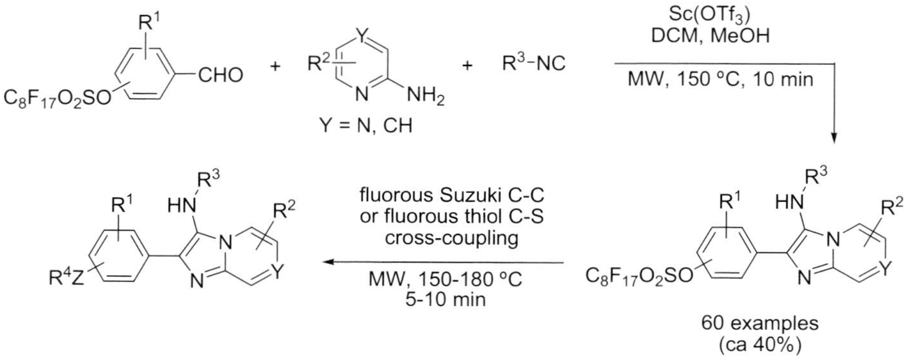

Scheme 7.85 Fluorous phase synthesis of imidazo[1,2-a]pyridine/pyrazine derivatives.

vating group in the subsequent cross-coupling reactions. After cooling to room temperature, the resulting fluorous-tagged condensation products were purified by simple solid-phase extraction (SPE) or recrystallization, before being subjected to various palladium-catalyzed cross-coupling reactions [97]. However, the yields for both the multicomponent reaction and the cross-coupling were rather low. Interestingly, the established microwave-optimized protocols have since been replaced by conventional heating in the context of a quick parallel library synthesis.

Regioselective Heck vinylations of enamides under controlled microwave irradiation have been studied by the Hallberg group using a palladium catalyst and fluorous bidentate 1,3-bis(diphenylphosphino)propane ligands (F-dppp) [98]. The Heck reaction gave identical yields of the products when using the non-fluorous and fluorous ligands, although the regioselectivity was slightly lower employing the fluorous ligand. The use of microwave heating led to enhanced reaction rates as compared to

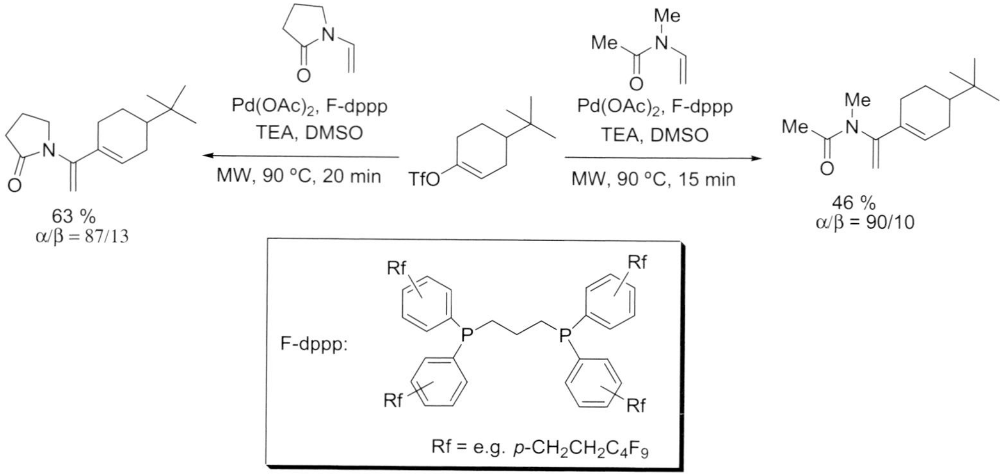

Scheme 7.86 Vinylation of enamides employing fluorous ligands.

classical heating. Vinylations of both cyclic and acyclic enamides with vinyl triflates were accelerated under microwave heating using fluorous ligands (F-dppp) and palladium(II) acetate as catalyst (Scheme 7.86). Related fluorous ligands were reported by Curran and coworkers [134].

Fluorous ligands introduce an ease of purification in that the tagged phosphine ligand, the palladium catalyst complexed ligand, and the oxidized ligand can be completely removed by direct fluorous solid-phase separation (F-SPE) prior to product isolation. Similarly, an example of a fluorous palladium-catalyzed microwave-induced synthesis of aryl sulfides has been reported, whereby the product purification was aided by fluorous solid-phase extraction [91].

In a more recent report, Larhed and coworkers have presented microwave-mediated fluorous reaction conditions for palladium-catalyzed aminocarbonylations [100]. A set of aryl halides was reacted with carbonyl hydrazides and molybdenum hexacarbonyl [Mo(CO)$_6$] as a source of carbon monoxide, employing fluorous triphenylphosphine (F-TPP) as ligand and the perfluorocarbon liquid FC-84 as a perfluorinated solvent (Scheme 7.87; see also Scheme 6.46 c).

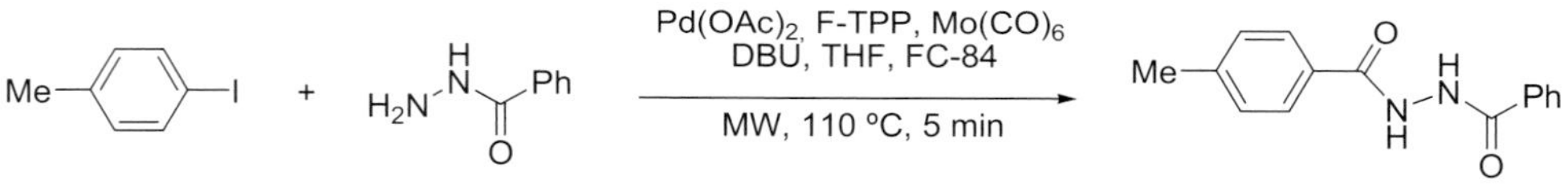

Scheme 7.87 Fluorous phase hydrazidocarbonylation.

The authors demonstrated the recyclability of the fluorous reagents, which showed no significant loss of efficiency in facilitating the model reaction shown in Scheme 7.87. After each run, the organic layer was separated and the perfluorinated liquid was applied to the next reaction mixture. Performing six cycles of the reaction afforded the corresponding product in 64–79% yield (Fig. 7.6).

An interesting approach involving the use of a fluorous dienophile as a diene scavenger under microwave conditions has been investigated by Werner and Curran [101]. The classical Diels–Alder-type cycloaddition of diphenylbutadiene with maleic anhydride as dienophile was accelerated under microwave heating, such that it

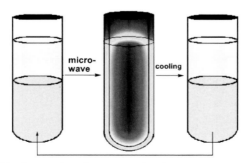

Fig. 7.6 Recycling of the fluorous ligand in aminocarbonylations (reproduced with permission from [100]).

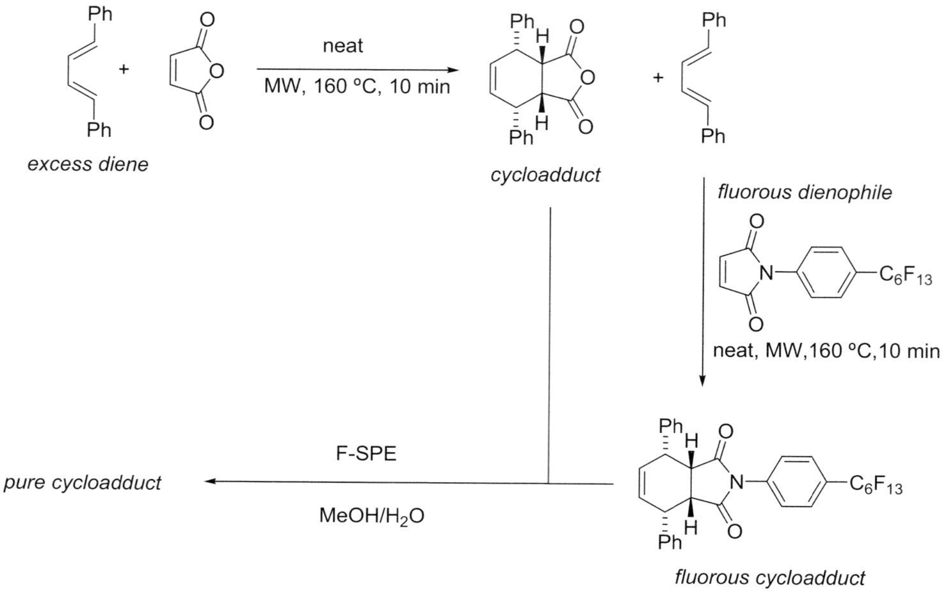

Scheme 7.88 Diene scavenging utilizing a fluorous dienophile.

could be achieved in 10 min at 160 °C. A structurally related fluorous dienophile scavenged the excess diene (1.5 equivalents) under the same microwave conditions and aided the purification of the final cycloaddition product (Scheme 7.88). The desired compounds were isolated in 79–93% yield and 90–93% purity after fluorous solid-phase extraction (F-SPE). Subsequent elution of the F-SPE column with diethyl ether furnished the fluorous Diels–Alder adduct.

7.4
Grafted Ionic Liquid-Phase-Supported Synthesis

Ionic liquids are particularly applicable for use in microwave-mediated liquid-phase reactions as they efficiently couple with microwaves resulting in very rapid heating profiles (see Section 4.3.3.2). Furthermore, ionic liquids are immiscible with a wide range of organic solvents and provide an alternative non-aqueous two-phase system. An interesting approach involves the attachment of ionic liquid components to organic substrates.

The very first report on the use of ionic liquids as soluble supports was presented by Fraga-Dubreuil and Bazureau in 2001 [102]. The efficacy of a microwave-induced solvent-free Knoevenagel condensation of a formyl group on the ionic liquid (IL) phase with malonate derivatives ($E^1CH_2E^2$) catalyzed by 2 mol% of piperidine was studied (Scheme 7.89). The progress of the reaction could be easily monitored by ^1H and ^{13}C NMR spectroscopy, and the final products could be cleaved from the IL

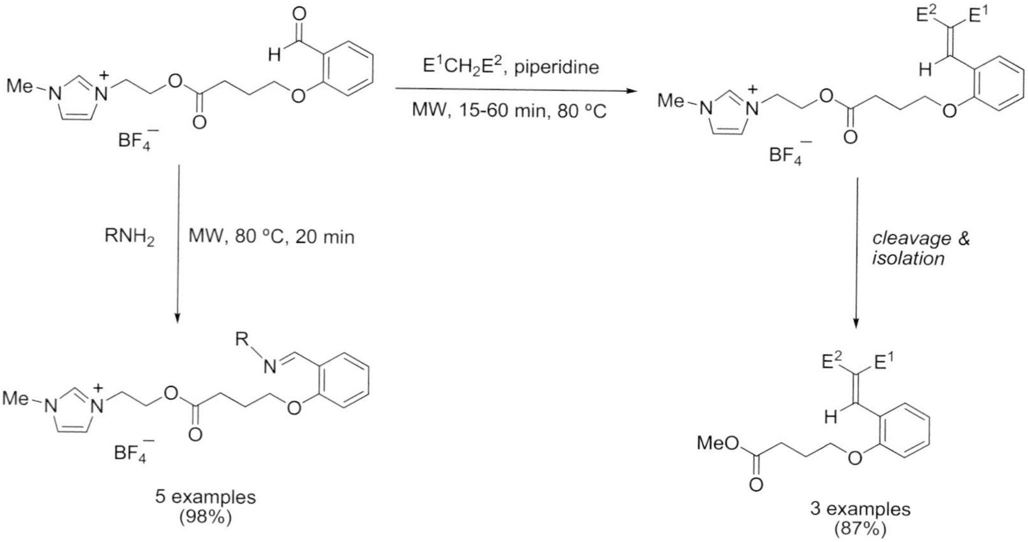

Scheme 7.89 Ionic-liquid-mediated Knoevenagel condensation.

phase and extracted in high yield without any need for additional purification. More importantly, the reaction times were short (15–60 min) and it was possible to reuse the ionic liquid in another cycle of the synthesis. This methodology was also instrumental in providing grafted aldimines within short microwave irradiation times. This enabled the regioselective introduction of amines onto the IL phase as precursors for some conventional regioselective 1,3-cycloadditions with imidates for the preparation of diversely substituted imidazoles on the ionic liquid phase (not shown).

This technique was extended by grafting the ionic liquid moiety to polyethylene glycol (PEG) units prior to attaching the benzaldehyde, as presented by the same group in a more recent study [103]. The corresponding imidazoles were treated with several ω-chlorinated alcohols to efficiently form the desired ionic liquids under microwave irradiation (Scheme 7.90). Since it is not only the length of the alkyl spacer attached to the imidazolium cation that drastically influences the physical properties (viscosity, hydrophobicity) of the resulting polyethylene glycol (PEG) ionic

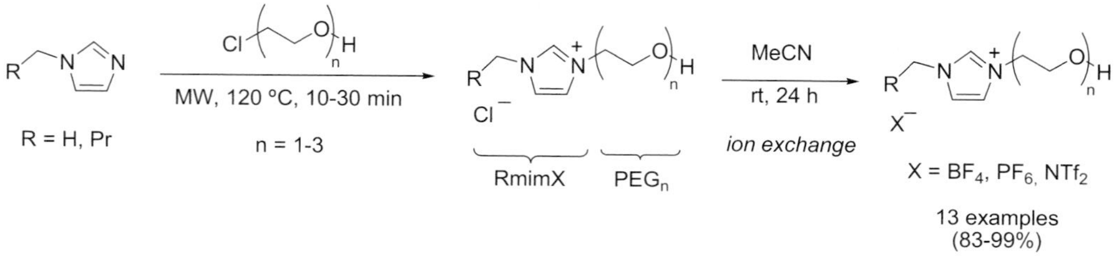

Scheme 7.90 Preparation of polyethylene glycol–ionic liquid phases.

liquid phases (PEG-ILPs) but also the counter-ion, a series of novel ionic liquids has been prepared by simple anion metathesis exchange. The synthesized imidazolium chlorides were treated with inorganic salts to afford the respective PEG-ILPs with tetrafluoroborate (BF_4^-), hexafluorophosphate (PF_6^-), and bis[(trifluoromethyl)sul-fonyl]amide (NTf_2) anions. The prepared compounds showed varying viscosity and miscibility with organic solvents and could be efficiently used for microwave-assisted reactions.

In subsequent work, it was demonstrated that this ionic liquid phase-bound organic synthesis methodology is also applicable to the formation of several hetero-cyclic scaffolds. Fraga-Dubreuil and Bazureau described the generation of a small 4-thiazolidinone library utilizing their previously described ionic liquid phase (ILP) (see above) in a one-pot, three-component condensation [104]. Equimolar amounts of the ILP, a primary amine, and mercaptoacetic acid were irradiated at 100 °C to afford – through an initial imine formation – the desired ILP-grafted thiazolidinones (Scheme 7.91). The cyclization step was monitored by ^{1}H NMR and required a rather long time as compared to other microwave-mediated transformations. The products were cleaved by amide formation, utilizing either primary amines or pyrro-lidine (not shown). This ester aminolysis usually requires rather harsh conditions, but was performed within 10–20 min at 100 °C under microwave irradiation employing potassium *tert*-butoxide base as a catalyst. The use of β-substituted pri-mary amines in the cleavage step required a higher temperature of 150 °C.

In a related study, the group of Bazureau applied their polyethylene glycol-grafted ionic liquid phases (ILPs) to the preparation of 2-thioxotetrahydropyrimidinones [105]. After the initial formation of acrylate-bound ILPs utilizing acryloyl chloride in refluxing dichloromethane, several primary amines were attached in a Michael addi-

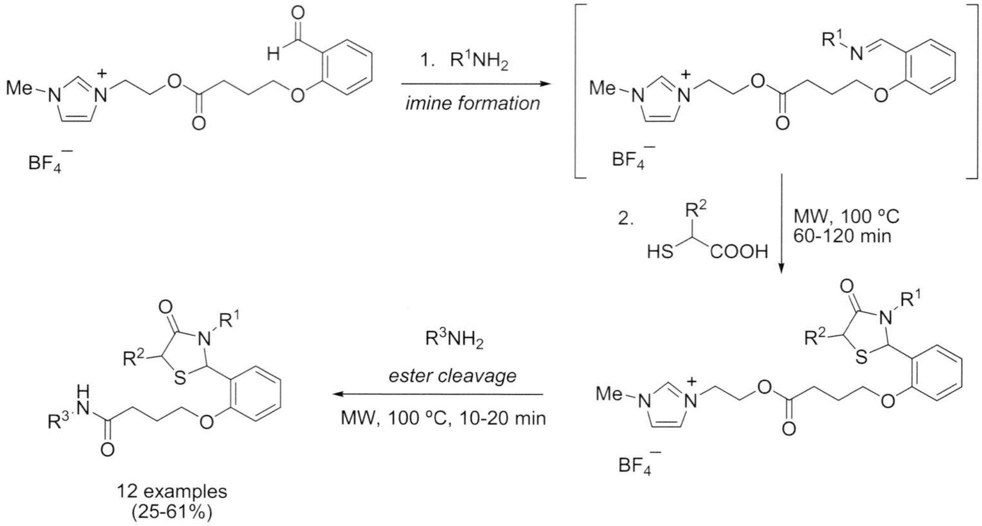

Scheme 7.91 Generation of a 4-thiazolidinone library utilizing grafted ionic liquids.

tion type reaction at ambient temperature. Subsequent treatment of the ILP-bound β-amido esters with alkyl isothiocyanates in dry acetonitrile gave the final intermediate (Scheme 7.92). Finally, cyclative cleavage was achieved under microwave irradiation employing 2 equivalents of diethylamine. The resulting mixtures were extracted with chloroform and, after evaporation of the solvent from the extracts, the crude products were purified by flash chromatography to afford the desired 2-thioxotetrahydropyrimidin-4(1*H*)-ones.

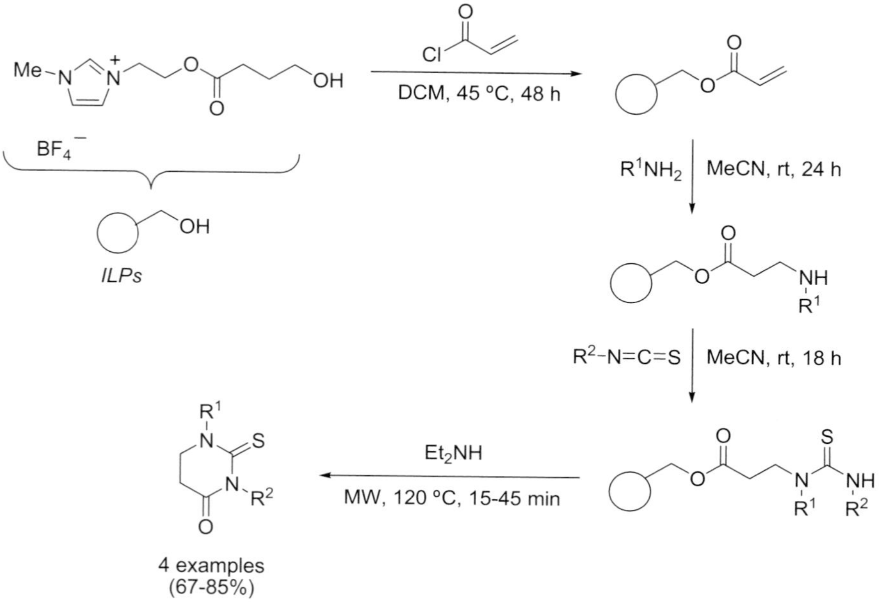

Scheme 7.92 Preparation of 2-thioxotetrahydropyrimidin-4(1*H*)-ones by ionic liquid phase organic synthesis.

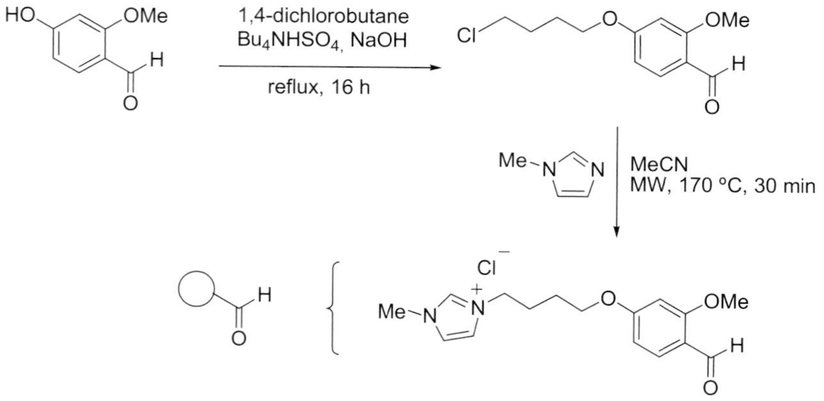

Scheme 7.93 Preparation of an *N*-imidazolium-based soluble AMEBA linker.

In a recent study, the group of Buijsman presented a microwave-mediated preparation of a different *N*-imidazolium-based ionic analogue of the well-known AMEBA solid support (Scheme 7.93). With this soluble support, a set of various sulfonamides and amides was prepared, and furthermore the use of this novel linker in the synthesis of a potent analogue of the antiplatelet drug tirofiban was presented [106].

7.5
Polymer-Supported Reagents

Besides traditional solid-phase organic synthesis (SPOS), the use of polymer-supported reagents (PSR) is attracting increasing attention from practitioners in the field of combinatorial chemistry [107, 108]. The use of PSRs combines the benefits of SPOS with the advantages of solution-phase synthesis (Fig. 7.7). The most important advantages of these reagents are the simplification of reaction work-up and product isolation, the work-up being reduced to simple filtrations. In addition, PSRs can be used in excess without affecting the purification step. By using this technique, reactions can be driven to completion more easily than in conventional solution-phase chemistry.

The polymer material involved in the development of polymer supports for immobilization of catalysts and/or reagents is generally a functionalized polystyrene/divinyl benzene co-polymer, as described in Section 7.1. Cross-linking with divinyl benzene imparts the polymer support with mechanical strength, making it resistant towards detrimental conditions such as vigorous stirring. Chemical functionality is introduced onto the polymer support by physical adsorption or by chemical bonding. These cross-linked poly(styrene-co-divinyl benzene) resins (1–2% cross-linking) are stable, have high loading capacity (>1 mmol/g) and high swelling properties, and are compatible with a variety of non-protic solvents. It is generally recognized that introducing a reactive species on a polymer support leads to site isolation. By increasing the percentage of cross-linking, the polymer chain mobility can be restricted, and thereby the interaction of reactive sites on a polymer chain can be restrained. The supported substrate molecule acts as a flexible spacer group to facilitate the interaction of reactive sites. This concept in terms of a polymer-supported catalyst is advantageous, resolving the need for a higher catalyst loading capacity. Moreover, the reactivity is improved when the catalyst bearing reactive sites can act in a co-operative manner towards a substrate molecule.

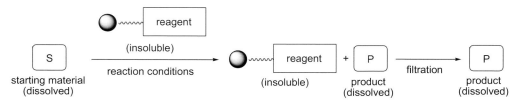

Fig. 7.7   Principle of polymer-supported reagents in organic synthesis.

Microwave-assisted organic transformations using polymer-supported reagents and/or a catalyst (see also Section 7.6) have been widely exploited in the chemical community. Microwave heating involves rapid instantaneous heating in closed vessels, often leading to high internal pressures, to temperatures well above the conventional boiling points of the solvents used (see Chapter 2). This kind of flash heating is adapted to the overall stability of the polymer backbone used and the supported species. On the contrary, conventional heating methods, involving prolonged exposure to higher reaction temperatures, often lead to degradation/decomposition of the material. This feature of microwave heating, together with the possibility of automating an entire synthetic process involving polymer-supported catalysts/reagents, makes it popular. The frequently observed increased reaction rates under microwave irradiation may conceivably be rationalized in terms of the selective absorption of reactive sites on a polymeric material (see Section 2.5.2). Typically, considering metal-impregnated catalysts, the metal particles would be heated preferentially over the polymer support [109]. Since microwave heating is influenced by the dielectric properties of the substrate, some temperature differences may arise when the absorption characteristic of the polymer support differs from that of the supported reagent and/or catalyst. In heterogeneous systems, the difference in the coupling abilities of the active sites and of the support together with the reaction medium often leads to localized heating or the development of "hotspots". Equally important is the interaction of a solvent with microwaves and its ability to diffuse into the polymer support for catalytic applications (see Table 7.1). Most functional groups on the polymer-supported system are within a polymer bead of diameter of ca. 100 μm, and the reactants in the solution enter the beads and react in the gel phase. Under microwave heating, the migration of reactants is influenced by the different swelling factors of some polymers with polar monomer units [110]. This is rate-determining and sometimes selectivity can arise because a more bulky group diffuses more slowly to the reactive sites [111].

Considering the characteristics of polymeric materials as supports for catalysts and/or reagents, the key factors that have made their application increasingly popular are ease of separation after the reaction, recyclability, and the possibility of reinstating the catalytic activity post-reaction.

A microwave-induced synthesis of 1,3,4-oxadiazoles has been described by Brain and coworkers employing a soluble polymer-supported Burgess' reagent (Scheme 7.94) [112]. The use of this supported Burgess' reagent offers a mild procedure for the cyclodehydration of 1,2-diacylhydrazines. Applying soluble polyethylene glycol (PEG)-supported Burgess' reagent [113], only 40% conversion was obtained after conventional reflux for 3 h in tetrahydrofuran. However, under microwave heating the reaction reached completion within 2–8 min. 1,2-Diacylhydrazines with a variety of substituted aromatic rings or heteroaromatic rings were cyclodehydrated with the soluble polymer-supported Burgess' reagent. The corresponding 1,3,4-oxadiazoles were obtained in varying yields but generally high purity within 2 min of microwave irradiation.

In more recent work, Brain and Brunton employed a different polystyrene-supported dehydrating agent (Fig. 7.8) for the synthesis of 1,3,4-oxadiazoles under ther-

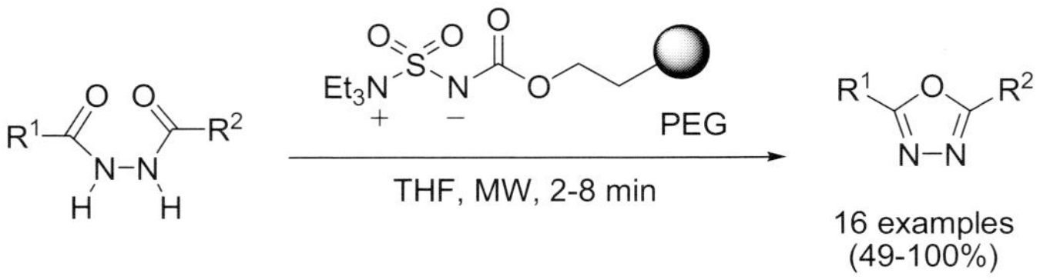

Scheme 7.94 Synthesis of 1,3,4-oxadiazoles utilizing a polymer-bound Burgess reagent.

mal and microwave conditions (Scheme 7.94) [114]. Addition of a homogeneous base such as guanidine improved the cyclodehydration reaction and allowed it to proceed to completion under thermal and even more markedly under microwave conditions. However, this necessitated subsequent purification steps for the removal of the base for the clean and quantitative isolation of the desired products. On the other hand, the use of a polymer-supported phosphazene base (Fig. 7.8) circumvents the need for additional purification of the reaction mixture for product isolation. Microwave heating proved advantageous over the thermal protocol, leading to a dramatic reduction in reaction times together with high yields and purities of the resulting 1,3,4-oxadiazoles. The same phosphozene base has also been used in amination reactions [99].

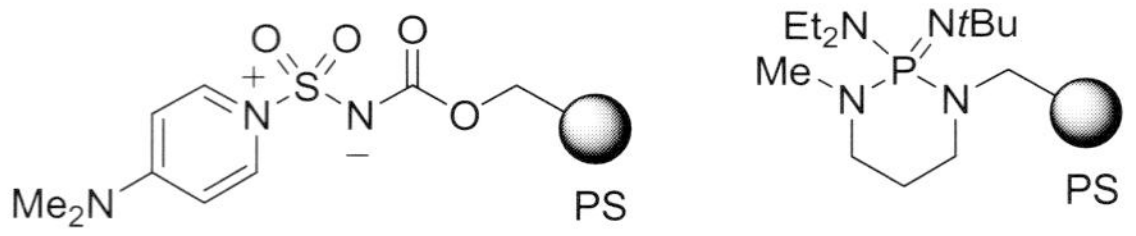

Fig. 7.8 Polystyrene-bound dehydrating agent (left) and polymer-supported phosphazene base PS-BEMP (right) utilized for oxadiazole synthesis.

The microwave-assisted thionation of amides has been studied by Ley and co-workers using a polymer-supported thionating reagent [115]. This polymer-supported amino thiophosphate serves as a convenient substitute for its homogeneous analogue in the microwave-induced rapid conversion of amides to thioamides. Under microwave conditions, the reaction is complete within 15 min, as opposed to 30 h by conventional reflux in toluene (Scheme 7.95). The reaction has been studied for a range of secondary and tertiary amides and GC-MS monitoring showed that it proceeded almost quantitatively. More importantly, this work was the first incidence of the use of the ionic liquid 1-ethyl-3-methylimidazolium hexafluorophosphate

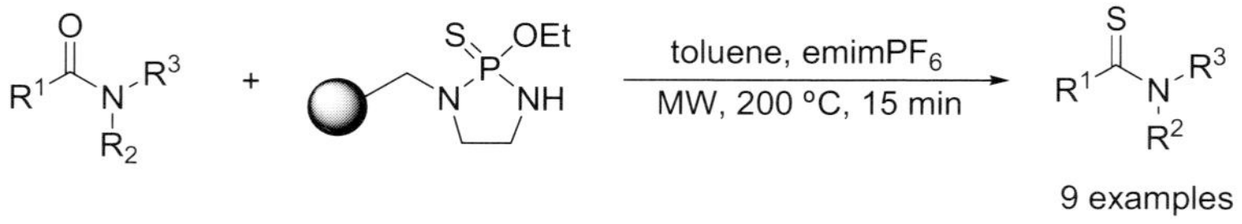

9 examples

Scheme 7.95 Thionation of amides utilizing a polymer-bound aminothiophosphate.

(emimPF$_6$) to dope non-polar solvents such as toluene to assist the heating under microwave irradiation (see Section 4.3.3.2).

A microwave-induced one-pot synthesis of alkenes by Wittig olefination has been reported by Westman using a polymer-supported triphenylphosphine [116]. The conventional solution-phase Wittig reaction often requires long reaction times and the triphenylphosphine unavoidably forms the triphenylphosphine oxide by-product in solution. To circumvent this, a similar reaction protocol using a solid-supported triphenylphosphine has been developed. Microwave heating firstly speeds up the actual chemistry and the products can be obtained free from organophosphorus contamination by simple filtration. The preparation of the actual Wittig reagent with *in situ* formation of the corresponding ylide has been realized in high yields within minutes, as opposed to several days by conventional methods.

Olefinations using a solid-supported triphenylphosphine were accomplished within 5 min of microwave heating at 150 °C using potassium carbonate as the base and methanol as the solvent (Scheme 7.96). A variety of aldehydes and organic halides were studied with the solid-supported phosphine, yielding the corresponding alkenes in excellent purities.

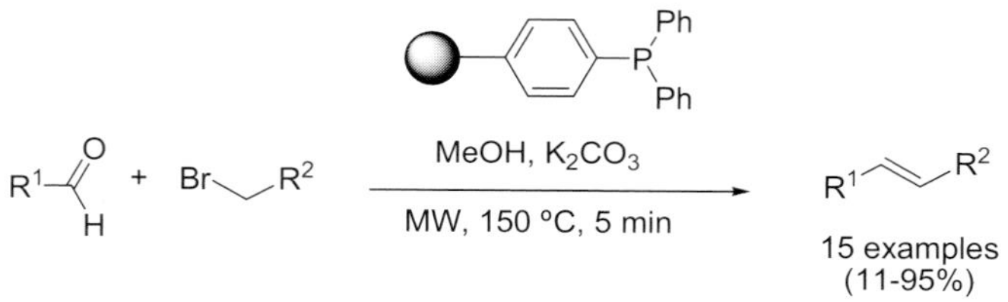

Scheme 7.96 Wittig olefinations using resin-bound triphenylphosphine.

Desai and Danks have reported on rapid solid-phase transfer hydrogenation utilizing a polymer-supported hydrogen donor (ammonium formate derivative) for the reduction of electron-deficient alkenes in the presence of Wilkinson's catalyst [RhCl(PPh$_3$)$_3$] under microwave heating [117]. Conventional hydrogen donors such as ammonium formate sublime at high temperatures and also release toxic gases such as ammonia. The polymer-supported formate, on the other hand, is not subject to these shortcomings, and it could easily be used for automated robotic synthesis.

In the work described by these authors, the formate utilized, which was immobilized on Amberlite® IRA-938, was admixed with the requisite substrate in a mini-

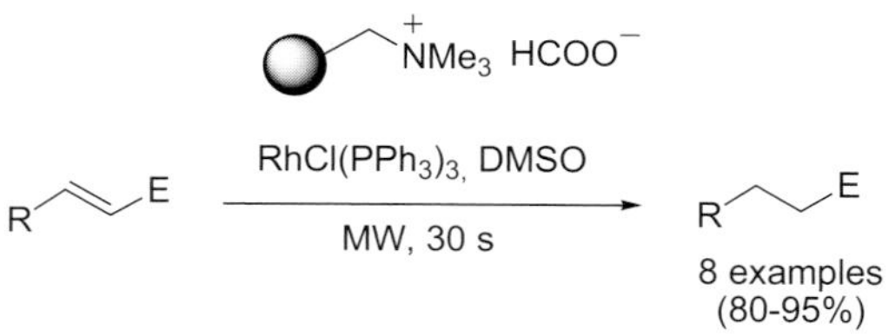

Scheme 7.97 Transfer hydrogenations using a resin-bound formate.

mum amount of dimethyl sulfoxide and the reaction mixture was irradiated in a sealed vessel for 30 s (Scheme 7.97). After cooling, the mixture was diluted with dichloromethane, washed with water, and dried. Evaporation of the solvent furnished the successfully hydrogenated compounds in high yields.

Other authors have described the application of a recyclable polymer-supported bromine chloride resin for regio- and chemoselective bromomethoxylation of various substituted alkenes under microwave conditions [118]. The immobilized brominating reagent utilized (perbromide resin, 1.6 mmol g^{-1} Br) was generated from a commercially available chloride exchange resin by simple bromination at room temperature (Scheme 7.98). The short-term microwave irradiations were carried out in refluxing methanol in a modified domestic oven equipped with a mounted reflux condenser. Bromomethoxylation was only observed across isolated double bonds, whereas conjugated double bonds remained unaffected under the reaction conditions. The crude products were easily recovered by filtration and evaporation of the solvent. The polymer-supported bromine resin could be recycled and regenerated by successive washings with methanol, acetonitrile, and chloroform and finally passing a solution of bromine in tetrachloromethane through it.

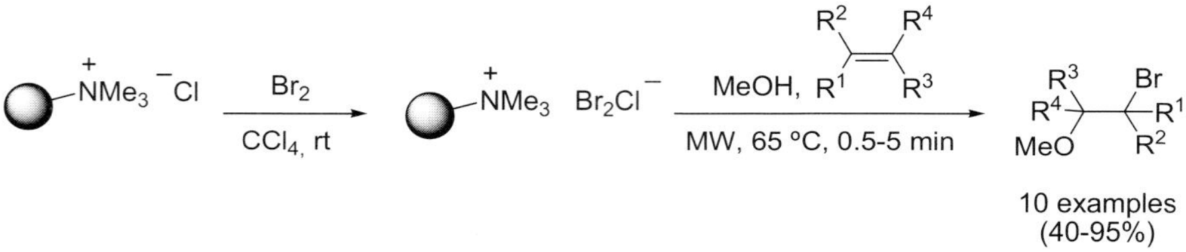

Scheme 7.98 Regio- and chemoselective bromomethoxylation.

The conversion of isothiocyanates to isonitriles under microwave conditions has been studied by Ley and Taylor using a polymer-supported [1,3,2]oxaphospholidine [119]. The use of 3-methyl-2-phenyl-[1,3,2]oxaphospholidine in solution is less favored [120] due to the associated toxicity and instability of the phosphorus-derived reagent, as well as the need to isolate the products from a complex reaction mixture by vacuum distillation. This drawback has been resolved by attaching the active [1,3,2]oxaphospholidine to a polymer matrix.

Commercial Merrifield resin was treated with aminoethanol and the resulting precursor was condensed with bis(diethylamino)phosphine in anhydrous toluene to generate the desired polymer-supported reagent. When using the novel solid-supported reagent, the conversions of the isothiocyanates could be carried out in a parallel fashion under microwave irradiation conditions, and the corresponding isonitriles were isolated in high purities (Scheme 7.99). This approach avoided any exposure to the highly toxic isocyanides during isolation and also limited the toxicity through the use of an immobilized phosphorus-derived reagent.

To demonstrate the feasibility of this method for library generation, the isocyanides produced were subjected to an Ugi three-component condensation with various primary amines and carboxybenzaldehyde. The resulting 2-isoindolinone derivatives were obtained in high to excellent yields.

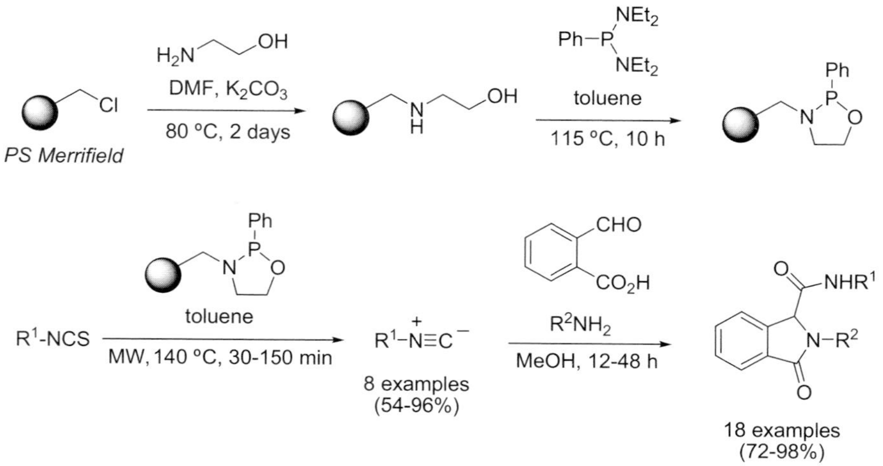

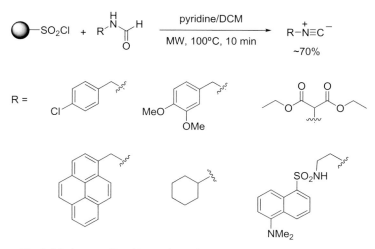

Scheme 7.99 Conversion of isothiocyanates to isonitriles utilizing a polymer-supported [1,3,2]oxazaphospholidine.

Bradley and coworkers have discussed a solid-phase mediated synthesis of iso-nitriles from formamides, utilizing polystyrene-bound sulfonyl chloride as a suitable supported reagent [121]. This commercially available polymer-supported reagent offers an efficient method for isonitrile generation under microwave irradiation, requiring only a simple filtration and acidic work-up. Six formamide derivatives were converted into the corresponding isonitriles in good purities utilizing 3 equiva-lents of the polymer-supported reagent (Scheme 7.100). However, increasing the degree of substitution of the formamide was found to have a detrimental effect on the purity of the desired products. The isonitriles were isolated by filtering off the

Scheme 7.100 Solid-phase mediated isonitrile synthesis.

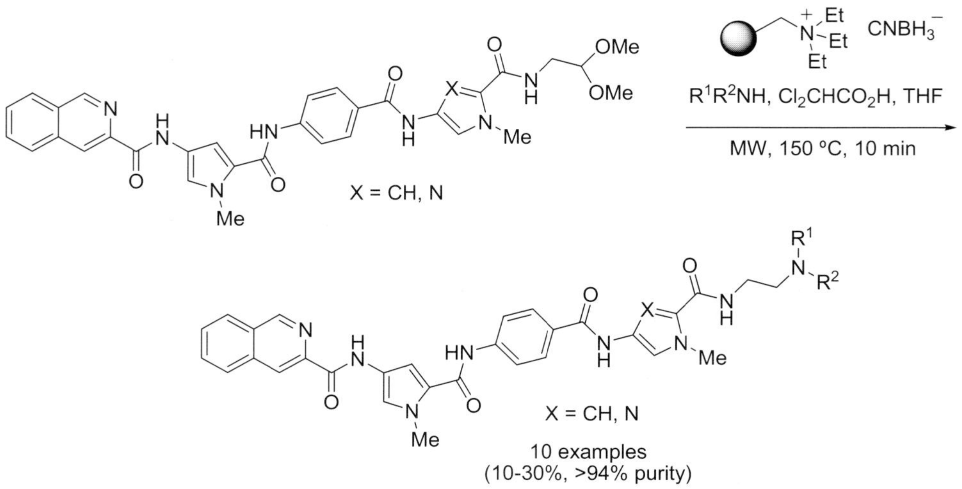

Scheme 7.101 Reductive amination of tetrameric isoquinolines.

resin, pouring the filtrate into water, and extracting with dichloromethane. Evaporation of the solvent furnished the desired products in approximately 70% yield. Furthermore, the sulfonyl chloride resin could be quantitatively regenerated by treatment of the sulfonic acid resin formed with phosphorus pentachloride (PCl$_5$) in N,N-dimethylformamide at room temperature. Solid-phase mediated isonitrile synthesis under microwave irradiation led to much faster reactions, thereby allowing rapid access to this important class of compounds, which are suitably predisposed for a broad range of subsequent syntheses.

In a recent study, the group of Moser has described the use of a polymer-bound borohydride in reductive aminations of tetrameric isoquinolines (Scheme 7.101) [122]. These tetrameric isoquinolines, which represented lead compounds in a search for antibacterial distamicyn A analogues, were prepared from the appropriate

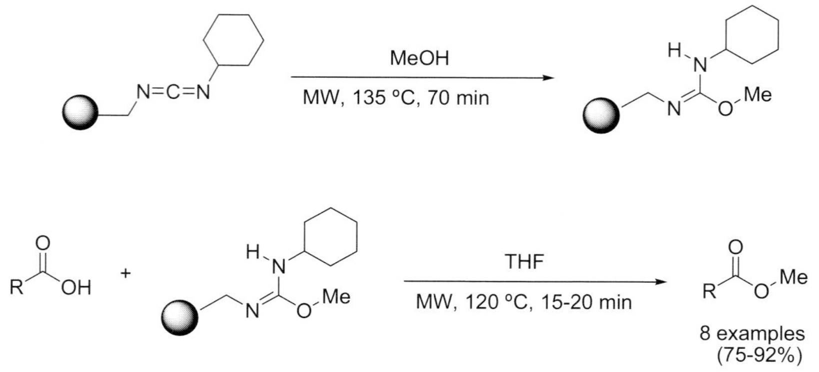

Scheme 7.102 Polymer-supported O-alkylation of carboxylic acids.

isoquinoline, imidazole, and pyrrole building blocks by standard amide bond formations. The final derivatization by reductive amination in the presence of Merrifield res-
in-supported cyanoborohydride was efficiently accelerated by microwave irradiation.

The group of Linclau has demonstrated the effective *O*-alkylation of carboxylic
acids using a polymer-supported *O*-methylisourea reagent [123]. Under conventional
conditions, complete esterifications were observed only after refluxing for several
hours in tetrahydrofuran, and the acidic work-up required limited the scope of
applicable substituents. In contrast, employing microwave heating led to complete
esterifications within 15–20 min, with only 2 equivalents of the polymer-bound
O-methylisourea being required (Scheme 7.102). Furthermore, the polymer-supported reagent simplified purification as the pure methyl esters were obtained by
evaporation of the solvent after a simple filtration step.

Sauer and coworkers have presented the use of polystyrene-bound carbodiimide
for convenient and rapid amide synthesis [124]. An equimolar mixture of 1-methy-
lindole-3-carboxylate, the requisite amine, and 1-hydroxybenzotriazole (HOBt) in
1-methyl-2-pyrrolidinone (NMP) was admixed with 2 equivalents of the polymer-
bound carbodiimide and irradiated for 5 min at 100 °C (Scheme 7.103). After cooling, the mixture was diluted with methanol and subjected to solid-phase extraction
utilizing silica carbonate. Evaporation of the solvent from the filtrate furnished the
desired compounds in excellent purities.

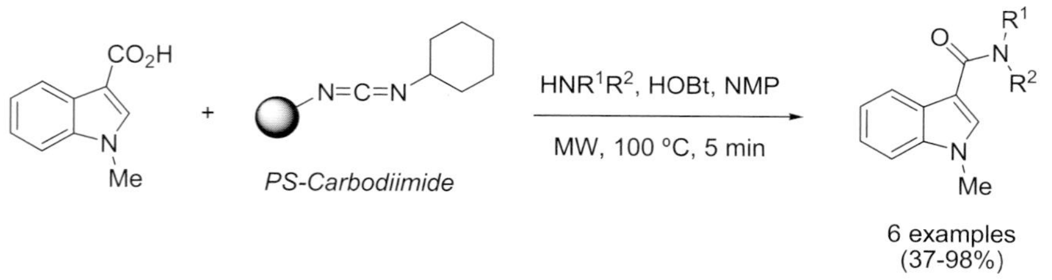

Scheme 7.103   Amide synthesis utilizing polymer-bound carbodiimide.

In a detailed investigation, Turner and coworkers have described the preparation
and application of solid-supported cyclohexane-1,3-dione as a so-called "capture and
release" reagent for amide synthesis, as well as its use as a novel scavenger resin
[125]. Their report included a three-step synthesis of polymer-bound cyclohexane-
1,3-dione (CHD resin, Scheme 7.104) from inexpensive and readily available starting
materials. The key step in this reaction was microwave-assisted complete hydrolysis
of 3-methoxy-cyclohexen-1-one resin to the desired CHD resin.

This novel resin-bound CHD derivative was then utilized in the preparation of an
amide library under microwave irradiation. Reaction of the starting resin-bound
CHD with an acyl or aroyl chloride yields an enol ester, which, upon treatment with
amines, leads to the corresponding amide, thus regenerating the CHD. This demonstrates the feasibility of using the CHD resin as a "capture and release" reagent for
the synthesis of amides. The "resin capture/release" methodology [126] aids in the
removal of impurities and facilitates product purification.

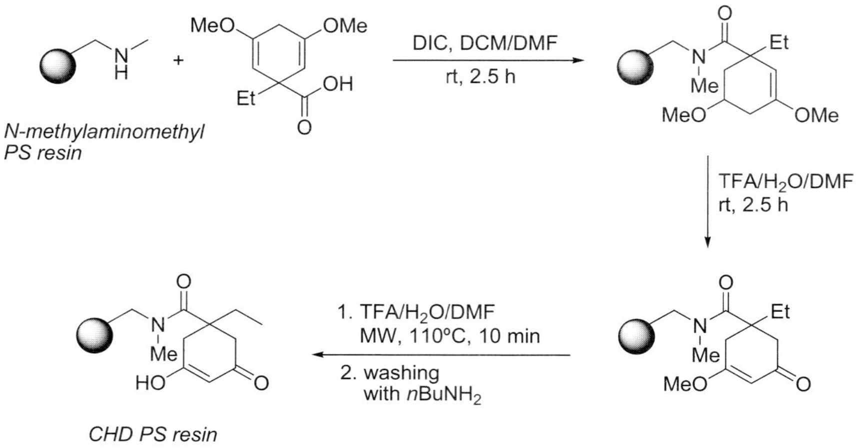

Scheme 7.104 Preparation of resin-bound cyclohexane-1,3-dione (CHD resin).

The synthesis of amides using the resin-bound CHD enol ester was found to be greatly accelerated under microwave heating, being accomplished in 30 min as compared to after overnight stirring at room temperature (Scheme 7.105), as described by the group of Turner [125]. Interestingly, kinetic monitoring has shown that the release of the amide into solution is apparently accelerated by microwave heating. The presence of an excess of amine proved to be necessary for complete release of the amide. Aromatic enol esters generally gave the corresponding amides in higher yields and purities than aliphatic enol esters, but lower yields in the case of aniline could be attributed to its lower nucleophilicity.

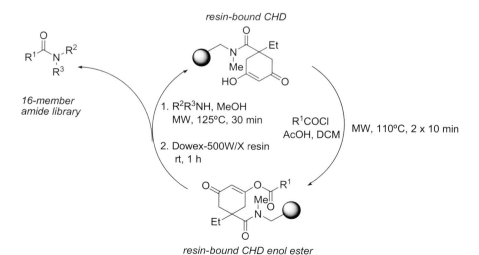

Scheme 7.105 Generation of an amide library utilizing resin capture–release methodology.

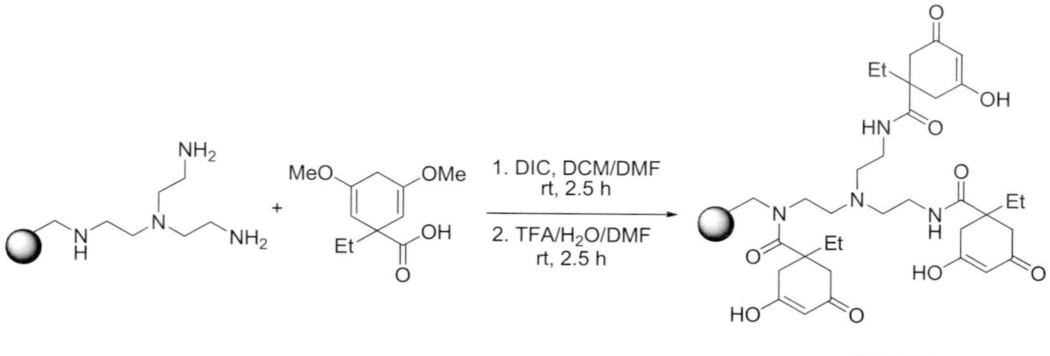

Scheme 7.106 Preparation of high-loading cyclohexane-1,3-dione scavenger resin (CHD-SR).

In order to explore the potential of CHD resins as scavenger materials, 1-ethyl-3,5-dimethoxycyclohexa-2,5-dienecarboxylic acid was anchored to a commercially available trisamine resin (Scheme 7.106), which yielded a high-loading cyclohexane-1,3-dione scavenger resin (CHD-SR) [125].

As part of the same study, the capacity of this novel resin to act as an allyl cation scavenger was demonstrated in a palladium-catalyzed O-alloc deprotection of O-alloc benzyl alcohol (Scheme 7.107) [125]. Benzyl alcohol was obtained in high yield with only trace amounts of by-product, thereby eliminating the need for further purification. The resulting C-allylation of the resin was evident from the presence of C-allyl signals in the relevant MAS-probe ^{1}H NMR spectrum.

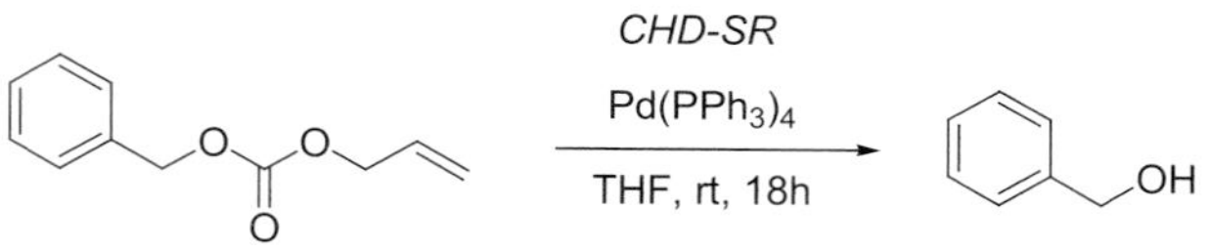

Scheme 7.107 Palladium-catalyzed deprotection of O-alloc benzyl alcohol employing the scavenging resin CHD-SR.

Another example of the "capture and release" strategy has recently been presented by Porcheddu and coworkers, utilizing polystyrene-bound piperazine for the generation of a pyrimidine library [127]. The authors disclosed a novel synthetic route to these valuable heterocycles, involving the condensation of polymer-bound enaminones with guanidines (Scheme 7.108). The enaminones utilized were readily prepared by the condensation of 6 equivalents of N-formylimidazole dimethyl diacetal and 3 equivalents of β-keto compounds with the polymer-bound piperazine in the presence of 10 mol% of camphorsulfonic acid (CSA). The mixture was heated in an open flask in a dedicated single-mode microwave reactor for 30 min to achieve full conversion of the substrates. After cooling, the functionalized resin was collected by filtration and subjected to the releasing cyclization. This step was performed by treating the polymer-bound enaminone with 1 equivalent of the appropriate guani-

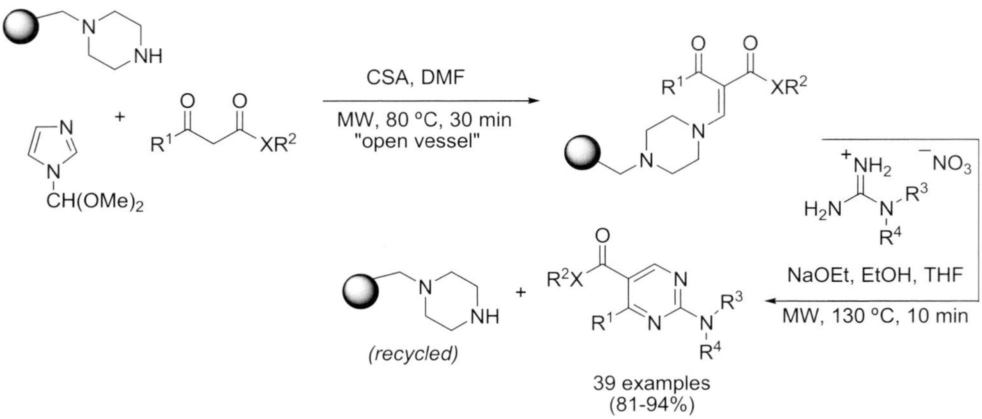

Scheme 7.108 Generation of a pyrimidine library through a capture and release strategy.

dine, which was liberated from its nitrate immediately prior to reaction. The mixture was irradiated in a sealed flask for 10 min at 130 °C to furnish the desired 2,4,5-trisubstituted pyrimidines. After the reaction, the piperazine resin could be collected by filtration and reused after several washings with ethanol. The products were isolated from the filtrate in high yields and excellent purity after aqueous work-up followed by evaporation of the solvent.

A recyclable solid-supported reagent has been developed for acylations by the group of Botta [128]. The solid-supported acylating agent could be easily applied in a parallel synthesis of the corresponding amides starting from amines (Scheme 7.109). The solid-supported reagent was removed by simple filtration after the reaction, the product was isolated by evaporation of the solvent from the filtrate, and the supported reagent could be reused at least twice before any decrease in the reaction yield was incurred. The 4-acyloxypyrimidines were prepared utilizing a domestic microwave oven by treating a solid-supported pyrimidine linker with various acyl chlorides. The solid-supported acyloxypyrimidines prepared in this way were used in the rapid and selective acylation of amines under microwave irradiation conditions (Scheme 7.109).

In a related and more recent study by the same group, syntheses of various solid-supported acylating agents for microwave-mediated transformations of amines, alcohols, phenols, and thiophenols have been presented [129]. In a microwave-mediated

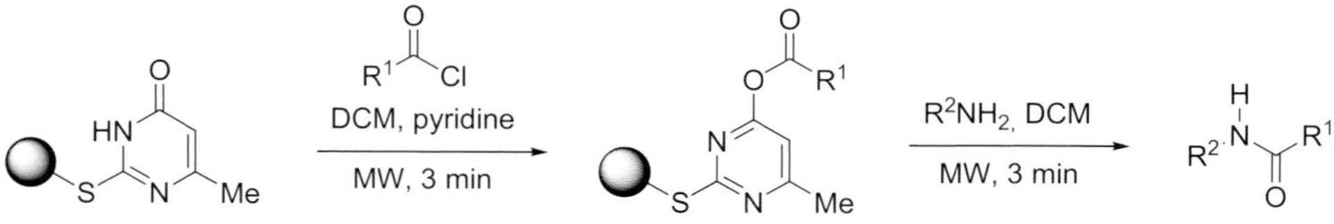

Scheme 7.109 Amide synthesis utilizing a reusable polymer-supported acylation reagent.

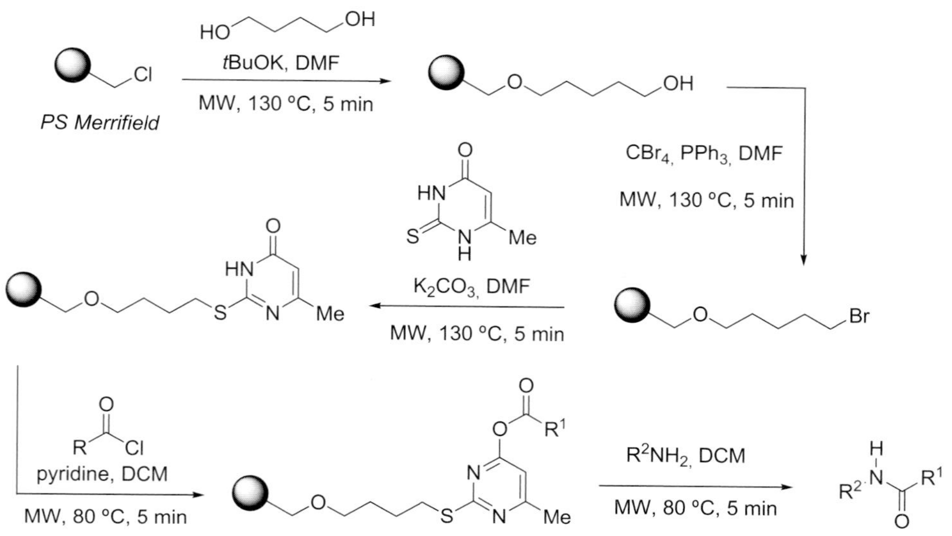

Scheme 7.110 Synthesis of a novel solid-supported reagent and its use in acylation.

procedure, Merrifield resin was first modified by attaching 1,4-butanediol to introduce a spacer unit. Bromination and subsequent reaction with commercially available 6-methyl-2-thiouracil, followed by treatment with an acyl chloride, afforded the desired polymer-bound pyrimidines (Scheme 7.110). The acylating ability of the resulting supported reagent was demonstrated by its reaction with benzylamine. Furthermore, the thiouracil building block utilized could be anchored to different solid supports to furnish alternative polymer-supported acylating agents [129].

Similarly, a solid-supported imide has been reported to serve as an acylating reagent under microwave conditions by Nicewonger and coworkers [130]. The starting imide was immobilized on aminomethyl polystyrene and in this case benzoyl chloride was chosen to prepare the acylating reagent (Scheme 7.111). Primary amines and piperazines were smoothly acylated at room temperature, but more hindered secondary amines required more time and higher temperatures, and anilines

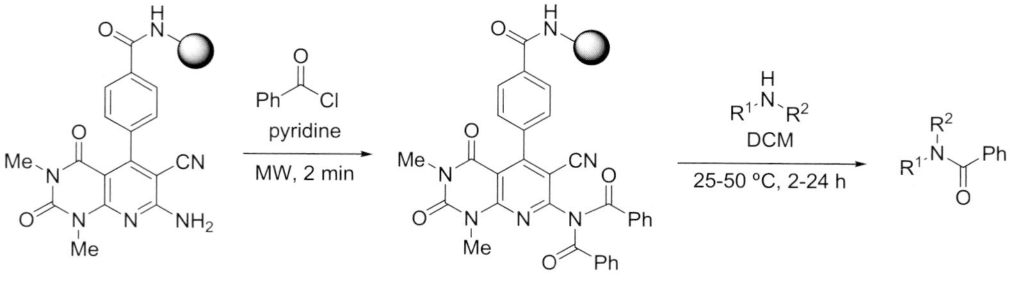

Scheme 7.111 Recyclable polymer-supported imides for amide synthesis.

could not be acylated at all. Above all, the resin-bound acylating agent was shown to be recyclable after washing with N,N-dimethylformamide and reactivation under microwave conditions.

The group of Swinnen has presented the use of a polymer-supported Mukaiyama reagent for the synthesis of amides [131]. The polymer-bound N-alkyl-2-chloropyridinium triflate utilized was initially synthesized from commercially available polystyrene Wang resin. To prove its effectiveness in even rather difficult coupling reactions, it was used in a microwave-accelerated amide synthesis involving sterically hindered pivalic acid (Scheme 7.112). The reaction did not require pre-activation of the carboxylic acid, and the reagents were simply dissolved in dichloromethane prior to addition of the resin. The mixture was then subjected to microwave irradiation at 100 °C for 10 min. Employing 3 equivalents of the acid furnished the desired product in 80% yield. Any remaining starting amine was removed by filtering the mixture through a sulfonic acid-derivatized SPE column.

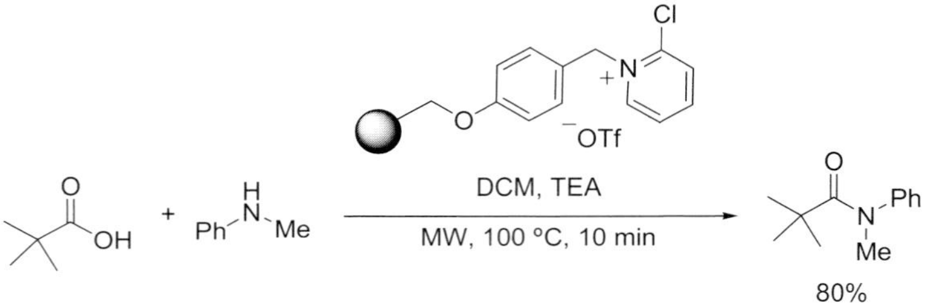

Scheme 7.112 Polymer-supported Mukaiyama reagent for amide synthesis.

As demonstrated by the Ley group, microwave heating has been applied in accelerating several slow intermediate reactions in the total synthesis of the natural product (+)-plicamine, as well as of a number of related spirocyclic templates [110, 132, 133]. The use of microwave heating in combination with polymer-bound reagents, catalysts, and scavengers proved beneficial in gaining rapid access to several key intermediates required for the synthesis of the target molecule. The initial precursor (Scheme 7.113) for the key methoxy-substituted intermediate was prepared by microwave irradiation (100 °C, 45 min) of a mixture of the starting tetracyclic lactam, trimethyl orthoformate, and a trifluoromethanesulfonate-functionalized polymer support (Nafion SAC-13). The required precursor intermediate was generated quantitatively, and was then further reduced to the desired key intermediate, bearing just one methoxy group on the unsaturated ring, using triethylsilane and Nafion SAC-13 under microwave heating for 30 min. This step was significantly more efficient under microwave irradiation, giving 95% conversion as opposed to only 43–56% yields under thermal heating. Alternatively, this key intermediate could also be accessed through a microwave-mediated process starting from the corresponding alcohol, most effectively in the presence of poly(4-vinyl-2,6-tert-butylpyridine) and methyl trifluoromethanesulfonate (MeOTf). Under microwave heating, the TFA-protected amide was hydrolyzed in quantitative yield by an ion-exchange hydroxide

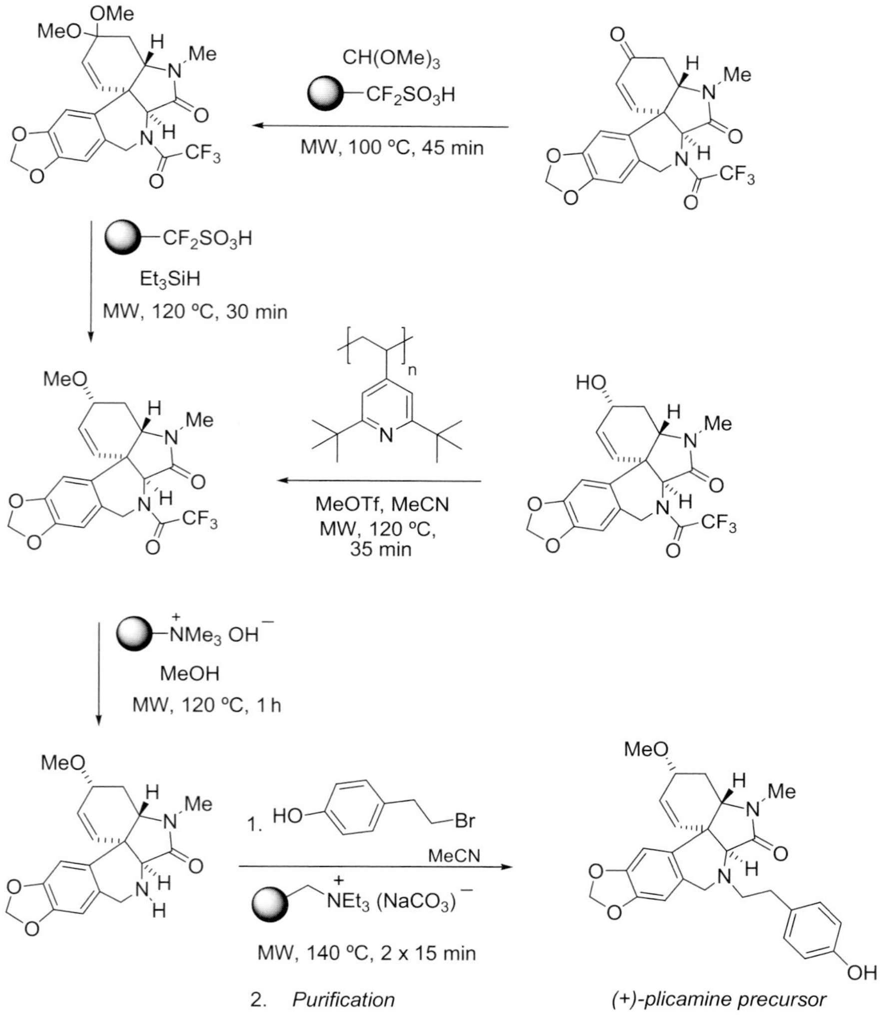

Scheme 7.113   Microwave-assisted preparation of a (+)-plicamine precursor.

resin within 1 h, as opposed to 6 h by conventional oil-bath heating. For access to the (+)-plicamine precursor, the deprotected intermediate was alkylated with the phenolic halide *p*-(bromoethyl)phenol. It was found that the conversion to the desired precursor could best be accomplished using a supported carbonate reagent under 2 × 15 min of pulsed microwave heating at 140 °C. The penultimate intermediate en route to the desired (+)-plicamine was isolated in high yield after purification, which was improved by microwave-assisted scavenging (see also Section 7.7).

7.6
Polymer-Supported Catalysts

Catalysts immobilized on polymeric supports have an important additional advantage over conventional homogeneous catalysts in that the spent catalyst can be removed after the reaction by simple filtration, as has already been discussed for polymer-supported reagents in general in Section 7.5. In many cases, the catalyst system may be regenerated and recycled several times without any significant loss of activity. Furthermore, transition metal catalysts immobilized on polymer resins have significant benefits in reducing metal contamination of the reaction mixture and the product.

7.6.1
Catalysts on Polymeric Supports

In a novel approach to improve the synthesis of valuable Biginelli compounds with different substitution patterns, the Kappe group has described the selective protection of dihydropyrimidinones [135]. This procedure was successfully catalyzed by a polymer-supported *N,N*-dimethylaminopyridine (PS-DMAP) (Scheme 7.114). This microwave-induced selective N3-acylation was applied to diversely substituted dihydropyrimidinone scaffolds, giving 30–97% isolated yields within 10–20 min of microwave heating. This methodology involving a polymer-supported DMAP is preferred to the solution-phase protocol due to ease of purification and work-up and the possibility of adaptation to a high-throughput format. The purification of the final products was improved by microwave-induced scavenging techniques (see Section 7.7).

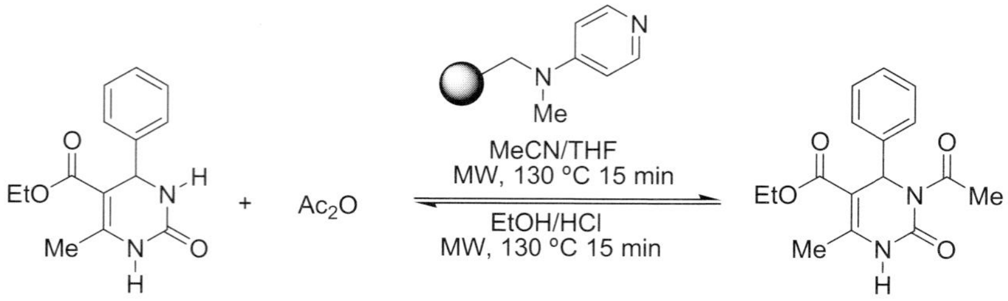

Scheme 7.114 N3-Acylations of dihydropyrimidines.

In a separate study, Öhberg and Westman applied the same PS-DMAP in a one-pot microwave-induced base-catalyzed reaction of *N*-aryl and *N*-alkyl amino acids (or esters) and thioisocyanates for the library synthesis of thiohydantoins (Scheme 7.115) [136]. Thiohydantoins are of interest due to their ease of preparation and the range of biological properties associated with this heterocyclic ring system. The use of PS-DMAP as the base in this reaction gave slightly lower yields compared to when triethylamine (TEA) was used, but it resulted in a cleaner reaction mixture and an easier purification procedure. Cyclizations of a number of *N*-substituted

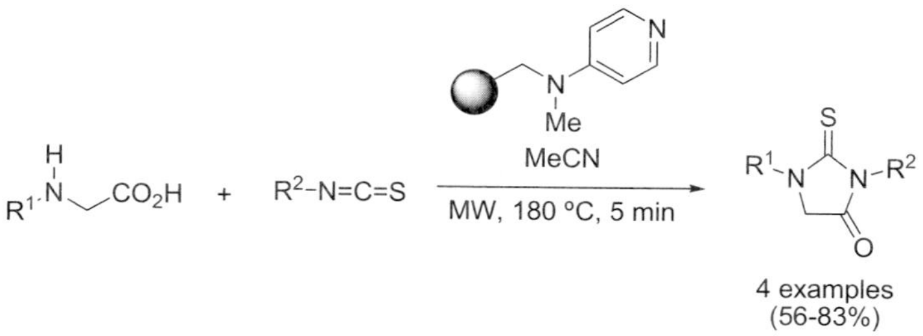

Scheme 7.115 PS-DMAP-catalyzed thiohydantoin synthesis.

amino acids with thioisocyanates gave satisfactory yields within 5 min of microwave heating.

Since metal catalysts are widely employed in organic transformations but are sometimes difficult to fully recover from reaction mixtures, the use of the corresponding polymer-bound species in microwave-mediated reactions to simplify the work-up is certainly of interest. Ring-closing metathesis (RCM) can be efficiently applied for the preparation of several small, medium, and macrocyclic ring systems. Microwave heating provides a highly desirable acceleration of such reactions using ruthenium-based catalysts. Ruthenium-based Grubbs' catalysts have been preferred in many studies involving ring-closing olefin metathesis due to their tolerance of a variety of common organic functional groups and low sensitivity to air and moisture (see Section 6.3.1).

In this context, the group of Kiddle has presented a rapid microwave-accelerated ring-closing metathesis reaction of diethyl diallylmalonate in the ionic liquid 1-butyl-3-methylimidazolium tetrafluoroborate (bmimBF$_4$) employing both homogeneous and immobilized Grubbs' catalysts in a sealed pressure tube utilizing a domestic microwave oven (Scheme 7.116) [137]. The authors observed that the polymer-bound Grubbs' catalyst gave significantly lower conversion (40%) as compared to the homogeneous analogue (80–100%). Nevertheless, the reaction could be readily accelerated by microwave heating without specific precautions. Similarly, notable accelerations in ring-closing metathesis have also been realized by Organ and coworkers using polymer-supported second-generation Grubbs' catalysts under microwave heating, with the further advantage of simple purification procedures (see also Scheme 6.74) [138].

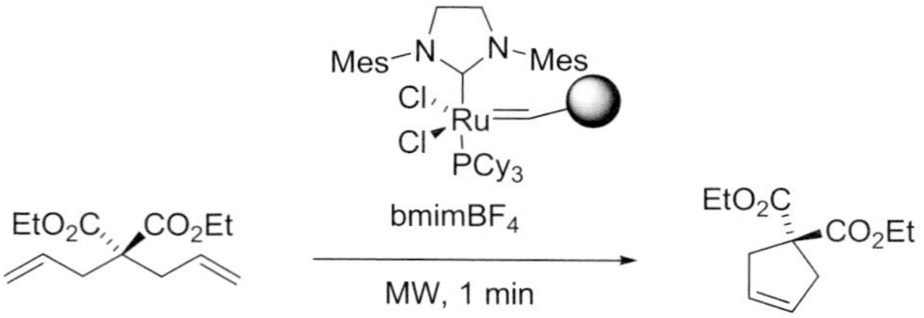

Scheme 7.116 Ring-closing metathesis utilizing ionic liquids.

Asymmetric catalysis provides access to several synthetically important compounds, and immobilized catalysts together with solid-supported chiral ligands have been equally instrumental. Chiral ligands immobilized on a solid support provide the advantage of being rapidly removable post-reaction while retaining their activity for further applications [139].

Microwave-assisted molybdenum-catalyzed allylic allylation has been studied using both free and polymer-supported bis-pyridylamide ligands [140]. The microwave-assisted catalytic reaction of 3-phenylprop-2-enyl methyl carbonate with dimethyl malonate $(CH_2(COOMe)_2)$ in the presence of N,O-bis(trimethylsilyl)acetamide (BSA) and a polymer-bound bis-pyridyl ligand (Scheme 7.117) proved to be rather slow, presumably because the insoluble resin-bound ligand renders the catalyst heterogeneous. However, when the reaction was performed with a twofold higher concentration of the reagents, it was complete after 30 min of microwave irradiation at 160 °C, and the product showed a branched-to-linear ratio of 35:1 and an enantiomeric excess of 97%. The polymer-supported ligand has obvious advantages over its unsupported analogue because after appropriate washing and vacuum drying, it could be reused at least seven times without any loss of activity.

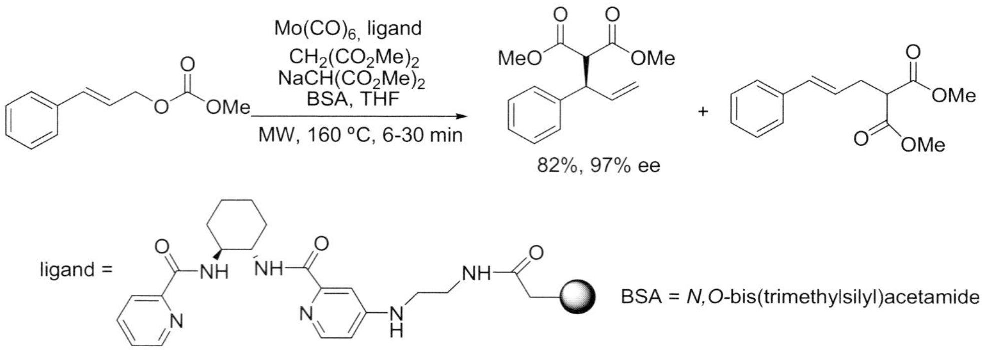

Scheme 7.117 Molybdenum-catalyzed allylic allylation.

The aforementioned polymer-supported bis-pyridyl ligand has also been applied in microwave-assisted asymmetric allylic alkylation [140], a key step in the enantioselective synthesis of (*R*)-baclofen (Scheme 7.118), as reported by Moberg and coworkers. The (*R*)-enantiomer is a useful agonist of the GABA$_B$ (γ-aminobutyric acid) receptor, and the racemic form is used as a muscle relaxant (antispasmodic). Under microwave heating, the enantioselectivity could be improved to 89% when using toluene as solvent (see also Scheme 6.52) [140].

In a more recent study, Wang and coworkers have discussed microwave-assisted Suzuki couplings employing a reusable polymer-supported palladium complex [141]. The supported catalyst was prepared from commercial Merrifield polystyrene resin under ultrasound sonification. In a typical procedure for biaryl synthesis, 1 mmol of the requisite aryl bromide together with 1.1 equivalents of the phenylboronic acid, 2.5 equivalents of potassium carbonate, and 10 mg of the polystyrene-

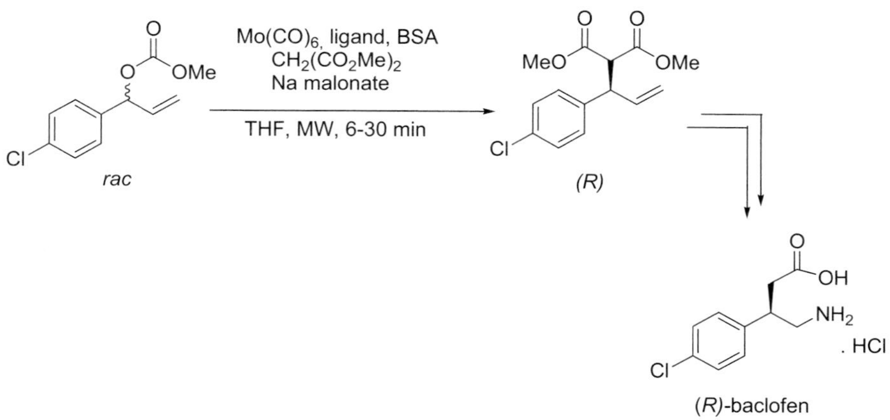

Scheme 7.118 Enantioselective synthesis of (R)-baclofen.

bound palladium catalyst (1.125 mmol Pd/g, 1 mol%) were admixed with 10 mL of toluene and 1 mL of water in a round-bottomed flask and irradiated for 10 min in a domestic microwave oven (Scheme 7.119). After cooling, the mixture was filtered and the catalyst was extracted with toluene and dried. In this way, the recycled polymer-bound catalyst could be reused five times without a loss of efficiency. To isolate the desired biaryl compounds, the aqueous layer of the filtrate was separated, and the organic phase was washed with water and then dried over anhydrous magnesium sulfate. After filtration and evaporation of the solvent, the crystalline biaryls were obtained in high yields. Further examples of Suzuki couplings involving immobilized palladium catalysts are illustrated in Scheme 6.28.

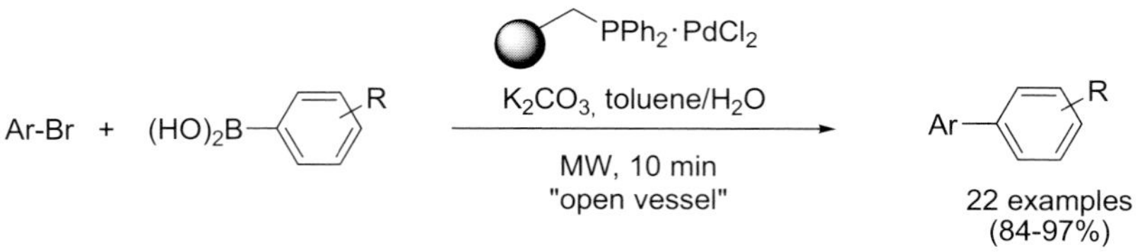

Scheme 7.119 Suzuki coupling utilizing a polymer-supported palladium catalyst.

In a related study, Srivastava and Collibee employed polymer-supported triphenylphosphine in palladium-catalyzed cyanations [142]. Commercially available resin-bound triphenylphosphine was admixed with palladium(II) acetate in N,N-dimethylformamide in order to generate the heterogeneous catalytic system. The mixture was stirred for 2 h under nitrogen atmosphere in a sealed microwave reaction vessel, to achieve complete formation of the active palladium–phosphine complex. The septum was then removed and equimolar amounts of zinc(II) cyanide and the requisite aryl halide were added. After purging with nitrogen and resealing, the vessel was transferred to the microwave reactor and irradiated at 140 °C for 30–50 min

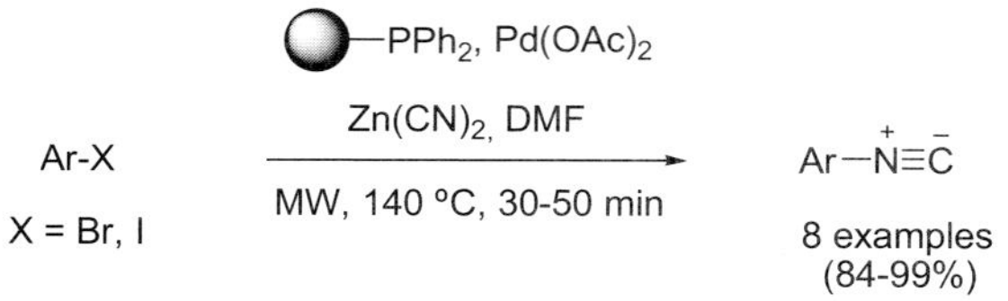

Scheme 7.120 Palladium-catalyzed isonitrile formation utilizing polymer-bound triphenylphosphine.

(Scheme 7.120). Finally, the resin was removed by filtration, and evaporation of the solvent furnished the desired benzonitriles in high yields and excellent purities.

Bergbreiter and Furyk have presented an oligo(ethylene glycol)-bound palladium(II) complex for microwave-mediated Heck couplings of various aryl halides [143]. This novel catalyst was prepared in a multistep procedure from poly(ethylene glycol) monomethyl ether via its methanesulfonyl intermediate, potassium thioacetate, and α,α'-dichloro-m-xylene. Final complexation was achieved by treatment of the 1,3-bis(MeOPEG$_{350}$thiomethyl)benzene with a solution of dichloro[bis(benzonitrile)]palladium(II) [Pd(PhCN)$_2$Cl$_2$] in acetonitrile. The Heck reactions were carried out in N,N-dimethylacetamide (DMA) utilizing 4 equivalents of alkene and either 2 equivalents of triethylamine or 3 equivalents of potassium carbonate as the base. The catalyst (0.01 mol%) was added as a solution in DMA immediately prior to subjecting the vessel to microwave irradiation. Heating at 150 °C for 10–60 min furnished the desired cinnamate products in moderate to good yields (Scheme 7.121). The catalyst could be recycled by performing the reactions in a 10% aqueous DMA/heptane mixture (1:2), with only a slight loss of efficiency after as many as four cycles [143].

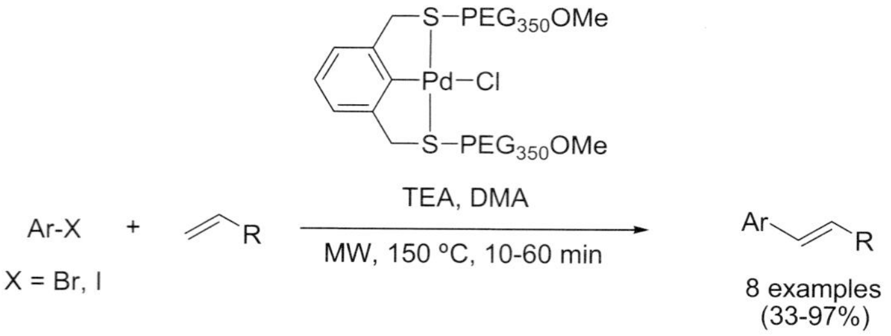

Scheme 7.121 Heck couplings utilizing an oligo(ethylene glycol)-bound SCS-palladium(II) complex as catalyst.

An application involving the use of a polystyrene-immobilized aluminum(III) chloride for a ketone–ketone rearrangement has been presented by Gopalakrishnan and coworkers [144].

7.6.2
Silica-Grafted Catalysts

Monoliths comprising cross-linked organic media with a well-defined porosity have emerged as useful supports for immobilizing catalysts. These supports offer advan-

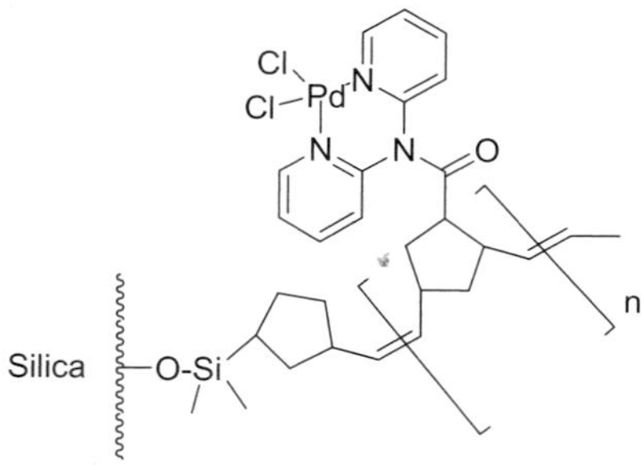

Fig. 7.9 Silica-grafted palladium catalysts.

tages of lower back pressure, enhanced diffusional mass transfer, and ease of synthesis, with many possibilities for the introduction of structural diversity. The group of Buchmeiser has presented the use of poly(N,N-dipyrid-2-yl-7-oxanorborn-2-en-5-yl-carbamino · PdCl$_2$)-grafted monolith supports for catalyzed Heck reactions [145]. Silica-based materials are equally useful as supports in slurry reactions under conventional as well as microwave conditions. Norborn-2-ene surface-functionalized silica has been used to graft N,N-dipyrid-2-yl-norborn-2-ene-5-ylcarbamide monomer, and "tentacles" of poly(N,N-dipyrid-2-ylnorborn-2-ene-5-ylcarbamide) were generated by controlled polymerization (Fig. 7.9).

The catalyst generated by palladium loading was utilized in microwave-assisted Heck reactions of iodoarenes with styrene, giving quantitative conversions to the corresponding C–C coupling products. The material could easily be removed by filtration and showed only minor leaching of Pd (< 2.5%) into the reaction mixture. These catalysts were successfully applied in the coupling of selected aryl iodides and aryl bromides under flow-through conditions in cartridges [145].

7.6.3
Catalysts Immobilized on Glass

Solution-phase organic coupling reactions in the presence of palladium catalysts are among the most extensively studied of organic transformations (see Section 6.1). Most of these reactions are, however, susceptible to poisoning of the palladium catalyst, with the associated disadvantage that relatively large amounts (1–5 mol%) of the catalyst have to be used for appreciable conversions. Organopalladium complexes in the form of homogeneous catalysts bearing phosphine ligands provide alternatives to overcome these disadvantages; however, their separation and recovery after the reaction can be difficult. Palladium metal on porous glass tubing has been investigated as a catalyst in reactions conducted in either a continuous or batchwise manner. In an early report, Strauss and coworkers described the Heck coupling of iodobenzene with allyl alcohol or styrene (Scheme 7.122) under microwave conditions in a pressure-tight batch reactor [146]. Although a reduced molar ratio of the catalyst (0.02 mol% palladium) was employed, >99% conversions with high regioselectivity were observed.

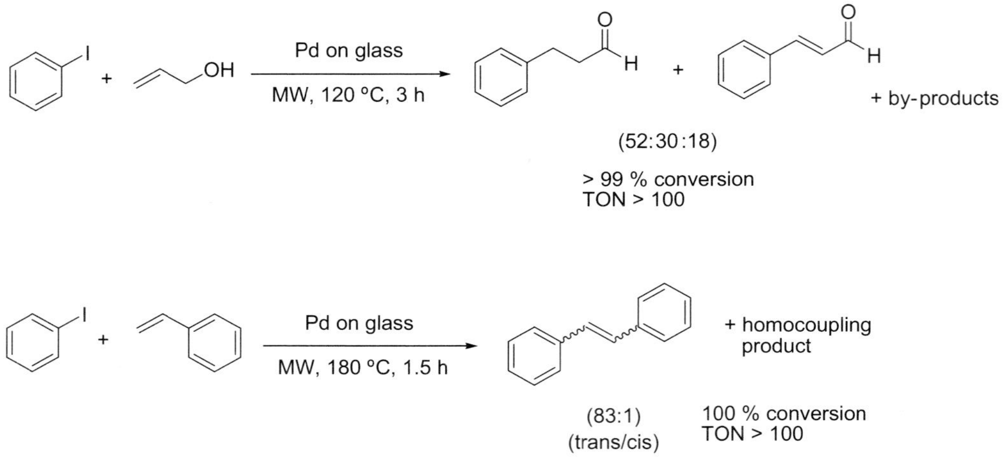

(52:30:18)

> 99 % conversion
TON > 100

(83:1)
(trans/cis)

100 % conversion
TON > 100

Scheme 7.122 Heck couplings utilizing palladium on porous glass.

In the same report, the Strauss group furthermore presented the effective use of palladium on porous glass to achieve quantitative conversions in couplings of phenylacetylene with iodobenzene and 4-bromobenzaldehyde. Additionally, satisfactory results were obtained for couplings of phenylacetylene with 4-bromoacetophenone and 2-bromopyridine [146].

Stadler and Kappe have reported on a palladium-doped microwave process vial, wherein palladium deposited on the inner glass surface was applied as a catalyst in heterogeneous C–P couplings leading to triphenylphosphines under microwave conditions (Scheme 7.123) [147]. The catalytic activity under microwave heating was, however, found to be somewhat lower (85% yield) than that using palladium-on-charcoal (98% yield) under identical reaction conditions (190 °C for 3 min; see Scheme 6.67). Nevertheless, the vials doped with the palladium deposit could be effectively reused several times without any significant loss of catalytic activity. Moreover, the need for catalyst filtration and additional reaction work-up is eliminated, simplifying the overall synthetic process.

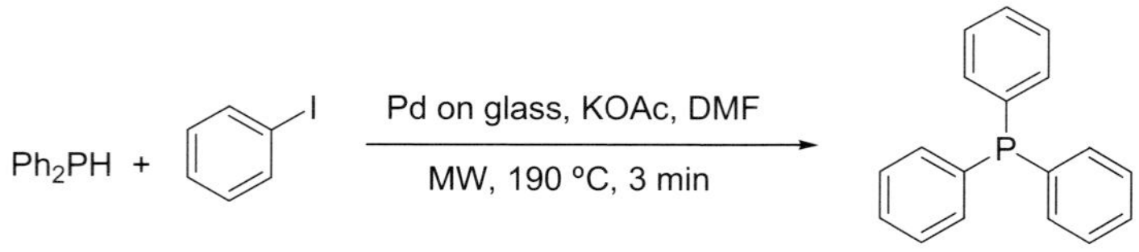

Scheme 7.123 Synthesis of triphenylphosphine by C–P coupling.

7.6.4
Catalysts Immobilized on Carbon

Carbon is inert in nature and has a high surface area, making it highly suitable as a support for catalysts. The surface characteristics and porosity of carbon may be easily tailored for different applications. Acid treatment is often applied to modify its surface chemistry for specific applications. Typically, active metal species are immobilized on carbon for catalytic applications.

Palladium-on-charcoal (Pd/C) is a catalyst widely employed in catalytic transfer hydrogenations. Microwave-assisted open-vessel reduction of double bonds and the hydrogenolysis of several other functional groups has been studied safely and rapidly by Bose and coworkers using 10% Pd/C catalyst and ammonium formate as the hydrogen donor (domestic microwave oven) [148]. Microwave-assisted selective hydrogenolysis of only an O-benzyl (Bn) group while retaining an N-benzyl group, together with reduction of an unsaturated ester to give a saturated side chain in a β-lactam (see Scheme 7.124 a), has also been reported. In a stereoselective preparation of β-lactam synthons under microwave-assisted catalytic transfer hydrogenation (CTH) conditions, unsaturation in a sugar moiety was successfully removed without disturbing the β-lactam ring.

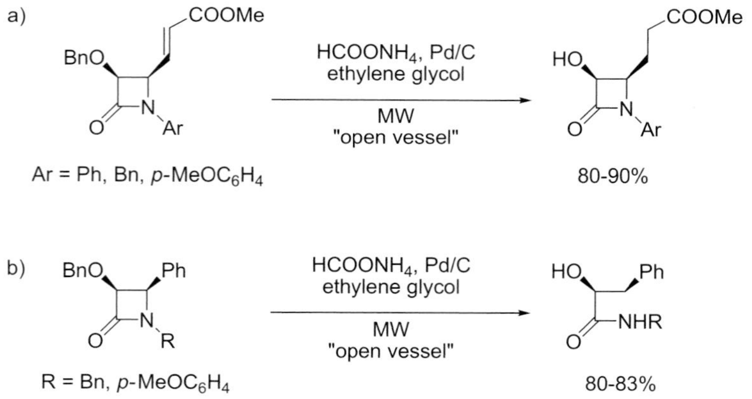

Scheme 7.124 Hydrogenolysis reactions catalyzed by palladium on charcoal.

In addition, microwave-assisted Pd/C-catalyzed transfer hydrogenolysis has been documented to cause rapid scission of 4-phenyl-2-azetidinones (Scheme 7.124 b). The hydrogenolysis conditions selectively deprotect the benzyloxy group at the C3 position to leave a hydroxy group with high yields within a few minutes. Microwave-assisted reduction of phenyl hydrazone has also been carried out using 10% Pd/C in the presence of ammonium formate, to give the corresponding amine in 92% yield within 4 min.

Related examples of microwave-assisted transformations involving palladium-on-charcoal are shown in Schemes 6.1, 6.19, 6.21, 6.46, 6.47, 6.67, 6.101, and 6.173. Examples of the use of nickel-on-carbon (Ni/C) are highlighted in Scheme 6.44. The use of graphite as a support in microwave synthesis has been discussed in Section 4.1.

7.6.5
Miscellaneous

The group of Choudhary has reported on a novel layered double-hydroxide-supported nanopalladium catalyst [LDH-Pd(0)], which showed superior activity in C–C coupling reactions compared to other supported catalysts, these ranging from acidic to weakly basic Pd/C, Pd/SiO$_2$, Pd/Al$_2$O$_3$, and resin-(PdCl$_4^{2-}$) [149]. In comparison with the homogeneous palladium(II) chloride catalyst, the LDH-Pd(0) catalytic system showed higher activity and selectivity, with excellent yields and high turnover frequencies (TOFs) in the microwave-assisted Heck olefination of electron-poor and electron-rich chloroarenes in non-aqueous ionic liquids (Scheme 7.125).

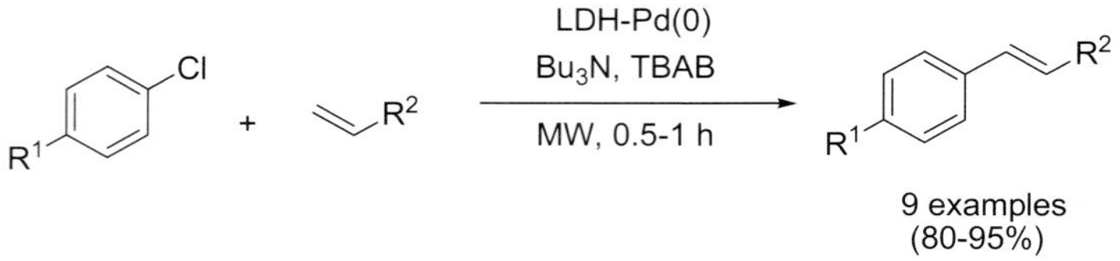

Scheme 7.125 Nanopalladium catalyst for Heck olefinations.

Under microwave heating, the Heck olefinations were achieved in 30–60 min, as opposed to 10–40 h by conventional heating. The recyclable heterogeneous LDH-Pd(0) catalytic system circumvents the need to use expensive and air-sensitive basic phosphines as ligands in the palladium-catalyzed coupling of chloroarenes. This novel Mg-Al layered double-hydroxide (LDH) support in the catalytic system stabilizes the nanopalladium particles and also supplies adequate electron density to the anchored palladium(0) species and facilitates the oxidative addition of the deactivated electron-rich chloroarenes.

Some early examples involving microwave-assisted solvent-free Sonogashira couplings using palladium powder doped on alumina/potassium fluoride as catalyst were described by Kabalka and coworkers (Scheme 4.4) [150]. In addition, this novel catalytic system has been used in microwave-assisted solvent-free Sonogashira coupling–cyclization of *ortho*-iodophenol with terminal alkynes, and similarly of *ortho*-ethynylphenols with aromatic iodides, to generate 2-substituted benzo[*b*]furans

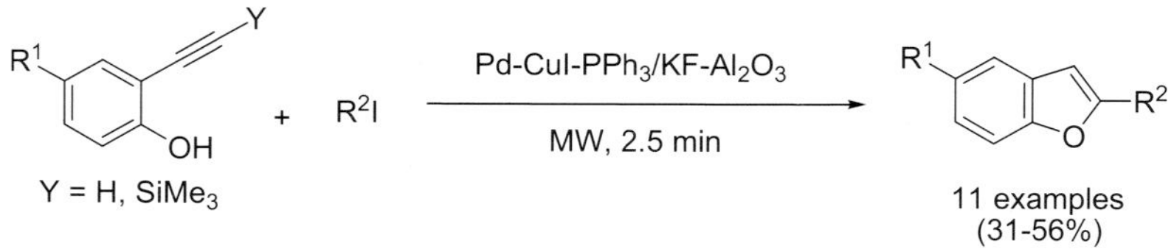

Scheme 7.126 Solvent-free Sonogashira coupling–cyclization.

(Scheme 7.126) [150]. All experiments within this report were carried out in a domestic microwave oven, utilizing septum-sealed round-bottomed flasks. The alumina in the reaction mixture acts as a temperature moderator in order to prevent the reagents from extensively reacting with the metal catalyst. After the reaction, the mixture was suspended in hexane, filtered, and the products were isolated by evaporation of the solvent from the filtrate and subjecting the residue to flash chromatography.

The group of Ley has reported on the use of palladium-doped perovskites as recyclable and reusable catalysts for Suzuki couplings [151]. Microwave-mediated cross-couplings of phenylboronic acid with aryl halides were achieved within 1 h by utilizing the supported catalyst (0.25 mol% palladium) in aqueous 2-propanol (Scheme 7.127). The addition of water was crucial as attempted transformations in non-aqueous mixtures did not proceed.

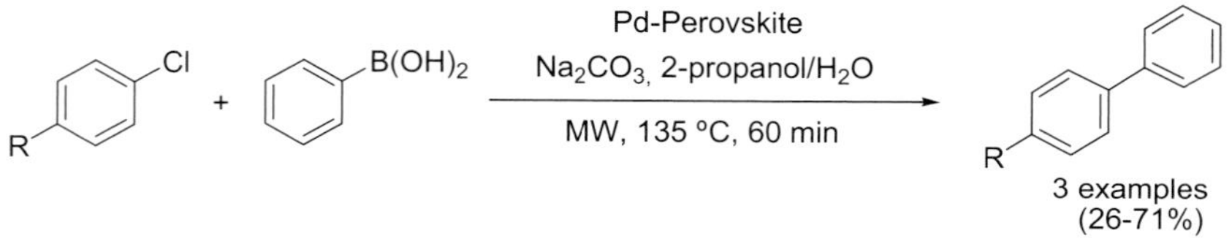

Scheme 7.127 Suzuki couplings utilizing palladium-doped perovskite as catalyst.

Similar Suzuki couplings have been performed by Hu and coworkers utilizing a poly(dicyclohexylcarbodiimide)/palladium nanoparticle composite [152]. This PDHC-Pd catalyst showed remarkable activity and stability under microwave irradiation. Near quantitative conversion (95% isolated yield) was obtained after 40 min of microwave heating of a mixture of iodobenzene with phenylboronic acid in dioxane. Re-using the immobilized catalyzed showed no significant loss of efficiency, as the fifth cycle still furnished a 90% isolated yield of the desired biphenyl.

A microwave-assisted Suzuki reaction performed in a microreactor device has been discussed in Section 4.5 [153]. More examples of microwave-assisted transformations involving immobilized catalysts are described in ref. [154].

7.7
Polymer-Supported Scavengers

The use of microwave heating in most examples of the syntheses discussed within this and the preceding chapter has the common purpose of speeding up the performed chemical transformation, in some cases to obtain a desired selectivity and often to improve the conversion to and yield of the desired products as compared to the conventional approach. The isolation of a clean and homogeneous product is also an integral part of any synthesis, and microwave heating has also been instrumental in improving many purification techniques. In this final section, we present specific examples of the use of microwave irradiation in combination with several functionalized polymers for scavenging and purification techniques.

The Kappe group has described microwave-induced N3-acylation of the dihydro-pyrimidine (DHPM) Biginelli scaffold [135] (Scheme 7.114) using various anhydrides. The process involved purification of the reaction mixture by a microwave-assisted scavenging technique. Volatile acylating agents such as acetic anhydride could simply be removed by evaporation, while other non-volatile anhydrides such as benzoic anhydride (Bz₂O) required an elaborate work-up suitable for a high-throughput format. Several scavenging reagents bearing amino functionalities with different loadings (Fig. 7.10) were applied to sequester the excess benzoic anhydride (Bz₂O) from the reaction mixture [135, 165].

In order to compare efficacies, the scavenging was studied under conventional (room temperature) and microwave conditions (Fig. 7.11). At room temperature, the polystyrene-based diamine (3.0 equivalents of the amine functionality) required 1–2 h for complete sequestration of the excess anhydride. In contrast, under sealed-vessel microwave heating, this was accelerated to completion within 5–10 min.

Comparable results were achieved with the silica-based diamine (3.0 equivalents of the amine functionality). At room temperature, the excess anhydride was sca-venged within 4 h. On the other hand, microwave heating at 100 °C in a sealed ves-

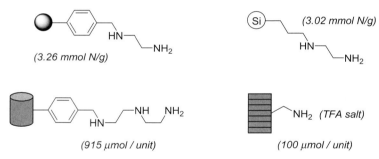

Fig. 7.10 Polymer-supported scavengers: polystyrene-bound diamine (top left), silica-based diamine (top right), Stratosphere Plugs (bottom left), and Synphase Lanterns (bottom right).

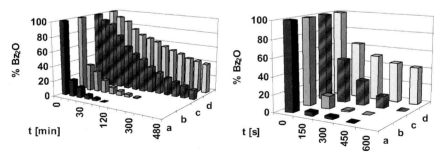

Fig. 7.11 Scavenging efficiencies of polymer-supported scavengers (a) polystyrene, (b) functionalized silica, (c) plugs, (d) lanterns] at room temperature (left) and under microwave heating (right).

sel significantly reduced the quenching time to 5–10 min. Scavenging using so-called Stratosphere Plugs, bearing a diethylenetriaminomethyl-polystyrene function, or Synphase Lanterns, bearing an aminomethyl-polystyrene function, proved somewhat less effective.

A parallel solution-phase asymmetric synthesis of α-branched amines has been reported by Ellman and coworkers based on stereoselective addition of organomagnesium reagents to enantiomerically pure *tert*-butanesulfinyl imines [156]. Microwave heating was utilized in two of the steps of this synthesis of asymmetric amines, both for the imine formation and for the resin capture (Scheme 7.128).

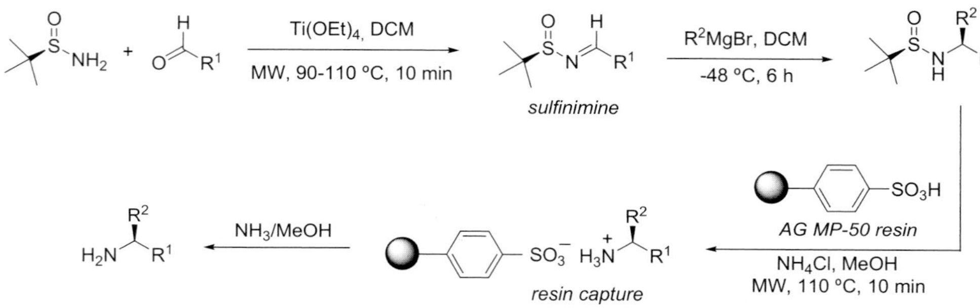

Scheme 7.128 Asymmetric synthesis of α-branched amines.

The initial condensation step in the synthesis of the sulfinimine substrates was catalyzed by titanium(IV) ethoxide as a Lewis acid under microwave heating. The excess of the Lewis acid was scavenged using a large amount of a support-bound diethanolamine (not shown in Scheme 7.128). In contrast to the reaction at room temperature, the imine was formed within 10 min of microwave irradiation, often in quantitative yield. After the Grignard addition step, microwave irradiations were utilized for acidic alcoholysis of the sulfinimide using macroporous sulfonic acid resin (AG MP-50). Under conventional reflux conditions, only partial sulfinyl group cleavage was observed in methanol. On the other hand, in some cases, complete consumption was achieved within 10 min under microwave heating at 110 °C. The microwave-assisted resin capture of amines allowed the preparation of a range of α-phenylethylamines and diphenylmethylamine derivatives in analytically pure form, in good overall yields and with high enantiomeric purity. This "resin-capture methodology" afforded the amine hydrochlorides without the need for chromatography or crystallization of any intermediates or products.

The group of Messeguer has presented the use of a high-loading polystyrene Wang aldehyde resin for the scavenging of excess amines in the preparation of piperazinium derivatives [157]. In the initial step, chloroacetyl chloride was reacted with a 50% excess of a primary amine at 0 °C for 30 min. After filtration and evaporation of the solvent, the residue was dissolved in dioxane, suspended with 3.5 equivalents of Wang aldehyde resin, and irradiated in a domestic microwave oven for 20–40 min, applying 4 min intervals. The resin was then filtered off and the resulting chloroacetamide was treated with 3 equivalents of the requisite pri-

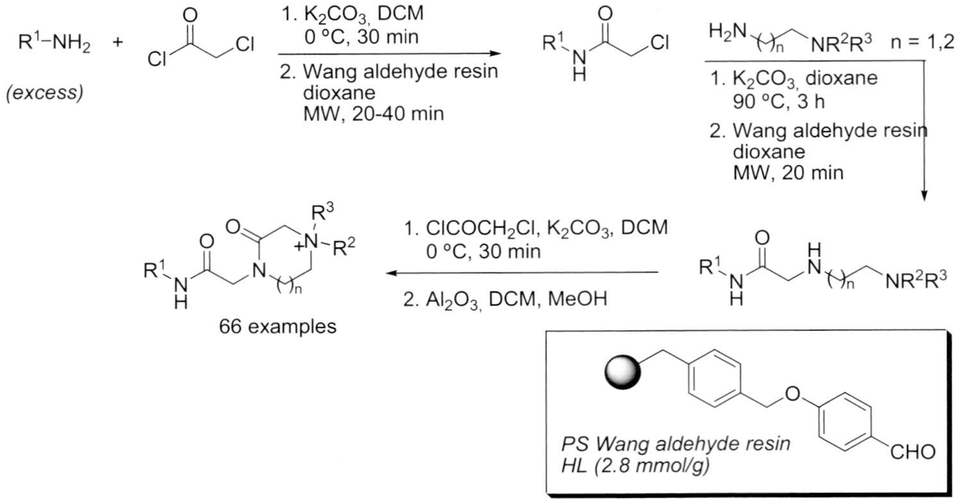

Scheme 7.129 Efficient amine-scavenging utilizing a polystyrene aldehyde scavenger.

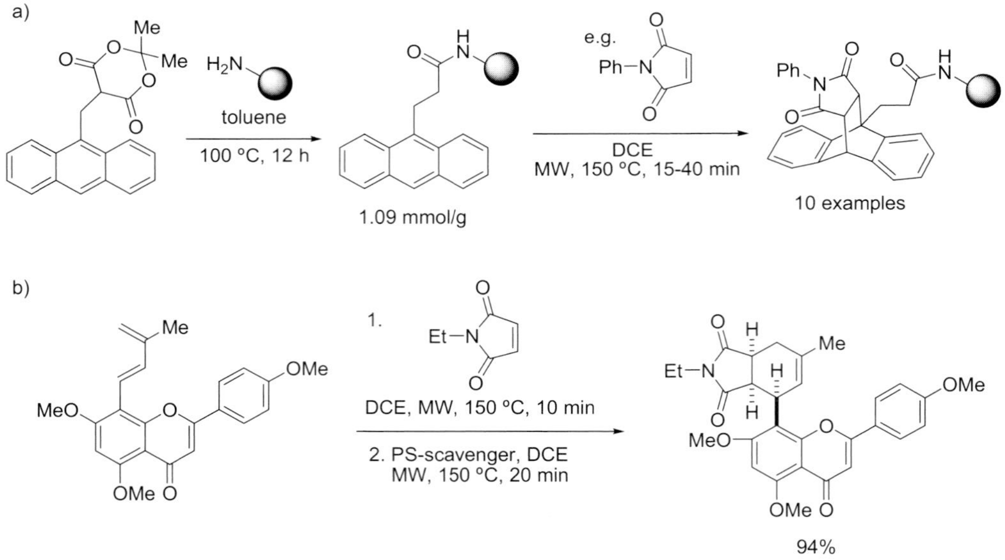

a)

b)

Scheme 7.130 Preparation and application of a novel polystyrene-bound anthracene as a dienophile scavenger.

mary diamine (Scheme 7.129). After the reaction, the solution was again suspended with the aldehyde resin and subjected to the microwave-assisted scavenging (domestic microwave oven). After five irradiation cycles for 4 min each, the resin was filtered off and the filtrate was concentrated to obtain the corresponding glycinamide. Further modification and cyclization steps furnished the desired heterocyclic scaffolds.

Finally, Porco and Lei have described the use of polymer-bound anthracene as an effective dienophile scavenger [158]. The scavenging agent utilized could be easily prepared by treatment of the commercially available corresponding Meldrum's acid derivative with aminomethyl polystyrene resin (Scheme 7.130 a). The reactivity of this novel polymer-bound scavenger was examined with a series of ten different dienophiles, including *N*-phenylmaleimide. Under microwave heating, the use of 2–3 equivalents of the resin proved sufficient to achieve effective scavenging of reactive dienophiles in less than 30 min. Rather unreactive derivatives, such as 1,4-naphthoquinone, required 40 min of irradiation in 1,2-dichloroethane (DCE).

To demonstrate the effectiveness of this scavenger, it was successfully applied in the synthesis of eight different flavonoid Diels–Alder cycloadducts (Scheme 7.130 b). The two-step microwave-mediated procedure furnished the desired compounds in high yields and excellent purities.

References

[1] K. C. Nicolaou, R. Hanko, W. Hartwig (Eds.), *Handbook of Combinatorial Chemistry*, Wiley-VCH, Weinheim, **2002**.

[2] F. Z. Dörwald, *Organic Synthesis on Solid Phase*, Wiley-VCH, Weinheim, **2002**.

[3] R. B. Merrifield, *J. Am. Chem. Soc.* **1963**, *85*, 2149–2154.

[4] R. Frank, W. Heikens, G. Heisterberg-Moutsis, H. Blöcker, *Nucl. Acids Res.* **1983**, *11*, 4365–4377.

[5] H. M. Geysen, R. H. Meloen, S. J. Barteling, *Proc. Natl. Acad. Sci. USA* **1984**, *81*, 3998–4002.

[6] R. A. Houghten, *Proc. Natl. Acad. Sci. USA* **1985**, *82*, 5131–5135.

[7] Á. Furka, F. Sebestyén, M. Asgedom, G. Dibó, *Highlights of Modern Biochemistry*, in *Proceedings of the 14th International Congress of Biochemistry*, Prague, Czech Republic, **1988**, *13*, p. 14, VSP, Utrecht; Á. Furka, F. Sebestyén, M. Asgedom, G. Dibó, *Int. J. Peptide Prot. Res.* **1991**, *37*, 487–493.

[8] A. Lew, P. O. Krutzik, M. E. Hart, A. R. Chamberlin, *J. Comb. Chem.* **2002**, *4*, 95–105.

[9] C. O. Kappe, *Curr. Opin. Chem. Biol.* **2002**, *6*, 314–320.

[10] P. Lidström, J. Westman, A. Lewis, *Comb. Chem. High Throughput Screen.* **2002**, *5*, 441–458.

[11] M. Larhed, A. Hallberg, *Drug Discovery Today* **2001**, *6*, 406–416.

[12] C. O. Kappe, A. Stadler, *Microwave-Assisted Combinatorial Chemistry*, in *Microwaves in Organic Synthesis* (Ed.: A. Loupy), Wiley-VCH, Weinheim, **2002**, p. 405–433; A. Stadler, C. O. Kappe, in *Microwave-Assisted Organic Synthesis* (Eds.: P. Lidström, J. P. Tierney), Blackwell Publishing, Oxford, **2005** (Chapter 7).

[13] B. Hayes, *Microwave Synthesis. Chemistry at the Speed of Light*, CEM Publishing, Matthews, NC, **2002**.

[14] N. K. Terret, *Combinatorial Chemistry*, Oxford University Press, New York, **1998**; M. Winter, R. Warrass, *Resins and Anchors for Solid-Phase Organic Synthesis*, in *Combinatorial Chemistry* (Ed.: H. Fenniri), Oxford University Press, New York, p. 117–138, **2000**.

[15] R. Pérez, T. Beryozkina, O. I. Zbruyev, W. Haas, C. O. Kappe, *J. Comb. Chem.* **2002**, *4*, 501–510.

[16] A. P. Combs, S. Saubern, M. Rafalski, P. Y. S. Lam, *Tetrahedron Lett.* **1999**, *40*, 1623–1626.

[17] B. M. Glass, A. P. Combs, Case Study 4–6: Rapid Parallel Synthesis Utilizing Microwave Irradiation, in *High-Throughput Synthesis. Principles and Practices* (Ed.: I. Sucholeiki), Marcel Dekker, Inc., New York, **2001**, Chapter 4.6, pp 123–128.

[18] B. M. Glass, A. P. Combs, Rapid Parallel Synthesis Utilizing Microwave Irradiation: Article E0027, *"Fifth International Electronic Conference on Synthetic Organic Chemistry"* (Eds.: C. O. Kappe, P. Merino, A. Marzinzik, H. Wennemers, T. Wirth, J. J. Vanden Eynde, S.-K. Lin), CD-ROM edition, ISBN 3–906980–06–5, MDPI, Basel, Switzerland, **2001**.

[19] I. C. Cotterill, I. Y. Usyatinsky, J. M. Arnold, D. S. Clark, J. S. Dordick, P. C. Michels, Y. L. Khmelnitsky, *Tetrahedron Lett.* **1998**, *39*, 1117–1120.

[20] C. M. Coleman, J. M. D. MacElroy, J. F. Gallagher, D. F. O'Shea, *J. Comb. Chem.* **2002**, *4*, 87–93.

[21] M. Erdélyí, A. Gogoll, *J. Org. Chem.* **2003**, *68*, 6431–6434.

[22] H.-M. Yu, S.-T. Chen, K.-T. Wang, *J. Org. Chem.* **1992**, *57*, 4781–4784.

[23] M. Erdélyi, A. Gogoll, *Synthesis* **2002**, 1592–1596.

[24] C. Lindquist, U. Tedebark, O. Ersoy, P. Somfai, *Synth. Commun.* **2003**, *33*, 2257–2262.

[25] H. J. Olivos, P. G. Alluri, M. M. Reddy, D. Salony, T. Kodadek, *Org. Lett.* **2002**, *4*, 4057–4059.

[26] A. Finaru, A. Berthault, T. Besson, G. Guillaumet, S. Berteina-Raboin, *Org. Lett.* **2002**, *4*, 2613–2615.

[27] H. Akamatsu, K. Fukase, S. Kusumoto, *Synlett* **2004**, 1049–1053.

[28] A. Stadler, C. O. Kappe, *Eur. J. Org. Chem.* **2001**, 919–925.

[29] A. Stadler, C. O. Kappe, *Tetrahedron* **2001**, *57*, 3915–3920.

[30] H. Yang, Y. Peng, G. Song, X. Qian, *Tetrahedron Lett.* **2001**, *42*, 9043–9046.

[31] V. A. Yaylayan, M. Siu, J. M. R. Bélangere, J. R. J. Paré, *Tetrahedron Lett.* **2002**, *43*, 9023–9025.

[32] J. Westman, R. Lundin, *Synthesis* **2003**, *7*, 1025–1030.

[33] G. A. Strohmeier, C. O. Kappe, *J. Comb. Chem.* **2002**, *4*, 154–161.

[34] M. Larhed, G. Lindeberg, A. Hallberg, *Tetrahedron Lett.* **1996**, *37*, 8219–8222.

[35] N.-F. K. Kaiser, U. Bremberg, M. Larhed, C. Moberg, A. Hallberg, *Angew. Chem. Int. Ed.* **2000**, *39*, 3596–3598.

[36] M. Alterman, A. Hallberg, *J. Org. Chem.* **2000**, *65*, 7984–7989.

[37] J. Wannberg, M. Larhed, *J. Org. Chem.* **2003**, *68*, 5750–5753.

[38] A. P. Combs, B. M. Glass, S. A. Jackson, *Methods Enzymology* **2003**, *369*, 223–231.

[39] P. Walla, C. O. Kappe, *Chem. Commun.* **2004**, 564–565.

[40] K. Weigand, S. Peka, *Mol. Diversity* **2003**, *7*, 181–184.

[41] W.-M. Dai, D.-S. Guo, L.-P. Sun, X.-H. Huang, *Org. Lett.* **2003**, 2919–2922.

[42] N. Kaval, W. Dehaen, E. Van der Eycken, *J. Comb. Chem.* **2005**, *7*, 90–95.

[43] D. Scharn, H. Wenschuh, U. Reineke, J. Schneider-Mergener, L. Germeroth, *J. Comb. Chem.* **2000**, *2*, 361–369.

[44] D. Scharn, L. Germeroth, J. Schneider-Mergener, H. Wenschuh, *J. Org. Chem.* **2001**, *66*, 507–513.

[45] M. D. Bowman, R. C. Jeske, H. E. Blackwell, *Org. Lett.* **2004**, *6*, 2019–2022.

[46] P. Grünefeld, C. Richert, *J. Org. Chem.* **2004**, *69*, 7543–7555.

[47] E. A. Yates, M. O. Jones, C. E. Clarke, A. K. Powell, S. R. Johnson, A. Porch, P. P. Edwards, J. E. Turnbull, *J. Mater. Chem.* **2003**, *13*, 2061–2063.

[48] P. Grieco, P. Campiglia, I. Gomez-Monterry, T. Lama, E. Novellino, *Synlett* **2003**, 2216–2218.

[49] P. Campiglia, I. Gomez-Monterry, L. Longobardo, T. Lama, E. Novellino, P. Grieco, *Tetrahedron Lett.* **2004**, *45*, 1453–1456.

[50] S. M. Miles, R. J. Leatherbarrow, S. P. Marsden, W. J. Coates, *Org. Biomol. Chem.* **2004**, *2*, 281–283.

[51] L. De Luca, G. Giacomelli, A. Porcheddu, M. Salaris, M. Taddei, *J. Comb. Chem.* **2003**, *5*, 465–471.

[52] R. E. Austin, J. F. Okonya, D. R. S. Bond, F. Al-Obeidi, *Tetrahedron Lett.* **2002**, *43*, 6169–6171.

[53] A. M. L. Hoel, J. Nielsen, *Tetrahedron Lett.* **1999**, *40*, 3941–3944.

[54] B. Henkel, *Tetrahedron Lett.* **2004**, *45*, 2219–2221.

[55] A. P. Frutos Hoener, B. Henkel, J.-C. Gauvin, *Synlett* **2003**, 63–66.

[56] G. J. Kuster, H. W. Scheeren, *Tetrahedron Lett.* **2000**, *41*, 515–519.

[57] B. Clapham, S.-H. Lee, G. Koch, J. Zimmermann, K. D. Janda, *Tetrahedron Lett.* **2002**, *43*, 5407–5410.

[58] H. M. S. Kumar, S. Anjaneyulu, B. V. S. Reddy, *Synlett* **2000**, 1129–1130.

[59] R. Schobert, C. Jagusch, *Tetrahedron Lett.* **2003**, *44*, 6449–6451.

[60] B. Martinez-Teipel, R. C. Green, R. E. Dolle, *QSAR Comb. Sci.* **2004**, *23*, 854–858.

[61] Y.-D. Gong, H.-Y. Sohn, M. J. Kurth, *J. Org. Chem.* **1998**, *63*, 4854–4856.

[62] B. Martin, H. Sekijic, C. Chassaing, *Org. Lett.* **2003**, *5*, 1851–1853.

[63] A.-M. Yu, Z.-P. Zhang, H.-Z. Yang, C.-X. Zhang, Z. Liu, *Synth. Commun.* **1999**, *29*, 1595–1599.

[64] S. Chandrasekhar, M. B. Padmaja, A. Raza, *Synlett* **1999**, 1597–1599.

[65] S. Weik, J. Rademann, *Angew. Chem. Int. Ed.* **2003**, *42*, 2491–2494.

[66] A. Bengtsson, A. Hallberg, M. Larhed, *Org. Lett.* **2002**, *4*, 1231–1233.

[67] N. E. Leadbeater, H. M. Torenius, H. Tye, *Mol. Diversity* **2003**, *7*, 135–144.

[68] L. Paolini, E. Petricci, F. Corelli, M. Botta, *Synthesis* **2003**, 1039–1042.

[69] N. Kaval, J. Van der Eycken, J. Caroen, W. Dehaen, G. A. Strohmeier, C. O. Kappe, E. Van der Eycken, *J. Comb. Chem.* **2003**, *5*, 560–568.

[70] C. G. Blettner, W. A. König, W. Stenzel, T. Schotten, *J. Org. Chem.* **1999**, *64*, 3885–3890.

[71] M. Xia, Y.-G. Wang, *J. Chem. Res. (S)* **2002**, 173–175.

[72] B. Sauvagnat, F. Lamaty, R. Lazaro, J. Martinez, *Tetrahedron Lett.* **2000**, *41*, 6371–6375.

[73] J. J. Vanden Eynde, D. Rutot, *Tetrahedron* **1999**, *55*, 2687–2694.

[74] M. Xia, Y.-G. Wang, *Tetrahedron Lett.* **2002**, *43*, 7703–7705.

[75] M. Xia, Y.-G. Wang, *Synthesis* **2003**, 262–266.

[76] A. Porcheddu, G. F. Ruda, A. Sega, M. Taddei, *Eur. J. Org. Chem.* **2003**, 907–912.

[77] C.-Y. Wu, C.-M. Sun, *Synlett* **2002**, 1709–1711.

[78] W.-J. Chang, W.-B. Yeh, C.-M. Sun, *Synlett* **2003**, 1688–1692.

[79] P. M. Bendale, C.-M. Sun, *J. Comb. Chem.* **2002**, *4*, 359–361.

[80] C.-L. Tung, C.-M. Sun, *Tetrahedron Lett.* **2004**, *45*, 1159–1162.

[81] M.-J. Lin, C.-M. Sun, *Synlett* **2004**, 663–666.

[82] W.-B. Yeh, M.-J. Lin, C.-M. Sun, *Comb. Chem. High Throughput Screen.* **2004**, *7*, 251–255.

[83] Y.-S. Su, M.-J. Lin, M.-C. Sun, *Tetrahedron Lett.* **2005**, *46*, 177–180.

[84] M.-J. Lin, C.-M. Sun, *Tetrahedron Lett.* **2003**, *44*, 8739–8742.

[85] W.-B. Yeh, M.-J. Lin, M.-J. Lee, C.-M. Sun, *Mol. Diversity* **2003**, *7*, 185–197.

[86] W.-B. Yeh, C.-M. Sun, *J. Comb. Chem.* **2004**, *6*, 279–282.

[87] M.-J. Lee, C.-M. Sun, *Tetrahedron Lett.* **2004**, *45*, 437–440.

[88] D. P. Curran, *Angew. Chem. Int. Ed.* **1998**, *37*, 1175–1196.

[89] M. Larhed, M. Hoshino, S. Hadida, D. P. Curran, A. Hallberg, *J. Org. Chem.* **1997**, *62*, 5583–5587.

[90] K. Olofsson, S.-Y. Kim, M. Larhed, D. P. Curran, A. Hallberg, *J. Org. Chem.* **1999**, *64*, 4539–4541.

[91] W. Zhang, Y. Lu, C. H.-T. Chen, *Mol. Diversity* **2003**, *7*, 199–202.

[92] T. Nagashima, W. Zhang, *J. Comb. Chem.* **2004**, *6*, 942–949.

[93] W. Zhang, C. H.-T. Chen, Y. Lu, T. Nagashima, *Org. Lett.* **2004**, *6*, 1473–1476.

[94] W. Zhang, T. Nagashima, Y. Lu, C. H.-T. Chen, *Tetrahedron Lett.* **2004**, *45*, 4611–4613.

[95] A.-L. Villard, B. Warrington, M. Ladlow, *J. Comb. Chem.* **2004**, *6*, 611–622.

[96] W. Zhang, P. Tempest, *Tetrahedron Lett.* **2004**, *45*, 6757–6760.

[97] Y. Lu, W. Zhang, *QSAR Comb. Sci.* **2004**, *23*, 827–835.

[98] K. S. A. Vallin, Q. Zhang, M. Larhed, D. P. Curran, A. Hallberg, *J. Org. Chem.* **2003**, *68*, 6639–6645.

[99] J. E. Moore, D. Spinks, J. P. A. Harrity, *Tetrahedron Lett.* **2004**, *45*, 3189–3191.

[100] M. A. Herrero, J. Wannberg, M. Larhed, *Synlett* **2004**, 2335–2338.

[101] S. Werner, D. P. Curran, *Org. Lett.* **2003**, *5*, 3293–3296.

[102] J. Fraga-Dubreuil, J. P. Bazureau, *Tetrahedron Lett.* **2001**, *42*, 6097–6100.

[103] J. Fraga-Dubreuil, M.-H. Famelart, J. P. Bazureau, *Org. Proc. Res. Dev.* **2002**, *6*, 374–378.

[104] J. Fraga-Dubreuil, J. P. Bazureau *Tetrahedron* **2003**, *59*, 6121–6130.

[105] H. Hakkou, J. J. Vanden Eynde, J. Hemelin, J. P. Bazureau, *Tetrahedron* **2004**, *60*, 3745–3753.

[106] M. De Kort, A. W. Tuin, S. Kuiper, H. S. Overkleeft, G. A. Van der Marel, R. C. Buijsman, *Tetrahedron Lett.* **2004**, *45*, 2171–2175.

[107] S. V. Ley, I. R. Baxendale, R. N. Bream, P. S. Jackson, A. G. Leach, D. A. Longbottom, M. Nesi, J. S. Scott, R. I. Storer, S. J. Taylor, *J. Chem. Soc., Perkin Trans. 1* **2000**, 3815–4195; S. V. Ley, I. R. Baxendale, *Nature Rev. Drug Disc.* **2002**, *1*, 573–586.

[108] A. Kirschning, H. Monenschein, R. Wittenberg, *Angew. Chem. Int. Ed.* **2001**, *40*, 650–679; C. C. Tzschucke, C. Markert, W. Bannwarth, S. Roller, A. Hebel, R. Haag, *Angew. Chem. Int. Ed.* **2002**, *41*, 3964–4000.

[109] S. Yoshida, M. Sato, E. Sugawara, Z. Shimada, *J. Appl. Phys.* **1999**, *85*, 4636–4638.

[110] I. R. Baxendale, A.-L. Lee, S. V. Ley, in *Microwave-Assisted Organic Synthesis* (Eds.: P. Lidström, J. P. Tierney), Blackwell Publishing, Oxford, **2005** (Chapter 6).

[111] P. Hodge, *Chem. Soc. Rev.* **1997**, *26*, 417–424 and references therein.

[112] C. Brain, J. M. Paul, Y. Loong, P. J. Oakley, *Tetrahedron Lett.* **1999**, *40*, 3275–3278.

[113] P. Wipf, S. Venkatraman, *Tetrahedron Lett.* **1996**, *37*, 4659–4662.

[114] C. T. Brain, S. A. Brunton, *Synlett.* **2001**, 382–384.

[115] S. V. Ley, A. G. Leach, R. I. Storer, *J. Chem. Soc., Perkin Trans. 1* **2001**, 358–361.

[116] J. Westman, *Org. Lett.* **2001**, *3*, 3745–3747.

[117] B. Desai, T. N. Danks, *Tetrahedron Lett.* **2001**, *42*, 5963–5965.

[118] G. Gopalkrishnan, V. Kasinath, N. D. Pradeep Singh, V. P. Santhana Krishnan, K. Anand Soloman, S. S. Rajan, *Molecules* **2002**, *7*, 412–419.

[119] S. V. Ley, S. J. Taylor, *Bioorg. Med. Chem. Lett.* **2002**, *12*, 1813–1816.

[120] T. Mukaiyama, Y. Yokota, *Bull. Chem. Soc. Jpn.* **1965**, *38*, 858.

[121] D. Launay, S. Booth, I. Clemens, A. Merritt, M. Bradley, *Tetrahedron Lett.* **2002**, *43*, 7201–7203.

[122] W. Hu, R. W. Bürli, J. A. Kaizerman, K. W. Johnson, M. I. Gross, M. Iwamoto, P. Jones, D. Lofland, S. Difuntorum, H. Chen, B. Bozdogan, P. C. Appelbaum, H. E. Moser, *J. Med. Chem.* **2004**, *47*, 4352–4355.

[123] S. Crosignani, P. D. White, B. Linclau, *Org. Lett.* **2002**, *4*, 2961–2963; S. Crosignani, D. Launay, B. Linclau, M. Bradley, *Mol. Diversity* **2003**, *7*, 203–210; S. Crosignani, P. D. White, B. Linclau, *J. Org. Chem.* **2004**, *69*, 5897–5905.

[124] D. R. Sauer, D. Kalvin, K. M. Phelan, *Org. Lett.* **2003**, *5*, 4721–4724.

[125] C. E. Humphrey, M. A. M. Easson, J. P. Tierney, N. J. Turner, *Org. Lett.* **2003**, *5*, 849–852.

[126] A. Kirschning, H. Monenschein, R. Wittenberg, *Chem. Eur. J.* **2000**, *6*, 4445–4450.

[127] A. Porcheddu, G. Giacomelli, L. De Luca, A. M. Ruda, *J. Comb. Chem.* **2004**, 105–111.

[128] E. Petricci, M. Botta, F. Corelli, C. Mugnaini, *Tetrahedron Lett.* **2002**, *43*, 6507–6509.

[129] E. Petricci, C. Mugnaini, M. Radi, F. Corelli, M. Botta, *J. Org. Chem.* **2004**, *69*, 7880–7887.

[130] R. B. Nicewonger, L. Ditto, D. Kerr, L. Varady, *Bioorg. Med. Chem. Lett.* **2002**, *12*, 1799–1802.

[131] S. Crosignani, J. Gonzalez, D. Swinnen, *Org. Lett.* **2004**, *6*, 4579–4582.

[132] I. R. Baxendale, S. V. Ley, C. Piutti, *Angew. Chem. Int. Ed.* **2002**, *41*, 2194–2197.

[133] I. R. Baxendale, S. V. Ley, M. Nessi, C. Piutti, *Tetrahedron* **2002**, *58*, 6285–6304.

[134] D. P. Curran, K. Fischer, G. Moura-Letts, *Synlett* **2004**, 1379–1382.

[135] D. Dallinger, N. Yu. Gorobets, C. O. Kappe, *Mol. Diversity* **2003**, *7*, 229–245.

[136] L. Öhberg, J. Westman, *Synlett* **2001**, 1893–1896.

[137] K. G. Mayo, E. H. Nearhoof, J. J. Kiddle, *Org. Lett.* **2002**, *4*, 1567–1570.

[138] M. G. Organ, S. Mayer, F. Lepifre, B. N'Zemba, J. Khatri, *Mol. Diversity* **2003**, *7*, 211–227.

[139] E. N. Jacobsen, A. Pfaltz, H. Yamamoto (Eds.), *Comprehensive Asymmetric Catalysis*, Springer, Berlin, **1999**.

[140] O. Belda, S. Lundgren, C. Moberg, *Org. Lett.* **2003**, *5*, 2275–2278.

[141] L. Bai, Y. Zhang, J. X. Wang, *QSAR Comb. Sci.* **2004**, *23*, 875–882.

[142] R. R. Srivastava, S. E. Collibee, *Tetrahedron Lett.* **2004**, 8895–8897.

[143] D. E. Bergbreiter, S. Furyk, *Green Chem.* **2004**, *6*, 280–285.

[144] G. Gopalakrishnan, V. Kasinath, N. D. Pradeep Singh, *Org. Lett.* **2002**, *4*, 781–782.

[145] M. R. Buchmeiser, S. Lubbad, M. Mayr, K. Wurst, *Inorg. Chim. Acta* **2003**, *345*, 145–153.

[146] J. Li, A. W.-H. Mau, C. R. Strauss, *Chem. Commun.* **1997**, 1275–1276.

[147] A. Stadler, C. O. Kappe, *Org. Lett.* **2002**, *4*, 3541–3543.

[148] B. K. Banik, K. J. Barakat, D. R. Wagle, M. S. Manhas, A. K. Bose, *J. Org. Chem.* **1999**, *64*, 5746–5753.

[149] B. M. Choudhary, S. Madhi, N. S. Chowdhari, M. L. Kantan, B. Sreedhar, *J. Am. Chem. Soc.* **2002**, *124*, 14127–14136.

[150] G. W. Kabalka, L. Wang, R. M. Pagni, *Tetrahedron* **2001**, *57*, 8017–8028.

[151] M. D. Smith, A. F. Stepan, C. Ramarao, P. E. Brennan, S. V. Ley, *Chem. Commun.* **2003**, 2652–2653.

[152] Y. Liu, C. Chemtong, J. Hu, *Chem. Commun.* **2004**, 398–399.

[153] P. He, S. J. Haswell, P. D. I. Fletcher, *Lab Chip* **2004**, *4*, 38–41.

[154] B. Desai, C. O. Kappe, in *Topics in Current Chemistry* (Ed.: A. Kirschning), Springer, Berlin, **2004**, Vol. 242, pp. 177–207.

[155] D. Dallinger, N. Yu. Gorobets, C. O. Kappe, *Org. Lett.* **2003**, *5*, 1205–1208.

[156] T. Mukade, D. R. Dragoli, J. A. Ellman, *J. Comb. Chem.* **2003**, *5*, 590–596.

[157] I. Masip, C. Ferrándiz-Huertas, C. García-Martínez, J. A. Ferragut, A. Ferrer-Montiel, A. Messeguer, *J. Comb. Chem.* **2004**, *6*, 135–141.

[158] X. Lei, J. A. Porco, *Org. Lett.* **2004**, *6*, 795–798.

8
Outlook and Conclusions

The diverse examples provided in Chapters 6 and 7 of this book should make it obvious that perhaps the overwhelming majority of chemical transformations can be successfully carried out under microwave conditions. This does not necessarily imply that dramatic rate enhancements compared to a classical, thermal process will be observed in all cases, but the simple convenience of using microwave technology will make this non-classical heating method a standard tool in the laboratory within a few years. In the past, microwaves were often used only when all other options to perform a particular reaction had failed, or when exceedingly long reaction times or high temperatures were required to complete a reaction. This practice is now slowly changing and due to the growing availability of microwave reactors in many laboratories, routine synthetic transformations are now also being carried out by microwave heating.

The benefits of controlled microwave heating, in particular in conjunction with the use of sealed-vessel systems, are manifold:

- Most importantly, microwave processing frequently leads to dramatically reduced reaction times, higher yields, and cleaner reaction profiles. In many cases, the observed rate enhancements may be simply a consequence of the high reaction temperatures that can rapidly be obtained using this non-classical heating method, or may result from the involvement of so-called specific or non-thermal microwave effects (see Section 2.5).
- An additional benefit of this technology is that the choice of solvent for a given reaction is not governed by the boiling point (as in a conventional reflux set-up) but rather by the dielectric properties of the reaction medium, which can be easily tuned, e.g., by the addition of highly polar materials such as ionic liquids.
- The temperature/pressure monitoring mechanisms of modern microwave reactors allow for an excellent control of reaction parameters, which generally leads to more reproducible reaction conditions.
- Because direct "in core" heating of the medium occurs, the overall process is more energy-efficient than classical oil-bath heating.
- Microwave heating can readily be adapted to a parallel or automatic sequential processing format. In particular, the latter technique allows for the rapid testing of new ideas and high-speed optimization of reaction conditions. The fact that a

Microwaves in Organic and Medicinal Chemistry. C. Oliver Kappe, Alexander Stadler
Copyright © 2005 WILEY-VCH Verlag GmbH & Co. KGaA, Weinheim
ISBN: 3-527-31210-2

"yes or no answer" for a particular chemical transformation can often be obtained within 5 to 10 min (as opposed to several hours in a conventional protocol), has contributed significantly to the acceptance of microwave chemistry both in industry and academia. The recently reported incorporation of real-time, *in situ* monitoring of microwave-assisted reactions by Raman spectroscopy will allow a further increase in efficiency and speed in microwave chemistry [1].

Apart from the traditional organic and combinatorial/high-throughput synthesis protocols covered in this book, more recent applications of microwave chemistry include biochemical processes such as high-speed polymerase chain reaction (PCR) [2], rapid enzyme-mediated protein mapping [3], and general enzyme-mediated organic transformations (biocatalysis) [4]. Furthermore, microwaves have been used in conjunction with electrochemical [5] and photochemical processes [6], and are also heavily employed in polymer chemistry [7] and material science applications [8], such as in the fabrication and modification of carbon nanotubes or nanowires [9].

So, *why is not everybody using microwaves?* One of the major drawbacks of this relatively new technology is equipment cost (Chapter 3). While prices for dedicated microwave reactors for organic synthesis have come down considerably since their first introduction in the late 1990s, the current price range for microwave reactors is still many times higher than that of conventional heating equipment. As with any new technology, the current situation is bound to change over the next few years, and less expensive equipment should become available. By then, microwave reactors will have truly become the "Bunsen burners of the 21st century" [10, 11] and will be standard equipment in every chemical laboratory.

References

[1] D. E. Pivonka, J. R. Empfield, *Appl. Spect.* **2004**, *58*, 41–46. For on-line UV monitoring, see: G. S. Getvoldsen, N. Elander, S. A. Stone-Elander, *Chem. Eur. J.* **2002**, *8*, 2255–2260.

[2] C. Fermér, P. Nilsson, M. Larhed, *Eur. J. Pharm. Sci.* **2003**, *18*, 129–132; K. Orrling, P. Nilsson, M. Gullberg, M. Larhed, *Chem. Commun.* **2004**, 790–791.

[3] B. N. Pramanik, U. A. Mirza, Y. H. Ing, Y.-H. Liu, P. L. Bartner, P. C. Weber, A. K. Bose, *Protein Science* **2002**, *11*, 2676–2687; B. N. Pramanik, Y. H. Ing, A. K. Bose, L.-K. Zhang, Y.-H. Liu, S. N. Ganguly, P. Bartner, *Tetrahedron Lett.* **2003**, *44*, 2565–2568; A. J. Bose, Y. H. Ing, N. Lavlinskaia, C. Sareen, B. N. Pramanik, P. L. Bartner, Y.-H. Liu, L. Heimark, *J. Am. Soc. Mass Spectrom.* **2002**, *13*, 839–850.

[4] B. Réjasse, S. Lamare, M.-D. Legoy, T. Besson, *Org. Biomol. Chem.* **2004**, *2*, 1086–1089; T. Maugard, D. Gaunt, M. D. Legoy, T. Besson, *Biotechn. Lett.* **2003**, *25*, 623–29; B. K. Pchelka, A. Loupy, J. Plenkiewicz, L. Blanco, *Tetrahedron Asymm.* **2000**, *11*, 2719–2732; G. Lin, W.-Y. Lin, *Tetrahedron Lett.* **1998**, *39*, 4333–4336; M.-C. Parker, T. Besson, S. Lamare, M.-D. Legoy, *Tetrahedron Lett.* **1996**, *37*, 8383–8386; M. Gelo-Pujic, E. Guibé-Pujic, E. Guibé-Jampel, A. Loupy, *Tetrahedron* **1997**, *53*, 17247–17252; M. Gelo-Pujic, E. Guibé-Jampel, A. Loupy, S. A. Galema, D. Mathé, *J. Chem. Soc., Perkin Trans.* **1996**, 2777–2780; J.-R. Carrillo-Munoz, D. Bouvet, E. Guibé-Jampel, A. Loupy, A. Petit, *J. Org. Chem.* **1996**, *61*, 7746–7749.

[5] Y.-C. Tsai, B. A. Coles, R. G. Compton, F. Marken, *J. Am. Chem. Soc.* **2002**, *124*, 9784–9788.

[6] P. Klan, V. Cvírka, in *Microwaves in Organic Synthesis* (Ed.: A. Loupy), Wiley-VCH, Weinheim, **2002**, pp 463–486 (Chapter 14).

[7] D. Bogdal, P. Penczek, J. Pielichowski, A. Prociak, *Adv. Polym. Sci.* **2003**, *163*, 193–263; F. Wiesbrock, R. Hoogenboom, U. S. Schubert, *Macromol. Rapid Commun.* **2004**, *25*, 1739–1764.

[8] S. Barlow, S. R. Marder, *Adv. Funct. Mater.* **2003**, *13*, 517–518; R. G. Blair, E. G. Gillan, N. K. B. Nguyen, D. Daurio, R. B. Kaner, *Chem. Mater.* **2003**, *15*, 3286–3293; M. Melucci, M. Gazzano, G. Barbarella, M. Cavallini, F. Biscarini, P. Maccagnani, P. Ostoja, *J. Am. Chem. Soc.* **2003**, *125*, 10266–10274.

[9] E. H. Hong, K.-H. Lee, S. H. Oh, C.-G. Park, *Adv. Funct. Mater.* **2003**, *13*, 961–966; T. J. Imholt, C. A. Dyke, B. Hasslacher, J. M. Perez, D. W. Price, J. A. Roberts, J. B. Scott, A. Wadhawan, Z. Ye, J. M. Tour, *Chem. Mater.* **2003**, *15*, 3969–3970; F. Della Negra, M. Meneghetti, E. Menna, *Fullerenes, Nanotubes, Carbon Nanostruct.* **2003**, *11*, 25–34; Y.-J. Zhu, W.-W. Wang, R.-J. Qi, X.-L. Hu, *Angew. Chem. Int. Ed.* **2004**, *43*, 1410–1424.

[10] The term "Bunsen burner of the 21st century" was originally coined by A. K. Bose (ref. 11), one of the pioneers of applying microwave heating to organic synthesis.

[11] A. K. Bose, B. K. Banik, N. Lavlinskaia, M. Jayaraman, M. S. Manhas, *Chemtech* **1997**, *27*, 18–24.

Index

Microwaves in Organic and Medicinal Chemistry. C. Oliver Kappe, Alexander Stadler
Copyright © 2005 WILEY-VCH Verlag GmbH & Co. KGaA, Weinheim
ISBN: 3-527-31210-2